RIPARIA

Riparia

Ecology, Conservation, and Management of Streamside Communities

Robert J. Naiman

Henri Décamps

Michael E. McClain

With a Foreword by Gene E. Likens

ELSEVIER
ACADEMIC
PRESS

AMSTERDAM • BOSTON • HEIDELBERG • LONDON • NEW YORK • OXFORD
PARIS • SAN DIEGO • SAN FRANCISCO • SINGAPORE • SYDNEY • TOKYO

Senior Acquisitions Editor	Nancy Maragioglio
Associate Acquisitions Editor	Kelly Sonnack
Project Manager	Kyle Sarofeen
Marketing Manager	Linda Beattie
Cover Design	Hannus Design Assoc.
Cover Photo	J.J. Latterell
Composition	SNP Best-set Typesetter Ltd., Hong Kong
Cover Printer	Hing Yip Printing Co., Ltd.
Interior Printer	Hing Yip Printing Co., Ltd.

Elsevier Academic Press
30 Corporate Drive, Suite 400, Burlington, MA 01803, USA
525 B Street, Suite 1900, San Diego, California 92101-4495, USA
84 Theobald's Road, London WC1X 8RR, UK

This book is printed on acid-free paper. ∞

Library of Congress Cataloging-in-Publication Data
Application submitted.

British Library Cataloguing in Publication Data
A catalogue record for this book is available from the British Library

ISBN: 0-12-663315-0

For all information on all Elsevier Academic Press Publications
visit our Web site at www.books.elsevier.com

Printed in China
05 06 07 08 09 10   9 8 7 6 5 4 3 2 1

Contents

Foreword xi
Preface xiii

1 Introduction 1
  Overview 1
  Purpose 2
  Hydrological Context 4
  Ecological Context 7
  Landscape Context 9
  Cultural Setting 11
  Rationale for Riparian Ecology 12
  Setting the Stage 18

2 Catchments and the Physical Template 19
  Overview 19
  Purpose 20
  Catchments and Hierarchical Patterns of Geomorphic Features 22
    Catchment Form and Channel Networks 23
    Catchment History 25
    Hierarchical Patterns of Geomorphic Features in Catchments 28
  Geomorphic Processes and Process Domains 32
    Headwater Erosion 33
    Channel Processes 34
    Floodplain and Channel Interactions 37
  Hydrologic Connectivity and Surface Water–Groundwater Exchange 40
    Surface Connectivity and Flooding 41
    The Dynamics of the Linked Surface–Subsurface Hydrologic System 45
  Conclusions 47

3 Riparian Typology 49
  Overview 49
  Purpose 49
  The Historical Context 51

Contents

Theoretical Basis for Classification 53
    Application of Ecological Information 55
        Inventory 55
        Classification 56
Emerging Classification Concepts 57
Geomorphic Classification 58
    Hierarchical Classification 59
    Rosgen's Classification 65
        Geomorphic Characterization (Aspect I) 67
        Morphologic Description (Aspect II) 69
    The Process Domain Concept 70
    The Hydrogeomorphic Approach 72
Biotic Classification 73
    Soils 73
    Plants 74
    Wildlife 75
Treating Complexity and Heterogeneity in Classification Systems 76
Attributes of an Enduring Classification System 77
Conclusions 78

4 Structural Patterns 79
  Overview 79
  Purpose 80
  Life History Strategies 81
  Morphological and Physiological Adaptations of Riparian Plants 82
  Reproductive Strategies 87
  Distribution, Structure, and Abundance 91
    Identification of Riparian Zones Based on Soils and Vegetation Type 91
    Biophysical Characteristics of Riparian Soils 92
        Organic Matter 93
        Moisture 94
        Fauna 96
    General Distributions of Aboveground and Belowground Communities . . . 97
        Lateral Zonation 97
        Longitudinal Zonation 100
    Successional and Seasonal Community Patterns 104
        Vegetative Succession 104
        Faunal Succession 112
    Density, Basal Area, and Biomass 112
  Biological Diversity 115
    Diversity Theory and Measurement 116
    Vegetative Diversity 119
        Site and Catchment Patterns 119
        Refuges 120
        Factors Controlling Species Richness 120

Faunal Diversity 121
    Diversity of Soil Organisms 122
    Aboveground Fauna Diversity 122

5 Biotic Functions of Riparia 125

Overview 125
Purpose 126
Water Use and Flux 127
Nutrient Fluxes 131
    Overview of Cycles and Processes 132
Production Ecology 135
    Growth and Metabolism of Riparian Trees 135
    Timing of Growth and Rates of Net Primary Production 138
    Litterfall 142
    Mortality Rates 145
    Root Production 147
Decomposition Dynamics 149
    Principles of Decomposition Dynamics 149
        Litter Quality 149
        Exogenous Nutrient Supply 150
        Temperature 150
        Oxygen Tension 150
    Nutrient Dynamics During Decay 151
    Factors Controlling Immobilization of Nitrogen 151
        Initial Litter Quality 151
        Exogenous Nutrient Supply 151
        Anaerobic Decay 152
        Temperature 152
    Mechanisms of Nitrogen Immobilization 152
        Nitrogen Accumulation in Microbial Biomass 153
        Nitrogen Accumulation in By-Products of Microbial Activity 153
    Decomposition of Riparian Litter 154
Information Fluxes 155
Microclimate 156
Conclusions 158

6 Biophysical Connectivity and Riparian Functions 159

Overview 159
Purpose 160
Patch Dynamics and a Landscape Perspective of Catchments 161
Nutrient Flows 163
    Riparian Zones as Buffers Against Nutrient Pollution from Upland
    Runoff 165
    Riparian Zones as Buffers Against High In-Stream Nutrient Levels 169
    Particle-Nutrient Considerations 172

Energy Flows and Food Webs 173
    Energy Flows Between Riparia and Adjointing Aquatic Systems 174
Large Animal Connections 179
    The Functional Grouping of Large Animal Interactions 180
    Pacific Salmon Influences on Riparian Ecosystems 183
Conclusions 186

7 Disturbance and Agents of Change 189
  Overview 189
  Purpose 190
  Major Categories of Change 192
    Human Demography 192
    Resource Use 192
    Technology Development 197
    Social Organization 197
  Riparian Disturbances 197
    Defining Anthropogenic Disturbance 198
    Understanding History: Basic Concepts and Approaches 198
        Legacies and Lag-Times 200
        Cumulative Effects 200
    Historical Examples of Riparian Alterations 201
    Pervasive Human-Mediated Changes 204
  Disturbance Ecology: Responses to Stress 206
  Ecological Consequences of Flow Regulation 209
    Theory 209
    Extent of Flow Regulation 211
    Effects on Native Species and Processes 213
    Alterations to Energy and Nutrient Budgets 216
    Basic Ecological Principles 219
  Consequences of Global Climate and Land Use Changes 221
    Climate Change 221
        Changes in Temperature Regimes 223
        Changes in Precipitation and Runoff Regimes 224
        Early Snowmelt Runoff 224
        Increased Precipitation Variability 224
        Reduced Discharge 225
    Can Riparia Adapt to Climate Change? 225
    Land Use Change 226
        Temperature Regimes 226
        Nutrient Enrichment 228
        Invasive Species 230
  Conclusions 232

8 Management 233
  Overview 233
  Purpose 234

Riparian Management: A Recent and Evolving Concern **235**

Economic Valuation of Riparia **237**

Social and Cultural Perspective **238**

Suitable Management Institutions **241**

Information Collection and Dissemination **243**

Riparian Management: A Process Linked to Catchment and to

River Management **246**

Riparian Benefits from Catchment Management **246**

Riparian Benefits from River Management **248**

Catchments and Rivers Benefit from Riparian Management **251**

Riparian Management: A Highly Specific Process **253**

Adaptiveness **254**

Sustainability **255**

Appropriateness **257**

Timber Harvest Practices **257**

Revegetation of Riverbanks **257**

Human Dimension of Riparian Management **259**

Shared Socioenvironmental Visions **260**

Social-Ecological Systems **261**

Communication Needs **266**

Conclusions **268**

Riparian Management as an Emerging Issue **268**

9 Conservation **269**

Overview **269**

Purpose **270**

Conserving Riparia for Biodiversity **271**

Conserving Riparia for Ecosystem Services **274**

Conserving Riparia for their Hydrologic Effects **276**

Riparian Conservation in a Management Context **277**

Human Benefits from Riparian Conservation **282**

Emergence of New Conservation Legislation **284**

Riparian Conservation for the Long Term **287**

Approaching Scientific Uncertainty in Ecological Problem Solving **287**

Integrating Ecological Knowledge with the Social and Ethical Aspects **288**

Conclusions **288**

10 Restoration **291**

Overview **291**

Purpose **292**

General Principles and Definitions **294**

Returning to More Natural Hydrologic Regimes **295**

Developing a Restoration Plan **301**

Getting Organized **302**

Identifying Problems and Opportunities **302**

Contents

Defining Goals and Objectives 303

Implementing, Monitoring, and Evaluating 304

Financial Incentives 305

Setting Priorities 306

Assessing the Ecological Integrity of Riparia 308

Specific Enhancements 314

Implementing Riparian Silvicultural Practices 314

Reintroducing Large Woody Debris 314

Designing Riparian Buffer Zones 317

Prescribed Grazing 319

Optimizing Riparia for Biodiversity 322

Conclusions 322

11 Synthesis **327**

Overview 327

Purpose 328

Riparia as Keystone Units of Catchment Ecosystems 329

Riparia as Nodes of Ecological Organization 331

A Unified Perspective of Riparian Ecology 332

Developing a Future Vision 333

Scenario Development 334

Forecasting, Prediction, and Projection 335

Emerging Perspectives, Tools, and Approaches 337

Emerging Perspectives 338

Emerging Tools and Approaches 341

Aboveground 341

Belowground 346

General Techniques 348

Principles for the Ecological Management of Riparia 350

Global Environmental Change 352

Conclusions 355

Bibliography **357**

Index **415**

Foreword

Despite years of study, the riparian zone remains a frontier for ecosystem study and landscape restoration and management. The riparian zone is potentially so important to ecological function in the overall catchment (watershed) as to be described by Robert J. Naiman, Henri Décamps, and Michael E. McClain as "the new important challenge we face at the present time." These experienced workers, the authors of this volume, are in a position to know. Yet, in many cases this zone may be so invisible to the causal observer that it becomes a part of the continuum from water to land. It is clear, however, that the serious researcher must assess the ecological functions of this zone, and also look *both* ways from the land–water interface (the shoreline) to gain insights about the overall functional processes of the entire catchment (e.g., Hynes 1975, Likens 1984). This boundary region can function as (1) a filter and/or modifier for organisms, water, and matter moving within the landscape; (2) a plane for the budgetary accounting of this flux of water, matter, and organisms; (3) an area of enhanced biological productivity, diversity, and aesthetics; (4) an area of specialized habitats, including specialized habitat for birds and other terrestrial biota and a spawning and nursery area for aquatic organisms; and (5) a zone for various and unique ecosystem functions, such as flood and erosion control, within the landscape.

The riparian zone can affect water quality and the functioning of aquatic ecosystems as well as the habitat diversity and functioning of terrestrial ecosystems. For example, by providing shade, inputs of coarse and fine particulate organic debris, and shoreline entanglements and complexity, riparian zones significantly enhance habitat quality in associated aquatic systems. These diverse and important features of land–water interactions depend on the unique structural dimensions and composition of this complicated ecotone or boundary between open water and the upland drainage basin.

Is the riparian zone a discrete, functioning ecosystem? Is the ecosystem concept robust enough to embrace this highly variable and variably bounded region of the larger landscape? Addressing such conceptual issues is challenging, and as valuable as the answers are, these questions are important for dissecting and understanding the overall structure and function of the landscape. For example, early conceptual views considered stream and river ecosystems as functionally inseparable components of the catchment (Bormann and Likens 1967), but only recently has this relation been quantified (e.g., Bernhardt et al. 2003). Likewise, what is the quantitative ecological and biogeochemical role of the riparian zone within a catchment?

Unfortunately, much of the riparian habitat in the United States and elsewhere has been degraded, seriously compromised, or threatened by human activity, such as land clearing, grazing, water withdrawal, waste disposal, and human habitation. Yet because riparian zones have critical functions in controlling hydrologic extremes, in cycling and retaining nutrients within the landscape, in reducing the flushing of biologically active contaminants to downstream estuaries and bays, in regulating the transmission of pests and diseases along rivers, in minimizing loss of eroded material from upland areas, and in providing unique habitats for organisms, there is great concern about the protection, restoration, and management of these ecologically vital areas. How are riparian zones formed and sustained, and how will human-accelerated environmental change (Likens 1991), including global climate change, affect their critical functions in the future?

Riparian areas are highly diverse, from broad forested bands in floodplains along lowland streams to narrow, intermittent areas along streams in highly incised, V-shaped valleys, and they are very dynamic (e.g., due to flooding). As such, there have been major technical challenges in attempts to develop a classification system for riparian areas (see, The Heinz Center Report, p. 148, 2002). Hundreds of scientific studies and papers have addressed the ecology and geohydrology of riparian zones, but possibly because these ecotones are so diverse and so dynamic, their boundaries so illusive, and because these areas have been so disturbed, they have been neglected for comprehensive study. This book is a welcome and valuable addition to the understanding and management of riparian systems throughout the world.

GENE E. LIKENS
Institute of Ecosystem Studies
Millbrook, New York

Preface

The intellectual roots of riparian ecology and management were formed several decades ago and are embedded in the development of catchment and floodplain perspectives. The pioneering achievements of G. E. Likens and F. H. Bormann at the Hubbard Brook Experimental Forest (Bormann et al. 1968, Likens et al. 1969, 1970), H. B. N. Hynes in Canada (Hynes 1975), and J. R. Karr and I. Schlosser (1978) in the midwestern United States provided visions of the importance of the land–water interface. With the emergence of ecosystem ecology as a legitimate field of inquiry, it was quickly realized that the choice of spatial boundaries, including riparian zones, had profound effects on the outcome and inferences of the results (Levin 1992). Indeed, it was realized that riparia were important locations where many processes changed or materials were transformed. Today, the frontiers in ecosystem science have evolved to place new emphasis on people–ecosystem interactions, spatial and temporal scale shifts, and cross-disciplinary linkages (Carpenter and Turner 1998)—and riparia are central to understanding and illustrating these issues.

The idea of a book on riparian systems was actually planted over 15 years ago as part of a small UNESCO-sponsored meeting in Toulouse, France. We were challenged to develop a program on land–water interactions, which eventually became "The Ecotone Programme," under the administrative leadership of the Man and the Biosphere Programme (MAB) and the International Hydrological Programme (IHP). The Ecotone Programme grew rapidly as researchers from approximately 25 countries shared their results and ideas in dozens of workshops and meetings, and in hundreds of publications, over the next decade. At the same time, the scientific community focused in many unexpected ways on the structure and workings of riparian systems— as mediators of land–water interactions—across the globe. Many researchers addressed basic science issues while others examined how knowledge from riparian systems could be applied to better catchment management—and the endeavor was highly successful on both fronts. When one compares the knowledge *ca.* 1986 with what is known today about riparian systems and their utility in resources management, the advances are astounding. And every year new discoveries continue to be made. The challenge of writing a book finally germinated a decade ago, but the daunting task of synthesizing the vast literature into a readable and understandable text was delayed until our intellectual courage could rise to the challenge.

What did we learn about land–water interactions, and especially riparian systems, in the early years? Perhaps, more than anything, we learned that even though riparian systems are the epitome of "heterogeneity," there are predictable patterns. As transitional semiterrestrial areas influenced by fresh water, they often exhibit strong biophysical gradients, which control energy and elemental fluxes and are highly variable in time and space. These attributes contribute to substantial biodiversity, elevated biomass and productivity, and an array of habitats and refugia. When riparian systems are properly managed, they make substantial and positive contributions to clean water as well as to ecosystem and human health.

We also learned that riparian systems are ideal places to illustrate the concepts of landscape ecology. A large variety of ecosystems—aquatic, semiaquatic, and terrestrial—may be found, side-by-side, along the hydrologic networks of entire drainage basins. These systems are typical of the patch–matrix–corridor model (Forman 1995). They are highly dynamic, to the point that their spatial organization interacts with their ecological processes in quasi-experimental ways. Not surprisingly, "landscape approaches" of riparian systems are becoming increasingly popular in the ecological literature. However—and this is a recurrent theme in this book—such approaches are not really "landscape approaches" if they lack the human and societal dimension. This means not only humans intervening on the spatial organization and ecological processes of riparian landscapes, but also humans interpreting, perceiving, and experiencing them. As a landscape, riparia are not merely places in the real world, they are also creations of the human mind (according to Lorzing's "The nature of landscape, a personal quest," cited by Rodieck 2002). Therefore, more than ideal places to illustrate the concepts of landscape ecology, riparia are ideal places to illustrate that landscapes are at the same time what we make, believe, see (or hear, smell, or feel), and know. Indeed, understanding the human and societal dimensions of riparia may be the new important challenge we face at the present time.

Riparian systems are associated with nearly all continental waters—lakes, streams, rivers, wetlands, springs, and estuaries. In this book, however, we largely narrow our focus to the riparian systems of small to medium-sized floodplain rivers where we elucidate fundamental patterns and processes that can be applied to other types of riparian systems. We describe heterogeneity at multiple scales of space and time, illustrate interactions among scales, and present conceptual models that integrate major system components. We illustrate how climatic and geological processes shape an array of physical templates, describe how disturbances redistribute materials, and illustrate how soils and subsurface processes form and are sustained on the major physical templates. Collectively, these processes strongly influence plant productivity and fluxes of channel-shaping large woody debris. Ultimately, the characteristics of riparian systems are an integration of climate (past and present), geological materials and processes, soil development and attendant microbial transformations, subsurface characteristics, plant productivity, animal activities, and large woody debris—and the active, continuous, and variable feedbacks among the individual components and the impacts of human societies that utilize them in so many ways.

In preparing this book, we conservatively examined over 5,000 professional articles, books, technical reports, and theses relating to riverine and riparian systems. One of our greatest challenges was identifying those synthesizing a topic or providing knowledge that could be applied to this book. Quite simply, we were stunned by the amount of literature, most of it generated since 1995, on riparia. We carefully examined articles written in English and French, and those written in German, Russian, Spanish, and Japanese if an English abstract proved interesting and useful. We suspect that there are many other articles that we were unable to read or discover.

We are greatly appreciative of the numerous individuals that freely shared ideas and information, and the dozens of organizations that supported our research efforts over the years. We are particularly grateful to the Andrew W. Mellon Foundation, the University of Washington, the Centre National de la Researche Scientifique (CNRS), the U.S. National Science Foundation, the Inter-American Institute for Global Change Research, the Ecosystems Center of the Marine Biological Laboratory (Woods Hole, Massachusetts), and the National Center for Ecological Analysis and Synthesis (NCEAS) of the University of California (Santa Barbara). The following individuals generously provided manuscripts, intellectual insights, advice, and technical assistance—and we are profoundly thankful: J. Aronson, E. A. Balian, V. Balta, K. Bartz, S. Bechtold, R. E. Bilby, P. A. Bisson, J. Braatne, S. Bunn, S. Carpenter, C. Dahm, M. Dixon, D. Drake, M. Duke, R. E. Edmonds, W. Elmore, J. Galloway, A. J. Glauber, S. Gregory, A. Gurnell, J. Helfield, F. Hughes, S. Jacobs, C. Johnson, G. Katz, C. Lake, B. Lassus, J. J. Latterell, S. Le-Floch, R. Lowrance, J. Makhzoumi, F. Malard, N. Minakawa, M. Molles, D. R. Montgomery, T. Moss, E. Muller, C. Nilsson, T. C. O'Keefe, H. Paerl, D. Peterson, N. Pettit, G. Petts, H. Piégay, G. Pinay, A. M. Planty-Tabacchi, G. Poole, K. Rogers, A. Rosales, A. Rosselli, B. Rot, J. Sabo, D. Sanzone, P. Shafroth, C. Simenstad, J. A. Stanford, J. C. Stromberg, E. Tabacchi, D. Terrasson, K. Tockner, L. Toth, M. G. Turner, J. V. Ward, C. E. Williams, and R. C. Wissmar. We extend our special thanks to Ava Rosales and Deanne Drake for assistance with the references, permissions, and collation of figures.

Robert J. Naiman
Seattle, Washington, USA

Henri Décamps
Toulouse, France

Michael E. McClain
Miami, Florida, USA

Introduction

Overview

- Riparian zones are transitional semiterrestrial areas regularly influenced by fresh water, normally extending from the edges of water bodies to the edges of upland communities. We introduce riverine riparia, place riparian systems in a landscape context, articulate the scope and purpose of the book, and present the environmental settings in which riparia occur.

- The book addresses the physical processes creating and maintaining riparia, the ecology of biotic communities inhabiting riparian zones, and the interactions with human cultures and management. Although special attention is given to vascular plants, the goal is to present a holistic perspective of floral and faunal assemblages in a system-scale perspective. Riparia are perceived as networks within catchments where biophysical processes linking terrestrial and aquatic systems converge within landscapes.

- Natural river systems are highly dynamic and characterized by multidimensional gradients. Constrained reaches with narrow riparian zones alternate along river courses with expansive alluvial floodplains. Riparian species are variously adapted to exploit the spatially and temporally dynamic habitat mosaic created by gradients in available materials and disturbance regimes. The resultant groundwater–surface water exchange pathways play major roles in structuring riparia and in determining their functional properties.

- Humans have shaped riparian landscapes since the beginning of human settlement in river valleys. Human activities have resulted in river systems characterized by reduced spatial and temporal dynamics, simplified gradients, truncated interactive pathways, and disconnected landscape components. A clear understanding of human perceptions of riparia is necessary for building a sustainable dynamic equilibrium between nature and culture.

- Riparia are an integral part of successful land management programs. A robust understanding of riparian ecology, as well as thorough monitoring and evaluation, is fundamental for successful planning and action. A sound scientific basis is requisite for effective conservation and rehabilitation of river riparia.

- Conceptually, riparia are closely linked to the main concepts of river systems ecology, namely the river continuum, the serial discontinuity, the flood pulse, and the hyporheic corridor concepts. However, perhaps the most effective perspective for understanding riparia is provided by the hierarchical patch dynamics concept.

Purpose

Riparian systems are transitional semiterrestrial areas regularly influenced by fresh water, usually extending from the edges of water bodies to the edges of upland communities. Because of their spatial position, they integrate interactions between the aquatic and terrestrial components of the landscape. They are dynamic environments characterized by strong energy regimes, substantial habitat heterogeneity, a diversity of ecological processes, and multidimensional gradients (Naiman et al. 2005). They are often locations of concentrated biodiversity at regional to continental scales. Two schematic representations of river corridors (Ward et al. 2002) may be used to visualize how riparian systems are spatially arrayed in alternating sequences of constrained and floodplain channels (Figure 1.1) and in braided-to-meandering channels (Figure 1.2). In Figure 1.1, lateral and vertical hydrologic exchange is concentrated near the river in constrained reaches, whereas it extends laterally and vertically in increasingly larger floodplain reaches. Figure 1.2 distinguishes a variety of lotic, semi-lotic, and lentic surface waters in a schematic river corridor from a braided to a

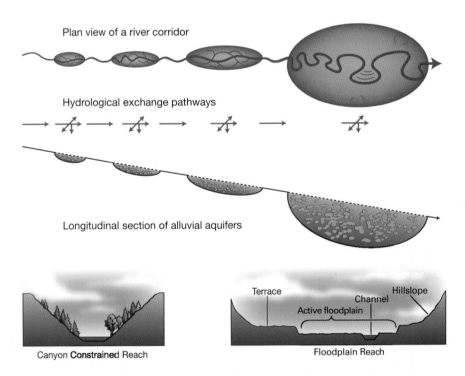

Figure 1.1 Schematic configuration of a river corridor as an alternating sequence of constrained and floodplain reaches (after Ward et al. 2002).

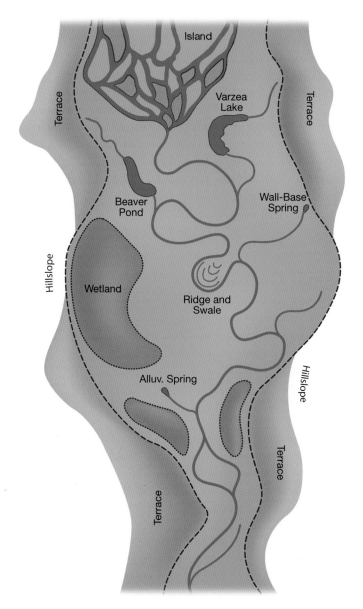

Figure 1.2 Surface water bodies and basic geomorphic features of a schematic river corridor in a braided-to-meandering transition zone. (after Ward et al. 2002).

meandering zone. These water bodies may be interconnected by surface waters during floods, but they are connected only by groundwater between floods.

Riparian systems form networks within *catchments* (i.e., drainage basins delineated by *watersheds* that are the *topographic divides or boundaries of the catchments*). The great river basins of the world are composed of hundreds of subbasins or tributary catchments, all of which contribute water and materials to riparian zones and to the main river course reaching the ocean. Lakes and all manner of wetlands (e.g., fens, bogs, and swamps) and associated groundwater aquifers may occur within the sub-catchments—and all have riparia of some type. In this book, we focus on riparian zones associated with running waters, largely excluding nonriverine riparian systems and wetlands associated with estuaries. This is for pragmatic reasons only. The professional literature on riparian systems is expanding exponentially, making a more focused treatment necessary.

Riparian zones are widely defined in terms of local conditions, and many people perceive riparia simply as plant communities growing on stream banks. Our approach is more expansive, examining riparia as dynamic, three-dimensional biophysical structures set in complex river corridors and cultural matrices from headwaters to the sea. Hence, we first provide a common background or context to define the hydrological, ecological, landscape, and cultural settings within which riparia occur. We then present the scope and purpose of the book, with a focus on some commonly assumed functions and tenets of riparian systems.

Hydrological Context

Rivers are indeed the arteries of continents, draining catchments that vary in size, geomorphic setting, biotic assemblage, and climate. It is well known that oceans recharge the hydrologic cycle, sending moisture inland to feed rivers in dynamic waves of precipitation falling as rain or snow or both, depending on locality, elevation, and time of year. Rivers gather the water to form complex surface (*epigean*) and subsurface (*hypogean*) channel networks that begin on or near the catchment divides (e.g., watersheds).

The Amazon, the world's largest river, provides a well-known example of these processes. It begins as headwater streams in the high Peruvian Andes and quickly descends ~5,000 m through precipitous channels. It then meanders across the northern Brazilian lowlands for more than 5,000 km to the Atlantic Ocean, a course where the elevation change is less than 50 m and the riverbed is actually below sea level in places owing to the erosive power of the massive river discharge (Sioli 1984, Junk 1997, McClain et al. 2001). The magnitude, variability, and characteristics of water and materials supplied from the Andes determine many aspects of the geomorphology, biogeochemistry, and ecology of the mainstem Amazon and the broad floodplains (see Figure 1.3). The diverse productive forests of the riparian floodplains become, in turn, important energy sources for heterotrophic communities in the main stem river (Junk 1997).

Another example is Triple Divide Peak in Glacier National Park, Montana where the waters of the three great rivers of North America begin. On the north slope of the peak, waters run via the Waterton and then Saskatchewan rivers to Hudson's Bay in the Arctic Ocean. On the west and south flanks, runoff courses down the Flathead River to the Columbia River and the Pacific Ocean. And on the east side, small spring brooks emerging from the talus slopes at the base of the peak initiate the mighty Missouri-Mississippi River that sends sediment plumes far out into the Atlantic Ocean's Gulf of Mexico. In a similar manner, three great rivers of Europe—the Danube, Rhône, and Pô—have glacier-fed headwaters in the Swiss Alps that historically had expansive floodplains in the lower reaches. The huge river systems in Asia— the Ganges, Lena, Mekong, Yangtze, Yellow—also coalesce in great mountain ranges,

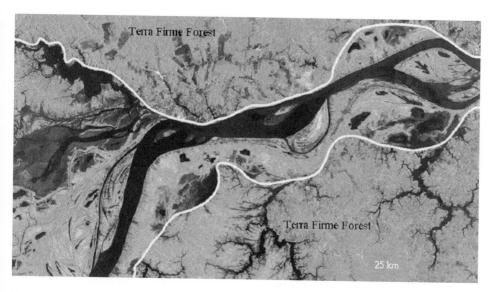

Figure 1.3 Radar image of the Amazon River and its floodplain to the west of Manaus, Brazil. White lines mark the approximate limit of the floodplain, with upland "terra firme" forests beyond. The dynamics of the river and floodplain are evident in the complex arrangement of scroll bars, floodplain channel, and lakes. The image is 75 km wide, the center of the image is Lat. 3.2°S, Long. 60.5°W, and was taken on October 20, 1995 by the JERS-1 satellite.

drop through deep gorges, and flow across plains or tundra, dissipating erosive energy in extensive floodplain forests and enormously complex deltas. The great rivers of Africa—the Nile, Congo, Zambezi—and South America—the Amazon, Orinoco, Parana, São Franciso, Magdalena—all have their specific characteristics shaped by their physiographic settings and the nature of their riparia. Sadly, most of the world's great rivers and their riparia are substantially altered from historic conditions, owing to centuries of flow control with dams, revetments, and abstractions (Dynesius and Nilsson 1994, Meybeck 2003).

Rivers move much more than just water. Erosion of the uplands and subsequent downstream transport and deposition of sediment in aggraded valleys is a natural attribute of all rivers. Headwaters typically are steep gradient and floods may move very large boulders, but as the slope of the channel decreases or aggrades in the valleys, sediment deposition and lateral reworking result in a dynamic biophysical mosaic across the floodplain (Figure 1.4). Every year rivers expand and contract in response to seasonal changes in runoff and to the shape of the valley. For example, in bedrock canyons, very little expansion is possible, whereas in aggraded reaches, floodwaters may penetrate into and flow across extensive floodplains for many kilometers.

Part of the water that collects in the catchment, including meltwater from snow and glaciers, infiltrates the substrata, forming shallow and deep aquifers that eventually contribute to surface runoff during periods of low discharge. If precipitation or meltwater volume exceeds infiltration capacity, water runs off the land in superficial channels. Periodically, runoff associated with flooding can be very erosive. The general pattern of river hydrology is high runoff during wet weather or snowmelt in the headwaters with base flow maintained by the more attenuated groundwater flow paths. In

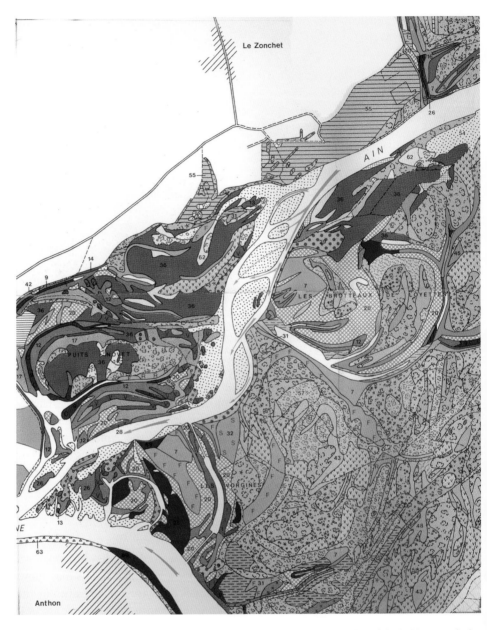

Figure 1.4 The vegetation mosaic of the lower Ain River, France: an illustration of the habitat complexity and diversity created by river meander migration (after Pautou and Girel 1986).

all cases, above and below ground, transport of dissolved and particulate materials by river flow is always occurring, and therefore the catchment is constantly being reshaped. In general, water-mediated erosion and transport results in bank or bottom cutting of the channel in one place and filling in another. It is the processes of cut-and-fill alluviation that are the primary formative processes of riverine landscape diversity, including riparian features.

The dissolved ion content and the erodibility of transported materials in runoff reflects the geology of the catchment, also affecting riparian characteristics through the delivery of nutrients and other important chemical elements. Some rivers are dilute because they drain uplands composed of granitic, basaltic, or other bedrocks that may produce sediments but few dissolved ions. However, vegetation distribution and abundance within the catchment and processes such as fire and herbivory that alter vegetative vigor, productivity, and succession also influence the chemical signature of rivers, especially as regards riverine export of plant growth nutrients and dissolved organic matter (Likens and Bormann 1974, Leopold 1994, McClain and Richey 1996). In general, rivers tend to increase their dissolved solids load with distance from the headwaters. However, that is not always the case. For example, the Madison River in Montana begins in the active volcanic areas of Yellowstone National Park and is fed by geysers and underground channels laden with sulfurous salts. But, farther downstream, tributary waters draining the more inert bedrocks of the Rocky Mountains substantially dilute the salt load. Streams draining rain forests in the active volcano belt of Costa Rica likewise are enriched by geothermal groundwater, whereas others are not, thus presenting a wide array of stream chemistries and communities within a local area (Pringle et al. 2000). In all cases, flooding tends to increase sediment loads and to decrease ion concentrations.

Thus, the basic physical character of rivers is dynamic in all three spatial dimensions (longitudinal, lateral, vertical) with concomitant influences on riparia. Expansion and contraction occur throughout the longitudinal course of the river (the first spatial dimension) in relation to precipitation and geomorphology (Naiman 1992). Precipitation and geomorphology also influence the lateral extent of the floodplain (the second spatial dimension). The third spatial dimension (vertical) is equally complicated. One segment of the channel may be fed largely by upwelling groundwater, whereas at other locations surface runoff may penetrate into bed sediments (alluvium) accumulated over millennia by cut-and-fill alluviation. During flooding, surface flow may recharge groundwater aquifers and spill out over the floodplains, eroding or depositing sediment in accordance with the energy dynamics of water interacting with geomorphic features. During dry periods, flow in the channel may be maintained by groundwater draining alluvial and karstic aquifers. Thus, rivers are not merely conduits for runoff from headwaters to the oceans. Rather, rivers are dynamic multidimensional pathways along which aquatic-terrestrial linkages vary spatially (in three spatial dimensions) and temporally (often considered as a fourth dimension; Ward 1989). Anthropogenic influences contribute greatly to this variation, as river valleys have been foci for human settlements and commerce for millennia.

Ecological Context

Riparian and riverine plants and animals are variously adapted, often uniquely so, to exploit the dynamic nature of river systems (Junk et al. 1989, Décamps 1996, Naiman and Décamps 1997, Naiman and Bilby 1998). For example, cottonwood (*Populus* spp.) seeds deposited along alluvial rivers in western North America germinate only during a brief period of suitable moisture content on fine sediment as floodwaters recede. Hence, seed release has to coincide precisely with the flood recession so as not to wash the seeds away or dry the substratum so quickly that seedling roots cannot grow fast enough to stay in contact with the capillary fringe of the water table.

Additionally, healthy rivers and their associated riparian zones are complex interconnected corridors that allow biota to disperse and adapt to particular conditions at particular locations. Fish and other aquatic and semiaquatic vertebrates and their prey, along with plants, microbes, and organic detritus, compose complex food webs within the habitat complex of the stream network, both above and below ground. Populations cluster in favored locations where resources support enough reproduction to sustain them, with gene flow maintained by immigration and emigration.

Dispersal is a natural feature of all populations in the struggle for living space and in the acquisition of resources needed to complete life cycles. River networks are ideal corridors for dispersal of individuals or propagules. Some organisms have life stages that are spatially dispersed along the river corridor. For example, migrating birds use riparia as navigational aids, stopover sites, and brood-rearing habitat. Conservation of many birds, such as sage grouse—greater sage grouse (*Centrocercus urophasianus*) and Gunnison's sage grouse (*C. minimus*)—requires ecologically functioning riparian areas (NRC 2002). Additionally, riparia offer unique habitat for many species, including the adult stages of numerous invertebrates, amphibians, reptiles, birds, and mammals that spend much of their life in water. One well-known example is beaver (*Castor canadensis*) that not only use riparia as habitat but also shape its community composition and the spatial-temporal dynamics of the vegetation (Naiman and Rogers 1997). Finally, plants adapted to flooding grow on the banks and on floodplains in complex vegetative arrays associated with variation in soils and local hydrology.

It is also crucial to consider riparia as systems where conservation and development need to be integrated, particularly where riparian resources and biodiversity are essential for livelihoods (Salafsky and Wollenberg 2000). But this is not an easy task; there are major deficiencies in linking ecosystems and management institutions and in rules governing the use of riparian areas (Berkes and Folke 1998). Often, management does not reflect the complexity and multiple functions of riparia. New institutions with adaptive comanagement approaches are necessary for a successful integration of conservation and development, as convincingly suggested in protected areas in the Ganges River Floodplain in Nepal (Brown 2003). There, the protection of emblematic, rare, and endangered species such as the Bengal tiger (*Pantera tigris*) and the Asian one-horned rhinoceros (*Rhinoceros unicornis*) is balanced against human needs.

Many variations on this general theme occur as organisms exploit the spatially and temporally dynamic mosaic of habitats within the interconnected pathways of rivers. In tropical regions, such as in South America and in Africa, fish life histories are tuned to the predictable flooding that provides access to floodplain lakes and riverine wetlands where food resources are seasonally abundant (Welcomme 1985). Indeed, the floodplains produce many times more fish biomass than the main river channels. This biomass production in turn supports a wide variety of higher consumers, including humans. Aboriginal populations focused on floodplains, locating villages in strategic locations for exploiting floodplain fisheries and other biotic resources, particularly edible plants as well as rushes and trees for building shelters.

Habitat for riverine and riparian organisms is a constantly changing mosaic, biophysically dynamic in space and time, and the biota are uniquely adapted to the dynamics of the system (Salo et al. 1986). Flow networks often encompass lakes and groundwater aquifers that are just slow-flowing environments within river continua embedded in the terra firma. Traditionally, ecologists have focused research on either purely terrestrial or aquatic attributes and processes, often attempting to segregate physical and biological attributes. Today it is well recognized that the key to understanding riverine and riparian networks is to integrate functional processes driving linkages between terrestrial and aquatic components across multiple biophysical gradients, from watershed divides to the oceans. This is riparian ecology.

How important are riparia in a catchment context and across biomes? Some studies provide preliminary support for the generality of riparian controls on river ecosystem structure and function, thus integrating landscape and food web ecology (Polis et al. 1997). Insights have been provided, for example, on marine nutrients from salmon (*Oncorhynchus* spp.) improving the growth of riparian trees (Helfield and Naiman 2001) and riparian animals such as river otter (*Lutra canadensis*) (Ben-David et al. 1998) and brown bear (*Ursus arctos*) (Hildebrand et al. 1999). Key issues concerning riparia include their potential role as keystone units of catchment ecosystems, which include acting as nodes of ecological diversity and providing clean water and flood control. A crucial issue is knowing how to integrate the complex multidimensionality into management decisions about riparian systems, especially when most are already culturally modified.

Landscape Context

Riparia form dendritic networks and, as such, may be the dominant structuring attribute that organizes catchments and landscapes. For example, riparian vegetation may act as buffer zones along rivers in various ways. Riparia minimize downriver flooding by physically slowing the water, absorbing it or increasing the rates of evapotranspiration. Riparia trap sediments and therefore influence downriver sedimentation. Finally, riparia constitute habitat for rare or uncommon species, and these species may move along the unique dendritic networks of riparian vegetation. Additionally, within a given climatic/geomorphic setting, fluvial dynamics and groundwater and surface water interactions have important impacts on the structure and function of riparia at the landscape scale. Throughout the book we examine this potentially unifying theme.

Landscape ecology, the study of interactions between spatial patterns and ecological processes in the context of spatial heterogeneity, holds the potential for developing a truly holistic perspective of riparian systems, one that rigorously integrates structure, dynamics, and function in a catchment context (see Sidebar 1.1; Tockner et al. 2002). Several decades have passed since it was first fully acknowledged that the character of the catchment basin, including riparian areas, fundamentally influences biotic patterns and processes in streams and rivers (Hynes 1975). Nowadays, river corridors—inclusive of riparia—are considered major components of viable landscapes (Malanson 1993, Forman 1995). A thorough analysis of riparian ecology in a landscape context may be attained in several ways, but using a hierarchical patch dynamics perspective has proven to be most useful (Townsend 1996).

Sidebar 1.1 Toward a Landscape Perspective of Riparia

A landscape perspective of riparian systems is frequently advocated in the professional literature, even if the meaning of such a perspective may differ between authors. This perspective is often an ecological one: Riparian systems are viewed as multiscaled nested hierarchies of interactive terrestrial and aquatic elements—that is, homogenous units (or patches) observable within a landscape at a given spatial scale (Poole 2002). According to Forman (1995), land mosaics along rivers appear as corridors where the interactions between water table, land surface, soil type, and slope determine the richness of vegetation and habitat. Observed patterns result from hydrologic flows, particle flows, animal activities, and human activities.

Such a perspective allows one to answer questions such as: How do patterns composed of patches and boundaries influence ecological processes? How, in turn, do ecological processes influence spatial organization? What are the causes and consequences of spatial heterogeneity at various scales? Thus, landscape ecology focuses on models and theories about spatial relations, on building a science for action, and/or on interdisciplinary approaches (Turner et al. 2001). Importance is given to the effect of spatial configuration on ecological processes, and the areas investigated are larger than those traditionally studied in ecology. Another aspect is the consideration given to humans and society, particularly as landscapes are comprised of both nature and culture—objective and subjective representations of the environment.

Thus, a landscape ecology of riparia is underpinned by two key ideas:

1. Spatial configuration influences the relationships developed by living beings between themselves and their environment, requiring one to understand how spatial organization of the environment shapes processes that drive the dynamics of populations, communities and ecosystems (Turner et al. 2001).
2. "Nature" cannot be divorced from "man and society," requiring one to be open to other disciplines often better qualified to study spatial organizations and humans (such as geography, history, anthropology, economy, and sociology). This also requires one to incorporate symbolic and aesthetic values and to remember that every landscape has witnessed a culture and therefore has a memory as well as an environmental *savoir faire* created and recreated with time (Nassauer 1997).

Our philosophy is that a landscape perspective can aid one enormously in understanding the causes and consequences of the current transformation of riparia, so long as it is inserted into a plurality of approaches where ecologists share their principles with landscape architects and designers and with the society who participates in the creation and the cultural representation of riparia.

A hierarchical patch dynamics perspective addresses the fundamental attributes of riparia, particularly the dynamics of heterogeneity in space and time, visualizing interactions between structure and function at scales ranging from microhabitats to landscapes. It also provides a framework for linking riparian ecology to key concepts underpinning river ecology, namely the river continuum, serial discontinuity, flood pulse, and hyporheic corridor concepts. This framework suggests a complex, dynamic, and nonlinear functioning for riparia involving a full range of interactions between the biophysical components, and thereby shaping the emergent ecosystem-scale characteristics.

In general, vegetation—whether upland or riparian—is the key moderator of cut-and-fill alluviation. Forests, shrub lands, and grasslands intercept and retain runoff and increase infiltration. However, evapotranspiration by vegetation is a primary feedback to the atmosphere that can deplete soil moisture, tap near-surface aquifers, and even withdraw significant amounts of stream flow from the channel. Vegetation moderates soil conditions as leaf litter is decomposed by soil microfauna, changing uptake trajectories of nutrients used by plants for growth. Nutrient cycles in the soil-vegetation complex of uplands determine the ion contents of runoff and thereby influence the production dynamics of riparian forests. Riparian forests, in turn, create microclimates through shading and transpiration and thereby influence stream and floodplain temperature patterns as well as nutrient cycles. Moreover, riparian trees and other vegetation eroded into river channels vastly alter water and material flow paths. Hence, both living vegetation and wood debris deposited in the riparian corridor change the ability of the water to transport sediments, and this changes channel shape over time, especially in expansive floodplain reaches that are heavily forested (Naiman et al. 2002, Gregory et al. 2003). In any case, riparian-derived wood strongly interacts with the bed sediment characteristics—and the load, the water volume, and the slope of the channel—to determine channel geomorphology over multiple time scales.

Fires, drought, mass wasting, wind throw, herbivory, and other natural disturbances, coupled with human interventions such as logging, urbanization, farming, and damming, alter vegetative patterns and soil–plant nutrient exchange at a variety of scales. This has direct consequences for ecological processes in rivers—such as productivity, biodiversity, sediment transport, and live and dead wood recruitment—as well as for riparia, which also are influenced by interactions with upland vegetation (e.g., seeds, leaves, and other organic matter recruitment) and grazing, nutrient fluxes, and other interactions with terrestrial animals and, finally, with humans.

Cultural Setting

As any landscape, riparia are both "natural" and "cultural." People see them differently according to the social group they belong to, as well as according to where they are from. And the way people see riparia may change with time. As Han Lorzing reminds us, a landscape is not merely a place of the real world; it is also a creation of the human mind (*in* Rodieck 2002). Riparia are at the same time factual (landscapes that we know), man-made (landscapes that we make), perceived (landscapes that we see, or we hear, smell, or feel), and emotional (landscapes that we believe).

This book is about riparian ecology. Nevertheless, as authors, we are conscious that ecology does not tell the whole story and that history, for example, may be more reliable than theory when people make decisions (Jackson 1994). This is not to expect historical knowledge to provide recipes or strategies for ecological management, conservation, or restoration. This is to acknowledge that "over centuries cultural habits have formed which have done something with nature other than merely work it to death, that help for our ills can come from within, rather than outside, our shared mental world" (Schama 1995). Such a shared mental world changes in time and space. In presently developed countries, reading books and looking at drawings, paintings, photographs, or films influence our "mental world." Everywhere, what people think should be a "natural" riparian landscape is strongly influenced by their "cultural" history, which differs between social groups and countries.

As eloquently suggested by Joan Nassauer (1997), landscapes more apt to be protected are those that are appreciated—in other words, those that satisfy our cultural or aesthetic aspirations. By incorporating principles that refer to ecological health to the cultural or aesthetic aspirations, we can obtain culturally sustainable landscapes. Such landscapes require sustained attention to the dynamics of ecological functions. They also require recognition of the limits and uncertainties of knowledge, leading, for example, to the protection of remnants of ecosystems even if we ignore why it may be interesting or useful to protect them. In addition, sustained attention to change must be remarkable in the sense that it must indicate an intention to care for riparia for the long term. Thus, a riparian landscape has a better chance of being cul turally sustainable if its ecological functions are known and if signs of intention fo long-term care are apparent.

Change probably characterizes the best examples of riparia: ecologically, culturally and scientifically. Ecologically, riparian landscapes change because they are highly dynamic ecological systems, independent of those who care for them. Culturally, the perception of riparian landscapes changes continuously in time and space because social groups evolve to view them differently. Scientifically, the perception of riparian landscapes is also changing because knowledge of their structure and function i improving—at a particularly high rate during the last two decades. These character istics make the study of riparia a fascinating topic in a period of accelerated envi ronmental and societal change.

Rationale for Riparian Ecology

Riparius is a Latin word meaning "of or belonging to the bank of a river" (*Webster New Universal Unabridged Dictionary*). The term *riparian* generally replaces the Lati and normally describes biotic communities living on the shores of streams and lakes Herein we use the term *riparian* as an adjective and the term *riparia* as a singular o plural noun to encompass the biotic assemblages of the aquatic-terrestrial transition zones associated with running waters. Riparian communities consist not only o higher plants, but also the flora and fauna, including those associated with the soil/sediment system. It is not, however, possible to present an even treatment of the various biotic communities. The best information is available for vascular plants, with a surprising paucity of data on other groups.

From this perspective, riparian ecology is very much focused on the ecology of rive floodplains, which can contain fluvial lakes and wetlands connected to river channel by surface and groundwater flow paths. Examples of the habitat mosaic of the ripar ian zone are shown in Figures 1.5, 1.6, and 1.7. Riparia clearly encompass the transition zone or ecotone between aquatic and terrestrial components of landscape (e.g., Junk et al. 1989, Naiman and Décamps 1990, NRC 2002). However, functiona processes are not limited to lateral and longitudinal vegetation gradients related to superficial water hydrology. The vertical dimension, a main determinant of so wetness, is a primary attribute determining the presence of hydrophilic vegetation and the nutrient and energy sources that are carried in groundwater flow paths coursing through riparia via alluvial aquifers (Hynes 1983, Stanford and Ward 1988).

Riparian zones are multidimensional systems shaped by some basic principles:

1. Water saturation gradients are determined by topography, geologic materials, and hydrodynamics.
2. Biophysical processes are driven by dynamic water saturation and energy gradients.
3. Surface and subsurface entities provide feedbacks that control organic energy and material fluxes.

Figure 1.5 The Garonne River downstream Toulouse, France. In most places the natural riparian forest has been replaced by poplar plantations and the riverbed has been dredged, lowering the river base flow level, which in turn affects the characteristics of the riparian forests (Photo: G. Pinay).

Figure 1.6 The Tagliamento River in northeast Italy is the last river in the Alps retaining a high degree of hydrogeomorphic dynamics. It is a braided river fringed by a ribbon of riparian forest along its entire length (Photo: K. Tockner).

Figure 1.7 Riparia are complex systems, especially where there are large variations in the physical environment and where natural populations of herbivores remain abundant—as along the semiarid Sabie River in Kruger National Park, South Africa (Photo: R. Naiman).

4. Biotic communities are structured or arrayed in space and time along gradients in three dimensions: longitudinal, lateral, and vertical.

Indeed, a basic premise is that dynamic interactions between ground and surface waters determine the persistence and productivity of riparian communities. Riparia are characterized by often large, complex biophysical gradients and are structured by antecedent geomorphic conditions, flood dynamics, and animal activities.

However, in an increasingly human-dominated world, riparia must be viewed in a landscape context—that is, as natural-cultural systems (see Sidebar 1.1). Such a perspective of riparian ecology extends what is currently known into a broader, more holistic synthesis. While surface and subsurface patterns and processes act as key drivers for sustaining riparian goods and services, it is human perceptions and cultural representations of landscapes that shape the dynamic complexity of contemporary riparian systems (Figure 1.8). These must be more resilient owing to a better understanding of land–water interchange at broad spatial scales, interactions of multiple drivers, slowly changing but powerful variables, and thresholds (see Sidebar 1.2). We develop this rationale in the chapters that follow.

In Chapter 2, riparia is placed in the context of catchments to address the geomorphic and hydrologic environments and processes. Chapter 3 examines geomorphic and biotic classifications from a hierarchical perspective, linking the physical

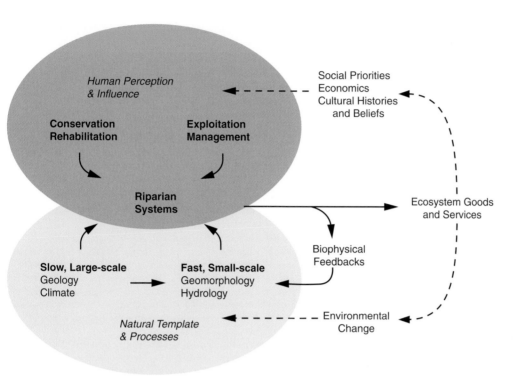

Figure 1.8 A synthesis of the approach to riparia adopted in this book, whereby the maintenance of riparian ecosystem goods and services depends on integrated cultural and natural factors. Riparian ecosystems evolve in response to large-scale, long-term physical variables like geology and climate, which determine the broad physical template and catchment characteristics. Smaller-scale, short-term physical variables determine the disturbance regime of riparia and strongly influence ecosystem function. Riparian biological communities also influence these small-scale physical variables through a number of feedbacks discussed in subsequent chapters. In many parts of the world, humans exert a controlling influence over riparian systems through their exploitation or conservation efforts. The nature of these influences depends directly on the perceptions of human societies toward riparia, where perceptions are influenced by social priorities, economics, cultural history, and beliefs. These factors are, themselves, ultimately controlled by the nature and value of ecosystem services provided by riparia. Humans also influence riparia indirectly by altering underlying physical variables like climate, sediment supplies, and hydrologic regime.

environment to biotic characteristics. Chapters 4 and 5 examine the biotic patterns and functions of riparian vegetation. Connectivity between riparian habitats, food webs, and the role of animals are the topics of Chapter 6, which demonstrates how riparia are controlled by interactive, nonlinear biophysical pathways at various scales of organization (single organism to entire catchments). In Chapter 7, human-driven environmental change is examined as a primary determinant of pattern and process in contemporary riparia. Human perceptions and management of riparia are covered in Chapter 8 and conservation and restoration issues are addressed, respectively, in Chapters 9 and 10. These chapters consider natural and cultural interactions in riparian settings by describing how human perception and management of riparian areas are interrelated. We conclude in Chapter 11 with a synthesis of riparian pattern and process and propose unifying principles for a robust riparian ecology.

Sidebar 1.2 Challenges for Riparian Science

Global assessments and many national assessments point to water as a critical limiting resource in the very near future. Globally, the supply of fresh water per person is declining, jeopardizing food production and the water supply for fish, wildlife, and natural ecosystems. Riparian systems, because they occupy the land–water boundary, will be the theater in which conflicting demands for water are played out in coming decades; hence, riparian science must provide the scientific underpinnings for water resource decisions. What should be the priorities for riparian science when faced with a world of increasing water shortages?

A truly integrated science of terrestrial and freshwater ecosystems is sorely needed. At present, the boundaries among subdisciplines of ecology as well as agency structures and political demarcations are not fostering science that seamlessly addresses complete, spatially extensive systems of land and water. There are promising beginnings in landscape ecology in the trend among aquatic ecologists to look beyond the shoreline and in the chapters of this volume. But a fully integrated framework for systems of land and freshwater has not yet coalesced, despite pressing needs and numerous calls. Building such a science is crucial and it must embody several key elements.

An integrated riparian science must embrace the full range of terrestrial–aquatic linkages to develop a much richer understanding of the bidirectional fluxes between these ecosystems. Ecologists historically have placed greater emphasis on fluxes of matter and energy from the land to the water than on fluxes from water to land. This emphasis was a response to urgent problems of eutrophication as well as recognition of the importance of hydrologic flow paths as integrators of land and water. However, understanding the magnitude and importance of fluxes of matter, energy, and information from the aquatic to the terrestrial systems should be enhanced in riparian science. This aspect of riparian systems is essential for understanding the aggregate behavior of land–water mosaics at broad scales.

Future riparian science must also tackle more directly the complex interactions among multiple drivers. As in other areas of ecology, driving variables that interact are often considered singly for simplicity's sake, despite recognition of multivariate causation and interactions among drivers. Riparian systems respond to myriad drivers that interact in very complex ways over multiple scales to determine system state and behavior. For example, flow regimes are determined by interactions among climate, geomorphic templates, land uses, vegetation patterns, and hydrologic modifications (e.g., levees or dams). Understanding how these drivers interact, and how their interactions vary in space and time, is critical to predicting the future state of riparian systems. Riparian scientists must embrace and fully explore this complexity.

Riparian science must strive to uncover ongoing but subtle changes that are difficult to perceive but that may have profound effects on riparian systems. The "big" changes, such as flood control, land-use conversion, or large nutrient inputs are conspicuous and have received considerable attention. However, other changes, though important, are more difficult to detect. For example, tree species composition has changed dramatically in riparian forests throughout most of the world, which may in turn influence ecosystem process rates and consumer populations. The abundance of coarse woody habitat in the riparian forests and littoral zone of lakes is slowly diminishing as rural residential development increases. In turn, aquatic communities may be disrupted in response to changes in habitat structure. The search to quantify and understand rates of change should be intensified, and the effects of slow but long-term changes must receive increased attention from both scientists and managers.

We challenge riparian science to identify thresholds, the conditions under which qualitative changes in the state and behavior of riparian systems are likely, and to determine whether such changes are likely to be irreversible. Ecological systems, in general, have a tremendous propensity to produce surprises. When integrated land–water mosaics are considered, with multiple drivers operating over a variety of scales, surprises are likely. The tangible consequences of qualitative shifts in riparian systems are likely to generate effects that cascade throughout natural ecosystems and human populations.

Change in riparian systems is eternal and inexorable. The future of viable riparian systems depends largely on trends in human population, demand for water, policies that allocate scarce water, and drivers such as changing climate and spatial configuration of watersheds. Restoration, in the sense of return to a baseline state, is neither imaginable nor practical. Instead, riparian science will be urgently needed in support of deliberate reorganizations of land-water interfaces. Such manipulations will be undertaken to use scarce water more efficiently to meet conflicting demands for social needs and support of ecosystems. Riparian systems must be more resilient, both to cope with unexpected extreme events such as storms and floods and to sustain their flows of ecosystem services in times of scarcity. Management for such resilience requires a stronger scientific foundation, which must include better understanding of land–water interchange at broad spatial scales, interactions of multiple drivers, slowly changing but powerful variables, and thresholds.

Monica G. Turner and Stephen R. Carpenter, University of Wisconsin, Madison

Table 1.1 Some Commonly Assumed Functions or Tenets of Riparian and Floodplain Ecology with Possible Alternative Explanations

Commonly Assumed Functions or Tenets of Riparian and Floodplain Ecology	Possible Alternative Explanations
Floodplains are areas of high biodiversity and production (Naiman and Décamps 1997, McClain et al. 2003).	Biota of riparia simply are transitional assemblages, mostly not in their optimal location along regional gradients; local gradients in riparia are so steep that production is extremely variable and overall not very high.
Riparia act as nutrient filters through interception of pollution-laden runoff (Lowrance et al. 1995, NRC 2002).	Riparia are nutrient sources for the river, owing to decomposition of leaves and other organic matter (as well as from lateral erosion).
Riparian zones are crucial habitat for long-range migrations of terrestrial animals such as neotropical birds (NRC 2002).	Birds and other animals may migrate preferentially through riparia simply in relation to spatial orientation with little regard to riparia per se.
Riparian soils form in ways similar to terrestrial soils (NRC 2002).	Riparian soils may be more allogenic than upland soils owing to continuous saturation.
Riparian trees are the keystones species for river ecosystems (Stanford et al. 1996, NRC 2002).	Riparian trees simply grow along wetness gradients and structures or habitats associated with wood debris do not contribute to the system more than other plant species might do.
Grazing and browsing are harmful to riparia (Donahue 1999).	Herbivory by large animals strongly and positively influences productivity and succession of riparia.
Floodplains harbor the most endangered (transformed) ecological systems on earth (e.g., by stream-flow regulation, urbanization, overgrazing, and exotic species invasions) (Tockner and Stanford 2002; Tockner et al. 2005).	Ecological systems associated with floodplains are highly resilient and display a great adaptive capacity.
Riparian landscapes are cultural representations of the reality as well as the reality itself (Décamps 2001).	Humans are just the main cause of destruction of "natural" riparia.
Riparia develop adaptive cycles with behavioral phases during which there are changes in resilience requiring a timing of management interventions (Gunderson and Holling 2002).	Resilience is a constant characteristic depending on intrinsic factors of the riparia under consideration.
Robust emergent properties can unify a new perspective for appreciation and management of riparia as functional units of river-scale catchments (this book).	There may be no consistency or universality of process outcomes owing to the very nonlinear nature of processes and responses in riparia.

Riparia are dynamic aquatic-terrestrial transition zones that are keystone elements of catchments. Nonetheless, there is little direct experimental support for this idea (but see Wallace et al. 1997 and Nakano et al. 1999). This book examines the data that support or refute this and other postulates, such as those listed in Table 1.1. These postulates result in a unified new perspective for the appreciation and management of riparia as functional units of riverine and catchment ecosystems. While riparia are of importance in relation to some of the most important issues of this new century—global change, sustainable development, environmental health, and equity—the postulates are also of importance in relation to human perceptions of riparia—perceptions highly variable in space and time as well as among social groups.

Setting the Stage

This book examines the body of theory, empirical evidence, and emergent properties of contemporary riparian ecology. Viewing riparia in the context of the ecology of large river catchments is needed to fully understand river-riparian-upland linkages and to compare processes and patterns across biomes. Such understanding is crucial for billion dollar (US$) river restoration projects, as proposed for the Kissimmee (Florida), Trinity (California), and Yakima (Washington) rivers in the United States and for sections of the Rhine and Danube rivers in Europe. Indeed, a robust riparian ecology also is needed to guide all aspects of river management, conservation, and restoration. Attaining a universal understanding and appreciation for riparian ecology in a catchment or landscape context means having healthy riverine systems for the long term. We ascribe herein to the convention that healthy riparia are key to providing sustained riverine goods and services, such as clean drinking water and productive wildlife and fisheries—which are important indicators of the quality of human life.

All large rivers of the world have expansive floodplains characterized by aquatic-terrestrial transition zones, and they are many times larger on an aerial or volumetric basis than the river channels themselves. This is the case of the Amazon River, whose several-million-year legacy of cut-and-fill alluviation has shaped various long-term succession states (Salo et al. 1986). This is also the case for other tropical rivers where natural fisheries are mainly produced on the river floodplains, not in the river channel (Welcomme 1985) and where the biomass of aquatic and semiaquatic organisms appears to be several, if not many, orders of magnitude higher on the floodplains than in the main river channel (Lowe-McConnell 1987). This is also the case for northern temperate rivers where rearing habitats in floodplains are crucial for sustained natural productivity of salmon and other fish (Reeves et al. 1998).

Clearly, a landscape perspective—including cultural perceptions and activities—has much to offer riparian ecology. We argue that riparia are one of the most powerful systems in the world to understand how spatial organization interacts with ecological processes. At the same time, because of the long history of human activity in floodplains, riparia provide useful examples to understand why human environmental perception is important when managing, conserving, or restoring human-dominated landscapes (Figure 1.8).

Catchments and the Physical Template

2

Overview

- The purpose of this chapter is to describe geomorphic and hydrologic processes influencing riparian system development and maintenance, paying special attention to the catchment context in which these processes operate and how they vary over space and time.

- Catchments are areas of the land surface in which all runoff drains to a single point on a stream or river channel. They are bounded by drainage divides and range in size from hundreds of square meters to millions of square kilometers.

- Catchment drainage networks may have dendritic, palmate dendritic, or trellised forms, depending on the nature of underlying geology. These networks vary in drainage density and gradient, which affect riparia by impacting flood intensity and stream power, respectively.

- In the system developed by Strahler (1964), groundwater-fed streams are ranked as first order and streams increase in order downstream as streams of equal order combine. Riparian ecosystems change in function with increasing stream order, and in any catchment the great majority (>80 percent) of length occurs in low-order streams (orders 1–3).

- Evidence for past climates, tectonic activities, and human alteration are recorded in the alluvial features of catchments, such as terraces and buried paleochannels. These features continue to affect modern-day catchment processes by constraining river movements or influencing hydrologic connectivity between river channels and adjoining subsurface or floodplain environments.

- River channel networks are highly structured and change in a systematic fashion from headwaters to mouth. These systematic changes can be grouped into hierarchies of habitats and alluvial features that reveal underlying processes controlling riparian form and function.

- The most basic geomorphic processes in catchments are erosion, transport, and deposition. These processes operate across all time and space scales but vary in relative importance along drainage networks. Erosive processes dominate in headwater regions, whereas deposition processes dominate near the bottom of catchments draining to the ocean or enclosed basins. Transport dominates in the mid-reaches of river systems.

- Erosion scours and eliminates riparian habitats and occurs when the shear stress imposed by flowing water exceeds the shear strength of the material over which it flows. The dominant forms of erosion include down-cutting and lateral movement of channels and scouring of channels and floodplains.

- Rivers transport materials as suspended load, dissolved load, and bed load; during transport, materials are sorted according to mean grain size. Riparian substrates generally become progressively fine-grained downstream.

- Deposition occurs where flow velocities decrease, as in river sections where gradient is reduced, in the shadow of boulders and large woody debris accumulations, and on floodplains. Depositional processes also sort sediments according to grain size such that coarser materials remain in and near river channels and riparian substrates become progressively finer across floodplains with distance from channels.

- Hydrologic processes strongly influence riparian habitats as the transport medium for sediments, but the presence or absence of water by itself is also an important control on riparian form and function.

- Surface water and groundwater of river corridors are linked, forming a single hydrologic system across the valley fill; this system also connects to the regional groundwater system. Water and its dissolved load are continually exchanged between the river, riparian aquifer, and regional aquifer.

- Flooding is a key process that distributes surface water to riparian environments and sets up gradients that drive surface water–groundwater exchanges. Four characteristics of floods are especially important to riparian and floodplain ecosystems: *magnitude*, *frequency*, *timing*, and *duration*.

- The rate at which water enters into and flows through subsurface sediments depends on the steepness of the hydraulic gradient and the hydraulic conductivity of alluvial sediments. Subsurface water flow is enhanced along paleochannels that may link modern channel flows to distal areas of the floodplain.

Purpose

Physical and biological processes are closely coupled in riparian systems, but physical processes ultimately control the structure and function of riparian biological communities. Tectonics, geology, and climate are the main forcing functions driving a hierarchy of physical processes that scale down to individual microhabitats and span time scales from seconds to eons (see Table 2.1). Tectonic activity determines base topography and topographic gradients, geology determines resistance of the landscape to physical and chemical erosion, and climate supplies the water and energy to erode the landscape and move accumulated sediments and solutes down river systems. Tectonics and geology are fundamental to the long-term evolution of river systems,

Table 2.1 Geomorphic and Hydrologic Variables Influencing Riparia

Geomorphic Variables	Hydrologic Variables
Catchment Topography	Climate
• Size	• Precip magnitude
• Form	• Precip distribution (time and space)
• Gradient	Discharge
Geology	• Flow depth
• Drainage density	• Flow velocity
• Material shear strength	• Erosive shear stress
• Mean grain size	Flooding
Tectonics	• Frequency
• Continental drift	• Magnitude
• Uplift/subsidence	• Duration
	Groundwater
	• Hydraulic conductivity
	• Hydraulic gradient

Geomorphic Processes	Hydrologic Processes
Erosion	Flow
• Mass movements	• Surface
• Down cutting	• Subsurface
• Hillslope steepening	Infiltration
• Lateral channel migration	Exfiltration
• Channel scouring	Evapotranspiration
Transport	
• Suspended	
• Dissolved	
• Bed load	
Deposition	
• Channel bar formation	
• Overbank deposition	
• Point bar formation	

while climate is fundamental to both long- and short-term processes. These fundamental variables blend in countless mixtures to produce a broad spectrum of river basins, ranging from high-gradient, arid catchments underlain by friable sandstones to low-gradient, humid catchments underlain by highly resistant granites.

This chapter examines the key geomorphic and hydrologic processes shaping river–riparian corridors in the context of the larger catchments within which they occur. Catchments display predictable patterns of geomorphology and hydrology based on their age, form, size, tectonic setting, lithology, and climate regime. Predictable longitudinal patterns are observed as one moves downstream from headwaters to estuaries, while predictable temporal patterns are observed as one considers catchment development through time. These patterns are linked to a clear hierarchy of controlling processes from the scale of whole catchments to the scale of individual stream reaches. Whereas climate, geology, and topography exert controlling influences on catchment form and dynamics at the scale of the whole river catchment, slope, flood frequency, and substrate type better explain geomorphic and hydrologic processes within individual stream and river segments.

Beyond the first-order control of channelized surface water and sediments in determining riparian form and structure, there is an important second-order physical control in the flow of surface and subsurface water laterally through floodplain and hyporheic zones. Flow onto the floodplain builds physical substrates for further plant colonization, but these surface and subsurface flows also connect the complex habitats of riverine landscapes in other important ways: They supply nutrients and their duration and timing influence ecosystem productivity. We begin this chapter by providing an overview of catchments, discussing their main characteristics and the factors that influence their form and evolution over time. We then describe the hierarchy of geomorphic processes that shape catchment features and the riparian environments that develop in them. Finally, we address specific hydrologic processes, including how they link riverine corridors and control the disturbance regimes that are fundamental to riparian form and function.

Catchments and Hierarchical Patterns of Geomorphic Features

A *catchment* is simply defined as an area of the land surface in which all runoff drains to a single point on a stream or river channel. The area is determined by topography, and one catchment is separated from another by a topographic ridge called the *drainage divide* or *watershed*. In reality, a catchment may be delineated upstream of any point along a stream or river, but it is customary to define catchments above confluences in stream and river networks. The term *watershed* is commonly used in the United States to mean catchment or drainage basin. For very large rivers, the catchment is generally referred to as a *river basin*.

Catchments range in size from as little as $100 \, m^2$ to greater than 6 million km^2, which is the area of the Amazon River basin in northern South America. In fact, the lower limit of catchment size could be further reduced to include any area (even a few m^2) that drains to a single point, but if we specify that a catchment must drain to a recognizable channel, then the minimum size is set by certain properties of the landscape. The drainage area above *channel heads* (or the most upstream point on stream channels) is determined by local valley slope, hydrology, and erosion rates. Channel initiation begins at the threshold where runoff becomes responsible for sediment transport, as opposed to landslides and debris flows at smaller scales (Montgomery and Dietrich 1992). The scale of large river basins is determined by the configuration of the Earth's continents (which reflects past tectonics) and the placement of the continental divide relative to smaller mountain ranges and the coastline. The scale of the Amazon River basin is made possible because the Andes Mountains and the continental divide lie along the western margin of the continent, and highlands to the north and south route Amazon runoff waters to the distal Atlantic coast, more than 6,000 km downstream of the river's source. Like several of Earth's largest rivers, the valley of the Amazon mainstem flows along an ancient rift valley that fixes the river's position and collects runoff from all major tributaries of the basin. The Mississippi River of the United States also flows along an ancient rift valley collecting runoff waters from the vast plains stretching between the Appalachian and Rocky Mountains.

Catchments are not, however, clearly identifiable everywhere. Geomorphic regions produced by continental glaciation or characterized by karst topography may have no discernable midsized catchments (Omernik and Bailey 1997), although the larger river basins draining these areas can be clearly delineated. Glaciated terrain is common along the northern boundary of the United States (North Dakota, Minnesota, Wisconsin, and Michigan), and karst features obscure basin boundaries in southern Florida and western Texas and Oklahoma, as well as in central France.

Catchment Form and Channel Networks

Catchments assume myriad forms depending mainly on topography and underlying bedrock. In low-relief regions underlain by massive geologic strata of similar erosional resistance, catchments develop a dendritic channel pattern that is self-similar, meaning that the geometry of the network is independent of scale (see Figure 2.1a). The self-similar geometric relationships of river systems were first quantified by Horton (1945) and have been further elaborated by a number of investigators (Rodriguez-Iturbe and Rinaldo 1997, Peckham 1999). In more mountainous zones, where sub-catchments are confined to elongated valleys, palmate and pinnate dendritic channel patterns develop where a number of shorter tributaries connect to the larger stream or river flowing along the axis of the valley (see Figure 2.1b). The influence of bedrock geology is particularly evident in catchments that cut across geologic strata of differing erosional resistance. Alternating beds of more and less resistant rock are common where the land surface intersects large synclines or anticlines. In these basins, stream flow is focused along less resistant beds, occasionally cutting through intervening resistant beds to link adjoining channel systems, and a so-called trellised drainage pattern develops (see Figure 2.1c).

The size and shape of catchments—when combined with climate—influence flood magnitudes and sediment loads and are therefore relevant to the structure and function of riparia. Larger catchments are clearly subject to larger flood magnitudes. Basins containing palmate and pinnate channel networks may be subject to intensified flash flooding when heavy runoff from smaller tributaries becomes focused at confluences of the axial river (Dunne and Leopold 1978). Likewise, round-shaped catchments experience greater flood magnitudes than elongated basins because flows from multiple tributaries converge more quickly (Brooks et al. 1997). Other physical characteristics of catchments that influence flood magnitudes are *drainage density* and *relief ratio* (Schumm and Hadley 1961). Drainage density is calculated as the sum of all stream segments in a basin divided by the area of the basin (with units of km/km^2). Basins with higher drainage densities are subject to more intense flooding, again because travel distances to the basin outlet are shorter. Relief ratio is the difference in elevation from the highest to lowest parts of the basin divided by the length of the basin along the main river axis. The gradient of individual stream channels also can be determined, and in both cases more intense flooding is associated with the steeper gradient channel or the catchment of higher relief ratio (Dunne and Leopold 1978).

Streams and rivers along a channel network are assigned an *order* or *linkage number* based on the tributaries feeding them. The most widely used system is that of Strahler (1957, 1964), which assigns an order of 1 to streams fed only by groundwater exfiltration and receiving no tributary inputs. The confluence of two first-order streams forms a second-order stream, the confluence of two second-order streams forms a

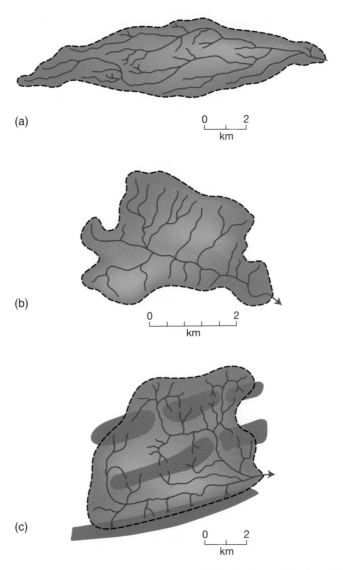

(a)

0   2
km

(b)

0   2
km

(c)

0   2
km

Figure 2.1 Examples of different drainage basin patterns, which develop as a function of topography and underlying bedrock. (a) A dendritic pattern develops in low-relief terrain underlain by similar bedrock. (b) A pinnate pattern develops in elongated mountain valleys. (c) A trellised pattern develops in areas of alternating more and less erodible substrates. (From Dunne and Leopold 1978.)

third-order stream, and so on (see Figure 2.2). Although this system is purely operational and does not consider process variables, it has proven useful in categorizing and grouping streams of shared geomorphological and ecological properties (Vannote et al. 1980, Montgomery 1999). Distinct riparian environments develop along streams and rivers of different orders (Gregory et al. 1991), so it is informative to consider the relative lengths of riparian environments occurring within each stream and river order. For example, in the ~20,000 km^2 Moisie River basin of Quebec, 87 percent (or ~28,000 km) of the total stream length occurs in first- through third-order streams, while seventh- through ninth-order rivers account for only 2.2 percent of length (or 711 km) (Naiman et al. 1987). As a point of reference, if extended in a continuous

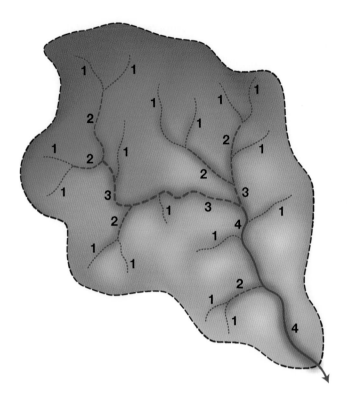

Figure 2.2 Stream ordering rules according to Strahler (1964). A higher-order stream is formed by the confluence of streams of equal but lesser order (from Dunne and Leopold 1978).

line, the ~32,000 km of total river length in the Moisie catchment would stretch 80 percent of the distance around the Earth's equator!

Catchment History

Over long time scales (10,000 to >1 million years), river basin processes are most influenced by tectonics and climate variability. The vestiges of this history are preserved mainly in the geomorphic features of river networks and are manifest today in complex assemblages of alluvium that still influence river dynamics and add complexity to the interaction of basin hydrology with riparian communities. Tectonics linked to plate movements cause regional uplift that reshapes the boundaries and gradients of rivers. Tectonic forces also produce folding, deformation, and faulting that leave complex mixtures of lithologic units within catchments. Climate changes linked to astronomical phenomena (Milankovich cycles) influence basins on 10,000- to 100,000-year time scales and include glacial/interglacial cycles and the rise and fall of sea level. On much shorter time scales (years to decades), climate may vary in response to the El Niño Southern Oscillation (ENSO) phenomenon, which may lead to extreme dry or wet runoff years, depending on location.

Evidence of past environments is found in alluvial complexes, which are the total preserved record of alluvial systems in catchments (Lewin 2001). In fact, the alluvial complexes of river basins are important tools in the reconstruction of past tectonic activity and climate variability (Maddy et al. 2001). Climate change is felt in catchments mainly as coupled changes in temperature, precipitation, and discharge. Temperature changes directly affect basin water balances, but they also indirectly affect river channel processes through their influence on plant communities and feedbacks linked to changes in evapotranspiration rates and substrate stabilization (Vandenberghe 1995).

Two historical features of catchments that have special significance for riparian structure and function are floodplain terraces and buried paleochannels. Terraces form when an increase in the gradient and/or power of a river system causes the channel to erode and incise the existing floodplain. The result is a new, narrower floodplain and more constrained river channel. River gradients are increased by both tectonic uplift and drop in sea level. Episodic and sustained regional tectonic uplift may produce large-scale flights of multiple terraces, resulting in dozens of terraces extending over hundreds of meters of vertical relief (Vandenberghe 1995, van der Berg and van Hoof 2001). With time, rivers may fill incised valleys and bury once-abandoned terraces to expand active floodplains to previous widths, but buried terraces remain as subsurface features influencing groundwater flows. These features are especially well developed in coastal river basins or the lower portions of large rivers, where fluctuations in sea level change the gradient of river systems and lead to repeated changes in the balance of erosion and deposition. Because sea level rises and falls in somewhat regular cycles, the net effect on alluvial complexes may be a mixture of old and new terraces, where multiple old terraces may be buried beneath more recent accumulations of sediments that either enhance or retard lateral flows of subsurface water. An increase in the erosional power of rivers independent of changes in gradient may also lead to terrace formation. This may occur during high discharge periods linked to melting glaciers or it may occur because of increases in precipitation. Increases in precipitation on well-vegetated catchments will increase the discharge of rivers without immediately producing concomitant increases in sediment load. The erosive power of the river is thus enhanced, and river incision occurs.

Buried paleochannels add special complexity to river corridors because they often act as preferential flow paths for subsurface flows connecting even the distal portions of floodplains and valley bottoms. *Avulsion* is the term describing the abandonment of an old river channel and the formation of a new one. In meandering systems, many sections of channel may be abandoned and left as oxbow lakes when the necks of meanders are cut off. *Crevasse splays* are accumulations of coarser sediments where river flows have broken through existing channels to deposit transported sediments in fan-like deposits within generally finer floodplain sediments. During severe periods of climate change, such as transitions between glacial and interglacial periods, channel forms may change completely and leave whole networks of abandoned channels.

An excellent example of the complexity of interplay between climate, discharge, and preserved alluvial formations is seen in a 250+ km section of the lower valley of the Mississippi River extending from the confluence of the Missouri and Ohio rivers to south of Memphis, Tennessee (Blum et al. 2000). There, an intricate alluvial complex formed at the end of the last glacial stage (see Figure 2.3a). High-amplitude fluctuations of flood discharge punctuating intervening times of slackwater produced a mosaic of braided coarse-grained channel deposits, crevasse-splay fans, fine-grained loess deposits, and sandy and silty fills. At least three episodes of large-scale, cross-

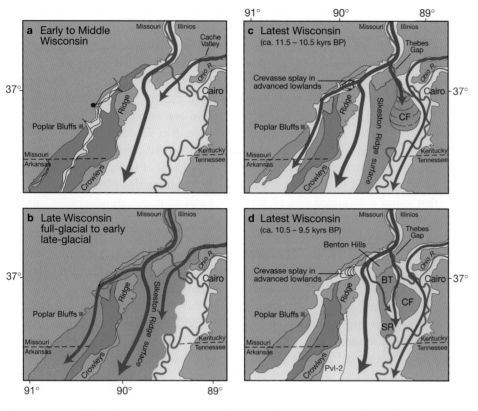

Figure 2.3 Model for the evolution of the lower Mississippi River at the end of the last Pleistocene (Wisconsin) glaciation. Where available, dates are reported as before present (BP). CF is the Charleston Fan. SR and BT mark remnant braided river sections from the period of full glaciation. See discussion in text. (From Blum et al. 2000.)

valley flooding occurred over this period. During the early stages of deglaciation, the Mississippi River overflowed its channel to develop new alluvial features (especially crevasse-splay deposits) in a much older paleochannel to the west (see Figure 2.3b). The river also broke through the Thebes Gap at this time (see Figure 2.3c), producing the Charleston Fan deposit. Later, the river abandoned the far western alluvial features, which were subsequently covered by fine loess deposits (see Figure 2.3d). Flows concentrated in the eastern channels formed braided channels of coarse-grained sediments. Finer-scale high-magnitude flood features linked to the breeching of glacial lakes also influenced the formation of this alluvial complex (Teller 1990). With the complete melting of the ice sheets, the Mississippi River developed into the meandering system it is today. Cut-and-fill processes in the modern river have reworked many of the paleo-alluvial features, but many others remain intact in the subsurface and in surrounding valley sediments (Blum et al. 2000). These abandoned channels and depositional features now crisscross the Mississippi valley bottom, providing subsurface preferential flow paths for the movement of upland groundwater to the river and for subsurface exchanges between the river and adjacent riparian and floodplain environments.

The past activities of humans may also leave a historical mark on catchments that influence modern riparian processes. Although human use and misuse of riparia are treated in detail in Chapters 7 and 8, it is useful to present here a few of the past human impacts that are relevant to basin geomorphology. River margins have been important to humans since our species evolved, and rivers have likely felt the effects of human activities just as long. Three kinds of activities have had the greatest impact: mining, deforestation, and the construction of artificial structures (Thieme 2001). Mining is an ancient human activity dating back to at least the early bronze age (ca. 5000 BP) when tin and copper mines were active across parts of Europe and the Mediterranean region (Mighall et al. 2002). In the Rio Tinto of southwest Spain, mines have operated nearly uninterrupted since 4500 BP (Hudson-Edwards et al. 1999). Mines were often located on hillsides, but the spoils of these operations washed into and accumulated in adjacent stream valleys. Additionally, placer mining, especially that using pressurized jets of water to wash alluvial sediments into sluicing devices, mobilizes large quantities of sediment that far exceed the river's transport capacity. An example of the legacy of such mining is found in rivers of the Sierra Nevada of California and Nevada, which were heavily impacted by gold miners in the mid to late 1800s. Inputs of sediments to the Bear River actually aggraded the river bed as much as 5 m and eventually caused the channel to shift locations (James 1991). Subsequent channel incision occurred in the Bear River and many other Sierra Nevada rivers, but large accumulations of coarse sediments persist in the subsurface and in adjoining terraces.

Land use change and its impacts on river systems is the focus of much current research, but the effects of land use change on river geomorphology reach back to prehistoric times. Increases in erosion rates attributed to deforestation can be difficult to quantify in any single time, but the accumulated effects of deforestation over centuries and millennia result in marked effects on catchments. For example, accumulations of sediment in the Drama basin of Macedonia can be linked directly to more than 7,000 years of land use conversion (Lespez 2003). There, significant aggradation of the river can be pinpointed to a period of agricultural expansion and settlements in the Late Bronze Age spanning 3600 to 3000 years BP. Peaks in aggradation appear to occur when deforestation expanded from low-lying areas to surrounding hillslopes where soils were less stable and more susceptible to erosion (Lespez 2003). Similar patterns are noted in the San River of the Eastern Carpathians in eastern Poland (Kukalak 2003). The initiation of construction of artificial structures likewise substantially alters valley geomorphology; this is considered in greater detail in Chapters 7 and 8.

Hierarchical Patterns of Geomorphic Features in Catchments

River channel networks are highly structured and change in a systematic fashion from headwaters to mouth. In most natural rivers, flow and channel size increase downstream in proportion with the size and flow of tributaries joining the main channel. Short-term and seasonal variability in river flows (i.e., flashiness) tends to decrease from headwaters to mouth as the variability of local climatic influences is integrated into the river flow. Because geomorphic features are strongly dependent on river flow and variability in that flow, they also change in a systematic and hierarchical fashion downstream.

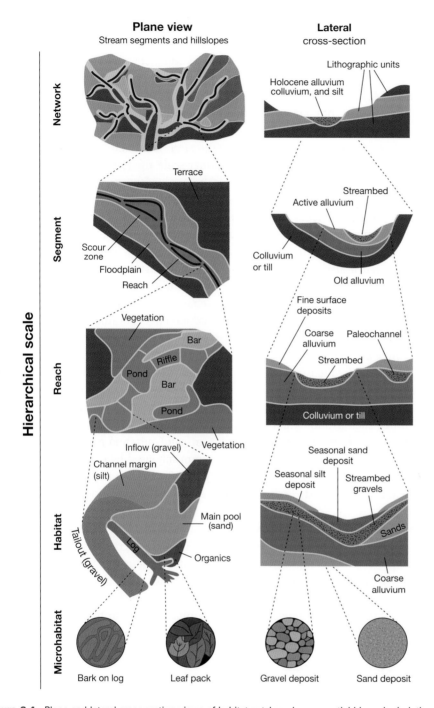

Figure 2.4 Plane and lateral cross-section views of habitat patches along a spatial hierarchy in lotic systems developed. Compiled by Poole (2002) and based on Frissell et al. (1986) and Dent et al. (2001).

Ecologists have found it useful to define a riverine hierarchical scaling system that consists of, from largest to smallest scale, the river network, river segments, river reaches, individual habitats, and microhabitats (see Figure 2.4) (Frissell et al. 1986). These scales are nested, in the sense that the river network is made up of multiple river segments, with each river segment consisting of multiple river reaches, and so on. At the scale of the entire *river network*—which extends from headwaters to mouth and may cover thousands to millions of square kilometers—only the coarsest features of the alluvial complex are discernable, and individual riparian features are certainly not visible. Major lithotopographic units may be defined (Montgomery 1999), and channel forms (e.g., braided or meandering) and major depositional features (e.g., alluvium-filled valley bottoms) can be distinguished in lower reaches of the river.

Floodplain riparian features emerge at the *river segment scale*, which may range from one to tens of kilometers in extent. Finer-scale riparian features of headwater streams are not easily discerned at this scale, but major features of downstream floodplains include oxbow lakes, back swamps, crevasse splays, longer channels abandoned because of avulsion, patches of floodplain forest, and so on. Other features, such as terraces and the larger point bars and cut banks of river channels may also be visible (see Figure 2.4).

The *river reach scale* considers short sections of rivers and streams, generally ranging from tens to hundreds of meters. It is at this scale that riparian features of headwater streams come into view, as well as channel features such as pools and riffles in mountain streams, channel bars and longitudinal bars in braided systems, and point bars and cut banks in small meandering systems. In very large rivers, the extent of this scale may be smaller than the width of the river channel, but features may nevertheless be defined by focusing attention on riverbanks—where levees may be discernable—or by considering the interior of floodplains.

The *river habitat scale* considers areas of 10 m or less on a side and discerns fine-scale features such as individual accumulations of large woody debris, communities of riparian vegetation, and different patches of sand, silt, and gravel.

The *microhabitat scale* is the finest scale commonly used by ecologists and considers areas of 1 meter or less. This is generally too fine a scale to discern geomorphologic features and instead describes the size and nature materials making up the features.

Geomorphologists recognize hierarchies of scale in catchment features as well and group features within a process framework. A recent hierarchical classification—which simplifies more complicated earlier systems—groups the geomorphic features of catchments into four levels (Lewin 2001) (see Figure 2.5):

- Level IV is the *Alluvial Complex*, which encompasses the entire geomorphic record of a catchment extending from headwaters to mouth and surface deposits to bedrock. This is the scale at which features reflect tectonic and climatic processes operating over geologic time.
- Level III is the *Architectural Ensemble*, which features local alluvial complexes at a scale similar to the *River Segment* scale of ecologists. This scale will still encompass the total alluvial accumulation in a valley bottom, including the floodplains of larger rivers. It also potentially preserves long histories of climate change and consequent changes in channel patterns and major sediment types.
- Level II is the *Form Unit*, which is generally tied directly to local sediment supply/energy environments. Channel versus overbank flows and hyporheic zones are clearly distinguished at this scale, and fine-scale hydrologic pathways may be delineated. Form units generally preserve a record of contemporary processes

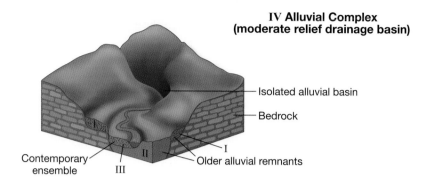

**IV Alluvial Complex
(moderate relief drainage basin)**

Isolated alluvial basin

Bedrock

Contemporary
ensemble

Older alluvial remnants

III

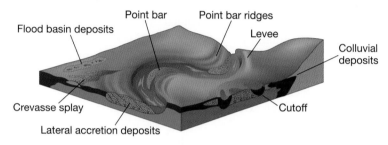

III Architectural Ensemble (meandering river)

Point bar   Point bar ridges

Flood basin deposits   Levee

Colluvial
deposits

Crevasse splay   Cutoff

Lateral accretion deposits

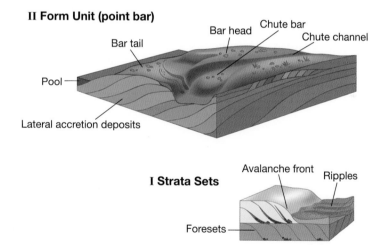

II Form Unit (point bar)

Bar head   Chute bar   Chute channel

Bar tail

Pool

Lateral accretion deposits

Avalanche front   Ripples

I Strata Sets

Foresets

Figure 2.5 Four hierarchical levels of geomorphologic classification proposed by Lewin (2002). Levels represent different levels of geomorphic processes as described in the text.

operating over decades to centuries at most. In highly dynamic systems, seasonal features may also be preserved.

• Level I is the *Strata Set*. As the name suggests, this level recognizes individual sedimentary surfaces in larger alluvial accumulations. Strata sets preserve a record of individual flood events, which may leave depositional or erosional features. Geomorphic variability at this scale would determine fine-scale variability in subsurface flow paths, and depending on the composition of the sedimentary substrate, it may determine the fine-scale availability of certain mineral-bound nutrients to riparian vegetation.

Geomorphic Processes and Process Domains

The fundamental work of rivers is to transport water, sediments, and solutes from land to sea. The processes through which this work is done physically create and destroy riparia, and they supply or deplete riparian water and nutrients. The most basic geomorphic processes in catchments are erosion, transport, and deposition. These processes operate across all time and space scales in catchments, but they do vary in relative importance in predictable ways. At the scale of large river basins, three broad geomorphic zones are commonly defined: an *Erosional Zone*, *Transfer Zone*, and *Depositional Zone* (Schumm 1977) (see Figure 2.6). The Erosional Zone includes headwater regions of the river basin where erosion dominates over deposition because—over long time scales—the rate at which river water removes sediments exceeds the rate at which sediments are supplied by physical weathering of basin lithologies. The Transfer Zone extends across lowland regions linking uplands to the sea (or the lake, in the case of endoreic basins). There, processes of erosion and deposition maintain a dynamic equilibrium such that over long time scales, the net geomorphic effect is simply to transfer sediments. Rivers are often quite dynamic through this zone, however, which corresponds to reaches where meandering rivers cut back

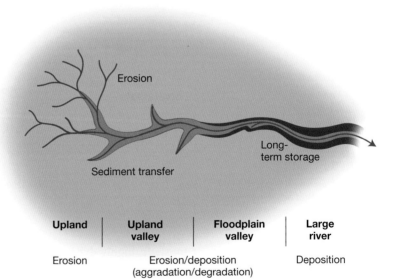

Figure 2.6 Diagram of the principle zones of distinct sediment behavior in a catchment. Headwater areas are predominantly erosional, whereas mid-reaches are predominantly transfer zones where erosion and deposition nearly balance. The lowest sections of rivers near sea level are predominantly depositional (from Church 2002 and modified from Schumm 1977).

and forth across broad floodplains. The depositional zone includes deltaic portions of rivers reaching the coastline, alluvial fans in more arid regions characterized by ephemeral river flows, and lacustrine deltaic areas of endoreic basins (such as the Lago Titicaca basin of the Andean Altiplano). The depositional zone represents a portion of the basin where—again, over long time scales—deposition rates exceed both erosion rates and the capacity of rivers to transport suspended materials. They form very near the *base level* of the river, or the very lowest elevation to which the river flows (sea level in most cases).

Geomorphic processes may also be grouped at the basin scale according to the size, frequency, and duration of ecological disturbances that they impose. Disturbances constitute geomorphic events—such as mass movements or floods—that alter habitats and thereby influence ecosystem structure and function. From this perspective, catchments are subdivided into hillslopes, channels, and floodplains. Montgomery (1999) terms these *Process Domains* and contends that they represent the most effective means of understanding the linkages between geomorphology and riparia. The ecological significance of these domains is treated in detail in subsequent chapters, but for the moment, these linked domains also serve as a useful framework for discussing processes that mobilize and transport sediments through catchments. Hillslopes of headwater valleys are the main sources of primary sediment to drainage systems and are predominantly erosional environments. Channels link headwater regions to the downstream portions of basins and are environments where sediments continually cycle via the processes of erosion → transport → deposition. Floodplains are important storage sites of fluvial sediments on their way to the sea, and the processes that control the storage and remobilization of these sediments generate the largest conterminous riparian environments in catchments. Along the punctuated continuum from hillslopes to lowland floodplains, riparian substrates are spontaneously created, gradually built up, left undisturbed for long periods, gradually deconstructed, or spontaneously destroyed. The following sections illustrate that the course of events in any one place along the drainage system depends on flow dynamics, the supply of sediments and large woody debris from upstream, and local vegetation. Each of these variables is highly dynamic in natural systems and easily, and often deliberately, perturbed by human actions. Our foci in this chapter are natural processes; human impacts are treated in Chapter 7.

Headwater Erosion

The process of erosion moves materials into streams and rivers from surrounding hillslopes and riverbanks. Erosion may be physical or chemical in nature; physical erosion is the mechanical abrasion that mobilizes sediments and organic materials, while chemical erosion is the dissolution of thermodynamically unstable minerals and soluble organic matter that mobilizes solutes into streams and rivers. Given the geomorphic focus of this chapter, we will focus only on physical erosion, as it supplies the raw material for the initial construction of riparian substrates and it also provides the recurring forces that disturb and thus reset riparian substrates.

The ultimate sources of sediments to streams and rivers are mass wasting of hillslopes and down cutting of river channels into underlying bedrock. Mass wasting predominates in the erosional zone and includes landslides, debris flows, and rock falls (Schumm 1977). Mass wasting occurs when the weight of material on a slope exceeds the strength of that material to hold itself up or, more precisely, when shear stress exceeds shear strength. The weight of the material applies a shear stress that increases with slope steepness. For rock slopes, the opposing force, shear strength, is a function of lithology and structure. Structure describes the spacing and orientation of discontinuities in rock mass like joints and fractures. Failure of rock slopes is more likely to occur when rocks are more fractured, especially if fractures are continuous and oriented downslope. Lithology determines the ease with which rocks fracture and the susceptibility of rocks to chemical weathering. For soil- or colluvium-covered slopes, shear strength is a function of internal friction (mechanical resistance to movement), cohesion between soil/sediment particles, and pore pressure. Vegetation is also an important component of shear strength on slopes covered with soil, as the root systems of plants bind soil particles.

Shear stress on hillslopes builds up over time as slopes become steeper, and rivers drive this steepening through valley incision, or down-cutting. Down-cutting is most rapid in mountainous areas where river gradients are steepest. As slopes become steeper and approach the threshold at which failure will occur, the most important variable becomes moisture content. Water increases hillslope weight (and shear stress) without a concomitant increase in shear strength. Slopes are far more likely to fail following significant rain events that saturate hillslope materials and lubricate joint surfaces. Mass wasting of this sort is most active in humid mountainous regions such as the western flank of the Cascade Mountains in the Pacific Northwest of the United States and the eastern flank of the Andes Mountains in northwestern South America. Actual fluvial erosion occurs once newly eroded sediments move off hillslopes and into the river channel. Unconsolidated sediments are entrained in stream flow as a direct function of flow velocity and as an indirect function of sediment grain size.

In addition to promoting erosion by over-steepening adjoining hillslopes, down-cutting of river channels also contributes sediments directly to the river system by eroding underlying lithologies. As with mass movements on hillslopes, channel down-cutting occurs when the shear stress of the river exceeds the shear strength of the underlying rock. Rivers cut downward and thereby reduce shear stress, until the two forces balance. At this point the river is said to be *graded*, and no further down-cutting will occur unless gradient increases. As noted in the previous section on basin history, both tectonic uplift and falling sea level can cause such an increase in gradient. Across the geomorphic zones of Schumm (1977), erosion of bedrock occurs almost exclusively in the Erosional Zone 1, whereas erosion in the Transfer Zone and Depositional Zone (Zones 2 and 3) mostly remobilizes sediments previously deposited by the river.

Channel Processes

Headwater streams occupy that part of a catchment where hillslopes are directly connected to channels (Church 2002). Primary sediments delivered to headwater streams by mass movements or down-cutting are entrained in the fluvial system and transported downstream. Channel transport is controlled by the stream energy and the size of sediments to be transported (Morisawa 1968). Stream energy reflects the total

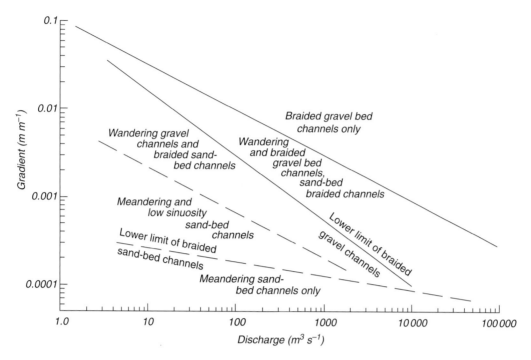

Figure 2.7 Fields of river channel morphologic pattern as a function of gradient and discharge. The relationship with sediment grain size is reflected in the descriptions of the fields. Braided gravel-bed channels occur in the highest energy settings, whereas meandering sand-bed channels occur in low-energy settings (from Church 2002).

erosive and transporting energy of flowing water and depends on velocity, depth, viscosity, and density. Deep and rapid flows are capable of mobilizing and transporting larger sediment grains. Velocity and depth are, in turn, influenced by the characteristics of the channel cross-section and bed materials; larger bed materials create greater roughness and thereby slow velocity. Viscosity varies as a function of water temperature, and density varies with temperature and with the dissolved and particulate load of the river. An increase in both viscosity and density increases the energy of streams and rivers. As stream or river energy increases, progressively larger particles are mobilized. Materials are thereby sorted, with finer material moving into transport before coarser material. Similarly, as energy decreases, larger material is deposited first, whereas finer material is carried further downstream. In this way, streams and rivers are continually acting to sort materials along and within their channels, resulting in characteristic deposits such as gravel bars, sand banks, and silt-filled side channels and backwaters. The relationships among river gradient, discharge, channel form, and sediment grain sizes are illustrated in Figure 2.7. Downstream patterns of these and other channel characteristics are shown in Figure 2.8.

Sediments move down rivers as either bed load or suspended load, and there is continual exchange between the two. Sediments of the bed load roll, bounce, and slide along the river bottom, moving more slowly than the flow of the water. The bed load consists of sediments small enough that the tractive force of the river will move them but too large for the buoyancy force of the river to maintain them in suspension. The *competence* of a river or stream at any moment or at any point along the channel is

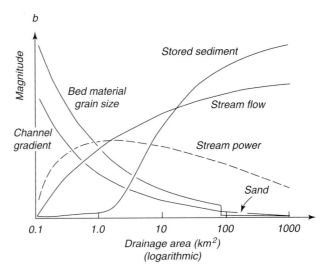

Figure 2.8 Illustration of downriver trends in key geomorphic and hydrologic parameters in catchments. Channel gradient and bed material grain size decrease exponentially downstream, while stream flow and stored sediment increase. Stream power generally peaks in early mid-reaches where the combined effects of stream gradient and flow volume are maximized (from Church 2002).

defined as the largest diameter of sediment grain (from sand grain to boulder) that can be moved in the bed load. Bed load transport is especially sensitive to the *flow regime* of the stream or river. Under a subcritical or low-flow regime, bed-load transport is minimal, with only individual grains moving across the bed. In contrast, under a critical or high-flow regime, bed-load transport is maximal and bed material moves downstream as a mass, which may take the form of moving dunes when sand grain sizes predominate.

The suspended load of a stream or river consists of sediment grains small enough that their settling velocity is less than the buoyant velocity of the flow, which is determined by the degree of turbulence. Turbulence generally increases with depth in channels (except at the bed surface, where flow becomes laminar), so there is some separation of grain sizes in depth profile. Larger sand grains are more abundant in the more turbulent lower depths of the channel, whereas finer grain sizes like silt are more evenly distributed across depths. All suspended sediments, however, move downstream at the same rate as water flow, and total suspended sediment load is closely correlated with discharge.

Stream and river energy varies greatly in space and time over drainage systems as a function of climate regime and variability, topography, substrate type, and channel and valley form. As energy regimes change, so does the competency of rivers and streams. Enhanced flow mobilizes and entrains sediments formerly stored in channel reaches, and reduced flow deposits sediments transported from upstream. Climate regime and variability control the temporal variability of fluvial flow regimes across all scales of catchments, although the manifestation of that variability varies depending on where a river reach lies in the larger basin. Sections of headwater streams and small rivers with limited contributing areas respond rapidly to each significant pre-

Figure 2.9 Example of sediment accumulation behind large woody debris on the Queets River, Washington. Accumulations like this enhance aquatic habitats and provide new substrates for the colonization of riparian plants (Photo: J. Latterell).

cipitation event, whereas downriver sections of large rivers, with large contributing areas, respond more slowly and may show no detectable response to rain events in individual upstream subbasins.

Using the channel-reach classification proposed by Montgomery and Buffington (1997), spatial variability in small fluvial systems may be observed in the alternating zones of rapid and slow flow in steps, pools, and riffles. Similar variability in flow regimes of larger streams, with rapid flow in chutes between boulders and slow flow on the downstream side of mid-channel boulders or in eddies formed in the shadow of boulders on the stream bank. In all channel sizes, accumulations of large woody debris create areas of high and low flow. Under stable flow conditions, sediments fractionate out of suspension according to grain size while passing through these sections, with enhanced deposition in low-flow zones. When flow increases, however, higher flow regimes bridge low-flow zones, and previously deposited sediments are resuspended and carried downstream. Significant bed-load movement may occur only during high-flow events in these stream reaches, and thus sediments move downstream in an episodic or sporadic manner. In the shadow of large woody debris piles and very large boulders, deposited sediments may remain shielded from high-flow events such that sediment deposition continues uninterrupted, being even enhanced during high-flow events (Abbe and Montgomery 2003). Under these conditions, sediment deposits may emerge as sandy banks when stream levels fall, providing fresh substrate for colonization by riparian vegetation (see Figure 2.9).

Floodplain and Channel Interactions

Valley form exerts a strong control on the competency of rivers by either constraining flow to a narrow channel or opening up broad valley bottoms to allow channel migration (see Figure 1.1). Constrained channel reaches tend to have a steeper gradient and be fast flowing when compared to unconstrained reaches. Thus, reach-scale patterns of erosion and deposition appear, with accumulation of sediments focused in lower-energy, unconstrained reaches. Ultimately, most major river systems emerge into lowland regions where valleys remain wide and channels are not again constrained. Unconstrained upstream and downstream reaches tend to accumulate significant amounts of sediments and to develop alluvial floodplains that host a great diversity of riparian substrate types. Rivers take on a meandering form in unconstrained low-gradient sections such as these. Meandering rivers, like the smaller streams described above, exhibit temporal and spatial variability in energy regimes, which leads to alternating periods of channel erosion and deposition. Unlike small, constrained streams, however, meandering rivers also interact strongly with their adjoining floodplains during flood events.

Focusing initially on channel processes, distinct spatial patterns in energy regimes are established in the bends of meandering river sections. Rapid flow is concentrated on the outside of bends and slow flow on the inside, and a secondary helical circulation develops to accentuate the gradient between inside and outside flows. An erosional zone, the cut bank, forms at the outside channel margin, and a depositional zone, the point bar, forms at the inside channel margin (see Figure 2.10). This combined process is responsible for the migration of meander belts across floodplains. As they grow, meanders migrate both perpendicular to as well as in the direction of dominant river flow. Meanders may grow to the point that their necks are cut off by the advancing cut banks of upstream and downstream meanders. At this point, the abandoned meander is left as an oxbow lake, and the process of meander development begins again. With time, meander belts may work themselves from one side of the valley bottom to the other, and rework virtually all the surface alluvial sediments. This process generates a highly complicated floodplain geomorphology consisting of low-lying ridge and swale topography and abundant oxbow lakes that may be completely filled with sediment or still flooded.

In the process of cut and fill, rates of erosion and deposition are more or less in equilibrium, so this is not a process that builds floodplains. Floodplains are built by the deposition of sediments from water that spills over the river channel margins and spreads out laterally. This process creates a number of other geomorphic structures important to the development of riparian substrates (see Figure 2.11). Natural levees form along the margin of riverbanks where floodwaters deposit the coarsest fraction of the sediment load. Rivers lose competency immediately upon overflowing their banks, and thus the coarsest sediment fractions settle out rapidly. Finer sediments are carried to progressively more distal sections of the floodplain, producing a clear gradient in substrate size. Crevasse splays develop where a river breaks through its levee prior to general flooding. In this situation, the water and sediment load are rapidly attenuated by the floodplain, and a distinct fan-shaped deposit develops. As the valley bottom aggrades and the alluvium becomes thicker, some areas of the floodplain more

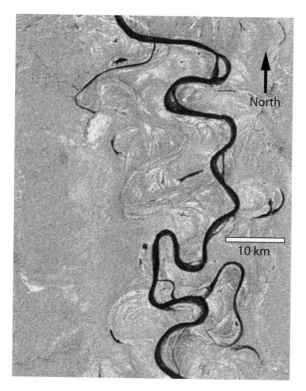

Figure 2.10 Meander belt of the Ucayali River in the Peruvian Amazon illustrating the 20- to 30-km wide zone of reworked sediments. Multiple oxbow lakes are apparent where meanders have been cut off from the main channel. 1996 JERS-1 Synthetic Aperture Radar (SAR) image.

distant from the river margin will lag behind because of less frequent sediment deposition. Eventually, these areas may become swamps as groundwater fills the valley alluvium to greater depth.

Very few detailed mass balances have been constructed for the exchange of materials between large meandering rivers and their floodplains. One exception is the detailed assessment of the mainstem Amazon River's interaction with its floodplain conducted by Dunne et al. (1998). That study quantified all major exchange pathways and concluded that the mean lateral flux of sediments (1,570–2,070 Tg yr^{-1}) between the river and adjoining flood plain over an 1,800-km section of the river exceeded the downstream flux (1,200 Tg yr^{-1}) by ~500 Tg yr^{-1}. The difference in fluxes represented sediment accumulating on the floodplain and in channel bars.

Hydrologic Connectivity and Surface Water–Groundwater Exchange

Riparian environments are created, structured, maintained, and destroyed by water and the solutes and sediments it carries. While our discussion thus far has focused on

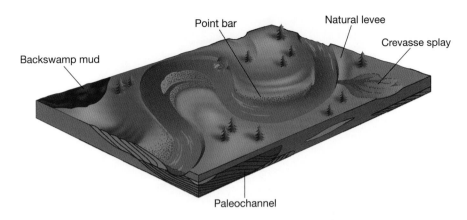

Figure 2.11 View of the main depositional features of floodplains featuring natural levees, point bars, crevasse splays, backswamp muds, and paleochannels. Paleochannels are also important as preferred pathways of subsurface flow in floodplains (from Ward et al. 2002).

geomorphic processes, we have repeatedly referred to the importance of hydrologic regimes, flow characteristics, and flooding as fundamental controls on geomorphology. Beyond its control on geomorphic processes, the presence or absence of water by itself is an important control on riparian form and function. Water table levels and groundwater flow paths are particularly important because nearly year-round access to water is a characteristic trait of riparian vegetation (Ward 1989, Amoros and Bornette 2002). In this section we describe the flow paths of water through catchments and the underlying hydrologic controls that influence riparia, whether directly or indirectly through geomorphology.

Surface water and groundwater of river corridors are linked, forming a single hydrologic system across the valley fill; this system also connects to the regional groundwater system (see Figure 2.12). Water and its dissolved load are continually exchanged among the river, riparian aquifer, and regional aquifer. As we have seen before, the rate and magnitude of this exchange depends on local climate, valley form, discharge, riverbed and riparian substrate material, channel configuration, and reach–scale variability in river gradient. Surface water–groundwater exchanges are key because they may link the river channel to distal portions of the floodplain, even during periods of low flow when surface water flows are entirely restricted to the river channel (Stanford and Ward 1992, Amoros and Bornette 2002). At the same time, riparian zones receive groundwater inflows from regional aquifers, and this water might differ considerably from river water in physicochemical properties. The following sections describe the fundamental processes that control hydrologic linkages in river corridors, surface water–ground water exchanges, and subsurface flows.

Surface Connectivity and Flooding

The flow of a river at any point along its length is revealed in its discharge, which is the volume per unit time (e.g., m³/s) of water moving down the channel (Hornberger et al. 1998). Discharge is calculated as the product of the cross-sectional area of the

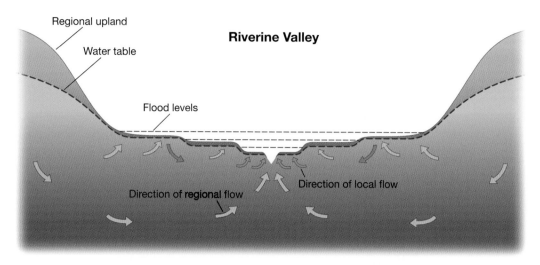

Figure 2.12 The multiple interactions of alluvial and regional groundwater systems in alluvial valleys. Pathways are shown as a function of different river levels as indicated by dashed lines above the active channel (from Winter et al. 1998).

wetted river channel (m^2) and the mean velocity of the flow (m/s). If discharge remains constant through a given river section, variations in channel cross-sectional area and flow velocity are strongly correlated. A decrease in river width will cause the flow to become deeper and faster, whereas an increase in width will cause flow to slow. An increase in the gradient of a river section will cause velocity to increase, leading to a decrease in wetted channel area. An in-depth treatment of catchment hydrology can be found in key texts such as Dunne and Leopold (1978), Black (1996), and Hornberger et al. (1998).

Hydrographs reveal the seasonal and interannual variability in stream and river flows, and they take on characteristic forms depending on the size and shape of the catchment and the local climate (see Figures 2.13a–c). Peaks on a hydrograph correspond to flood events that may scour riverbanks or transport sediments onto floodplains. Small rivers or streams, and thus the riparian environments bordering them, are sensitive to individual precipitation events and have dynamic hydrographs characterized by a large number of peaks (floods) over a year. In perennial rivers, floods may be distributed through the year but are usually concentrated in the rainy season (see Figure 2.13a). Rivers in arid regions are often ephemeral and can be without surface flow for considerable periods. Their hydrographs reveal only occasional floods (see Figure 2.13b). Mountain rivers and streams may be strongly influenced by spring snowmelt (see Figure 2.13c). Large rivers and their riparian environments are less sensitive to individual precipitation events because the scale of the basin surpasses the scale of the storm and because the flow of the river integrates the flow of a large number of upstream tributaries, some of which may be in flood while others are not. This is revealed in the more damped hydrographs characteristic of large rivers.

Floods in headwaters initiate *flood waves* that propagate downriver and accumulate in downstream river sections. Thus, the same flood event will affect riparia in distinct ways. This downstream accumulation is reflected in an example from the Potomac

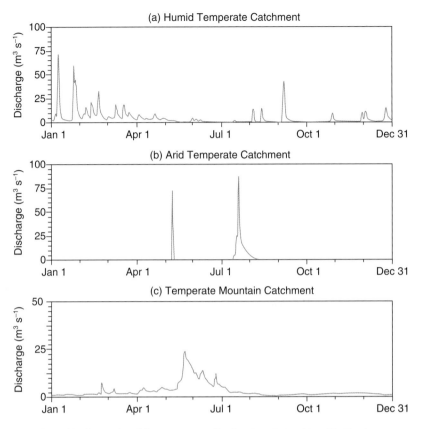

Figure 2.13 Annual hydrographs of three streams reflecting the relationship with climate zones. (a) Perennial flow patterns develop in humid climate zones. (b) Ephemeral flow patterns develop in arid climate zones. (c) Snowmelt flow patterns develop in high mountain climate zones. (From Hornberger et al. 1998.)

River at Washington, DC (see Figure 2.14a). Although it peaks approximately $1-1\frac{1}{2}$ days after the peaks in upstream sections of the Shenandoah and South Rivers, flow in the Potomac River peaks at a discharge in excess of 4,000 m³/s, which far exceeds the individual flood peaks of its tributaries and indicates accumulation of flow. But while the total discharge in the Potomac River exceeds those of its tributaries, the area-normalized discharge is actually lower (see Figure 2.14b). This illustrates the extent to which flood waves are attenuated as they move down river systems. Attenuation is a result of friction and temporary storage, which depends on the shape and roughness of the channel and the connectedness of the channel to its floodplain. Rivers that have been straightened and isolated from their floodplains propagate flood waves with very little attenuation and cause more severe flood events in downstream sections, significantly impacting riparia adapted to undisturbed hydrologic regimes. In such highly engineered systems, flooding must be controlled by systems of dams.

Rivers overtop their banks when the volume of floodwater exceeds the channel volume. This overbank flow spreads out across the floodplain, filling low-lying areas first and extending inland until the entire volume of overbank floodwater is accommodated. An important variable in determining the area of floodplain affected by

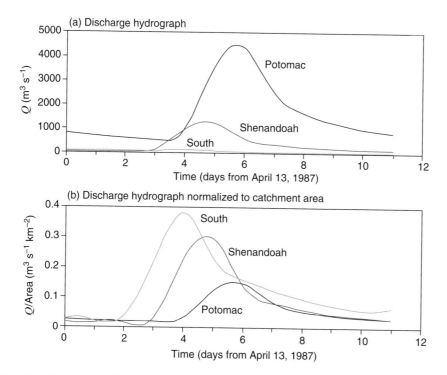

Figure 2.14 Hydrographs of three nested catchments. Discharge is shown to increase downstream in the river network (a), while the area-normalized discharge decreases downstream (b). (From Hornberger et al. 1998.)

river waters is the preexisting moisture condition on the floodplain when the river overtops its banks. River waters flooding a dry floodplain will extend to the maximal area allowed by its volume and will infiltrate floodplain sediments. Direct rainfall, rising groundwater, and local tributaries may, however, saturate the floodplain before the river crests (Mertes 1997). Where this occurs, river waters will be constrained to a smaller proportion of the floodplain, and significant geomorphological and biogeochemical effects may result. This local (or perirheic) water is generally sediment poor and may have solute chemistries distinct from those of the river. In either case, when river levels drop, surface water will flow quickly back into the river—usually via a system of floodplain channels—and groundwater in the alluvial aquifer will exfiltrate slowly back into the river.

Four characteristics of floods are especially important to riparian and floodplain ecosystems: *magnitude, frequency, timing,* and *duration*. Magnitude refers to the maximal discharge associated with an individual flood and therefore the intensity and severity of the event. The range of flood magnitudes for a particular river section depends mainly on climate and the catchment area upstream; flood magnitudes are greatest in large river basins in humid regions (Costa 1987). Variations in flood magnitude within any basin are generally expressed as recurrence intervals. Recurrence intervals are calculated from historical records by ranking (from smallest to largest) the natural-log-transformed discharge of recorded floods and calculating an exceedance probability for floods of each magnitude. Exceedance probability is calculated as the rank (r) divided by the total number of recorded floods (n) + 1 [i.e., $r/(n + 1)$] (See Hornberger et al. 1998 for a detailed description of the procedure.) The recurrence interval is then calculated as 1 divided by the exceedance probability. In this way, a flood with an exceedance probability of 0.01 would translate into a recurrence interval of 100 years. Truly extraordinary floods like the Upper Mississippi flood of 1993 occur when there is a convergence of conditions that promote flooding. The summer of 1993 experienced 1 in 200 year rainfall amounts over large areas of the Mississippi River headwaters. Flooding was further intensified by above-normal soil moisture levels, below-normal evapotranspiration, and the orientation of major rainfall along the river channel (Kunkel et al. 1994).

Flood duration is the amount of time that an area of riparian zone or floodplain is flooded seasonally or during individual flood events. Flood duration varies as a function of topography; low-lying areas close to channels are first to be flooded and last to be drained and thus experience the longest duration of flooding within any floodplain or riparian zone. Duration of flooding, or, conversely, duration of dry periods, strongly influences the nature of floodplain vegetation. In the very predictable flood regime of the central Amazon floodplain, there is a clear zonation of plant communities according to flood duration (see Figure 2.15). Floodplain areas lower than

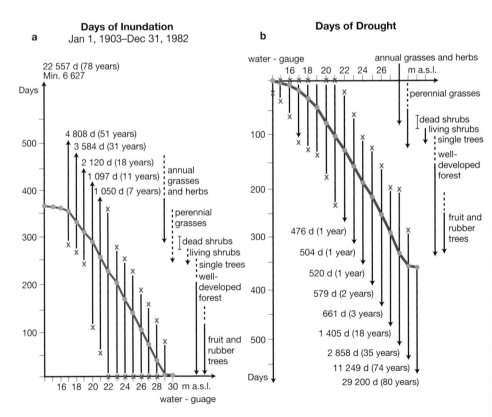

Figure 2.15 The zonation of plant communities on the Amazon floodplain as a function of flooding characteristics. More elevated sections of the floodplain (*x* axis) are inundated fewer days during the year (*y* axis). Patterns range from grass communities in the longest hydroperiod areas of the floodplain to well-developed forests in the shorter hydroperiod areas (from Junk and Piedade 1997).

19.5 masl experience more than 300 days of flooding during an average year and are colonized only by annual plants. Areas above 19.5 masl are colonized by perennial plants, transitioning from grasses to flood-tolerant shrubs (at 20.5 masl or 270 days flooding) and eventually to flooded forests (at 22 masl or 230 days flooding; Junk and Piedade 1997).

The Dynamics of the Linked Surface–Subsurface Hydrologic System

Water flows downhill, and in the linked surface water–groundwater systems of river corridors, hills are reflected in the topography of the continuous surface plane formed by the river surface and adjoining water table (i.e., hydraulic surface). The shape of the water table in the regional aquifer will generally follow the surface topography. Likewise, in the river corridor, this hydraulic surface has a primary downstream gradient that conforms to the valley gradient, so both river water and riparian groundwater flow down the valley. There are also a number of secondary gradients that develop because of microtopography in the river and water table surfaces. Small "topographic highs" develop in rivers where water piles up behind obstructions in the channel (such as boulders, large woody debris dams, or channel bars), at the head of riffles, or in meander bends (see Figure 2.16). Each of these highs establishes gradients into channel sediments, and flow moves from the surface to subsurface. As discharge increases and river water levels rise (but do not overflow the channel), a gradient is established into adjoining subsurface sediments, and infiltration occurs. Conversely, when the water level falls, an opposite gradient forms and subsurface water will exfiltrate. Water that infiltrates into the subsurface along a microtopographic gradient may travel for tens to hundreds of meters downstream before eventually reentering the river channel or being lost to the regional aquifer (Edwards 1998). On these excursions through the valley subsurface, water interacts with fill material, the roots of riparian vegetation, and subsurface organisms. It also mixes with regional groundwaters to form a unique hydrochemical and biological area known as the *hyporheic zone* (see Chapter 6).

The rate at which water enters into and flows through subsurface sediments depends on the steepness of the hydraulic gradient and the hydraulic conductivity of the sediment. These relationships are expressed by the equation $R_f = k(dh/dl)$, where R_f is the rate of flow (m s^{-1}), k is the hydraulic conductivity (m s^{-1}), and dh/dl is the hydraulic gradient or the change in head over distance (both in m). At the microtopographic highs described above, an increase in gradient forces localized surface water flow into the subsurface, although the rate of infiltration will vary greatly depending on the grain size of bed sediments. An increase or decrease in gradient also forces infiltration or exfiltration as river water levels change. This exchange flow is distributed along large sections of the river channel but again is enhanced in areas of the riverbank characterized by larger mean grain sizes (Malard et al. 2002). In the valley-wide system, flow is ultimately controlled by the primary, down-valley hydraulic gradient. As in the river, however, there is secondary localized topography that reflects the topography of the ground surface. This is especially clear in terraced valleys, but it also occurs under more low-lying floodplain features such as ridge and swale lines.

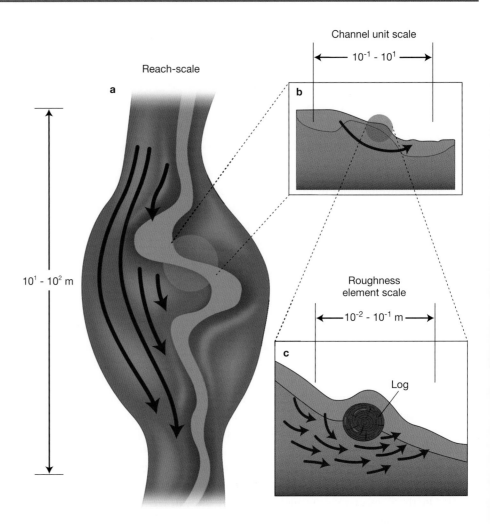

Figure 2.16 Examples of subsurface or hyporheic flow paths at spatial scales ranging from less than 1 meter to greater than 100 meters. Reach-scale flow paths develop primarily as a function of channel sinuosity, while finer-scale surface–subsurface exchanges develop as a function of channel topography (channel unit scale) and obstructions in the channel (roughness element scale) (from Edwards 1998).

Although hydraulic gradients must be set up for surface water–groundwater exchanges to occur, the most important variable in determining the spatial distribution and variability of flow rates is mean grain size. Hydraulic conductivity is extremely sensitive to mean grain size. A one order-of-magnitude change in mean grain size—say from 0.1 mm (fine sand) to 1 mm (coarse sand)—results in a greater than one order-of-magnitude change in hydraulic conductivity and thus flow rate. The mean grain size of channel and valley fill sediments ranges over multiple orders of magnitude and is differentially distributed, or sorted, according to the varying flow regimes of the surface water system. In a basin-scale longitudinal dimension, coarser sediments are concentrated in headwater regions, especially those that are lithologically and topographically prone to the production of gravel- and boulder-sized weathering products. In a lateral dimension on lower floodplain sections of rivers, coarser sediments accumulate in the active channel bed (where flow regimes are greatest).

Linear deposits of coarse-grained sediments may also crisscross the floodplain along paleochannels (Poole et al. 1997). The heterogeneous distribution of mean grain sizes results in preferred pathways of subsurface water flow, whether at the margin of rivers or distal portions of the floodplain. Paleochannels constitute the most important preferred pathways away from the active channel, while preferential subsurface flow adjacent to the active channel is likely to occur in the basal deposits of point bars (the largest grain sizes in the profile) and in gravel bars.

Conclusions

The diversity of form and function in riparian communities may be largely attributed to the spatial and temporal diversity of the physical environment within which they live. In this chapter we have examined the complex ways that geomorphic and hydrologic processes combine to create and destroy riparian habitats and how hydrologic flows link riparia along and across valley bottoms. Understanding riparia requires a catchment perspective because catchment size and form, coupled with climate and underlying lithographic units, exert fundamental control over the suite of geomorphic and hydrologic processes influencing riparia at any point in the drainage system. Catchment alluvial complexes are highly structured and reflect systematic changes in controlling variables over space and time. In large part these changes are predictable over the length of drainage systems from headwaters to mouth and from the time of initial channel formation to eventual abandonment and burial, producing a clear hierarchy of riparian habitat types.

It is useful to group geomorphic processes into process domains or hydrogeomorphic zones that reflect the dominant variables controlling riparian substrate dynamics. Gradients from dominantly erosional to depositional environments create the complexity and diversity of hydrogeomorphic zones and extend both longitudinally and laterally along short sections and the full extent of fluvial corridors. For example, headwater streams in mountains are dominantly erosional but contain environments ranging from recently deposited landslides to scoured bedrock surfaces. Sediment movement in these headwater streams is episodic and linked to major storms that destabilize hillslopes and transport sediments downstream in clear pulses. In low-relief mid-reaches of catchments, sediment transport is more consistent, but gradients between erosional and depositional environments are equally steep. Cut-and-fill channel migration exerts a primary control on riparian habitat creation and destruction, and overbank floodplain deposits extend riparian habitats across valley bottoms along gradients of grain size and inundation regime.

Surface water and groundwater systems of river corridors form a single hydrologic system that interconnects riparian habitats and regulates the exchange of solid and dissolved materials between habitats. Flow velocity and depth are the main factors determining the power that a stream or river has to erode and transport sediment materials, and each of these factors increases with distance downstream. Microtopographic features and the hydraulic conductivity of alluvial deposits drive exchanges between surface water and groundwater components of the system. Once in the subsurface, groundwater flows will be preferentially directed along paleochannels, and these subsurface features may provide important connectivity between the river channel and distal portions of the floodplain.

We have now considered the geomorphic and hydrologic templates controlling and sustaining riparian habitats, and it should be clear that proper management, conservation, and restoration of riparia require a basic understanding of these fundamental processes. Next we examine the approaches taken to classifying complex stream and riparian environments.

Riparian Typology

3

Overview

- Classifications or typologies are intellectual constructs in which objects with similar relevant attributes are grouped together to meet the purposes of the classifier. They involve collection of biophysical data at appropriate spatial scales, selection of appropriate variables, and application of a methodology for grouping variables into reasonably homogenous units.

- Classification is essential for scientific and managerial activities, such as developing inventories, interpreting data, extrapolating information from specific sites to larger or to other areas, setting strategic objectives or standards, and evaluating the uniqueness of areas. Classification of river corridors has a long history, but there is no universally accepted riparian classification system.

- Information on riparian systems can be in the form of inventories, plant associations, indicator species, biophysical processes, and assessments of biotic integrity. Each has a distinct purpose and therefore differs in the appropriate application(s).

- Geomorphic and climatic processes shaping abiotic and biotic features provide a foundation for understanding the dynamics of riparian systems. This chapter examines geomorphic and biotic classifications from a hierarchic perspective, linking the physical environment to biotic characteristics. Specifically, we discuss hierarchical perspectives, the Rosgen classification, the process domain concept, and the hydrogeomorphic approach. Biotic approaches to classification include soils, plant communities, and fauna.

- Riparian complexity and heterogeneity make it difficult to develop a robust classification system, but these can be somewhat accommodated with well-designed and thoughtfully analyzed investigations.

Purpose

Typologies or classifications use similarities of form and function to impose order on a variety of natural stream morphologies. Basically, they are intellectual constructs in which objects with similar relevant attributes are grouped together to meet the purposes of the classifier. This may seem like an easy task; however, riparian forests and wetlands associated with streams are often biophysically dynamic and diverse, and thereby challenging to classify. There is no universally accepted riparian classification system, but there are several channel classification schemes that encompass riparian areas. Each of the many channel classifications in common use has advantages and disadvantages in geological, engineering, and ecological applications, and no single classification can satisfy all possible purposes or encompass all channel types. Furthermore, channel classification cannot substitute for focused observation and clear thinking about channel processes (Montgomery and Buffington 1997, 1998). Channels are complex systems that need to be interpreted within their local and historical context. Classification merely provides one of many tools that can be applied to particular problems.

Classifications or typologies describe biophysical attributes in a brief and effective form that can be used within or among ecoregions. Classifications involve the collection of data on the environment at an appropriate spatial scale or scales, the selection of variables that are most appropriate to the classification, and the application of a methodology for grouping those variables into an understandable typology. A well-designed classification should be useful as a management tool, and many classifications are designed with management applications in mind. An enduring classification system should incorporate as many biophysical aspects of riparian formation and processes as possible, include a temporal or historical perspective on the dynamics of the fluvial corridor. Such a system should also be applicable at a range of spatial scales and provide consistent and reproducible results (Gurnell et al. 1994).

Classification is essential for scientific and managerial activities, such as developing inventories, interpreting data, extrapolating information from specific sites to larger or to other areas, setting strategic objectives or standards, and evaluating the uniqueness of areas. Classification provides a necessary framework for research and development of class-specific management strategies and monitoring programs, and it also provides a way of simplifying complex information and can have important educational values (Quinn et al. 2001).

In general, riparian typology has roots in two broad disciplines. The first is a physically based geomorphic approach that classifies the structure and dynamics of river corridors. The second is a biologically based approach that inventories or classifies plant communities on various geomorphic surfaces or physical templates associated with river corridors. In reality, the combination of these two approaches produces the most effective classification systems, and that is evident in many of the recently developed approaches (see Innis et al. 2000, Quinn et al. 2001). Recent classifications are increasingly multidisciplinary, smaller scale, hierarchically structured, and applied mostly in North America, Europe, and Australasia.

Traditionally, wetlands and riparian zones have been combined in classification schemes because of their "semiaquatic" characteristics. Both have an intimate association with the hydrologic regime, existing as ecotones between fully terrestrial and fully aquatic environments. Nevertheless, inherent differences between wetlands and riparia are significant. They differ substantially in spatial context, disturbance regimes, hydrologic setting, and ecological organization. Terrestrial and aquatic processes and species are intimately associated in wetlands, whereas riparian zones are more terrestrial and generally do not include the in-stream environment. Riparian systems are highly connected, spatially linear systems, whereas freshwater wetlands are relatively lightly connected and spatially circular or oblong entities. Natural riparian systems are often shaped by high-energy, system-resetting disturbance regimes, whereas wetlands are relatively less physically dynamic and the disturbance less catastrophic. Basically, riparian plants are often disturbance limited, whereas wetland plants are limited by competition. Despite these fundamental ecological differences, riparian zones were historically treated as a type of wetland (e.g., Cowardin et al. 1979). The emergence of riparian ecology, largely independent of wetland ecology, has led to an increasing separate treatment of the two.

This chapter explores the historic development of stream channel and riparian classifications and the application of information to riparian typology. We examine riparian classifications based on physiography, scale-dependent hierarchy, and biological features, and survey recent advances that promote an ecosystem approach to typology.

The Historical Context

The classification of river corridors, with its emphasis on the stream channel rather than riparia, enjoys a long history (reviewed by Wasson 1989, Naiman et al. 1992b, Montgomery and Buffington 1998, Naiman 1998). The strong relations between in-channel processes, communities, and the adjacent riparian vegetation make these early stream classifications relevant here (see Table 3.1).

River classification debuted in the 19th century with recognition of differences between mountain and lowland channels (e.g., Surell 1841, Dana 1850) and the broad geomorphic delineations of channel types (e.g., Powell 1875, Gilbert 1877, Davis 1890). Later, at the beginning of the 20th century, starting with Léger's (1909) evaluation of the biogenic capacity of streams to produce salmonids, more detailed observations of channel patterns were considered. Léger's classification is interesting because he considered the biophysical stream environment as well as the riparian vegetation; with subsequent modifications, this approach is still used in France. Later, well-known classifications based on stream order, on the nature of the channel bed, and on channel patterns and processes served a variety of purposes, forming the foundations for the hierarchical classifications that emerged near the end of the 20th century. Contemporary with the geomorphic classifications, Huet (1949), Illes and Botosaneanu (1963), and Verneaux (1973) provided classifications based on faunal zonation of fish and aquatic insects from headwaters to the sea. Although they and subsequent researchers implicitly recognized the connection between in-stream and riparian characteristics, it was not until the 1960s that explicit connections were made between stream and riparian systems in Hynes's (1970) exceptional treatise, *The Ecology of Running Waters*.

Many early attempts at stream channel classification recognized that biotic zonation patterns generally correlated with gradient or other abiotic features, such as temperature or water chemistry. Huet (1954), for one, recognized the importance of larger spatial scales by incorporating valley form. Later, classifications using stream order

Table 3.1 An Abbreviated Chronology (1875 to 2002) of Interesting and Important Stream Channel and Riparian Typologies

Reference and Year	Contribution
Powell 1875	Possibly the first broad delineation of channel types
Gilbert 1877	Similar to Powell; a broad delineation of channel types
Davis 1890	Concept of young to mature stream types
Léger 1909	First examination of the biophysical stream environment that characterizes the riparian vegetation
Gilbert 1914	Recognized differences between bedrock and alluvial channels in terms of sediment transport capacity
Horton 1945	Concept of stream orders; later modified by Strahler (1957)
Leopold and Wolman 1957	Differentiated straight, meandering, and braided channel patterns based on relationships between slope and discharge
Schumm 1977	Classified alluvial channels based on dominant modes of sediment transport
Frissell et al. 1986	Developed a hierarchy of spatial scales reflecting differences in processes and controls on channel morphology
Rosgen 1994	Recognized nearly 50 channel types based on patterns, entrenchment, width-to-depth ratio, sinuosity, slope, and bed material size
van Coller et al. 1997	Use of statistical techniques to relate riparian community type to physical channel features
Montgomery 1999	Concept of process domains at the catchment scale; accounts for the influence of geomorphic processes on aquatic and riparian systems
Lewin 2001	Geomorphic classification of alluvial systems into a hierarchy of four spatial levels containing distinct typologies
Church 2002	Delimitation of riverine landscapes and habitats based on thresholds in flow regime, the quantity and caliber of sediment, and topographic setting.

and linkage number were successful in describing patterns of in-stream ecological processes such as primary production and detritus dynamics (e.g., Minshall et al. 1983, Naiman et al. 1987). In spite of widespread recognition of distinct biotic zones along rivers, there were many early critics because key physical parameters change gradually, as well as discontinuously, along the stream continuum (e.g., slope and width), and the biological communities and processes respond to changes in the physical template.

There are two general limitations to these historic systems. First, the reliance on species as indicators of ecological zones means that the biotic zonation schemes are only valid in basins with similar zoogeographic, geologic, and climatic histories. Despite relating physical factors to biotic patterns, these schemes failed to construct a conceptual framework for the classifications that could transcend regions. Second, for both physical and biotic zonation systems, there were no features relating geologic and climatic processes, which regulate the physical features of stream corridors, to the classification system. Therefore, these efforts were ineffective at relating catchment-scale processes to dynamic changes in channel features (Naiman et al. 1992b, Montgomery and Buffington 1997). However, the application of landscape ecology concepts (such as patches, boundaries, and connectivity) to rivers has become a useful approach to overcome some of this difficulty (Décamps 1984, Strayer et al. 2003).

Theoretical Basis for Classification

Studies of boundaries, of which riparia are one broad type, are an important and rapidly evolving part of contemporary ecology. Ecological boundaries can be classified in many ways (Strayer et al. 2003). Nevertheless, there are only four main classes of boundary traits: (1) origin and maintenance, (2) spatial structure, (3) function, and (4) temporal dynamics. This makes the task of classification somewhat easier (see Table 3.2).

Riparian boundaries arise in various ways. In practice, riparian boundaries are a mixture of investigative and tangible boundaries involving the imposition of human order onto a real natural structure (Cadenasso et al. 2003). Boundaries may arise because of discontinuities between patches (consequential) or they may cause discontinuities between patches (causal). Existing boundaries may have arisen from forces still in operation (contemporary) or from forces no longer operating at that site (relict). Likewise, endogenous and exogenous forces—and the interactions between endogenous and exogenous forces—may tend to maintain or destroy a boundary through time, and the control of boundary characteristics by endogenous and exogenous forces may shift over time. Collectively, it is the relative mix of boundary attributes and forces that shape the heterogeneity of natural riparian systems.

There is an enormous range in the spatial structure of ecological boundaries. At least 11 attributes have been recognized (Strayer et al. 2003). Some of the important attributes for riparia include grain, extent, dimensionality, geometry, sharpness,

Table 3.2 Some General Attributes of Ecological Boundaries

Key Questions	General Attributes
How did the boundary originate, and how is it maintained?	• Causal or consequential • Contemporary or relic • Endogenous or exogenous origin • Endogenous or exogenous controls (maintenance or suppression)
What is the spatial structure of the boundary?	• Grain size • Extent • Thickness and dimensionality • Geometry of adjacency • Abruptness, steepness • Patch contrast • Integrity (perforation versus unbroken) • Geometric shape and tortuosity
What are the functions of the boundary?	• Interactive or noninteractive • Transformation • Transmission • Absorption • Amplification • Reflection • Neutral
How does the boundary change over time?	• Changes in *any* structural or functional property • Age and history

Modified from Strayer et al. 2003.

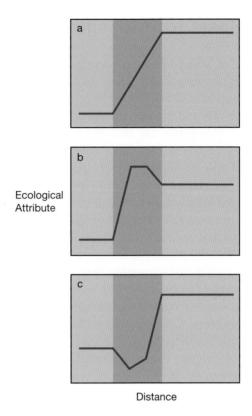

Ecological
Attribute

Distance

Figure 3.1 (a) A noninteractive boundary that averages the properties of the adjacent patches. (b,c) Two examples of interactive boundaries where a specific property is amplified or attenuated. The gray area is the boundary, which has a finite width (adapted from Strayer et al. 2003).

contrast, integrity, and tortuosity. For example, as different grain sizes are used, the same physical structure may appear to be very different or may not appear at all. Other boundary attributes, such as length, tortuosity, and sharpness, depend on grain size—and this is an important consideration when making comparisons. Some boundaries, such as riparia, have a thickness and therefore have the same dimensionality as the patches they separate.

Conditions in the boundary may be interactive or noninteractive. Boundary conditions may be a simple average of conditions in adjacent patches, or they may reflect interactions that occur along the respective boundaries (see Figure 3.1). Many kinds of mechanisms can create interactive boundaries; indeed, much of the literature on the positive and negative aspects of boundaries has been focused on identifying, quantifying, and managing these interactions.

The boundaries themselves affect ecological processes and can be classified by the general mode of processing that occurs. They may transform materials at aerobic–anaerobic boundaries (e.g., ammonia–nitrate reactions); they may transmit materials, energy, or organisms; or they may absorb them into their structure. Other types of boundaries may amplify ecological phenomenon, of which denitrification is a good example. In other cases, boundaries may be reflective, returning materials and organisms to patches where they originated, or they may be neutral.

In any case, boundaries change properties and position over time. Boundaries may become sharper or more diffuse; conduits may appear or disappear. For example, vegetation growth makes the riparian boundaries more or less permeable depending on biotic assemblage and successional stage. Riparian boundaries will shift position depending on the lateral dynamics of the river. Age and history of a boundary are important, if only to determine the functional properties and the local conditions around the boundary. For example, oxidation of iron seeping from groundwater can cause the progressive formation of ferricretes (a layer of hardened iron oxides) along aerobic–anaerobic boundaries that vary with age, affecting the permeability of other materials (Phillips 1999).

Collectively, riparia share most of the attributes found in all types of boundaries. Understanding the general properties and behavior of boundaries provides a robust theoretical underpinning for the classification of riparian boundaries that follows.

Application of Ecological Information

A variety of data sources support classification, including routine field surveys, laboratory determinations, acquisition of remotely sensed information, and thematic mapping of topography, geology, and soils. These different data sources, of course, present different problems in terms of spatial resolution, generalization, and data standards. In developing an operational scheme, spatial units for data handling must be defined that represent a compromise between the natural boundaries representative of riparian patches and the environmentally arbitrary boundaries that often define the spatial units within which environmental data sets are collected. In achieving an effective riparian classification, a range of technical issues are associated with the assembling, storing, and retrieving of vast amounts of information and the integration of these data in a manner that produces valid, understandable results. It is also important to recognize the long-standing dilemma whereby bottom-up and top-down classifications may not produce the same result (Gurnell et al. 1994).

Information on riparian systems can be in the form of inventories, plant associations, indicator species, biophysical processes, and assessments of biotic integrity. These are distinct in their nature and therefore the application(s) for which they are appropriate. In general, inventories generate comprehensive lists of ecological attributes; geomorphic and plant associations collate that information into recognizable groups; indicators are ecological attributes that respond in a known way to disturbance; biophysical processes are hydrogeomorphic and ecological mechanisms associated with distinctive communities; and assessments compare ecological attributes in relation to specific criteria such as a "desired state" (see Table 3.3). Riparian indicators and assessments are not considered further in this chapter, but a comprehensive discussion of them can be found in Innis et al. (2000), Montgomery and MacDonald (2002), and, to some extent, in Chapters 9 and 10. Biophysical processes are discussed throughout the text as drivers of riparian community groups and as products of those groupings.

Inventory

An inventory is a list of measurable biophysical features (e.g., species, nutrient contents, and so on). Inventories normally apply only to a narrow window of space and

Table 3.3 Differentiating Among Inventories, Classifications, Indicators, and Assessments

Term	Definition
Inventory	A list of observable or measurable physical, chemical, and biological features. The results of an inventory apply only to a narrow window of space and time.
Classification	Groupings based on important common attributes with the categories depending on the objectives of the classification. Classifications use the "snapshots" generated by inventories.
Indicator	An organism or ecological community strictly associated with particular environmental conditions. The presence of the organism or community is indicative of these conditions. The use of indicators requires a comprehensive understanding of the systems and the interactions occurring therein.
Assessment	An integrated statement regarding the current system state and the factors contributing to that state. Often used to evaluate ecological integrity through conclusions drawn from rigorous analyses of inventories and classifications.

Adapted from Innis et al. 2000.

time, in essence providing a detailed snapshot of extant conditions. Some inventory procedures have been standardized because they may form a basis for riparian classification and assessment.

One of the most widely used riparian inventory approaches was developed by the U.S. Department of Agriculture's Forest Service to evaluate streamside habitats (Platts et al. 1987). This approach provides a well-organized, comprehensive list of potential parameters (many tested for accuracy and precision), including detailed instructions in the proper measurement of parameters and a discussion of statistical considerations. Although it does not address selection and integration of parameters for building an inventory, the approach has been used by many researchers that do.

The U.S. Department of the Interior's Bureau of Land Management (BLM) generated two widely used riparian inventory manuals. One is a two-stage approach that consists of an extensive inventory that produces a broad-scale classification, which is followed up by a site-specific, intensive inventory (Meyers 1989). The second builds on Meyers' approach and previous BLM manuals to generate a multi-stage, interdisciplinary approach for ecological site inventories (Leonard et al. 1992). Subsequent inventory approaches (e.g., Briggs et al. 1997) employ well-established methods to gather quantitative information and to select attributes that provide a maximum amount of information. Collectively, the inventories provide part of the foundation for a classification system.

Classification

Classification or typology provides a means of applying the biophysical "snapshots" generated by an inventory. This is accomplished by grouping spatially explicit areas based on common biophysical attributes. The attributes used and the categories created depend on the objective of the classification; the same area could be classified differently using different sets of attributes or criteria (Kondolf 1995). As highly complex systems, riparian zones are not only distinct from the surrounding landscape but are usually distinct from each other—and riparian systems change characteristics along the length of the river (e.g., Quinn et al. 2001). The natural variation makes comparison between and within riparian systems problematic unless a cohesive classification is available.

Emerging Classification Concepts

Ideally, a classification system should be based on a hierarchical ranking of linkages between the geologic and climatic settings, the physical riparian habitat features, and the biota. The geomorphic and climatic processes that shape abiotic and biotic features provide a conceptual and practical foundation for understanding the structure and processes of riparian systems (Montgomery and MacDonald 2002; see also Chapters 4 and 5). Furthermore, an understanding of process allows riparia to be viewed in a larger spatial and temporal perspective. It also allows us to infer the direction and magnitude of potential changes due to natural and human disturbances. A riparian classification system, based on patterns and processes and how they are expressed at different temporal and spatial scales, is one of the pillars of successful management (e.g., Kovalchik 2001, Quinn et al. 2001, Snelder and Biggs 2002).

Conceptually, individual classification units can be thought of as an integrated collection of *ultimate* and *proximate* controls on riparian system characteristics. These terms generally correspond to higher and lower levels of a hierarchical ranking of controlling factors. *Ultimate controls* refer to a set of geologic factors that act over large areas ($>1\,\mathrm{km}^2$), are stable over long time scales ($>10^4$ yr), and dictate the range of conditions possible in a drainage network. These include physical characteristics, such as regional geology and climate, and biotic characteristics, such as zoogeography (Moyle and Li 1979, Biggs et al. 1990). *Proximate controls* refer to local geomorphic and biotic processes important at small scales ($<10^2\,\mathrm{m}^2$), which can change channel characteristics over relatively short time periods ($<10^4$ yr). Proximate controls are constrained by ultimate controls and include such physical processes as discharge, temperature, hillslope erosion, lateral channel migration, and sediment transport, and the biotic processes of reproduction, competition, disease, and herbivory—all of which may be influenced by an equally diverse array of human impacts. Within this conceptual framework, management strategies for maintaining important physical and ecological structures may be tailored to local conditions.

The idea of ultimate and proximate controls arose from hierarchy theory and has been used to construct a continuum of habitat sensitivity to disturbance and recovery time (Frissell et al. 1986; see Figure 3.2). In this scenario, microhabitats are most sensitive to disturbance and catchments are least sensitive. Furthermore, individual events affecting smaller-scale habitat characteristics generally do not affect larger-scale system characteristics (however, collectively they can have an impact), whereas large disturbances directly influence the smaller-scale features of channels. For example, on a small spatial scale, deposition at one site may be accompanied by scouring at a nearby site, and overall the reach or segment does not appear to change significantly. In contrast, a large-scale disturbance (such as a debris flow in mountainous regions) initiated at the segment level is reflected in all lower levels of the hierarchy (reach, habitat, microhabitat). On a temporal scale, siltation of microhabitats may alter biotic communities over the short term. However, if the disturbance is of limited

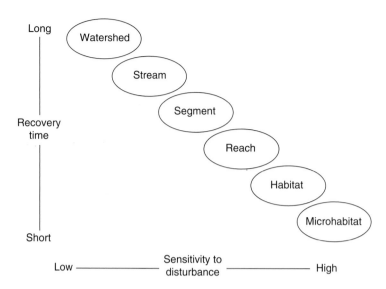

Figure 3.2 Relationship between recovery time and sensitivity to disturbance for different spatial scales associated with stream systems (from Naiman et al. 1992,). See Table 3.4 for definition of spatial scales.

scope and intensity or if the stream has a high transport capacity, the system may recover quickly to pre-disturbance levels.

Tailoring management strategies to stream types implies that the classification system includes the physical and biotic characteristics of the riparian system, as well as the disturbance regime creating and maintaining those characteristics. The most useful classification systems are able to categorize the types and frequencies of disturbance that may affect riparia as well as predict adjustments to the biophysical characteristics.

Geomorphic Classification

Most geomorphic approaches to classification have focused on stream channels rather than on the adjacent banks, but they still provide useful insights into the creation and maintenance of distinct riparian landforms. Two of the more popular approaches—hierarchical classification (Frissell et al. 1986) and Rosgen's classification (Rosgen 1994)—largely address system structure. More recent classifications— the process domain concept (Montgomery and Buffington 1997, 1998; Montgomery 1999) and the hydrogeomorphic approach (Brinson 1993, Hauer et al. 2002)—also incorporate physical processes and some evaluate channel responses to changing conditions (e.g., Montgomery and MacDonald 2002). Riparian corridor classifications formulated at the landscape scale provide additional insights based on the shape of the riparian boundary and speculation about how a specific corridor type might affect stream and riparian functions (Forman 1995). Although presented separately here, in reality, each of the geomorphic classifications shares some common origins and perspectives.

Hierarchical Classification

Inherent linkages among different ecosystem components at various scales have intrigued researchers for decades. Ecologists often have applied the concept of hierarchical classification as a means of identifying the appropriate scale at which to assess a given set of channel dynamics and to understand how the juxtaposition of biophysical patches at that scale may influence ecosystem dynamics (Poole 2002). Fluvial geomorphologists, on the other hand, have used the concept to understand bottom-up and top-down linkages across multiple spatial and temporal scales. The first comprehensive attempt at an element-based hierarchical classification of streams described the elements (i.e., stream network, segment, reach, habitat unit, and microhabitat) and provided the approximate size of the element and the frequency of disturbances that create and destroy them (Frissell et al. 1986). The structure contained within a given element influences its function, character, and operation within the river corridor.

Recent stream classification systems in North America use a hierarchical perspective linking large regional scales (ecoregions) with small microhabitat scales (see Table 3.4). This approach is especially useful since stream processes occur at scales spanning ~16 orders of magnitude (10^{-7}–10^{8} m spatially and 10^{-8}–10^{7} yr temporally; Minshall 1988). Several classification systems have been developed using nested landscape or channel features (Warren 1979, Frissell et al. 1986, Hawkins et al. 1993, Rosgen 1994, Montgomery 1999). Hierarchical classifications have two basic features:

1. They can be broadly applied (e.g., global, national, regional scales), but they are flexible enough to be modified for subregional purposes. When using hierarchical classifications, it is important to understand the proportional roles of controlling factors in determining the long-term and short-term riparian characteristics. This is because the relative importance of factors changes with spatial and temporal scale.
2. A hierarchical approach requires fewer variables at any one level than many other classification systems. Within most geographic regions, managers and scientists need only one or two spatial and temporal scales for classification purposes. Ultimately, however, the scope of the issue, or the nature of the question being considered, should determine the appropriate scale(s) of resolution.

One of the first efforts toward a hierarchical stream classification system, Warren (1979) described 11 levels that ranged from regional (>10^{2} km^{2}) to microhabitat (<1 m^{2}). Levels were defined largely by four variables: substrate, climate, water chemistry, and biota. Warren did not propose a classification system *per se,* but his contribution to the conceptual evolution of classification is noteworthy because he presented an explicit theoretical structure for a complex hierarchical system. He stressed assessment of site potential (i.e., all possible developmental states and performances that a system may exhibit while still maintaining its integrity as a coherent unit) rather than current conditions. Evaluating potential system states assists in distinguishing natural variability from human disturbance.

Within a few years, Warren's approach was extended by incorporating spatially nested levels of resolution (e.g., catchment, stream, valley segment, reach, habitat unit,

Table 3.4 Some Events or Processes Controlling Stream Habitat on Different Spatio-Temporal Scales

System Level	Linear Spatial Scale[a] (m)	Evolutionary Events[b]	Developmental Processes[c]	Time Scale of Continuous Potential Persistence[a] (yr)
Stream system	10^3	Tectonic uplift subsidence; catastrophic volcanism; sea-level changes; glaciation; climate shifts	Planation; denudation; drainage network development	10^6–10^5
Segment system	10^2	Minor glaciation, volcanism; earthquakes; very large landslides; alluvial or colluvial valley infilling	Migration of tributary junctions and bedrock nick-points; channel floor incision; development of new first-order channels	10^4–10^3
Reach system	10^1	Debris torrents; landslides; log input or washout; channel shifts, cutoffs; channelization, diversion, or damming by humans	Aggradation/degradation associated with large sediment-storing structures; bank erosion; riparian vegetation succession	10^2–10^1
Habitat or channel unit system	10^0	Input or washout of wood, boulders, and so on; small bank failures; flood scour or deposition; thalweg shifts; numerous human activities	Small-scale lateral or elevation changes in bed forms; minor bed-load resorting	10^1–10^0
Microhabitat system	10^{-1}	Annual sediment and organic matter transport; scour of stationary substrates; seasonal macrophyte growth and cropping	Seasonal depth and velocity changes; accumulation of fines; microbial breakdown of organics; periphyton growth	10^0–10^{-1}

From Frissell et al. 1986, with permission.

[a]Space and time scales indicated are approximate for a second- or third-order mountain stream. Caution is advised in using absolute spatial scales for the hierarchy. Depending on the specific situation, for example, a channel reach may be tens to hundreds of meters long while a habitat unit may be less than one meter to several meters long. Perhaps a better spatial index that preserves geomorphic similitude is scaling by channel width because there is no absolute association of channel size with stream order.

[b]Evolutionary events change potential capacity—that is, extrinsic forces that create and destroy systems at that scale.

[c]Developmental processes are intrinsic, progressive changes following a system's genesis in an evolutionary event.

and microhabitat; see Table 3.5 and Figure 2.4). An important conceptual advancement, Frissell and his colleagues (1986) articulated patterns within each hierarchical level as well as the origins and processes of development. The following discussion of hierarchical classification, summarized from Montgomery and Buffington (1998), provides insights into the physical basis for such an approach.

There are several spatial scales to be considered, ranging from large geomorphic provinces to small channel units. Geomorphic provinces consist of regions with similar landforms that reflect comparable hydrologic, erosional, and tectonic processes over areas greater than 1,000 km^2 (see Table 3.6 and Figure 3.3). Major physiographic,

Table 3.5 Habitat Spatial Boundaries Conforming to the Temporal Scales of Table 3.4

System Level	Capacity Time Scale[a] (yr)	Vertical Boundaries[b]	Longitudinal Boundaries[c]	Lateral Boundaries[d]	Linear Spatial Scale[a] (m)
Stream system	10^6–10^5	Total initial basin relief; sea level or other base level	Drainage divides and sea coast or chosen catchment area	Drainage divides; bedrock faults, joints controlling ridge valley development	10^3
Segment system	10^4–10^3	Bedrock elevation; tributary junction or falls elevation	Tributary junctions; major falls, bedrock lithological or structural discontinuities	Valley sideslopes or bedrock outcrops lateral migration	10^2
Reach system	10^2–10^1	Bedrock surface; relief of major sediment-storing structures	Slope breaks; structures capable of withstanding <50 years flood	Mean annual flood channel; mid-channel bars; other flow-splitting obstructions	10^1
Habitat or channel unit system	10^1–10^0	Depth of bed load subject to transport in <10-yr flood; top of water surface	Water surface and bed profile slope breaks; location of genetic structures	Same as longitudinal	10^0
Microhabitat system	10^0–10^{-1}	Depth to particles immovable in mean annual flood; water surface	Zones of differing substrate type, size, arrangement; water depth and velocity		10^{-1}

From Frissell et al. 1986, with permission.
Scaled to approximate a second– or third–order mountain stream; see cautions in Table 3.3, footnote a.
Vertical dimension refers to upper and lower surfaces.
Longitudinal dimension refers to upstream–downstream extent.
Lateral dimension refers to cross-channel or equivalent horizontal extent.

climatic, and geological features bound geomorphic provinces and impose broad controls on channel processes. Watersheds within a geomorphic province tend to share roughly similar relief, climate, and lithologic assemblages. Although geomorphic provinces identify broad areas likely to host comparable watersheds, the concept remains too general for predicting specific channel attributes or disturbance responses. Geomorphic provinces do, however, provide a general context for investigating and interpreting channel processes, and therefore channel response. This is not surprising since it has been suspected for decades that stream channel form can be predicted along the length of the river within geographic regions (Leopold et al. 1964).

Catchments define natural systems for routing sediment and runoff into and through channel networks (see Chapter 2, Table 3.6, and Figure 3.3). Although the appropriate scale of catchment-level classification ultimately is site specific, drainage basins 50–500 km^2 provide practical units for examining the influence of catchment processes on channel morphology and disturbance regimes (Montgomery et al. 1995). However, catchment-level classification of channel networks neglects fundamental differences in sediment production and transport processes of finer-scale valley morphologies.

Table 3.6 Hierarchical Levels of Channel Classification and Associated Spatial Scales

Classification Level	Spatial Scale
Geomorphic province	$1,000\,km^2$
Catchment	$50–500\,km^2$

Valley segment
   Colluvial valleys
   Bedrock valleys
   Alluvial valleys
	$10^2–10^4\,m$

Channel reaches
   Colluvial reaches
   Bedrock reaches
   Free-formed alluvial reaches
      Cascade reaches
      Step-pool reaches
      Plane-bed reaches
      Pool-riffle reaches
      Dune-ripple reaches
   Forced alluvial reaches
      Forced step-pool
      Forced pool-riffle
	$10^1–10^3\,m$

Channel units
   Pools
   Bars
   Shallows
	$10^0–10^1\,m$

From Montgomery and Buffington 1998.

    Valley segments define portions of the drainage network exhibiting similar valley scale morphologies and governing geomorphic processes at a scale of 10^2 to 10^4 m (see Table 3.6 and Figure 3.3). For example, valley segments in mountain drainage basins are classified into colluvial, bedrock, and alluvial valley types based on valley fill, sediment transport processes, channel transport capacity, and sediment supply (Montgomery and Buffington 1997, 1998). Colluvial valley bottoms lacking a well defined channel indicate insufficient hydraulic erosion to initiate and maintain a channel; these unchanneled valleys (regionally referred to as *hollows*, *swales*, or *head walls*) often extend upslope of the smallest channels. Channeled colluvial valleys below hollows indicate the emergence of fluvial transport. Nevertheless, the influence of fluvial processes on colluvial valley form and incision is often secondary to transport by periodic debris flows in steep landscapes. In contrast, the maintenance of colluvial valleys in low-gradient landscapes requires incision by streams; extensive networks of colluvial valleys in low-gradient landscapes are likely due to long-term climate change.

    The second type of valley segment, bedrock valley, typically is confined within solid riverbanks and lacks significant valley fill. Narrow valley bottoms result in relatively straight channels, although deeply incised bedrock meanders may occur. Channel floors in bedrock valleys consist of either exposed bedrock or thin, patchy accumulations of alluvium. Little sediment storage in bedrock valley segments indicate downstream transport of virtually all the material delivered to the channel, suggesting that transport capacity exceeds sediment supply over the long term.

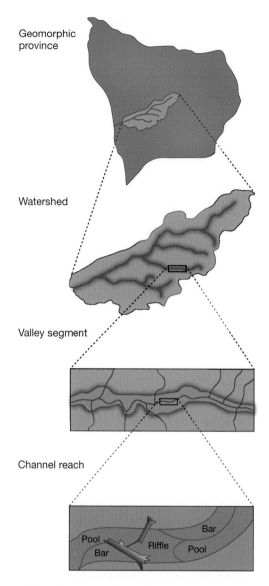

Geomorphic province

Watershed

Valley segment

Channel reach

Bar
Pool
Bar   Riffle   Pool

Figure 3.3 Geomorphic province, watershed, valley segment, channel reach, and channel unit (e.g., pools, bars, riffles) scales of classification illustrated for the Olympic Peninsula, a mountainous region in Washington (adapted from Montgomery and Buffington 1998).

The third segment type, alluvial valleys, transport and sort sediment loads supplied from upslope but lack the capacity to routinely scour the valley to bedrock. Channels in alluvial valley segments may support either narrow or wide floodplains. Thick alluvial deposits in unconfined valley segments imply a long-term excess of sediment supply relative to transport capacity. Both the specific channel morphology and degree of confinement reflect local channel slope, sediment supply, and hydraulic discharge.

Channel reaches exhibit similar bed forms over stretches of stream that are many channel widths in length (see Table 3.6 and Figure 3.3). Again, using the mountain drainage basin example, channels are classified into colluvial, bedrock, and alluvial reach types. Basic features of the colluvial and bedrock reaches are described above, whereas free-formed alluvial reaches exhibit a wide variety of morphologies and roughness configurations that vary with slope and position within the channel network. Montgomery and Buffington (1997) suggest that the ratio of transport capacity to sediment supply controls the roughness configurations that shape alluvial reach morphology, which can be categorized into five free-formed alluvial channel reach types: cascade, step-pool, plane-bed, pool-riffle, and dune-ripple. Transitional morphologies also occur, as this classification imposes order on a natural continuum.

Cascade reaches are found on steep slopes with high rates of energy dissipation. They are characterized by longitudinally and laterally disorganized bed material; typically cobbles and boulders confined by valley walls. Step-pool reaches consist of large rock fragments (clasts) organized into discrete channel-spanning accumulations that form a series of steps separating pools containing finer material. The stepped morphology of the bed results in alternating turbulent flow over steps and tranquil flow in pools. Plane-bed reaches are characterized by a relatively featureless gravel/cobble substrate that encompasses channel units (described later) and are often referred to as *glides*, *riffles*, or *rapids* (Bisson et al. 1982). Plane-bed reaches may be either unconfined or confined by valley walls and are distinguished from cascade reaches by the absence of tumbling flow and less roughness (i.e., ratio of the largest grain size to bankfull flow depth). Pool-riffle reaches are typically unconfined by valley walls and consist of a laterally oscillating sequence of bars, pools, and riffles resulting from migrating cross-channel flow that causes flow convergence and scour on alternating banks as well as opposing flow divergence on the opposite side that results in local sediment accumulation in discrete bars. Dune-ripple reaches are unconfined, low-gradient, sand-bedded channels that exhibit a succession of mobile bed forms with increasing flow depth and velocity.

The presence of other materials also can force the morphological expressions of alluvial reaches. External flow obstructions such as large woody debris and bedrock outcrops force local flow convergence, divergence, and sediment retention that respectively form pools, bars, and steps (see Figure 3.4). The morphologic impact of large woody debris, in particular, depends on the amount, size, orientation, and position of debris, as well as channel size and rates of debris recruitment, transport, and decay (Bilby and Bisson 1998). Flow obstructions can force specific channel morphologies on steeper slopes than is typical of analogous free-formed alluvial morphologies. In particular, large woody debris may force pool-riffle formation in otherwise plane-bed or bedrock reaches. It is important to recognize forced morphologies as distinct reach types because the interpretation of whether such obstructions govern bed morphology is essential for understanding how channels respond to altered conditions.

Channel units are morphologically distinct areas that extend up to several channel widths in length and are spatially embedded within a channel reach; they are the morphologic building blocks of a reach (see Figure 3.3). Channel units are classified as various types of pools, bars, and shallows (i.e., riffles, rapids, and cascades). Distinctions among the units focus on topographic form, organization, and areal density of clasts, local slope, flow depth, velocity, and, to some extent, grain size. Although there is a general association of specific channel unit morphologies with reaches, prediction of channel unit properties is complicated by site-specific controls, particularly in forest environments rich in large woody debris. While the channel unit is biologically

Figure 3.4 Piles of large woody debris in the Queets River, Washington. Large wood is instrumental in shaping the morphology—and thereby the classification—of alluvial rivers in forested regions and in the regeneration of riparian vegetation. (Photo: J Latterell.)

relevant, interpretation of the abundance, characteristics, and response potential of channel units depends on the context imposed by reach-level channel types.

Rosgen's Classification

Although not strictly a hierarchical system, the Rosgen (1994) classification is widely used in western North America by private landowners and resource managers. It is based on geomorphic and in-channel characteristics including channel gradient, sinuosity, width-to-depth ratio, bed material, entrenchment, channel confinement, soil erodibility, and stability. It also includes subcategories that may change over short temporal scales and are characterized by riparian vegetation, channel width, organic debris, flow regime, meander patterns, depositional features, and sediment supply. Rosgen's stream-type classification system has been used widely for site-specific riparian forest and fisheries management and for predicting geomorphic and hydrologic processes.

Rosgen's classification requires an explanation because of its wide use. It is based on a morphological arrangement of the aforementioned stream characteristics organized into relatively homogeneous stream types rather than being based on geomorphic or hydrologic processes. The Rosgen system assumes physical laws govern contemporary channel morphology, resulting in observable stream features and fluvial processes (such as water hydraulics and transported materials). A change in any one of the physical processes causes channel adjustments, which lead to changes in other fluvial processes, resulting in new channel features. There are, however, strong and well-founded objections to Rosgen's classification (e.g., Miller and Ritter 1996, Juracek and Fitzpatrick 2003). It is argued that Rosgen extended his classification beyond its use as a communications tool into the realm of predicting fluvial processes, and his system fails to demonstrate that the criteria upon which it is based have geomorphic significance.

Specific limitations of the Rosgen classification relate to spatial and temporal difference in geomorphic processes, including time dependence on when the reach was classified, uncertain applicability across physical environments, difficulty in identifying a true equilibrium condition as one transient state can be replaced by another, potential for incorrect determination of bankfull condition (bankfull flows in some stream types are difficult to delineate), and the uncertain process-scale significance of classification criteria (Juracek and Fitzpatrick 2003). We continue with a brief description of the Rosgen's classification, keeping in mind its limitations.

Rosgen (1994) recognized four hierarchical aspects to classification (see Table 3.7)

Table 3.7 Hierarchy of River Inventories

Level of Detail	Inventory Description	Information Required	Objectives
I	Broad geomorphological characterization	Landform; lithology; soils; climate; depositional history; basin relief; valley morphology; river profile morphology; general river pattern	To describe generalized fluvial features using remote sensing and existing inventories of geology, landform evolution, valley morphology, depositional history, and associated river slopes; relief and patterns used for generalized categories of major stream types and associated interpretations
II	Morphological description (channel types)	Channel patterns; entrenchment ratio; width-to-depth ratio; sinuosity; channel material; slope	To delineate homogeneous stream types that describe specific slopes, channel materials, dimensions, and patterns from "reference reach" measurements; provides a more detailed level of interpretation and extrapolation than Level I
III	Stream "state" of condition	Riparian vegetation; depositional patterns; meander patterns; confinement features; fish habitat indices; flow regime; river size category; debris occurrence; channel stability index; bank erodibility	To further describe existing conditions that influence the response of channels to imposed change and provide specific information for prediction methodologies (such as stream bank erosion calculations); provides for very detailed descriptions and associated prediction/interpretation
IV	Verification	Involves direct measurements and observations of sediment transport, bank erosion rates, aggradation/ degradation processes, hydraulic geometry, biological data such as fish biomass, aquatic insects, riparian vegetation evaluations, and so on	Provides reach-specific information on channel processes; used to evaluate prediction methodologies; to provide sediment, hydraulic, and biological information related to specific stream types; and to evaluate effectiveness of mitigation and impact assessments for activities by stream type

Modified from Rosgen 1994.

I. Broad geomorphic characterization
II. Morphologic description of the channel
III. Stream condition
IV. Verification

The first two aspects address physical channel characteristics, forming the basis of his system, and are discussed below. The third aspect addresses the stream state by describing existing conditions that influence the response of altered channels and provides specific information for prediction. The fourth aspect addresses verification of reach-specific information on channel processes, which is used to evaluate predictions. Those interested in additional information on Aspects III and IV are referred to the original publication.

Geomorphic Characterization (Aspect I)

The purpose is to provide a broad characterization that integrates the landform and fluvial features of valley morphology with channel relief, pattern, shape, and dimension. Aspect I combines the influences of climate, depositional history, and vegetative life zones on channel morphology. Generalized categories of stream types are initially delineated using descriptions of dominant slope range, valley and channel cross-sections, and plan-view patterns (see Figure 3.5 and Table 3.8).

The longitudinal profile serves to identify slope categories for stream reaches. For example, streams of type Aa+ have channel gradients greater than 10 percent with frequently spaced, vertical drop scour-pool features (see Figure 3.5 and Table 3.8). The cross-sectional profile also can be inferred at this broad level as can information concerning floodplains, terraces, structural control features, confinement, entrenchment, and valley versus channel dimensions. For example, the type A streams are narrow, deep, confined, and entrenched, while the width of the channel and the valley

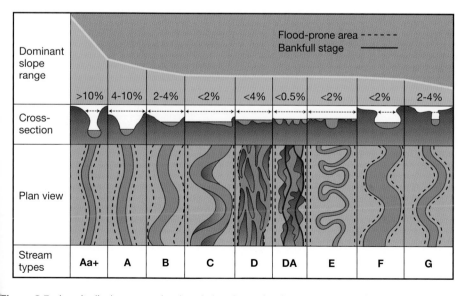

Figure 3.5 Longitudinal, cross-sectional, and plan views of major stream types (after Rosgen 1994).

Table 3.8 Summary of Delineative Criteria for Broad-Level Classification

Stream Type	General Description	Entrenchment Ratio	Width-to-Depth Ratio	Sinuosity	Slope	Landform/Soils/Features
Aa+	Very steep, deeply entrenched, debris transport streams	<1.4	<12	1.0–1.1	>10 percent	Very high relief; erosional, bedrock, or depositional features; debris flow potential; deeply entrenched streams; vertical steps with deep scour pools; waterfalls
A	Steep, entrenched, cascading step-pool streams; high energy/debris transport associated with depositional soils; very stable if bedrock or boulder-dominated channel	<1.4	<12	1.0–1.2	4–10 percent	High relief; erosional or depositional and bedrock forms; entrenched and confined streams with cascading reaches; frequently spaced, deep pools in associated step-pool bed morphology
B	Moderately entrenched, moderate-gradient, riffle dominated channel with infrequently spaced pools; very stable plan and profile; stable banks	1.4–2.2	>12	>1.2	2–3.9 percent	Moderate relief, colluvial deposition and/or residual soils; moderate entrenchment and width-to-depth ratio; narrow, gently sloping valleys; rapids predominate with occasional pools
C	Low-gradient, meandering, point-bar, riffle-pool, alluvial channels with broad, well-defined floodplains	>2.2	>12	>1.4	<2 percent	Broad valleys with terraces in association with floodplains and alluvial soils; slightly entrenched with well-defined meandering channel; riffle-pool bed morphology
D	Braided channel with longitudinal and transverse bars; very wide channel with eroding banks	N/A	>40	N/A	<4 percent	Broad valleys with alluvial and colluvial fans; glacial debris and depositional features; active lateral adjustment with abundance of sediment supply
DA	Anastomosing (multiple channels) narrow and deep with expansive well-vegetated floodplain and associated wetlands; very gentle relief with highly variable sinuosities; stable stream banks	>4.0	<40	Variable	<0.05 percent	Broad, low-gradient valleys with fine alluvium and/or lacustrine soils; anastomosed (multiple channel) geologic control creating fine deposition with well-vegetated bars that are laterally stable with broad wetland floodplains
E	Low-gradient, meandering riffle-pool stream with low width-to-depth ratio and little deposition; very efficient and stable; high meander width ratio	>2.2	<12	>1.5	<2 percent	Broad valley/meadow; alluvial materials with floodplain; highly sinuous with stable, well-vegetated banks; riffle-pool morphology with very low width-to-depth ratio
F	Entrenched meandering riffle-pool channel on low gradients with high width-to-depth ratio	<1.4	>12	>1.4	<2 percent	Entrenched in highly weathered material; gentle gradients, with a high width-to-depth ratio; meandering, laterally unstable with bank-erosion rates; riffle-pool morphology
G	Entrenched "gully" step-pool and low width-to-depth ratio on moderate gradients	<1.4	<12	>1.2	2–3.9 percent	Gully, step-pool morphology with moderate slopes and low width-to-depth ratio; narrow valleys or deeply incised in alluvial or colluvial materials, i.e. fans or deltas; unstable, with grade control problems and high bank erosion rates

From Rosgen 1994.

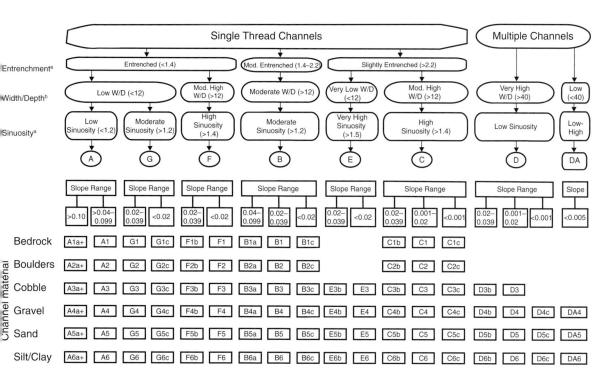

aValues can vary by ±0.2 units as a function of the continuum of physical variables within stream reaches.
bValues can vary by ±2.0 units as a function of the continuum of physical variables within stream reaches.

Figure 3.6 Key to classification of natural rivers (from Rosgen 1994).

is similar. The plan view morphology is simply the pattern of the river from above. For example, type A streams are relatively straight, whereas type C streams are meandering.

Morphologic Description (Aspect II)

After streams are separated into the major categories of A through G (see Figure 3.5 and Table 3.8), Aspect II is applied, separating them into discreet slope ranges and dominant substrate grain sizes. This results in 42 subcategories of stream types (see Figure 3.6). In reality, however, there is a normal range of values for each criterion, and this important observation is incorporated into Rosgen's classification system. This aspect recognizes and describes a morphologic continuum within and among stream types. The continuum is applied where values outside the normal range are encountered but do not warrant a unique stream type. For example, selected channel slopes in Figure 3.6 are sorted by subcategories of: a+ (>10 percent), a (4–10 percent), b (2–3.9 percent), c (<2 percent), and c– (<0.01 percent). A limitation of Aspect II is that no biophysical criteria are established for the arbitrary sorting of slopes into subcategories.

The emphasis on channel materials is equally important, as they are critical for sediment transport and hydraulic influences (such as channel roughness) and for the modification of the river's form, plan, and profile. Interpretation of biological function and stability also require this information. Using the "pebble count" method of Wolman (1954), with a few modifications for large bank materials and sand, the particle size distribution of substrate materials is determined easily in the field.

There is disagreement about whether the principal units of classification should be temporally stable (e.g., valley segments) or dynamic (e.g., stream types). Arguments for temporal stability suggest that a reach, once classified, is of little management value if it changes naturally over the time scale of land-use practices (Frissell et al. 1986). In contrast, a dynamic classification based on smaller, more dynamic units provides a more accurate description of present conditions in the reach (e.g., riparian width). Both perspectives may be useful for management assessments depending on the specific objectives.

The Process Domain Concept

The process domain concept (PDC) is a sharp departure from more structurally based classification perspectives because it is spatially oriented by type of physical process. The assumption underlying the PDC is that the influence of spatial and temporal variability in geomorphic processes on biotic systems is controlled by the disturbance regime (i.e., frequency, magnitude, timing, and duration of changes in the physical environment). Process domains are predictable *areas* of a landscape within which distinct suites of geomorphic processes govern physical habitat type, structure, and dynamics. The disturbance regimes associated with process domains dictate the geomorphic template upon which ecosystems develop (Swanson et al. 1988). Fundamentally, the concept of process domains provides a means of representing spatial differences in the disturbance regimes that structure habitat.

The most basic set of process domains includes hillslopes, hollows, channels, and floodplains (Montgomery 1999). However, each can be subdivided into finer-scale distinctions (e.g., types of channels or floodplains). Fundamentally, the PDC holds that one can define and map domains within a catchment characterized by different geomorphic processes, disturbance regimes, response potentials, and recovery times, and that the divisions have ecological significance (Table 3.9).

A typical mountain catchment, for example, has a number of distinct landscape–scale process domains (Montgomery 1999; see Figure 3.7). Convex hillslopes define zones of sediment production where rock becomes colluvial soil through processes such as tree throw, splintering by freeze-thaw cycles, or burrowing by fossorial mammals. Hollows are fine-scale, unchanneled valleys where topographic

Table 3.9 Effect of Spatial Scale on Geomorphic Influences on Ecosystems and the Biological Attributes Most Affected

Spatial Scale	Geomorphic Influences	Biological Attributes
Regional/Physiographic Province	Climate, topography, geology	Community type
Valley Segment/Channel Reach	Routing of sediment, water and organic matter	Community composition and species abundance
Channel Unit or Patch	Local factors/disturbance history	Habitat use by individuals

From Montgomery 1999, with permission.

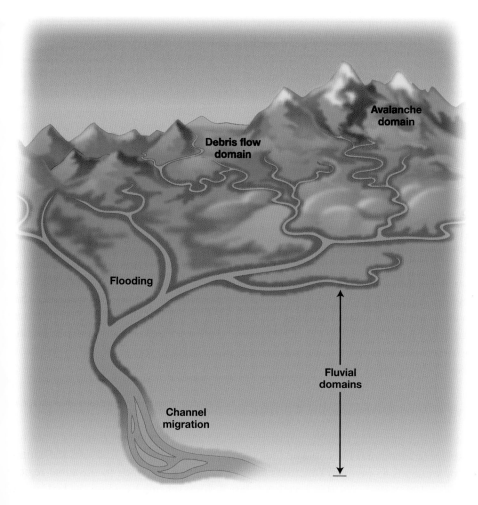

Figure 3.7 Typical coarse-scale riverine process domains for river systems in the mountainous Pacific Northwest region of North America (adapted from Montgomery 1999).

convergence concentrates sediment moved by soil creep, raveling, and biogenic transport. Consequently, soil profiles thicken through time in hollows. Topographic convergence also focuses surface and subsurface flows to hollows, which produces elevated soil moisture and leads to a cycle of gradual colluvial infilling followed by catastrophic evacuation by shallow landsliding every few thousand years. Immediately downslope of hollows there are ephemeral colluvial channels, steep alluvial channels, and low-gradient, alluvial floodplain channels—each of which has distinct process domains. Collectively, the physical channel types can be readily mapped in the field, and with appropriate methods can be predicted from digital topography (Dietrich et al. 1992).

There is strong correspondence between process domains and riparian vegetation. Relations among stream gradient, floodplain geometry, and riparian forest diversity illustrate relationships between identifiable landscape units and riparian forest characteristics (e.g., Hupp 1982, Hupp and Osterkamp 1996) and valley-bottom geomorphic surfaces characterized by contrasting flow-driven disturbance frequency and intensity have distinct vegetative communities (e.g., Osterkamp and Hupp 1984, van Coller et al. 1997). Additionally, riparian vegetation composition and structure is known to differ among confined and unconfined valley segments (Gregory et al. 1991). In wide, unconfined valley bottoms with well-developed floodplains, disturbances associated with floods and lateral channel migration can lead to a complex riparian forest, whereas confined alluvial channels that are not free to migrate laterally typically have only narrow riparian vegetation or terrestrial associations that extend to the channel banks—and the canopy may be fully enclosed by late successional species. Confined channels recently disturbed by flooding or debris flows can be lined by distinct riparian or early successional communities. Well-defined longitudinal corridors of riparian or early successional vegetation in confined channels indicate either synchronous catastrophic disturbance or a frequent disturbance regime.

Overall, the spatial and temporal variability of geomorphic processes governing habitat quality, availability, disturbance, and recovery impart either a lateral zonation or a patchy character to riparian habitats. It is well known that patterns of habitat availability define conditions within which communities and community interactions develop. At the catchment scale, process domains identify distinct process zones that divide (and thereby classify) channel networks into channel reaches or valley segments dominated by different disturbance regimes and environmental characteristics. Hence, the process domain shapes not only the physical channel type; it ultimately assists in shaping the riparian community.

The Hydrogeomorphic Approach

The hydrogeomorphic (HGM) approach is a collection of concepts and methods for developing functional indices. The indices are subsequently used to assess the capacity of wetlands to perform functions relative to similar wetlands in the same region (Brinson 1993). Basically, it is a wetland classification scheme based on the hydrologic and geomorphic settings. The HGM was initially designed for use in the context of the Clean Water Act (Section 404 Regulatory Program permit review) to consider alternatives, minimize impacts, assess unavoidable project impacts, determine mitigation requirements, and monitor the success of mitigation projects. However, a variety of other potential applications for the HGM have been identified, including assessing the functions of wetlands associated with riparian floodplains (Hauer et al. 2002).

The HGM classification identifies groups of wetlands using three criteria that fundamentally influence how wetlands function: geomorphic setting, water source, and hydrodynamics. Geomorphic setting is the landform and the position in the landscape. Water source refers to the primary water source, such as precipitation, overbank floodwater, or groundwater. Hydrodynamics refers to the level of energy and the direction of water movement. Using these criteria, Brinson (1993) identified five hydrogeomorphic wetland classes at the continental scale; two additional ones were added later and regional scale subclasses have been developed. Regional guidebooks are available from the U.S. Army Corps of Engineers Environmental Laboratory at http://www.wes.army.mil/el/wetlands/wlpubs.html. Strictly speaking, the HGM at this stage of development is not a riparian classification. However, the application of the HGM to gravel-bed, alluvial riverine floodplain wetlands is currently underway for the northern Rocky Mountains (Hauer et al. 2002).

Biotic Classification

Biotic communities serve as integrators of ecological conditions expressed over different time and space scales and, therefore, can serve as one approach to classification. Several systems have been developed, and all are based on community-level patterns seen in riparian soils, plants, and animals. In general, biotic classifications assume a predictable relationship between the biota and the geomorphic and hydrologic templates.

Ultimately, biogeographic factors restrict the geographic scope of classification schemes based on assemblages of plants or animals. However, spatial variability in physical and biotic factors shaping community dynamics can also limit geographic scope. It is well known that environmental disturbance regimes vary with climate and geology (Poff and Ward 1989). In systems where seasonal flow patterns are predictable, riparian communities may be persistent and resilient. However, in systems with highly variable and unpredictable flow patterns, communities can exhibit sharp temporal fluctuations in structure.

Additional biotic factors further compound species–habitat relationships. In riparian habitats where competition and trophic interactions are important, fluctuations in physical and chemical conditions modify competition as well as alter the intensity and direction of plant–herbivore and predator–prey interactions. Further, variability in productivity between riparian sites also contributes to wide-ranging diversity patterns. Moreover, various human activities produce major alterations in community composition (e.g., species introductions, chemical pollution, harvest) without altering physical habitat structure.

Soils

Upland and riparian soils are derived from a diversity of parent materials and, in the case of riparian soils, are largely distributed and influenced by water. While the parent material of upland soils is chiefly the rock underlying the site, the mineral component of riparian soils originates as stream-deposited sediment in stratified bands of varying textures. Thus, riparian soils tend to be more heterogeneous in mineral characteristics than their upland counterparts. Periodic sediment deposition to riparian areas is accompanied by the import and flushing of organic matter and nutrients from the site by water. This increases the heterogeneity of riparian soils by producing a bare soil surface in some areas and rich deposits of organic matter and nutrients in others. Additionally, there are a wide variety of site-specific factors—moisture, fluctuating water tables, litterfall, and decomposition—that subsequently contribute to soil development. A compounding factor is the periodic disturbance of riparian sites by floods, which tend to leave riparian soils poorly developed.

Therefore, even though upland soils are widely used in classifications systems, it should not be surprising that there is no site-level classification system specifically directed toward riparian soils. Most riparian soils belong to the Entisol, Inceptisol, or Histosol orders (USDA Soil Survey Staff 1998). Entisol is a catchall category for soils with little pedogenic development (i.e., soil-forming processes, primarily weathering and mineral translocation but not direct vegetative and physical disturbance). Entisols

73

include the suborders aquent (regular or permanent saturation within the soil profile), fluvent (fine-textured alluvial soils), and psamment (sandy alluvial soils). Inceptisols have more development than Entisols but are still found mostly in locations where recent (i.e., Holocene) colluvial or fluvial processes reset soil development. There are various suborders of Inceptisols, reflecting mostly climatic influences. Histosols are organic soils—bogs and muskeg. Nevertheless, if one also includes other substrate characteristics such as sediment grain size and geologic origin of particles, riparian soils eventually may prove to be a useful perspective for site-level classification.

Plants

The distribution of plant species in temperate and tropical riparian systems is closely associated with variations in physical templates. The vegetative patterns have been described most often in relation to the strong vertical and lateral gradients associated with flooding and water availability. Thereby, hydrologic characteristics are often inferred as major determinants of the plant distributions. Recently, however, a number of researchers have recognized the importance of geomorphic processes in shaping vegetative patterns (e.g., Gregory et al. 1991, Montgomery 1999) incorporating them into their classifications (Hupp and Osterkamp 1985, 1996; van Coller et al. 1997).

Several classification systems based on riparian vegetation patterns use community type as the fundamental classification unit. Community type is defined either by present vegetative composition or potential climax vegetation (Swanson et al. 1988). Inferences are drawn regarding environmental gradients and successional relationships between community types. Stratification of community types is based on overstory or understory vegetation. The understory (herbs and shrubs), because of its higher turnover rate, is a better indicator of current soil and hydrologic conditions, whereas the canopy is a better integrator of longer temporal patterns. As with other biotic classification systems, the most valuable riparian classification schemes center on relationships to physical environmental factors.

Taking riparian classification one step further, many authors acknowledge the need for classifying riparian zones with respect to conservation value or ecological potential (Slater et al. 1987, Harris 1988, Swanson et al. 1988, Baker 1989, Gregory et al. 1991, Gurnell et al. 1994, Quinn et al. 2001). For example, species richness, rarity, and frequency of occurrence can be used to determine the conservation value of different stream segments (Slater et al. 1987).

Classifying vegetative communities in relation to geomorphically distinct river segments or valley types incorporates concepts from landscape ecology, especially the hierarchical relationships of different landscape elements. This approach has been successfully employed by several researchers, most notably Hupp and Osterkamp (1985, 1996), Harris (1988), and van Coller et al. (1997). The long-term investigations of Hupp and Osterkamp established the intrinsic affinity between bottomland riparian vegetation and discrete fluvial landforms across a broad array of sites in the United States. Harris, working in the Sierra Nevada Mountains of California, limited classification units to the stream segment scale and addressed the importance of larger-scale factors in determining smaller-scale patterns. The geomorphic-vegetation units differed in their sensitivity to management, yet were useful for purposes of resource inventory, detailed ecological studies, and prediction of human-induced alterations. Harris' classification system was an important step forward in coupling contrasting landscape processes to biotic resources and in attempting to predict the sensitivity of riparian systems to disturbance. Van Coller et al. (1997), working along a semi-arid, bedrock-controlled river in South Africa identified six riparian vegetation types using TWINSPAN analysis, relating them to dominant geomorphic channel types. This particular study is instructive because, once the vegetation–fluvial geomorphology classifications and links are established, it allows researchers to develop scenarios of vegetation responses to progressive sedimentation from land use in the catchment.

There are three methods for assessing the integrity of riparian systems that, although not classification systems themselves, possess some important attributes of classification systems. The first method is the USDA Forest Service's Integrated Riparian Evaluation Guide (USDA 1992). It presents a three-level procedure including analytical methods, extensive field methods, and intensive site characterization. An inventory of dozens of physical (e.g., hydrologic, geomorphic), chemical (e.g., soil attributes), and biological parameters sets the stage for various classifications, including soil and vegetative communities. The actual ecological evaluation is limited to a vegetative comparison with the potential natural community.

A related approach, the Riparian Evaluation and Site Assessment (RESA, Fry et al. 1994), is modeled after the U.S. Natural Resources Conservation Service's Land Evaluation and Site Assessment (LESA). RESA assists in determining appropriate land use in agricultural areas, ranks riparian sites for management activities, and prescribes riparian buffer widths. The RESA approach utilizes a semiquantitative index that incorporates biophysical information and integrates upland conditions with site characteristics.

The second method, the System for Evaluating Rivers for Conservation (SERCON, Boon et al. 1998), integrates riparian attributes into an overall river evaluation. SERCON is a weighted index that scores various attributes within the "conservation criteria" of physical diversity, naturalness, representativeness, rarity, species richness, and several special features. Although not widely used currently, SERCON is expected to gain popularity as it is closely associated with the well-known River Habitat Survey in the United Kingdom.

The third method, the Riparian, Channel, and Environmental inventory (RCE; Petersen 1992), assesses the biophysical conditions of small stream systems in lowland agricultural settings. It is a multi-metric index that generates a score by integrating 16 characteristics of the stream channel and the riparian zone. The index places strong emphasis on physical attributes of the riparian zone, such as bank stability, sediment retention and shading, which reflect riparian functions important to the stream channel. Fundamentally, the RCE describes in-stream conditions, which it seems to accomplish fairly adequately.

Wildlife

The use of animals as a riparian classification tool has received less attention, especially in comparison to the use of plants. Even though there are hundreds of studies of animals using riparian systems (e.g., Raedake 1988, Naiman and Rogers 1997), only a few attempt to use animals for developing a riparian typology. Animals are, of course, highly mobile, but the potential exists for using habitat specialists for delineating riparian types. For example, a novel method using butterflies as indicators of riparian condition in the southwestern United States may be useful for classification (Nelson and Anderson 1994). An index of butterfly riparian quality, based on species richness and disturbance susceptibility for individual taxa, is generated for each species using attributes that reflect changes in the entire riparian system. These attributes include species mobility, larval host-plant form and specificity, and riparian dependency. The success of the method lies in the fact that the indicator is restricted in its habitat due to the unique nature of the riparian zone in an arid landscape.

Treating Complexity and Heterogeneity in Classification Systems

Riparia exhibit both biocomplexity as well as heterogeneity. *Biocomplexity* refers to the array of species and biophysical linkages, whereas *heterogeneity* refers to the spatial and temporal patterns of species and communities (Naiman et al. 2005). Since riparian communities are not static systems, the expression of their biophysical characteristics makes classification more of an art than a science.

Riparia respond continuously in time and space to a complex array of hydrologic (e.g., water regimes, hydraulic shear stress, sediment deposition, erosion, deposition of large woody debris) and biotic (e.g., animal activities, plant production) influences, producing a broad array of community types. Further, riparian community patterns integrate a diverse array of landscape elements and processes operating on numerous spatial and temporal scales. As we saw in Chapter 2, these include longitudinal, lateral, and vertical gradients in geomorphic features (e.g., gravel bars, terraces, islands), surface and subsurface flows of water and nutrients, and disturbance regimes (e.g., floods, drought, fire, wind). Fluvial actions (e.g., erosion, transport, deposition) are the dominant agents of riparian change and constitute one suite of the natural disturbance processes primarily responsible for sustaining a high level of heterogeneity in riparian systems (Poff et al. 1997, Ward et al. 2002). It is the hydrological connectivity—the flux of matter, energy, and biota via water—in combination with animal activities that largely sustain riparian heterogeneity (Naiman et al. 2005). Changes in just one of these controlling factors can shift riparian assemblages from one class to another in a short period. Although individual features of riparia (e.g., a specific vegetative patch type) may exhibit dynamic transitions, their relative abundance within a catchment tends to remain in quasi-equilibrium over decades to centuries.

How can one accommodate riparian complexity and heterogeneity in establishing a classification system? One approach is to use ordination techniques, such as TWINSPAN, to group communities based on similarity of attributes. Van Coller et al. (1997) and others have used this approach in effectively linking riparian community types to geomorphic channel features. For example, they were able to identify six plant community types on four channel types (either bedrock anastomosing, mixed anastomosing, pool-riffle, and braided) composed of 12 morphological units (e.g., macro-channel bank, lateral bar, and so forth) on the Sabie River in South Africa. A similar approach used on the Queets River, Washington, groups communities onto three general geomorphic templates (Balian and Naiman 2005). The templates—in this case active floodplain, young terrace, and mature terrace—are identified from aerial photos and digital elevation models. Balian and Naiman used the templates as the basis for measuring riparian tree production, eventually extrapolating production to the entire 57-km^2 floodplain. Whatever the approach, acquiring a system-level perspective requires classifying riparia, with its inherent heterogeneity, in a manner that can be extrapolated to larger spatial scales.

Attributes of an Enduring Classification System

Like any system that organizes many criteria into categories, the ones described here were developed with specific objectives in mind. They can be transferred to other situations with different objectives but only after considerable thought and, possibly, technical modifications. However, the underlying principles are basically sound and adaptable to new situations.

Although specific riparian classification systems are not numerous or diverse, there is a consensus developing on the fundamental attributes of an enduring classification system. These attributes relate to the ability to encompass broad spatial and temporal scales (including a historical perspective), to integrate structural and functional characteristics under various disturbance regimes, to convey information about underlying mechanisms that control riparian features, and to accomplish this at low cost and at a high level of uniform understanding (Naiman 1998, Juracek and Fitzpatrick 2003). No existing classification system adequately meets all of the ideal attributes, and it is likely that no single classification system ever will. The concepts of Frissell et al. (1986) and Rosgen (1994) are regarded as important intellectual advancements, but they do not provide the level of understanding of physical processes needed to predict riparian responses to specific types of catchment disturbances (such as debris flows or hillslope failures). Consequently, specific links between physical and biological processes within these classification systems remain poorly defined. Physical, process-based approaches that offer more promise in meeting these attributes are currently being used to address a wide variety of stream-related management issues in mountainous (Montgomery and Buffington 1998, Montgomery 1999, Montgomery and MacDonald 2002) and other regions (Quinn et al. 2001, Snelder and Biggs 2002). Nevertheless, the generally narrow perspective provided by all existing classification systems places constraints on their broad effectiveness. Generally, with the exception of the approaches previously noted, there is often no explicit consideration of the underlying geomorphic context or the potential response of the channel segment to disturbance.

Despite these caveats, hierarchical classification systems have been useful in making scientists and resource managers aware of the diversity of channel types and the need for a variety of management prescriptions for habitat protection and conservation (see Chapters 8 and 9). This is especially important in regions with numerous categories of channel segments, and where pervasive land use provides an economic base for industries employing tens of thousands of people. For example, the evolving classification system used as part of the State of Washington Forest Practices Regulations (Montgomery and Buffington 1998, Montgomery 1999) allows resource managers and scientists to consider alternative riparian-related forestry practices (e.g., silvicultural techniques, cutting patterns) tailored to specific channel configurations. Narrowly defined techniques and regulations are normally effective across only a few channel types. Simple prescriptive management, such as riparian zones of fixed width, is less effective than management techniques adapted to local topography and to natural disturbance regimes.

Riparian classification is fundamental for designing new approaches for resource management. Several authors have made recommendations for general improvement of classification systems (e.g., Goodwin 1999, Montgomery and MacDonald 2002). The recommendations include basing classifications on processes, controlling factors, temporal change, and thresholds; treating classifications as hypotheses; diagnosing channel condition; and incorporating stream size as a criterion. These recommendations relate to the placement of logging access roads; the decisions on when, where, and how much tree harvest should occur; and the development of silvicultural restoration techniques and system models, all of which require adherence to channel type. The most effective stream and riparian models include aquatic and terrestrial disturbance regimes, unique species mixtures, spatial and temporal heterogeneity, and microclimate gradients—all of which vary by channel type. Further, emerging silvicultural techniques for riparian tree species account for genetic vitality, stand development, and system complexity—factors that are specific to channel types (Berg 1995).

Conclusions

Even though the search for an ideal classification system continues, the fundamental principles of an ideal system are reasonably well articulated. Perhaps the best approach is for resource managers to adapt guiding principles using adaptive management for specific situations (Holling 1992, Barrington et al. 2001). The task is difficult because it requires a holistic, long-term perspective. Once in place, riparian classification provides a solid foundation for making resource decisions that affect the environmental quality of stream corridors for decades. When complete and properly applied, a riparian classification offers the potential to improve management decisions. However, if incomplete and/or improperly applied, the use of a riparian classification may become problematic, resulting in inappropriate action or no action when activities are needed.

Structural Patterns

4

Overview

- This chapter summarizes life history strategies exhibited by riparian vegetation, provides descriptions of community structure, illustrates how community assemblages typically vary across the catchment and through time, and analyzes biological diversity in terms of theory and actual patterns.

- Life histories reflects strategies for acquisition of high-quality food and shelter (habitat), growth dynamics, reproduction, and defense against being consumed. Riparian plants have specific morphological, physiological, and reproductive adaptations suiting them for life in high-energy and wet environments. Adaptations enable some plants to grow on large woody debris, other species to establish upon mineral soils in the floodplain or grow in saturated or flooded soils, and seeds or plant fragments well suited for survival. In addition, many riparian plants are specifically adapted to cope with flooding, sediment deposition, physical abrasion, and stem breakage.

- Collectively, plant life history strategies include a suite of coadapted characteristics, which enhance reproductive success under specific environmental conditions. Some primary strategies include the balancing of sexual with asexual reproduction, the mode of seed dispersal, the optimizing of seed size, the timing of dormancy, and seed longevity.

- Patterns in the distribution, structure, and abundance of riparian plant communities are influenced by general biophysical templates, and these patterns and templates are repeated throughout the world. The precise details (e.g., history, species, phenology, productivity) differ regionally, but the underlying patterns and templates, as well as their determinates, remain relatively predictable.

- Responses of riparian vegetation to soil conditions (organic matter, moisture, nutrient availability) and to mycorrhizal associations underpin general patterns of community structure and succession, as well as the expression of ecological characteristics, such as abundance and biomass.

- Even though riparian communities appear to be disproportionately species rich compared to their surroundings, it is still not known how many species are present in most catchments. Even though no studies fully quantify what percent of all life utilizes riparian corridors, available evidence suggests that for some taxa, the biodiversity is quite high. Biodiversity in riparian corridors is best documented for vascular plants and vertebrates, with the general patterns dependent on specific environmental settings.

Purpose

Riparian zones exhibit wide ranges of physical variability, resulting in community characteristics that are vividly expressed by a broad array of life history strategies and by successional and demographic patterns. Consequently, riparian zones are among the biosphere's most complex ecological systems (Naiman and Décamps 1990, 1997). The variability of biotic characteristics found in natural riparian zones reflects the inherent physical heterogeneity of drainage networks, processes shaping stream channels, and ecological characteristics of community assemblages. In effect, riparian communities are products of intense interactions among biotic and abiotic factors—whether they are occurring in the present or have occurred in the past. Further, as we will see, the biota themselves have strong, long-term influences on the geomorphic structures and processes by which they are shaped.

Interactions between biotic patterns and physical processes are quantitatively complex but generally easy to conceptualize (recall Figure 1.8). There are strong hierarchical interactions among hydrogeomorphic processes (e.g., catchment-scale processes), habitat dynamics, and riparian communities. Lithotopographic units, areas with similar topography and geology and where similar suites of geomorphic processes occur, have profound influences on the creation of habitat (Montgomery 1999, Naiman et al. 2000). The ultimate emergence of riparian community patterns is a reflection of the physical templates and their individual dynamics, as well as subsequent modifications of the physical environment by the biotic community. This concept is explored further in Chapters 5 and 6.

Riparia are often thought of as living plant communities, and much of the literature has focused on them to the exclusion of detritus, soils, fungi, and animals. Indeed, there is considerable information about riparian plants, especially the aboveground components. Unfortunately, little is known about the belowground components—the roots and their associated organisms. Likewise, there is only scattered information on the characteristics and spatial structure of detritus (large woody debris is an exception), soils, fungi, and many animals (some birds and large mammals are exceptions), especially invertebrates. As a consequence, even though this chapter is admittedly biased toward biotic groups and components where information is readily available, there remain substantial research opportunities and challenges on topics and biotic groups where only sparse information is available.

This chapter explores the complex—but often predictable—biotic patterns of riparian communities. Initially we summarize the variety of basic life history patterns exhibited by riparian plants. Then we move to descriptions of community structure, including animals that inhabit riparia above and below ground, illustrating how structure typically varies across the catchment and through time. Finally, since riparian communities are thought to be unusually diverse, we analyze this perception in terms of theory and actual patterns.

Life History Strategies

Life history strategies refer to the various ways that organisms use resources to carry out essential functions: acquisition of high-quality food and shelter (habitat), growth patterns, reproduction, and defense against being consumed (Grime 1979, Bazzaz 1996). Each requires a complex set of resources—mainly water, carbon, nitrogen, and phosphorus—that comprise the plant structures (e.g., leaves, stems, fruits, roots) used to carry out the different ecological functions. Variation in resource allocation occurs through inherent differences in the chemicals demanded by specific structures, the relative mass of different structures or organs, and the relative numbers of different structures produced. This variation occurs in individuals through evolutionary time, within and among populations, and especially among species.

At an evolutionary level, resource allocation involves balancing fecundity against probability of survival, as well as the effects of this balance on fitness (Grime 1979). At an ecological level, allocation includes the relationship between investment in one function and investments in others, such as the relationship between defense and growth. At a physiological level, allocation entails partitioning resources within the plant, as well as the consequences of this partitioning for resource gain or loss.

Much has been written about the life history strategies of organisms in stable (or predictable) and chaotic (or unpredictable) environments. In general, organisms in relatively stable or predictable environments are thought to live longer, become larger, reproduce slower, and invest more into individual offspring than organisms in chaotic or unpredictable environments (Grime 1979). In riparian communities, one finds nearly every conceivable strategy due to the great physical heterogeneity as well as to the high degree of *predictability* and *unpredictability* encountered, depending on spatial position in the riparian zone. Additionally, even though some parts of riparian zones are subject to rapidly changing environmental conditions, the changes are reasonably predictable within the life span of the organism and within the life spans of subsequent generations, and the life history strategy of individual species is responsive to those conditions. For example, within a seasonal or longer cycle, hydraulic characteristics may vary widely at the river edge. Nonetheless, the occurrence of specific physical conditions is reasonably predictable and the biotic community with its inherent life history traits reflects those conditions.

Examples of variability in riparian life history traits abound. Many short-lived organisms, such as springtails (collembola), are abundant in soils and on vegetation immediately after the spring flood, whereas the vegetation they use for habitat and food may live for many decades or centuries. Mobile birds and larger mammals may use the riparian zone when conditions are favorable but may move when conditions are better elsewhere. Some plants use the seasonally high river flows to distribute seeds and propagules, some store seeds in soils waiting for better conditions, and others have developed adaptations to use the fragments produced by physical damage to reproduce asexually. Indeed, there are a variety of strategies that can take advantage of the inherent physical heterogeneity, providing refugia for a good percentage of the regional pool of organisms (Salo et al. 1986).

Morphological and Physiological Adaptations of Riparian Plants

Riparian plants have numerous morphological, physiological, and reproductive adaptations suiting them for life in high-energy and wet environments (Mitsch and Gosselink 1993, Blom and Voesenek 1996, Naiman and Décamps 1997). Adaptations by riparian plants enable tree species to grow on large woody debris, pioneer species to root in bare mineral soils on the floodplain surface, obligate hydrophytes to grow in saturated or flooded soils, and seeds or plant fragments to survive burial. Additionally, many riparian plants are specifically adapted to cope with flooding, sediment deposition, physical abrasion, and stem breakage.

The active channel and floodplain can be harsh environments for plant colonization and establishment. Even though vegetation may initially colonize a variety of sites, it only establishes successfully on "safe sites." Safe sites have suitable conditions for germination (water and oxygen), are refuges from herbivores and floods, and have environmental conditions compatible with life history requirements (Harper 1977).

The classification of plants into guilds of similar life history strategies is useful for understanding riparian forest succession and species distribution. Functional adaptations of plants fall into four broad categories (Naiman et al. 1998b):

- **Invader:** Produces large numbers of wind- and water-disseminated propagules that colonize alluvial substrates.
- **Endurer:** Resprouts after breakage or burial of either the stem or roots from floods or after being partially eaten (e.g., herbivory).
- **Resister:** Withstands flooding for weeks during the growing season; also withstands moderate fires or disease epidemics.
- **Avoider:** Lacks adaptations to specific disturbance types; individuals germinating in an unfavorable habitat do not survive.

Categorizing the major riparian tree species in temperate North America and Europe illustrates the variety, as well as some of the similarities, of life history strategies among riparian plants. For example, willow (*Salix* spp.) is ubiquitous as a pioneer plant adapted to several types of disturbance. Willow seeds germinate and establish in postfire landscapes, individuals resprout following low-intensity fires that have not destroyed the root system, and adventitious roots appear when the stems are fragmented by floods, debris flows, or herbivory. These adaptive characters make them well suited as invaders, endurers, or resisters depending on local environmental conditions. In contrast, Sitka spruce (*Picea sitchensis*), a riparian tree of the Pacific coastal rain forest, is more restricted, colonizing elevated large woody debris and mineral substrates on the floodplain (Harmon and Franklin 1989). Once established, Sitka spruce is resistant to both flooding and sediment deposition. However, it is fire sensitive and, in response to that type of disturbance, it is classified as an avoider.

Note that those trees most often thought of as riparian species (black cottonwood, Sitka spruce, alder, willow, coast redwood) are also species categorized as resisters or endurers in response to specific riparian disturbances (Naiman et al. 1998b).

Many morphological adaptations of plants (e.g., adventitious roots, stem buttressing, root and stem flexibility, seed characteristics) are responses to anoxia in the soil, unstable substrate conditions, or reproductive requirements (see Sidebar 4.1).

Sidebar 4.1 Adaptations to Flooding in Riparian Plants

There are seven species of the riparian plant *Rumex* (sorrel) along the banks of the Rhine River (The Netherlands), and all show marked zonation patterns in response to river level fluctuations (see Figure SB4.1). Each species has morphological and physiological features allowing it to thrive in specific riparian areas (Blom et al. 1990). Flow in the Rhine River is seasonally variable, resulting in various degrees of submergence and water-logging of riparian soils, which influence the duration and intensity of soil anaerobiosis. The zonation of *Rumex* provides a remarkable example of how plants adapt to riparian environments. *R. acetosa* and *R. thyrsiflorus* are sensitive to floods and found only on the high point of the river levee. *R. obtusifolius*, *R. crispus*, and *R. conglomeratus* find optimal conditions in relatively wetter areas on the lower slope of the levee. Finally, *R. maritimus* and *R. palustris* thrive in permanently wet soils.

As the first effects of soil wetting appear, there are profound changes in the morphology of the roots system (Blom et al.1990). Roots adjust to the major decrease in oxygen exchange between the atmosphere and the soil as any remaining oxygen is soon exhausted by root and microbial respiration. In *Rumex*, there are at least three morphological root changes in response to anaerobiosis: (1) an increase in root branching, (2) the development of new adventitious roots, and (3) an altered vertical distribution of the laterals, with more roots concentrated in the upper soil layer near better oxygenated water. Concomitant with alterations in root morphology are changes in root anatomy. Improved porosity due to expansion of the intercellular spaces is an obvious adaptive response. The formation of large channels in the root cortex enhances diffusion of atmospheric or photosynthetic oxygen from the shoot to the roots so that aerobic metabolism can be maintained.

Species' distributions of *Rumex* can be explained by a differential ability to elongate petioles and stems, a process regulated by the hormone ethylene, in order to protrude above the water surface. Renewed contact between leaves and air after submergence stimulates the formation of a new aerenchymatous root system in wet-tolerant species but not in wet-intolerant species. Increased porosity in wet-tolerant species enable plants to longitudinally transport aerial and photosynthetic oxygen to the rhizosphere. Reproductively, *Rumex* employs two strategies. Some species delay flowering and seed production during periods of flooding, surviving as vegetative plants while others accelerate flowering during short dry periods to produce seeds between successive floods.

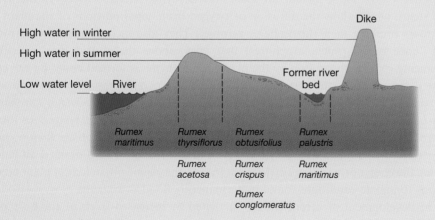

Figure SB4.1   The zonation of *Rumex* species with respect to water levels in the Rhine River riparian zone (modified from Blom et al. 1990).

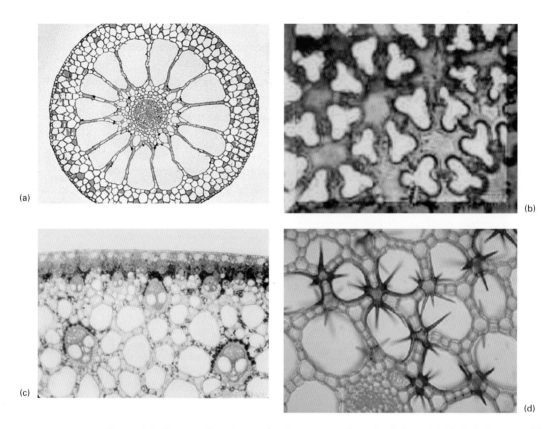

Figure 4.1 Cross-section micrographs of aerenchyma in wetland plants: (a) *Myriophyllum* stem (Photo: A. Roberts). (b) *Cyperus javanicus* (Photo: D. Webb). (c) Papyrus (*Cyperus papyrus*) leaves (Photo: D. Webb). (d) Aerenchyma with lignified inner hairs of pond lily (*Nymphaea* sp.); a section through the leaf stalk (Photo: D. Webb).

Metabolic depletion of oxygen in the rhizosphere (i.e., rooting zone) occurs rapidly when soils are flooded, and vascular plants have evolved several mechanisms and structural adaptations to anoxic environments (Mitsch and Gosselink 1993, Blom and Voesenek 1996). These include air spaces (aerenchyma—aerated tissue) in the roots and stems, which allow the diffusion of oxygen from aerial portions of the plant to the roots (making their rhizospheres aerobic). They also include adventitious roots and "knee roots" (pneumatophores) that grow above the anoxic zone—possibly enhancing oxygen absorption by the plant. Aerenchyma is formed either by cell collapse (lysigeny) or by the enlargement of intercellular spaces resulting from cell separation without collapse (schizogeny). Aquatic plants such as *Nymphoides*, *Luronium*, and *Littorella* species (and also corn, *Zea mays*) possess the typical lysigenous form, whereas schizogeny can be found in wetland plants such as *Caltha*, *Rumex*, and *Filipendula* (see Figure 4.1). Note, however, that flooding of well-drained alluvium during the cool, dormant season does not lead to the classic physiological plant adaptations to anoxic conditions found in permanently flooded wetlands.

The development of aerenchyma is mediated by increased levels of the hormone ethylene, which is synthesized when soil conditions turn anoxic (Kozlowski et al. 1991, Mitsch and Gosselink 1993). Aerenchyma are common among ethylene-producing species of the families commonly found in floodplains with poor soil drainage, such as Cyperaceae (sedges) and Juncaceae (rushes). Related to ethylene synthesis is a process known as *rhizosphere oxygenation*. It occurs in wetland plants where oxygen moves from roots to the adjacent soil, creating a minute oxidized zone. This process is believed to effectively mediate the toxic effects of soluble reduced ions found in anoxic soil, such as manganese (Mitsch and Gosselink 1993). The second structural adaptation to anaerobic conditions (to sediment deposition associated with flooding)—adventitious roots and pneumatophores—occurs in a number of floodplain trees (e.g., black cottonwood, willow, red alder, cypress, and coast redwood).

In the central Amazon basin, where floodplains are submerged for 50 to 270 days annually and where water depths can reach 10 m, the hundreds of thriving riparian plant species show strong morphological, phenological, and physiological adjustments to wet conditions, which would otherwise result in oxygen starvation (Parolin 2001). In plants examined to date, all produced adventitious roots, lenticels, and stem hypertrophy (i.e., height growth) and showed adaptive changes in leaf and root respiration, leaf chlorophyll, water potential, and photosynthetic assimilation under both waterlogged and drought conditions.

Flooding also stresses plants through erosion of the soil surface and abrasion by suspended sediment and debris. A number of trees and other plants, however, actually thrive in the active channels of rivers experiencing extreme variations in flow. A key adaptation is stem flexibility among the woody species (e.g., willow, alder, and cottonwood). Stem flexibility accommodates the shear stress associated with floods and debris flows. Additionally, major floods and debris flows in mountainous regions of North America and Europe occur during periods when deciduous trees are without leaves, thereby reducing the drag on the plant stem. In drier regions where minimal flow occurs throughout much of any year, and for periods of up to several years, other tree species are able to establish in the active channel. However, species that survive for longer periods must possess characteristics facilitating survival during occasional inundation by fast-flowing, debris-laden water. Trees such as paperbark (*Melaleuca*) in Australia are successful because they are able to recline or become prostate in floods, reducing the associated shear stress. Further, they have a multistemmed form, a modified crown with weeping foliage that folds during flooding, a thick and spongy bark to repel abrasive materials, roots that anchor into firm substrate beneath the channel floor to resist removal, roots that regenerate quickly when needed, and populations aligned in linear groves parallel to the direction of flow to provide protection (Fielding and Alexander 2001). Likewise, *Breonadia* growing on rock outcrops in the mid-channel of South African rivers possesses similar adaptations to resist the occasionally fierce flow event (van Collier et al. 1997), as do many other species living in high-energy environments (see Figure 4.2).

Reproductive adaptations take on many forms and are discussed more fully below. However, two morphological adaptations improve the reproductive abilities of plants in environments where heavy litter and silt accumulations are common during floods. By the second growing season, plant species with large seeds (e.g., *Rubus*) are the most successful in surviving litter accumulations, whereas an ability to spread laterally (e.g., *Carex*, *Festuca*) is the most successful strategy for silt accumulations. However, *Ranunculus* and *Viola* have the ability to use both strategies (Xiong et al. 2001). Those species with long-lived seeds (e.g., *Agrostis*, *Carex*, *Juncus*, *Ranunculus*) are of special importance for the recovery of riparian vegetation as they form seed banks, thereby

(a)

Figure 4.2 (a) *Breonadia salicina* is adapted to grow on bedrock in South African rivers where it is able to withstand substantial flooding or, if severely damaged, able to (b) resprout from the remaining stems and branches (Photos: R.J. Naiman).

(b)

Figure 4.2 *Continued*

contributing substantially to plant community development regardless of the type of accumulation. Nevertheless, in other situations, sprouting from buried plants may be more important than seeds in the revegetation process (van der Valk et al. 1983)

Reproductive Strategies

Stimulated by theoretical concepts of life history evolution, empirical studies of reproductive strategies and resource allocations have been major research foci in ecology. Theoretical concepts suggest that life history and competition should have major effects on reproductive allocation of resources. In newly opened habitat, the fitness of colonizing organisms is likely to depend on fecundity, whereas in crowded habitat, high fecundity may compromise the competitive ability of organisms to persist. For parallel reasons, organisms that reproduce once in a lifetime should have a higher reproductive allocation than organisms that reproduce multiple times, because it pays to maximize present reproduction if future reproduction has been sacrificed. Some studies support these predictions, especially if the comparisons are limited to congeneric species (Bazzaz 1996).

Plant life history strategies include a suite of coadapted characteristics, which enhance reproductive success under specific environmental conditions (Barbour et al.

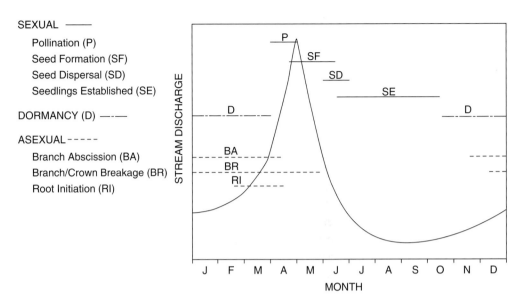

SEXUAL ———
  Pollination (P)
  Seed Formation (SF)
  Seed Dispersal (SD)
  Seedlings Established (SE)

DORMANCY (D) —·—

ASEXUAL - - - - -
  Branch Abscission (BA)
  Branch/Crown Breakage (BR)
  Root Initiation (RI)

Figure 4.3 Generalized timing and duration of reproductive events for riparian cottonwoods in relation to the annual pattern of stream discharge. Modified from Braatne et al. 1996.

1987). Some primary strategies include the balancing of sexual with asexual reproduction, the mode of seed dispersal, the optimizing of seed size, the timing of dormancy, and seed longevity. The illustrations below address seed dispersal and trade-offs between sexual and asexual reproduction.

An essential adaptation of riparian plants is a strategy where seed dispersal coincides with the seasonal retreat of floodwaters when moist seedbeds are available for successful germination and colonization. Seed dispersal by wind (anemochory) is one common strategy, and North American cottonwoods (*Populus* spp.) provide excellent examples of this reproductive phenology (Braatne et al. 1996). Cottonwood seeds borne by fluffy, cotton-like hairs are dispersed long distances by wind as well as water (hydrochory). Seed dispersal typically coincides with declining river flows following springtime snowmelt and storm flows, thereby increasing the probability of seeds landing in favorable microsites along the riverbanks (see Figure 4.3). Seed viability is short, generally lasting only 1 to 2 weeks under natural conditions. Once a seed becomes wet, viability is lost in 2 to 3 days if the microsite is not favorable. Germination is rapid on appropriate microsites. The root radicle emerges from the seed, enters the soil, and cotyledons begin to expand within 24 hours. Young roots are noted for their development of hairs that quickly attach to sand and silt particles to provide anchorage and absorption. Survivorship, however, is highly regulated by substrate texture and soil moisture, with the pattern of river discharge playing a key role in the establishment of young trees (see Figure 4.4).

The general pattern described here for cottonwood is repeated for other riparian plant species throughout the world. In Australia and elsewhere, the reproductive phenology of riparian tree species is timed to coincide with the unique seasonal hydrology and rainfall of specific rivers (Pettit and Froend 2001). Key hydrograph components include the timing and magnitude of flood peaks, the rate of decline of the recession limb, and the magnitude of base flows.

Figure 4.4 Successful seedlings and clonal saplings form arcuate bands that tract specific elevations along meandering rivers and especially at point bars at the end of meander lobes. The curving band of even-aged saplings and trees originates from discrete floods, providing a hydrologic history of the river. In the photo, sapling bands of narrowleaf cottonwoods (*Populus angustifolia*) and plains cottonwoods (*P. deltoides*) are thriving on a meander lobe of the Oldman River near Lethbridge, Alberta. (Photo: S. B. Rood; from Braatne et al. 1996.)

In addition to wind dispersal, seeds and vegetative fragments of many species rely on transport by flowing water (hydrochory). Bald cypress (*Taxodium distichum*) and water tupelo (*Nyssa aquatica*) in South Carolina swamp forests rely on water more than wind to disperse seeds away from parental trees (Schneider and Sharitz 1988). In flowering ash (*Fraxinus ornus*), typically a wind-dispersed species, dispersal downstream in southern France is much more extensive than in other directions, suggesting that water transports seeds during autumn flooding (Thébaud and Debussche 1991). In Sweden, there is a positive relationship between diaspore floating capacity (length of time a seed will float) and the frequency of species in the riparian vegetation (Johansson et al. 1996). Nevertheless, anemochory and hydrochory are not the only means for dispersal. Dispersal by animals (zoochory), especially by birds, may be even more important, but few empirical data exist for comparison (Naiman et al. 2003).

In addition to sexual reproduction by seeds, many riparian plant species reproduce by clonal growth (e.g., vegetative or asexual reproduction). Notable examples include redwood (*Sequoia*), willow (*Salix*), cottonwood or poplar (*Populus*), and ash. Multiple sprouts can result from burial during floods (Sigafoos 1964) and abrasion during floods can stimulate stump sprouts (Bégin and Payette 1991). Riparian cottonwood, for example, exhibit the trait of branch abscission, or cladoptosis, in which healthy branch tips are shed and develop adventitious roots on moist soils with the potential to grow into new but genetically identical trees (see Figure 4.5a; Stettler et al. 1996).

(a)

(b)

Figure 4.5 (a) Cottonwood (*Populus* spp.) occasionally shed branch tips in the spring, which upon contact with a moist surface develop adventitious roots and grow into genetically identical trees—a process known as *cladoptosis*. (Photo: R. F. Stettler.) (b) Broken or cut branches also form adventitious roots upon contact with moist surfaces, forming suckers at several points (Photo: S. McKay.) From Naiman et al. 1998.

Even though cladoptosis is commonly observed, the success of the branch tips in establishing new trees appears to be rare. Clonal establishment from broken branches appears to be more common, occurring in 91 percent of the saplings examined in British Columbia (see Figure 4.5b; Rood et al. 2003b).

A unique feature of clonal growth is that rooting at nodes of individual shoots may create physiologically distinct plants with independent fates. This developmental feature has important physiological and ecological implications because it affects how the plant functions as a single unit and interacts with its environment. Additionally, whether a species possesses the capacity to grow clonally, as well as the specific mechanism it employs (e.g., bulbs, rhizomes, stolons, fragmentation), presumably reflects phylogenetic inertia as well as adaptations to particular types of environments. Clonal growth is obviously advantageous where horizontal spread is favored over vertical growth. Beyond that, it is favored in environments where seed and seedling mortality are high (as for *Salix* and *Populus*), as well as in other stressful environments (Bazzaz 1996).

Distribution, Structure, and Abundance

Conceptually, the distribution, structure, and abundance of riparian communities are a reflection of "spatial and temporal heterogeneity"—a phrase frequently used in ecological sciences. Even though "heterogeneity" seems to imply a random or chaotic state, nothing could be further from the truth. Indeed, there are clear patterns in riparian communities guided by general biophysical templates, and these patterns and templates are repeated throughout the world. Of course, the precise details (e.g., history, species, phenology, productivity) differ regionally, but the underlying patterns and templates, as well as their determinates, remain relatively predictable.

The heterogeneity observed in riparian characteristics reflects the complex interactions occurring among hydrology, lithology, topography, climate, natural disturbances, and the life history characteristics of riparian organisms (Naiman et al. 2005). Yet biotic responses to the physical environment are often predictable, provided enough information is available about the available pool of organisms and the spatial and temporal scales over which environmental drivers operate. Annually predictable events (e.g., snowmelt) can often lead to directional selection, whereas widely variable timing to events (e.g., peak flows) leads to disruptional selection, resulting in high levels of genetic heterozygosity within populations with regard to timing of reproduction, flood tolerances, and so forth.

In this section we address patterns in the distribution, structure, and abundance of riparian plant communities. We discuss the responses of riparian organisms to soil conditions, describe general patterns of community structure and succession, and show how community characteristics are reflected in abundance and biomass.

Identification of Riparian Zones Based on Soils and Vegetation Type

As discussed in Chapter 3, the spatial extent of the riparian zone may be difficult to delineate precisely because of the biophysical heterogeneity associated with river

91

corridors. The actual width of riparian zones is related to the size of the stream, the stream's position in the drainage network, the hydrologic regime, and the local physical configuration. Delineating precise limits for the riparian zone therefore becomes dependent on choosing environmental characteristics that have strong influences on the plant community or other attributes that can be easily identified. Generally, this is accomplished by measuring the spatial extent of herbaceous plants adapted to wetted soils, the production of nutritional resources for aquatic systems, the local geomorphology, and identification of areas showing frequent sediment erosion or deposition (Naiman and Décamps 1997). In practice, the identification of "true" riparian plants, as we saw in Chapter 3, is a complex and somewhat contentious subject (Hauer and Smith 1998, Innis et al. 2000).

Biophysical Characteristics of Riparian Soils

Soil development within alluvial environments is highly variable. Frequent erosional and depositional disturbances from flooding create a complex mosaic of soil conditions in the active floodplain—and this fundamentally influences vegetation colonization and establishment (Oliver and Larson 1996). Well-drained soil or recently deposited mineral alluvium may be found adjacent to very poorly drained organic soils in abandoned channels or hillslope seeps. In an elegant paleo-study of the fluvial architecture of the upper Buntsandstein in Spain, García-Gil (1993) was able to elucidate the complex structure of riparian deposits and their attendant soils. She found that fluvial sedimentation took place in coexisting braided and meandering channels, interspersed with ephemeral discharge episodes. In Tasmania, Nanson et al. (1995) describe an equally complex vertical–lateral sequence of sedimentation on the Stanley River (see Figure 4.6). They found that, as the river stabilized, finer materials were transported, accreting as sand and silt on the floodplain surface. In addition to variable sediment deposition, movement pathways for subsurface water often created a patchwork of nutrient and oxygen concentration gradients below ground. In total,

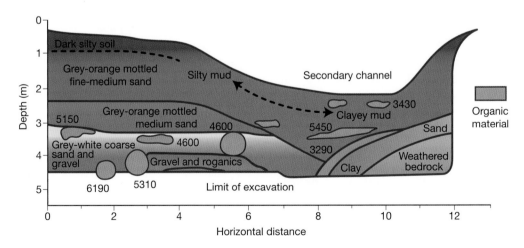

Figure 4.6 An example of the variability in sediment characteristics found within the riparian soils of the Stanley River, Tasmania. Radiocarbon dates are shown for logs and litter; dates of the logs are adjusted to give the age of the outmost growth rings. Modified from Nanson et al. 1995.

this heterogeneity in soil conditions is a major factor in determining plant productivity and diversity (Naiman et al. 2005).

Organic Matter

Soil texture and mineralogy affect most aspects of soil biogeochemistry and are reflected in strong correlations with soil organic matter (OM), nitrogen (N) mineralization, and moisture. River sediments are sorted according to the stream energies they are subjected to, and newly formed river bars evolve rapidly. In rivers with large sediment supplies in approximate equilibrium with transport capacity, rapid downcutting of the river channel and concomitant elevation of floodplain surfaces limit the amount of time that terrestrial areas are subject to frequent flooding (Church 2002). In forested catchments, trapping of large amounts of sediment by organic debris jams may lead to especially dynamic local patterns of aggradation and degradation (Montgomery et al. 1996, Edwards et al. 1999). Following any period of rapid change (i.e., normally days to months), the particle size distribution can be thought of as a fixed template relative to its influence on subsequent OM cycling (Bechtold et al. 2003).

Soil texture and mineralogy affect moisture and nutrient cycling through their influence on the size of pore spaces between particles and on the amount of surface area for adsorption of organic and inorganic substances (see Figure 4.7). The size of pore spaces between soil particles limits the movement of liquids and gases—most

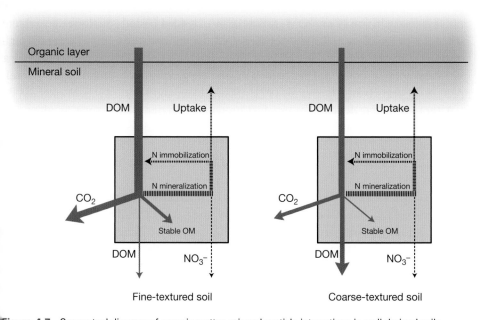

Figure 4.7 Conceptual diagram of organic matter–mineral particle interactions in well-drained soils. Adsorption of infiltrating dissolved organic matter (DOM) to fine particles increases short- and long-term retention of organic matter. Short-term retention of labile OM results in increased mineralization in fine-particle fractions. Long-term retention increases the pool of stable OM. Though it does not directly contribute to mineral N pools, fine particles and associated OM reduce leaching losses by enhancing ion exchange and water retention. Courtesy of J. Bechtold, University of Washington.

importantly, water and oxygen—through soil. In poorly drained soils, anoxic conditions can result in large accumulations of OM and high denitrification rates (Davidson 1995, Ludwig et al. 2001). In well-drained soils, substantial OM is preserved by adsorption to mineral particles and incorporation within stable aggregates (Christensen 1992), and often results in strong correlations between OM and soil texture (J. Bechtold, University of Washington, unpublished data).

Rates of carbon (C) and N mineralization increase with decreasing particle size (Christensen 1992), apparently because of the adsorption of percolating dissolved OM that would otherwise leach from the soil (Nelson et al. 1993). The net effect is an increase in OM in the fine-particle fractions. Although some OM may be retained for long periods by physical or chemical inhibition of decomposition in the anaerobic environment, mineralization of much greater amounts of loosely adsorbed OM may be especially rapid because of the concentration of microbes in fine-particle fractions. Similarly, temporary retention of mineralized N maximizes opportunities for microbial immobilization as well as vegetative uptake. The overall effect is improved efficiency of cycling through microbial or soil-vegetation pathways. High rates of N mineralization in fine-textured soils are strongly correlated with enhanced primary production (Pastor et al. 1984, Reich et al. 1997). Large OM pools in "older" riparian soils and, conceivably, correlations between texture and OM result from increased return of OM in litterfall.

Just as mineral soils undergo long-term changes in their capacity to retain OM as minerals, weather, aeolian, colluvial and fluvial processes redistribute mineral particles, influencing soil OM dynamics over shorter time scales. This is particularly important in early successional riparian systems in which soils weather rapidly, are physically unstable, and initially have few fine particles. Characteristic of these systems is a rapid initial increase in soil N and OM, followed by extended periods of slower accumulation to a point of equilibrium between OM inputs and respiration, leaching and erosion (J. Bechtold, University of Washington, unpublished data). These changes are particularly rapid in floodplains, which reach near-maximum N and organic C levels in 20 to 50 years (van Cleve et al. 1971, 1993, Luken and Fonda 1983, Walker 1989).

Although fluvial deposition and fast-growing, N-fixing plants such as alder can contribute large inputs of C and N to young floodplain soils, retention of fine sediments is also important in maintaining soil organic matter and nutrients. Soil particle size distribution is strongly reflected in soil C concentrations (Bechtold and Naiman 2005). In contrast, site age may only weakly correlate with soil C, and that relation could be explained entirely in terms of fine-particle increases over time. This strongly suggests that stabilization of OM by fine particles, and not variation in OM inputs, is the primary mechanism controlling the size of mineral soil OM pools in riparian zones. The largest increases in C occur during the first decade of vegetative development, about the same time as increases in fine-particle concentrations occur, and may in fact be due to fluvial deposition.

Moisture

The ability of soils and sediments to hold water and the existence of tributary and groundwater flows are related to the accumulation of OM in soils, and are equally important as OM in determining vegetative distribution (Brinson 1990, Hupp and Osterkamp 1985). Additionally, distance from the river, microtopographic variation above ground, and variations in belowground hydraulic connectivity determine lag times between rising discharge in the main channel and arrival of water on site. Once water has arrived, the composition of the soil and the underlying alluvium (as well as the rate of evapotranspiration) determine how long the substrata remain saturated—and great heterogeneity may occur over an annual hydrograph (Hughes 1997). For example, when a flood peak travels rapidly over a floodplain with fine-grained alluvium, recharge of floodplain soil moisture may not occur in critical rooting zones, suggesting that floods of longer duration or in rapid succession are necessary for this to occur. Nevertheless, external water sources flowing subsurface toward the river channel can still allow vegetation to persist largely independent of the annual flow regime (Dawson and Ehleringer 1991, Thorburn and Walker 1994). Indeed, older cottonwood, river red gum (*Eucalyptus camaldulensis*), and other mature riparian trees apparently can use subsurface water rather than nearby river water, presumably because subsurface water in the alluvial aquifer or flowing from upslope is a more reliable water source than the river channel. Nevertheless, it is known that all riparian cottonwoods will use shallow alluvial groundwater linked to stream water, particularly in arid regions, if available. This is based on extensive studies of natural plant occurrence, population declines following river damming or dewatering, physiological water relations, isotopic composition of xylem water, and reestablishment of cottonwoods after flow augmentation (Rood et al. 2003a). As discussed in Chapter 2, the alluvial aquifer may be the primary source of main channel flow during drought, especially if the aquifer is recharged by river flow at upstream locations.

Any discussion of soil characteristics and their effects on riparian communities is confounded by interactions between the natural environmental gradients away from and along the river and the patchy mosaic of substrate types whose availability change on different temporal scales (van Coller et al. 2000). Substrate type—with its inherent nutrient and moisture characteristics—and the availability of subsurface water exert profound influences on the initial colonization as well as the long-term survival of plants. Along the semiarid South African Sabie River, the species-specific proportion of juvenile trees establishing and surviving the first few years of life depends on the type of substrate (see Table 4.1). Mature individuals, however, often show responses to changes in subsurface flow. Many researchers have demonstrated the key role of subsurface water to plant zonation, community structure, growth dynamics and survivorship (e.g., Stromberg et al. 1996, Harner and Stanford 2003). In general, the relation between individual plants and moisture depends on several interacting

Table 4.1 Proportion of Juveniles (Stem Diameter <3 cm) of Selected Riparian Tree Species Occurring on Contrasting Substratum Types

Substrate Type	Breonadia salicina	Syzigium guineense	Combretum erythrophyllum	Acacia robusta	Diospyros mespiliformis
Bedrock	97	40			2
Alluvial stones				17	49
Alluvial sand		12	96	25	8
Fine alluvial soil	3	48	4	40	23
Nonalluvial soil				18	18

Data adapted from van Coller and Rogers 1995.

factors that influence both water uptake and water demand. These include root architecture, ability of soil to hold moisture, growth of new roots to access water, ability of the plant to adjust physiologically and morphologically to changing moisture availability, and plant age and size.

In response to subsurface flow patterns, various chemical and biological transformations may occur in a predictable sequence within narrow redox ranges (see also Chapter 5), and these can affect soil and plant characteristics (Hill 2000). Where groundwater flows in a shallow subsurface path toward the stream, a sequence of lateral zones dominated by anaerobic respiration, denitrification, and sulfate reduction may occur between the upland perimeter and the stream. Reverse sequences of redox reactions can be associated with anaerobic groundwater discharging upward into oxidized riparian sediments resulting in increased respiration, nitrification, and sulfate formation as the water emerges at the surface (Hedin et al. 1998). Vertical patterns of increased reduction and oxidation are also linked to water-table fluctuations in riparian zones. Increased oxidation of the soils because of seasonal water-table declines may produce a vertical sequence of methane (CH_4) loss, accumulation of sulfate (SO_4^{2-}), and increased nitrate concentrations due to nitrification (Dahm et al. 1998).

Fauna

Riparian soils, however, are composed of more than just minerals, OM, and water. They are also home to active biotic communities that are integral to the long-term vitality of riparia. Here we refer mainly to the soil fauna rather than the microbial assemblages. Microbial communities, although numerous and diverse (and discussed in Chapters 5 and 6), are largely dormant in dense soils that sometimes form in poorly drained riparia, and are largely confined to the immediate microsite where they reside. Soil invertebrates, along with viable roots and their associated mycorrhizae, appear to be necessary to significantly stimulate microbial activity by providing new microsite surfaces or substrates and by oxygenating the soil. Faunal communities, like the physical soil processes, are highly dynamic over time and space (Lavelle 1997). Soil animals have evolved in an environment that imposes three major requirements on them: (1) the ability to move in a compact environment with a loosely connected porosity, (2) the ability to feed on low quality resources, and (3) the ability to adapt to the occasional drying or flooding of the porous space.

Soil invertebrates have a continuum of strategies from the smallest microfauna that colonize the water-filled pore space to macrofauna that alter the soil environment to suit their own needs. Three artificial size groupings have been distinguished: the microfauna, mesofauna, and macrofauna. The microfauna are <0.2 mm on average and mainly include the protozoans, nematodes, tardigrades, and rotifers living in the water-filled pore spaces. The mesofauna average 0.2–2 mm and include the collembolans, acarids, and the smaller oligochaetes that live in the air-filled pore space of soil and litter. The macrofauna are >2 mm, encompassing the termites, earthworms, and larger arthropods with the ability to dig through soil and create structures for their movement and habitat (e.g., burrows, galleries, nests, and chambers).

In general, soil invertebrates have limited ability to digest the complex organic substrates of riparian soil and litter, but many have developed symbiotic interactions with microflora that permit them to use soil resources. With increasing size, the relationhip between microflora and fauna gradually shifts from predation to mutualism of increasing efficiency. Additionally, the excrement of soil invertebrates is of utmost importance in the decomposition of riparian organic matter, in the formation and maintenance of soil structure and, over longer periods of time, in specific pedological processes called "zoological ripening of soils" (Bal 1982). Unfortunately, data are largely lacking for most riparian soil animals, especially beyond the larger taxonomic and functional groups.

General Distributions of Aboveground and Belowground Communities

Ecologists increasingly find that complex combinations of competitive and facilitative interactions influence distributions and abundances of riparian plants (Levine 2000). Distributional patterns in riparian aboveground and belowground communities vary with spatial position in the catchment as well as with latitude—a reflection of regional environmental regimes. A rich and comprehensive literature exists describing catchment processes and how they are distributed in space and time (summarized in Chapters 2 and 3). The lateral distribution of riparian biota has been linked to hydroperiods, landforms, and sediment types as well to competition and life history factors. While several studies now demonstrate these relationships for particular rivers in a lateral perspective, few studies address longitudinal variations in riparian biota (Cordes et al. 1997).

Most literature describing vegetation along rivers makes use of phytosociological entities. In phytosociology, a hierarchical division of plant communities is used, leading from class to order, to alliance, to association, and eventually to lower entities. Vegetation units are described by characteristic species, differential species, and accompanying species. The advantage of using phytosociological terminology is that long lists with hundreds of plant names can be avoided, and abiotic characteristics are implicitly known. Disadvantages are the many differences of opinion between phytosociological schools of thought and between scholars within schools (Higler 1993). In this section we address general community patterns laterally from the river channel to hillslopes and from headwaters to the sea using a phytosociological approach.

Lateral Zonation

Lateral distributions of vascular plants are related to often well-defined valley and fluvial landforms resulting from distinct hydrogeomorphic processes (Hupp and Osterkamp 1985, 1996; Rot et al. 2000). Among such landforms are depositional shelves, active channel bars, natural levees, interfluvial channels, and swales and terraces (see Chapter 2). These and other features represent increasingly higher levels or surfaces above the wetted channel, with correspondingly decreasing flow durations and flooding frequency (see Sidebar 4.2). Many well-known studies have related riparian vegetation to local topography, suggesting that elevation above the wetted channel is a major factor determining species patterns. However, as discussed later in this chapter, this has proven to be not entirely true.

Generally, vascular plants are distributed laterally according to the life forms described earlier (Fonda 1974, Higler 1993, Vadas and Sanger 1997, and others). Closer to the wetted channel one encounters earlier serial stages of vascular plants

Sidebar 4.2 Lateral Distribution of Riparian Vegetation

A physiographic approach to plant distributions often provides strong predictive understandings of riparian plant patterns based on a dynamic landscape perspective. For example, in Passage Creek, Virginia, the distribution patterns of woody vegetation within the bottomland forest are persistently allied to fluvial landforms, channel geometry, stream-flow characteristics, and sediment-size characteristics (Hupp and Osterkamp 1985). There are distinct species distributional patterns on four common landforms: depositional bar, active channel shelf, flood-plain, and terrace (see Figure SB4.2). Woody plants patterns develop more as a result of hydrologic processes associated with each fluvial landform rather than from sediment size characteristics, although certain species may be exceptions. Likewise, riparian plant communities (trees and herbaceous layer) along steep mountain streams can be differentiated by four landform classes (floodplain, low and high terraces, hillslope) based on elevation above the active channel (Rot et al. 2000). Ordination of 602 vegetation plots in western Washington shows strong separation between landform classes (54 to 67 percent) with incorrectly classified plots normally being most closely allied to adjacent landforms (see Figure SB4.3).

Figure SB4.2 Block diagram showing alluvial landforms. From the lowest, the features are CB = channel bed, DB = depositional bar, AB = channel-shelf bank, FB = floodplain bank, FP = floodplain, T_l = lower terrace, T_u = upper terrace, and HL = hillslope. From Hupp and Osterkamp (1985).

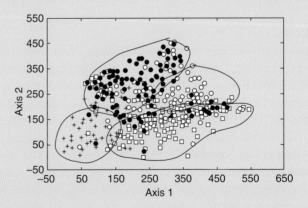

Figure SB4.3 Detrended correspondence analysis ordination of 602 vegetation plots by landform. Each plot was ordinated by species and percent cover. +, floodplain; □, low terrace; ○, high terrace; ●, slope (modified from Rot et al. 2000).

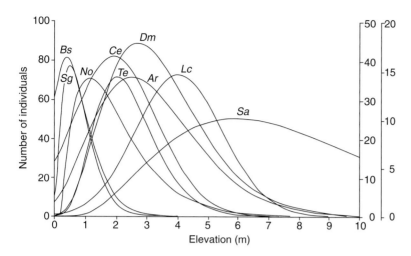

Figure 4.8 Distribution and abundance of mature individuals (stem diameter >3 cm) of selected tree species as a function of the elevational gradient of the Sabie River riparian forest. The species are *Breonadia salicina (Bs)*, *Syzygium guineense (Sg)*, *Nuxia oppositifolia (No)*, *Combretum erythrophyllum (Ce)*, *Trichilia emetica (Te)*, *Acacia robusta (Ar)*, *Diospyros mespiliformis (Dm)*, *Lonchocarpus capassa (Lc)*, and *Spirostachys africana (Sa)*. The three scales correspond to different species: 0–100 individuals for *Bs* and *Sg*; 0–50 individuals for *No*, *Ce*, and *Sa*; and 0–20 individuals for *Te*, *Ar*, *Dm*, and *Lc*. From van Coller 1993.

adapted to low nutrient environments and high light levels, whereas at slightly higher elevations away from the channel one encounters a larger proportion of woody, long-lived, shade-tolerant, and often water-intolerant vascular plants. In locations with well-defined physical changes in elevation, such as terraces, the plant assemblages will be distinct. However, in many situations, the elevation changes are more gradual or the local flora has broader environmental tolerances. In those cases, there are broad overlaps in lateral distributions. For example, in the diverse riparian forests of Southern Africa and Amazonia, there are often very broad lateral overlaps in species distributions because of the historical legacies of sites, the patchy distribution of microsites in space and time, and the broad physiological tolerances of the plants (see Figure 4.8; Salo 1990, van Coller 1993). Further, there are complex but predictable interactions between the overstory and the understory to consider. In a coastal Alaska example, well-drained sites are better for trees than for understory vegetation; the opposite is true for poorly drained sites, regardless of the lateral position from the wetted channel (Hanley and Brady 1997).

The lateral distribution of organisms other than woody vascular plants is less well documented, although studies of specific groups (e.g., nonvascular flora, herbaceous plants, invertebrates, amphibians, birds, mammals) generally are available locally. Additionally, it is not clear whether the distribution of these groups is directly related to the geomorphic surface (as for woody vascular plants), to another co-related environmental variable, or to a combination of variables. For example, bryophytes (mosses), liverworts, and lichens often compose 45 to 90 percent of the ground cover in the temperate rainforest (Hanley and Hoel 1996), and there are strong gradients in composition with increasing distance from small streams in the state of Oregon (Jonsson 1997). However, the lateral changes correspond to several environmental gradients—distance from stream; occurrence of boulders, deciduous stems, and large

Table 4.2 Physical Environmental Variables (of Nine Analyzed) that
Correlated Strongly with Ordination Patterns

Riparian Components Analyzed	Physical Factors
Tree species	Moisture,* nitrogen,** silt, elevation**
Shrub species	Elevation*
Pitfall invertebrates	Moisture,* nitrogen,* elevation**
Litter invertebrates	Clay,** elevation*
Litter invertebrates/dry weight	Nitrogen, silt, elevation*
Bird species (year averaged)	Elevation**
Major taxa	Moisture, nitrogen,* silt, elevation

From Catterall et al. 2001.
*P < 0.05; **P < 0.01.

woody debris; elevation in relation to the wetted channel; and stream size, among
others. Thus, bryophyte vegetation varies among different types of streams, and the
general distribution of this taxon cannot be represented by a single or a few stream
types.

Lateral patterns among contrasting taxa (e.g., plants, birds, insects) offer insights
into the different environmental factors influencing distributions. For example, in the
eucalypt forests of Australia, there are strong associations between local topography,
drainage, and the distributional patterns of riparian biota (Catterall et al. 2001).
Microtopographic position is generally associated with biotic composition; however,
nitrogen is occasionally important, as is moisture (see Table 4.2). Other environmental variables thought to be important (i.e., soil pH, percent of gravel, clay and sand,
and soil carbon) show little association with any of the biotic assemblages.

Longitudinal Zonation

Longitudinal variations in riparian vegetation can be explained in terms of both
Holocene and historical changes in climate, sea level, and the length of time surfaces
have been available for colonization (Nilsson et al. 1991). Like lateral zonation, variations in elevation, channel gradient, valley constraint, geomorphic processes, fire
regimes, sediment and litter loads, and substrate diversity also largely explain longitudinal patterns (Rot et al. 2000, Catterall et al. 2001). Spatially, many of these factors
covary along the river's length and are reflected in their influence on species distributions. Vegetative composition also reflects the different lag times involved in the
responses of abiotic and biotic elements to temporal change and is, therefore, in a
state of constant adjustment. Despite the strong abiotic influences on riparian vegetation, biotic influences such as competition, herbivory, and disease are significant in
shaping community patterns. There is evidence that some riparian plant species could
exist beyond their present range if competition, herbivory, or disease were reduced or
eliminated (Keddy 1989, Naiman and Rogers 1997).

Many regional ecologists, when asked, could easily describe how the biota changes
from headwaters to the sea, as well as articulate the mechanisms responsible for those
changes. Surprisingly, there are few comprehensive published accounts that empirically quantify this conspicuous landscape pattern for individual catchments. Nevertheless, it is possible to provide overviews for specific regions by assembling studies
describing vegetation associated with various stream sizes.

The mountainous coastal region of Pacific North America provides just such a vegetative example (Naiman et al. 1992a). The riparian forests range from compact alpine vegetation along steep, constrained, headwater channels to floodplain forests along unconfined, low-gradient, alluvial channels (see Figures 4.9a–c). Characteristic patterns of riparian forest responses to climate and disturbance include long, narrow patches of low-stature slide alder (*Alnus sinuata*) parallel to steep gradient avalanche channels; fan-shaped deciduous forest patches of red alder, willow, and cottonwood overlying alluvial and debris-flow deposits emptying from confined steep valley tributaries to higher-order valley floors; and multiple-aged valley forest patches of coniferous islands (Sitka spruce, western hemlock, western red cedar) set within a matrix of deciduous forest dominated by black cottonwood (*Populus trichocarpa*) and willow (*Salix* spp.). The resulting mosaic of riparian patches, often dominated by deciduous species, generally is set within a valley bottom dominated laterally by coniferous species. Similar patterns, but composed of different species, have been carefully described for rivers in southern France and in Italy (Pinay et al. 1990, Cattaneo et al. 1995).

Riparian fauna also show strong latitudinal patterns in community composition. Much of the understanding of wildlife associations with riparian habitat is based on research conducted in arid and semiarid regions where riparian zones act as "ribbons" of organization for the surrounding landscape because of the concentration of water, nourishment, and habitat as compared to the uplands (Malanson

(a)

Figure 4.9 Riparia exhibit strong gradients in species assemblages and structure from headwaters to the sea. These three photos depict a typical longitudinal profile one might encounter in the Pacific Coastal ecoregion of North America. (Photos: R. J. Naiman and J. J. Latterell.)

(b)

Figure 4.9 *Continued*

1993, Kelsey and West 1998). Contrasts between riparian and adjacent upland microclimates and vegetative communities are much less dramatic in mesic regions. Consequently, riparian wildlife communities are not strikingly different than those of adjacent upland areas. All riparian zones do, however, provide three important functions for wildlife. First, riparian areas increase wildlife species diversity by providing habitat for obligate riparian species, for species that seek edge habitats, and for species associated with early successional plant communities. Second, the riparian zones provide refugia for upland species when that habitat experiences a major disturbance (e.g., fire, deep winter snow). Finally, as we will see later, riparian zones can function as topographic landmarks to visually cue species during migration or dispersal.

The riparian wildlife community is composed of obligate and generalist species. Use of riparian habitat varies, depending on the life history characteristics of individual species and the availability of suitable habitat elsewhere—in some cases rendering the distinction between obligate and generalist discretionary. Riparian obligates are species considered so highly dependent on riparian and aquatic resources that they would disappear with the loss of this habitat from a drainage

(c)

Figure 4.9 *Continued*

basin. Amphibians often dominate the list of riparian obligates, in numbers and pro-portions of species present. Species utilizing both riparian and upland forests are considered riparian generalists. In the temperate rain forests of North America, there is a rich riparian wildlife community with nearly 400 species of amphibians, reptiles, birds, and mammals (see Table 4.3). Overall, 29 percent of those species are riparian obligates with the taxon-specific percentages ranging from 12 percent for mammals to 60 percent for amphibians.

Composition of riparian wildlife communities varies over space and time. As seen for the flora, riparian conditions and processes shape wildlife distributions at both local and landscape scales (Kelsey and West 1998). In mountainous regions, small streams (generally first, second, and third order) in upper basins tend to have high gradients, steep side slopes, nearly continuous forest canopy, narrow strips of riparian vegetation, and food webs based on allochthonous inputs to the stream (Naiman et al. 1992a). High densities of stream-breeding amphibians, a virtual absence of fish, and few obligate bird species distinguish riparian wildlife communities there (see Figure 4.10). Lower-basin rivers (generally sixth order and greater) have low gradients, extensive floodplains, open canopies, wide riparian zones, and productive autotrophic communities. Riparian zones along the large rivers provide habitat for ungulates, reptiles, raptors, herons, kingfishers, waterfowl, river otter (*Lutra canadensis*), and many other species. Mid-order streams function as transition areas between small streams and large rivers, where the overlap of species distributions creates different community patterns.

Table 4.3 Numbers, by Taxonomic Class, of Native Riparian Obligate and Upland Species Compared to the Total Number of Native Species in the Pacific Coastal Ecoregion

	Riparian Obligates	Upland Specialists	All Species	Percent Riparian Obligate
Amphibians	18	7	30	60
Reptiles	3	12	19	16
Birds	78	93	231	34
Mammals	13	31	107	12
Total	*112*	*143*	*387*	*29*

From Kelsey and West 1998, with permission.

Successional and Seasonal Community Patterns

Vegetative Succession

The earliest studies of vegetation dynamics in riparian zones did not refer to the concept of succession but nevertheless illustrated the process (Fitzpatrick and Fitzpatrick 1902). Many successional patterns in riparia are primary succession, but an equal number of successional patterns begin with plant fragments, propagules, or biomass remaining from previous communities. Avalanche, flood, wind, fire, drought, disease, litter accumulation, herbivory, and other physical influences on the vegetation leave unique biotic legacies that are displayed in various successional patterns (Nilsson et al. 1993, Naiman et al. 1998b). As we have seen, many riparian plants possess adaptations allowing them to recover and reproduce by stem flexibility, root suckering, and formation of adventitious from plant fragments. The amount of biotic material remaining to initiate secondary succession depends on the nature of disturbance. Disturbances, especially scouring floods, also prepare sites for primary succession by new species favored by the new conditions. In essence, the dynamics of riparian vegetative colonization and succession is a complex and often elusive process involving species-specific life history characteristics in combination with local geomorphology and climate.

Vegetative succession can be described as occurring in four stages (Oliver and Larson 1996):

1. **Establishment:** Plants colonize sites following a disturbance or formation of a new depositional landform. The establishment stage is the developmental period before the growing space is fully occupied.
2. **Stem Exclusion Stage:** After all the available growing space is occupied, species with a competitive advantage in size or growth expand into the space utilized by other plants, eliminating them from the community. New plants, for the most part, are excluded from colonizing by intra- and interspecific competition. A predictable vertical sorting of individuals (e.g., vertical stratification by height) in even-aged stands occurs during this stage with some individuals growing faster than others.
3. **Understory Initiation Stage:** Mortality in the overstory initiates understory development. Understory initiation is characterized by the invasion of shade tolerant herbs, shrubs, and trees. These can be the same species as those present during the stand initiation stage, but they grow slowly, creating a stand with multiple canopy layers.

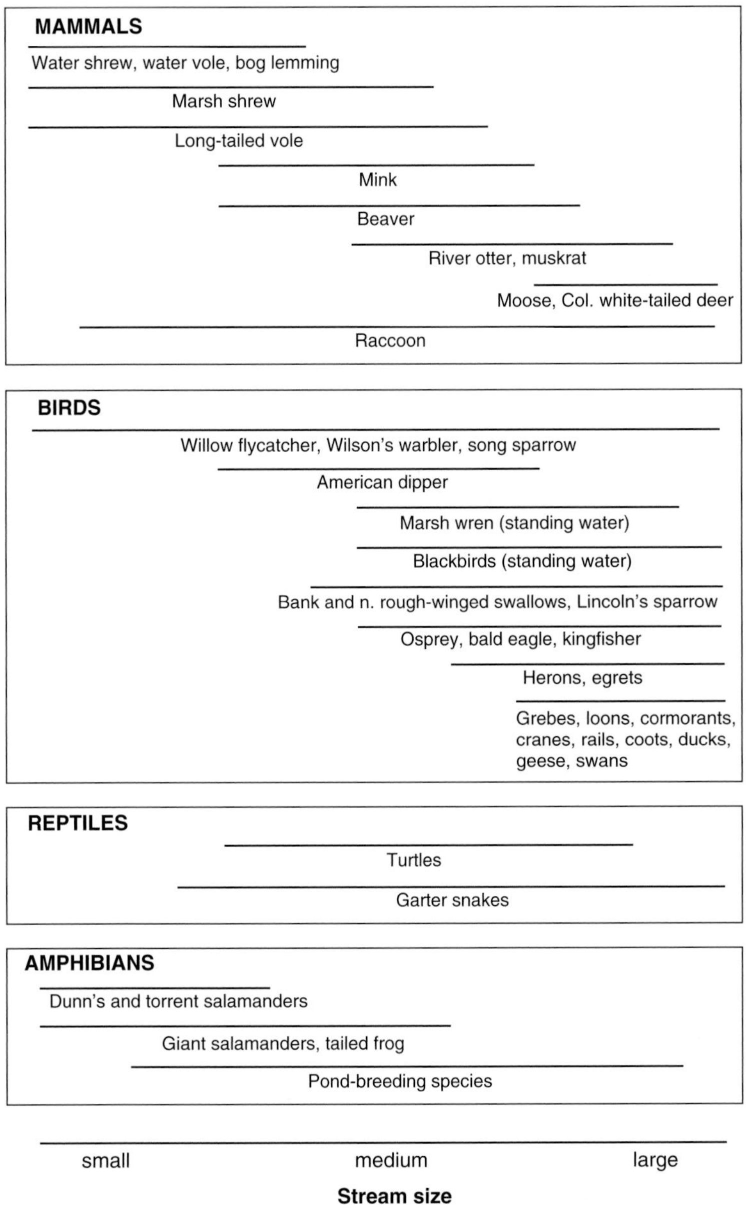

MAMMALS

Water shrew, water vole, bog lemming

Marsh shrew

Long-tailed vole

Mink

Beaver

River otter, muskrat

Moose, Col. white-tailed deer

Raccoon

BIRDS

Willow flycatcher, Wilson's warbler, song sparrow

American dipper

Marsh wren (standing water)

Blackbirds (standing water)

Bank and n. rough-winged swallows, Lincoln's sparrow

Osprey, bald eagle, kingfisher

Herons, egrets

Grebes, loons, cormorants, cranes, rails, coots, ducks, geese, swans

REPTILES

Turtles

Garter snakes

AMPHIBIANS

Dunn's and torrent salamanders

Giant salamanders, tailed frog

Pond-breeding species

| small | medium | large |

Stream size

Figure 4.10 Distribution of selected native riparian obligates as a function of stream size in the Pacific Coastal ecoregion of North America. From Kelsey and West 1998.

4. **Mature Stage:** Individual trees die as a forest stand ages, opening up canopy space, which can be occupied by advanced regeneration in the understory. When the trees that invaded immediately following the initial disturbance begin to die, the stand enters into an old-growth condition. This is an autogenic process whereby trees regenerate and grow without the influence of external disturbances—changes originate from the biota. Note, however, that allogenic processes (originating from the environment) are also important in shaping the forest. Riparian forests are described as mature if they have three-dimensional structural characteristics such

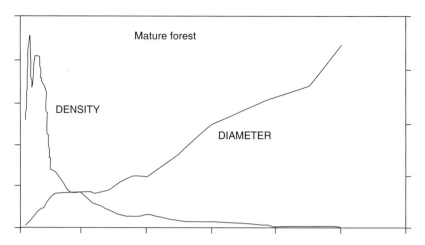

Figure 4.11 Crown density and crown diameter as a function of tree height for a mature Amazonian forest. Data are presented for all strata, including the herbaceous ground layer. From Terborgh and Petren 1991.

as large, living old trees; large, dead standing trees; massive fallen logs; relatively open canopies with foliage in many layers; and a diverse understory.

An illustration of succession can easily be seen in the floodplain forests of the Amazon, where the sizes of canopy tree crowns steadily increase through time (see Figure 4.11; Terborgh and Petren 1991). Normally, trees comprising the sunlit upper canopy have the broadest crowns. An average 50-m-tall tree can have a mean crown diameter of 3.5 m, casting shade over 0.1 ha; an area large enough to accommodate the crowns of a dozen 20-m-tall trees. Although few in number, the influence of emergent trees on the vegetation in their shadows is likely to be very great. Structurally, mature riparian forests have the greatest horizontal and vertical variation of all the successional stages, with both large and small trees growing in separate and intermixed patches.

The traditional Clementsian model of vegetative succession is probably not appropriate for riparian areas (Baker and Walford 1995). Clements' original idea of one community facilitating the development of a second community over time, with an eventual convergence toward a single climax vegetation, does not adequately describe the temporal dynamics and spatial mosaic of riparian communities. It may be more informative to think of riparian successional processes as having multiple quasi-stable states. In riparian settings, succession is not completely predictable because both autogenic and allogenic processes affect the successional trajectory. Floods and other physical events introduce sudden allogenic changes in substrate particle size or soil development, or reset patch age. However, between physical events, autogenic changes in surface sediment size are more gradual as soil develops and available nutrients increase. Collectively, these influences result in multiple states that persist for relative long periods (decades to centuries), hence they are quasi-stable conditions (Schumm 1977). More broadly, the actual riparian vegetation is a reflection of multiple influences from the catchment scale to the scale of local topography, influencing water and sediment regimes. The composition of contemporary vegetative communities reflects the different lag times involved in the response of abiotic and biotic elements to temporal change and is therefore in a constant state of adjustment.

Examples of multiple quasi-stable states are abundant for riparian vegetation. On the Garonne River in southern France, cyclical successional processes occur within the active floodplain where flood-induced erosion and deposition are common (see Figure 4.12). However, on terraces that no longer flood, the successional dynamics are not reversible, and internal autogenic forces dominate. One successional model, however, is never adequate for multiple locations on the same river because of strong longitudinal changes in the physical environment. On the Red Deer River in Alberta, for example, longitudinal variations in vegetation communities require at least two successional models: one typical of the balsam poplar-spruce communities of the upper river (see Figure 4.13a) and a second model more typical of the cottonwood sites of the lower river (see Figure 4.13b). The two models are distinguished by differences in species (there are also many common species) but also by different elevations with respect to the low water stage at which similar communities in the succession are found (Cordes et al. 1997).

Perhaps the most comprehensive general model of riparian succession on floodplains is that formulated by Hughes (1997). She organized floodplain components and processes as a broad spatio-temporal hierarchy spanning seven magnitudes of time and ten magnitudes of space (see Figure 4.14). Her synthesis of a global body of knowledge on floodplain biogeomorphology into a general model points to the commonality of processes and patterns in floodplain forests.

As these and other models demonstrate, vegetative succession in riparian forests is influenced by many interconnected biophysical variables acting from local to landscape scales (Dixon et al. 2002). Indeed, this is a highly complex subject because the multiscale biophysical influences on riparian vegetation are not necessarily linked in a causal hierarchy, making separation of the relative influences of fine- and broadscale processes difficult. Nevertheless, there are three local factors influencing riparian succession of particular interest: presence of large woody debris, sediment size distributions, and accumulations of litter.

Large woody debris is a common and important element in river corridors and, therefore, it should not be surprising that it affects the colonization and survivorship of riparian plants. Large woody debris has many important ecological functions since it often moves only during unusually high flows (Maser and Sedell 1994). These functions include acting as a substrate for colonization, protecting young plants from flood scour, providing nutrients and moisture at critical times, regrowing vegetation from living large woody debris, and diversifying habitat in the corridor. Large woody debris is abundant along alluvial rivers throughout forested regions, often reaching masses of >100 t/ha and >100 large woody debris piles/km of channel bank (Bilby and Bisson 1998, Gurnell et al. 2000, Pettit et al. 2005). If the large woody debris accumulations are large enough to continuously resist high flows, sediments accumulate in low shear-stress zones immediately downstream of piles, and vegetative islands begin to form and expand over time (Fetherston et al. 1995, Abbe and Montgomery 1996, Edwards et al. 1999). In fact, this may be a key mechanism initiating the heterogeneous patterning of riparian forests. Gurnell et al. (2001) provide a conceptual model of the important role of large woody debris in the formation of riparian islands for the Fiume Tagliamento in Italy (see Figure 4.15), which is nearly identical to processes

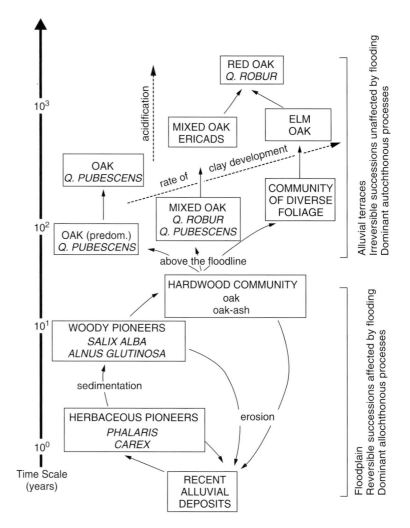

Figure 4.12 Multisuccessional pathway development of riparian vegetation along the Garonne River, France, in response to changing hydrologic conditions occurring over several centuries. From Décamps et al. 1988.

observed in the alluvial rivers along the Pacific Coast of North America and elsewhere.

The important role of sediment characteristics in establishing riparian vegetation is well known, especially for willow and cottonwood (Braatne et al. 1996, Scott et al. 1996). Sediment characteristics fundamentally affect the moisture and nutrient regimes that are so important for young plants, although too much silt accumulation or removal can adversely impact some plants after establishment (Langlade and Décamps 1995). Cottonwood and willow require bare, moist surfaces protected from disturbance for successful establishment. These habitats occur at predictable places in channels where floods can deposit fine materials, holding moisture and nutrients better than coarse sediments.

The interaction between sediment characteristics and vegetative communities, however, becomes quickly complicated for a number of reasons. Riverine corridors are notorious for their heterogeneity in the three-dimensional size sorting and depo-

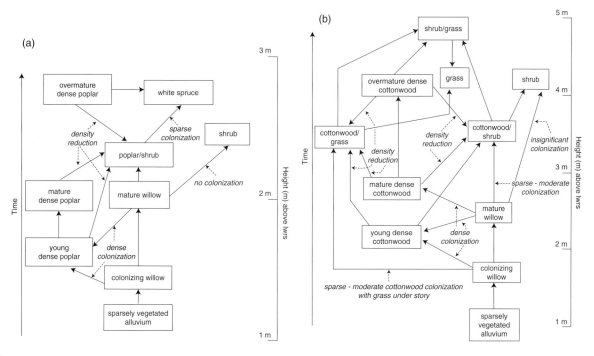

Figure 4.13 Successional models for sites in the (a) upper and (b) lower Red Deer River, Alberta. LWRS = low water reference site. From Cordes et al. 1997.

sition of sediment. A thin layer of fine sediment may be underlain by coarse cobbles, and vice versa, differentially affecting moisture and nutrient regimes for seedlings and saplings. Likewise, seasonal variations in the timing and pattern of flooding have profound effects on seedling survival and subsequent plant community structure. For example, on the Wisconsin River, there can be a 100-fold range in seedling density among years and a different dominant species each year depending on interactions between summer flow conditions, species-specific dispersal and germination phenology, and sediment characteristics (Dixon et al. 2002). The longer-term consequences of initial sediment characteristics on vegetative succession are not well documented, remaining as a fertile frontier in riparian ecology.

Less well known than the effects of large woody debris or sediment on primary succession are the effects of litter accumulations—yet litter has important impacts. The total production of plant litter, the proportion of leaf litter, and the decomposition rate of litter are greater in riparian corridors than in upland ecosystems (Xiong and Nilsson 1997). The accumulation and decomposition of plant litter have long been considered as complex and important factors in controlling both vegetative structure and ecosystem function.

Plant litter initially has an inhibitory effect on vegetation, although the magnitude varies with vegetative composition, latitude, litter type and quality, ecosystem, and target species (Xiong and Nilsson 1999). While some investigations show litter favoring vegetation development (e.g., protecting seeds from predators, buffering lethal frosts, conserving water during dry conditions, adding nutrients, and carrying plant diasporas during redistribution by water), a worldwide analysis suggests that the negative effects outweigh the positive ones during the first year. After the first year, the

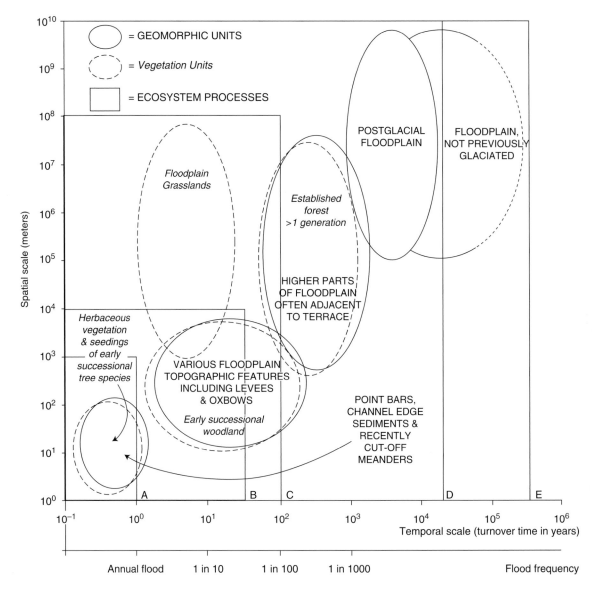

Figure 4.14 The organization of floodplain components and processes as a spatio-temporal hierarchy. (A) Primary succession of herbaceous vegetation and early successional woody species associated with the annual flood. (B) Primary and secondary floodplain succession associated with floods of medium magnitude and frequency. (C) Long-term floodplain succession associated with the widespread erosion and reworking of sediments by floods of high magnitude but low frequency. (D) Species migration upstream and downstream, local species extinction, long-term succession on terraces, emergence of life history strategies associated with climate and base-level change, and the influence of post-glacial relaxation phenomena on hydrologic and sediment inputs to the floodplain. (E) Species evolution and changes in biogeographic range associated with tectonic activity, eustatic uplift, and climate change. From Hughes 1997.

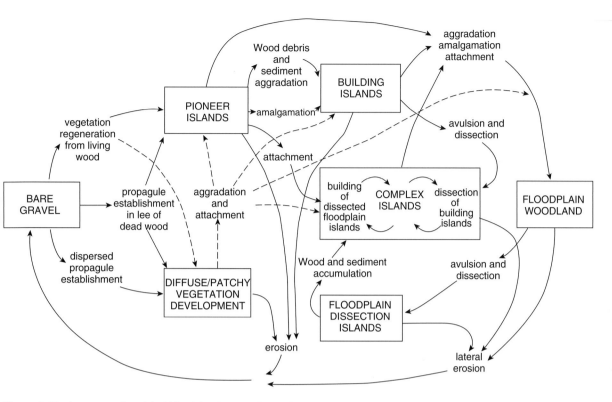

Figure 4.15 A conceptual model of island development for the Fiume Tagliamento in Italy. The solid arrows represent the most commonly occurring pathways. From Gurnell et al. 2001.

litter effects become mostly positive as the successful vegetation becomes established. It appears that litter-derived nutrients released during decomposition stimulate plant growth. There is also a trend for species richness to increase after the first year but at a slower rate than the overall vegetative biomass.

How does litter affect plant reproduction? There are two morphological strategies that improve the reproductive abilities of plants in environments where litter and silt accumulations are common. By the second growing season, plants with large seeds (e.g., *Ranunculus, Rubus, Viola*) are the most successful in surviving litter accumulations, whereas an ability to spread laterally (e.g., *Carex, Ranunculus, Festuca, Viola*) is the most successful strategy for silt accumulations (Xiong et al. 2001). Those species with long-lived seeds (e.g., *Agrostis, Carex, Juncus, Ranunculus*) are of special importance for the initiation of succession as they contribute substantially to plant community development regardless of the type of accumulation. Nevertheless, in other situations, sprouting from buried plants may be more important than seeds in succession (van der Valk et al. 1983).

As a cautionary note, most general models of riparian plant succession, and factors shaping successional processes, have focused on larger streams and floodplains. In contrast, relatively little is known about processes controlling successional processes or patterns along small streams. As with other biotic components that gradually change from the uplands to the sea, factors directing primary and secondary succession in riparian vegetation change too, especially if there are strong physical contrasts between stream sections. The important successional factors in uplands will differ in type and degree of influence with those of the lower reaches of the river corridor.

Faunal Succession

Faunal associations, as one might expect, show strong affinities with the state of vegetative succession. As riparian forests develop from an open canopy of grasses, forbs, and young woody plants to a multiple layer canopy of mature trees with organic soils, there are concomitant changes in the invertebrate, amphibian, reptile, bird, and mammal communities (Lavelle 1997, Kelsey and West 1998). For example, in the bird communities of coastal rivers of northwestern America, one finds understory seedeaters and insectivores requiring the dense shrubs for nesting and cover associated with early successional vegetation. As riparian vegetation matures and becomes structurally more complex, bird community composition changes to include canopy seedeaters, canopy and bark insectivores, species of aerial foragers adapted to canopies, and a variety of larger predators (see Figure 4.16).

The faunal assemblages are even reflected in the composition of soil invertebrate communities (Lavelle 1997), predatory spider communities (Moring and Stewart 1994), and insect communities (often ants, beetles, and butterflies; Williams 1993, Framenau et al. 2002, Sadler et al. 2004) associated with different physical templates and successional stages. The wildlife-habitat and entomological literatures are replete with other examples of faunal species and communities reflecting the characteristics of the riparian habitat and its associated resources.

Density, Basal Area, and Biomass

Often the density and basal area of riparian forests is as great or greater than that of upland forests because of the relatively favorable growing conditions. Although values vary widely within regions, largely due to successional stage and soil conditions, the variation is usually less than an order of magnitude. Riparian forests in warm and humid regions tend toward greater stem density and basal area than those in more arid regions and cooler latitudes. The aboveground biomass of mature riparian forests ranges between 100 and 300 Mg/ha, with few exceptions (Brinson 1990, Balian and Naiman 2005). Leaves represent 1 to 10 percent of the total. In the few empirical studies, belowground biomass tends to be less than the aboveground biomass, ranging from 5 to 120 percent of it. However, considerable caution should be exercised in broadly applying these generalities. This is because of the sparse database, the emphasis on mature forests rather than on earlier successional stages, the concentration of studies in the northern hemisphere, and the very few studies that include belowground biomass in measurements.

There is, however, emerging interest in better quantifying density, basal area, and biomass of riparian forests at contrasting successional stages. In the coastal rain forest of North America, where early successional stages are dominated by willow and alder and the mature forests are dominated by Stika spruce (age: ~>250 yr), stem density decreases from ~26,000 to 500 stems/ha as succession proceeds, basal area increases from ~16 to 69 m^2/ha, and biomass increases from ~18 to 542 Mg/ha (see Figure 4.17). Nearly 90 percent of the change for each parameter takes place in the first 20 to 40 years, with the values changing relatively little thereafter. In each case >87 percent of

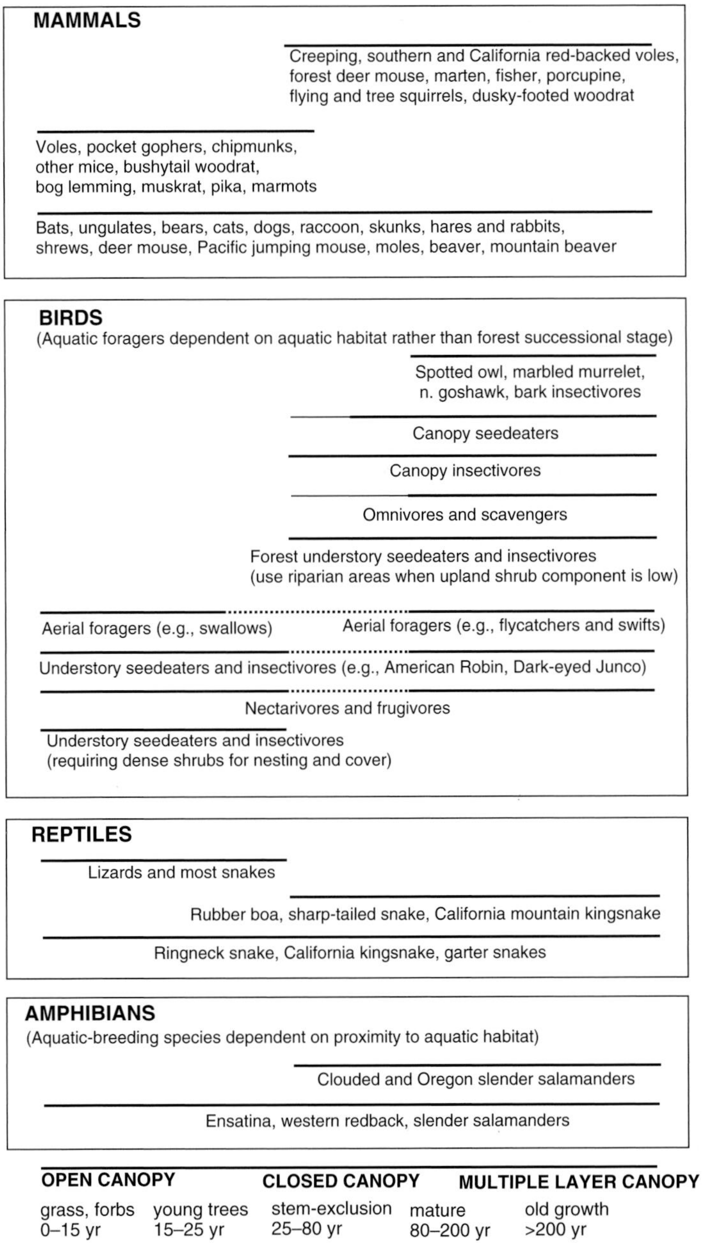

MAMMALS

Creeping, southern and California red-backed voles,
forest deer mouse, marten, fisher, porcupine,
flying and tree squirrels, dusky-footed woodrat

Voles, pocket gophers, chipmunks,
other mice, bushytail woodrat,
bog lemming, muskrat, pika, marmots

Bats, ungulates, bears, cats, dogs, raccoon, skunks, hares and rabbits,
shrews, deer mouse, Pacific jumping mouse, moles, beaver, mountain beaver

BIRDS
(Aquatic foragers dependent on aquatic habitat rather than forest successional stage)

Spotted owl, marbled murrelet,
n. goshawk, bark insectivores

Canopy seedeaters

Canopy insectivores

Omnivores and scavengers

Forest understory seedeaters and insectivores
(use riparian areas when upland shrub component is low)

Aerial foragers (e.g., swallows)   Aerial foragers (e.g., flycatchers and swifts)

Understory seedeaters and insectivores (e.g., American Robin, Dark-eyed Junco)

Nectarivores and frugivores

Understory seedeaters and insectivores
(requiring dense shrubs for nesting and cover)

REPTILES

Lizards and most snakes

Rubber boa, sharp-tailed snake, California mountain kingsnake

Ringneck snake, California kingsnake, garter snakes

AMPHIBIANS
(Aquatic-breeding species dependent on proximity to aquatic habitat)

Clouded and Oregon slender salamanders

Ensatina, western redback, slender salamanders

OPEN CANOPY		CLOSED CANOPY		MULTIPLE LAYER CANOPY
grass, forbs	young trees	stem-exclusion	mature	old growth
0–15 yr	15–25 yr	25–80 yr	80–200 yr	>200 yr

Figure 4.16 Distribution of selected mammals, birds, reptiles, and amphibians according to forest successional stage (thick line = primary habitat, thin line = secondary habitat, dashed line = gap in distribution) in the coastal rivers of northwestern America. From Kelsey and West 1998.

the variation can be explained by stand age, providing a strong predictive model for estimating stem density, basal area, and biomass over the lifetime of these forests. A caution, however, is that for the younger forest patches (<40 years old), there can be considerable spatial variability in density, basal area, and biomass for plots of approximately the same age.

Along the more arid river corridors of Montana, a similar pattern is evident as young willow and cottonwood develop into a more mature forest (age: ~60 yr).

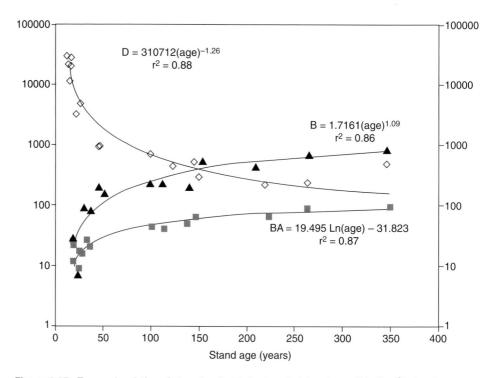

Figure 4.17 Temporal variation of stem density (D; ◇; stems/ha), basal area (BA; ■; m²/ha) and stem biomass (B; ▲; Mg/ha) with increasing plot age, all tree species within a plot are included. Trend lines estimate curve shape for each variable. Logarithmic scale is used to display the three variables in the same graph. Modified from Balian and Naiman 2005.

Density decreases from >10,000 to <1,300 stems/ha, aboveground biomass increases from 1 to 193 Mg/ha, and belowground biomass increases from 6 to 94 Mg/ha (Boggs and Weaver 1994, Harner and Stanford 2003). In terms of nutrients, biomass changes correspond to notable accumulations of N, P, and extractable K. Total ecosystem N increases from 3 to 8 Mg/ha, P increases from 7 to 9 Mg/ha, and K increases from 0.5 to 3.0 Mg/ha (Boggs and Weaver 1994). We return to density, basal area, and biomass in the next chapter when we examine vegetative growth and production.

Biological Diversity

River corridors normally possess highly diverse floral and faunal communities and attendant biological processes (Ward et al. 2002). These attributes emerge from the unique spatial organization—a mosaic of habitats continually changing in response to variable water flows above and below ground—and biotic responses to the locally variable topography and climate. As floods alter habitats, they also modify interactions and exchanges between habitats. The net effect is substantial small-scale heterogeneity (e.g., Naiman et al. 2005). Further, the inherent heterogeneity introduced by local topography affects the frequency and duration of inundation, and the local microclimate influences the ability of individual species to flourish, thereby adding even more physical as well as genetic heterogeneity. Collectively, heterogeneity at local scales offers a great variety of conditions for life, making riparian corridors focal points for diversity. Riparian diversity is further magnified at the catchment scale since riparian corridors extend from the highest to the lowest elevations, and the higher elevation riparia are spatially numerous, which tends to augment regional diversity.

Even though riparian communities are considered to be disproportionately species-rich compared to their surroundings (Naiman et al. 1993, 2000; Sabo et al. 2005), it is still not known how many species are present in most catchments. No studies fully quantify how many species utilize riparian corridors, but available evidence suggests that for some taxa the numbers are quite high (Tockner and Ward 1999). Biodiversity in riparian corridors is best documented for vascular plants and vertebrates.

Studies throughout the world indicate that riparian areas generally have high levels of both of plant and animal diversity (e.g., Salo et al. 1986, Junk 1989, Nilsson et al. 1989, Décamps and Tabacchi 1994). In the Pacific Northwest of the United States, 74 percent of all plant species within one catchment were found in the riparian corridor (Pollock et al. 1998). In another local area, 60 percent of 480 wildlife species occurred in wooded riparian habitats, and 68 species of mammals, birds, amphibians, and reptiles required riparian areas to satisfy a vital habitat need during all or part of the year (Raedeke 1988). In the Santa Monica Mountains in southern California, less than 1 percent of the land area is riparian, but approximately 20 percent of the native vascular plant species have their primary habitat there (Rundel and Sturmer 1998). Nilsson (1992) reports 13 percent (>260 species) of the entire Swedish flora of vascular plants occurring along the Vindel River. Junk and others (1989) report that all periodically flooded forests in the Amazon basin may have about 20 percent of the 4,000 to 5,000 estimated Amazonian tree species. Planty-Tabacchi and others (1996) report 1,396 vascular plant species along the Adour River riparian corridor in France, representing 30 percent of the French flora (see also Sidebar 4.3). An even more striking example is that the riparian corridor along the main channel of the Vindel River includes 60 percent of the vascular plant species in the entire river basin (Nilsson and Svedmark 2002). However, the information just presented may be slightly misleading because comparative studies are seldom undertaken in surrounding uplands. Do riparia have more species than nearby uplands? Examining previously published data from seven continents, Sabo et al. (2005) suggest that riparia increase regional, or γ-diversity, >50 percent by harboring different species assemblages, rather than more species, when compared to surrounding uplands.

Diversity, however, is much more than just the number of species. It also includes structural (habitat) diversity as well as functional (process) diversity (e.g., nutrient cycling processes, symbiotic relationships), and all three are tightly interconnected. In this section we will focus on compositional or species diversity, having already addressed structural diversity and leaving functional diversity for the next chapter.

Diversity Theory and Measurement

Three general theories of biodiversity explain local variation in species richness, all of which have the potential to at least partially explain biodiversity patterns in riparian corridors. These are the dynamic equilibrium model (Huston 1979, 1994), the theory of resource competition in a heterogeneous environment (Tilman 1982), and the intermediate disturbance hypothesis (Connell 1978).

The Amazon and Orinoco rivers, connected in their headwaters by the Casiquiare Channel, together deliver >20 percent of the global river discharge to the ocean. Tributaries draining from the Andes are dominated by white-waters (so named for their rich sediment and nutrient loads) that create the Várzea lake-wetlands in the lowland floodplains. In contrast, rivers draining the oldest Amazonian craton (areas with Podzol soils) are dominated by oligotrophic (poor in sediment and nutrients) clear- and blackwaters that create a relatively unproductive riparian system known as Igapó. Whether the tributaries originate in the Andes or in the Podzol areas, the rivers are characterized by seasonally variable water discharge and differential sediment production. The water and material fluxes shape a mosaic of fluvial environmental patterns in the various riparian reaches. Flooded riparian forest is the main vegetation type of the floodplains, and it is of remarkable importance for aquatic life and riparian people (Rosales et al. 1999). The flooded riparian forests are classified as seasonal or permanent Várzea or Igapó and are somewhat similar to the larger upland matrix that is basically forested. In extensive areas, however, riparian forests are clearly differentiated as gallery forests within widespread savanna uplands.

Historically, gallery forests were considered to have been especially important for the maintenance of overall forest diversity during drier and cooler Pleistocene periods. As the region gradually became wetter and warmer, the biota eventually re-dispersed, contributing to the wide genetic differentiation and geographic species distribution of plants in Amazonia. Additionally, the extensive riparian floodplains, with their lateral interactions with the main river channel, build and renovate diverse fluvial geomorphologic features that further influence biota distribution and diversity. A strong seasonality in flood pulse regulates the presence of biological and ecological processes, resulting in seasonal patterns of plant phenology, germination, and growth. The Amazon and Orinoco riparian forests, like riparian systems elsewhere, conserve nutrients as well as act as controls on nutrient fluxes. Finally, the riparian plant species have numerous, and interesting, physiological and anatomical adaptations to cope with prolonged periods of anaerobiosis.

In terms of species diversity, the Amazon and Orinoco riparian forests increase in diversity as one moves from an alpha (habitat), to a beta (landscape), and to a gamma (regional) perspective. For example, several studies indicate that flooding intensity is negatively correlated to tree species diversity at local levels. However, at a basin scale or regional level, the unique floristic composition of sectors with high flooding intensity increase the number of species in the riparian corridor. These variables are key factors in the planning of riparian conservation as they control uniqueness (levels of rarity, endemism, redundancy of specific types of river sectors), succession (young or mature stages), and terrestrialization (similarity with richer upland forest ecosystems).

Judith Rosales, Centro de Investigaciones Ecológicas de Guayana, Universidad Nacional Experimental de Guayana, Chilemex, Puerto Ordaz, Edo. Bolivar, Venezuela.

Huston's (1979, 1994) dynamic equilibrium model makes important predictions about site patterns of species diversity based on the interaction of productivity and disturbance. The dynamic equilibrium model suggests predictable relationships between productivity, disturbance, and diversity, and explains the effects of interactions between disturbance and productivity on patterns of species richness among competing species. Although the model describes a continuum in productivity–disturbance–diversity gradients, it can be reduced to four basic states to illustrate the interacting effects of productivity and disturbance on diversity:

1. In highly productive, undisturbed systems, diversity tends to be low because dominant species eliminate nondominant species through competitive exclusion. If the community is productive but disturbances are more frequent, then community diversity can increase because competitive dominants are killed or their growth is slowed by the disturbance, and they cannot competitively exclude other species.
2. Conversely, if a site is unproductive but disturbances are infrequent, diversity will be high because the rate of competitive exclusion is slow.

3. Unproductive and frequently disturbed sites tend to be low in diversity because frequent disturbances eliminate slow growing species before they have a chance to reproduce.
4. Productive, frequently disturbed sites tend to be high in diversity because growth rates are high, but disturbances are frequent enough that competitive exclusion does not occur.

The theory of resource competition in a heterogeneous environment predicts that species diversity is controlled primarily by resource availability and productivity, and the influence of disturbance is not emphasized (Tilman 1982). This theory makes several predictions about relationships between plant diversity, resource richness, and spatial heterogeneity. First, species richness is expected to be highest in relatively resource-poor habitats but should decline rapidly in extremely resource-poor habitats and decline slowly as resource richness increases. This results in a unimodal resource richness–species richness curve, similar to previously published models such as Grime's (1973) productivity-diversity model. Second, communities with the greatest diversity should have many codominant species, whereas more resource-rich, lower-diversity communities should be dominated by a few species, with most species being rare. Finally, for a given level of resource richness, increased spatial heterogeneity should lead to increased species richness, with the most marked effects in resource-poor habitats.

The intermediate disturbance hypothesis, modified for river corridors, predicts that biotic diversity will be greatest in communities subjected to intermediate levels of disturbance (Connell 1978, Ward and Stanford 1983). At low levels of disturbance, competitive interactions will result in lower diversity due to exclusion of species. Such communities often are dominated by large, relatively long-lived species. On the other extreme, high disturbance also will result in lower diversity due to exclusion of poor colonists or long-lived species. These communities tend to be dominated by small, short-lived species.

Together, the models of Huston, Tilman, and Connell suggest that potential productivity, disturbance, and spatial heterogeneity are key factors controlling local patterns of diversity, a suggestion supported by field and experimental observations and other theoretical models (see Pollock 1998).

It is important to clearly identify the spatial scale at which diversity is measured. Biodiversity exhibits patterns at seven scales, which are applicable to any biotic group or groups. The seven scales are known as microhabitat (point or internal-alpha) diversity, between-microhabitat (pattern or internal-beta) diversity, alpha diversity, beta diversity, gamma diversity, delta diversity, and epsilon diversity (Pollock 1998). The number of species found in a microhabitat (e.g., a quadrat) is considered *point diversity*, whereas the rate of change in species composition between microhabitats in a community is *between-microhabitat diversity*. The number of species found in a habitat patch is *alpha diversity*, whereas the rate of species turnover between different types of habitat patches is *beta diversity*. The term *gamma diversity* describes diversity levels over large areas but has been used to describe two different types of coarse-scale diversity patterns, leading to some confusion over the term (see Pollock 1998). *Delta diversity* refers to changes between geographic regions or along climatic gradients. Finally, *epsilon diversity* refers to the total number of species observed in a region or in the total number of gamma-scale landscapes sampled.

Point, alpha, gamma, and epsilon diversity are considered *inventory diversity* measures because they tabulate the number of species in a given area. Pattern, beta, and delta diversity are considered *differentiation diversity* measures because they measure the rate of change in the number of species between areas. Although these measures were initially designed to measure the diversity of plants at different spatial scales from quadrats to large geographic regions, there is no specific spatial scale attached to these definitions. In reality, most studies of diversity only examine several scales at most, and alpha, beta, and gamma diversity are the terms most commonly used, regardless of the spatial scales involved (Ward and Tockner 2001).

Field observations indicate that the predicted pattern of plant diversity as a function of scale in riparian corridors is hypothetically as follows: alpha and gamma diversity are high, beta diversity is moderate, and delta diversity is low. This expected pattern of diversity reflects the fact that in riparian corridors, a number of relatively small habitat patches in close proximity have high numbers of species (so alpha diversity is high). However, many of these species have wide ecological amplitudes, and species composition between habitat patches overlaps considerably, thereby lowering beta diversity. Gamma diversity is high because there is a wide array of habitat types. Species turnover rates may be low, but there are enough different habitats that species continue to accumulate. Delta diversity (e.g., changes in species composition between catchments) is low wherever there is very little endemism among riparian plants within a region. Finally, over large areas, species diversity slowly increases as the size of the area increases, primarily because of changes in climate. Therefore, epsilon diversity in riparian corridors is low if the area is climatically similar and increases as the area sampled encompasses a wider array of climate conditions. Few studies have compared the rate of increase in species in riparian corridors relative to uplands at such large scales. Most studies lending support to the idea that riparian corridors are floristically diverse have largely examined alpha or gamma diversity (Nilsson 1986, Kalliola and Puhakka 1988, Nilsson et al. 1989, Décamps and Tabacchi 1994, Pollock et al. 1998, Sabo et al. 2005), while little or no research has examined beta, delta, or epsilon diversity.

Vegetative Diversity

The contrasting patterns found at site and catchment scales, the roles of river islands and other refuges, and the relative influence of factors controlling species expression are key to understanding riparian plant diversity.

Site and Catchment Patterns

Biodiversity patterns in riparian corridors are structured along longitudinal, lateral, and vertical dimensions. Unfortunately, little is known about vegetative patterns and spatial changes in species composition of interconnected riparian systems within large catchments and, additionally, few data are available for the vertical dimension or for nonvascular plants.

There are, however, some studies partially addressing the longitudinal dimension. Many riparian corridors vary in species richness along the river's course, but some do not (e.g., Baker 1990). There appear to be three types of riverine reaches that fundamentally differ in terms of biodiversity: constrained river reaches increase in biodiversity downstream, whereas braided reaches have relatively low diversity and meandering reaches have high biodiversity (Ward 1998). In general, longitudinal studies show that plant diversity generally increases downstream, with the peak reached in the piedmont (transition zone between mountain and lowland domains) as the riparian zone widens, at major tributary deltas, or, in Sweden, at the position of the coastline during the Holocene (Nilsson et al. 1989, 1994, Planty-Tabacchi et al. 1996, Salo et al. 1986).

These catchment-scale patterns suggest a maximum diversity at an intermediate level of disturbance, which induces considerable spatial heterogeneity. However, other physical attributes (e.g., channel gradient, lithology, level of confinement) and local climate also modify site-specific species richness and can result in no net trend in longitudinal species richness. This has been illustrated in the United States in the state of Colorado and in Sweden. In Colorado, where the climate becomes progressively drier with decreasing elevation, species richness is slightly greater in subalpine riparian forests but longitudinal trends differ by plant group (forb, graminoid, shrub, tree)—resulting in no net trend (Baker 1990). In Sweden, species richness per site (alpha diversity) is generally higher along the main channel of the Vindel River than in the tributaries (Nilsson et al. 1994). Nevertheless, the tributaries also show certain floristic characteristics of their own, such as higher proportions of woody plants, geophytes, and native species. The main channel has its greatest species richness at intermediate altitudes, whereas tributaries have the least number of species at intermediate altitudes. In total, the diversity of the flora in the main channel corridor is not simply a reflection of the flora in the tributaries, implying that conservation strategies need to consider stream size as well as geomorphic complexity of the channels.

Lateral patterns in plant diversity also generally peak at intermediate levels of disturbance and moisture availability. Diversity is often low immediately adjacent to the river, increasing with local elevation and geomorphic complexity, and then declining slightly upslope of the riparian-upland interface (e.g., Gregory et al. 1991, Dixon and Johnson 1999, van Coller et al. 2000). Nevertheless, there are variations to this generalization. In the relatively species-rich uplands of western Australia, species richness peaks at the ecotone between the riparian and upland forests (Hancock et al. 1996). Along South African rivers, van Coller et al. (2000) combined two prominent conceptual frameworks—environmental gradients and patch hierarchy—to illustrate multivariate controls on lateral biodiversity patterns in a highly complex environmental setting. In central Amazonia, where river levels fluctuate up to 14 m and riparian plants are inundated 50 to 270 days annually, tree species richness and diversity are strongly limited by the water level and the duration of flooding (Ferreira and Stohlgren 1999). There is also a high (42 to 60 percent) overlap of species between habitats and, while flood duration may decrease local diversity, it also creates and maintains high landscape-scale diversity by increasing lateral heterogeneity.

Refuges

Riparian corridors act as refuges for organisms and processes under several circumstances, thereby conserving diversity. Two key circumstances are (1) the formation and disintegration of islands (short term) and (2) the providing of mesic conditions during

119

prolonged droughts (long term). River islands are often very dynamic, creating a continuum of disturbance and environmental conditions across the river landscape, thus providing a diversity of riparian habitats supporting a wide range of successional stages and ecological processes (Edwards et al. 1999, Ward et al. 2000). Unfortunately, there are scant empirical data on the biodiversity of river islands because so few natural ones remain. In contrast, there are better data on the role of riparian corridors during drought.

It is well known that riparian forests often act as safe sites for regional flora and fauna during widespread environmental shifts, such as decadal-long dry periods. However, the story also has validity at millennium time scales. Some present-day humid tropical zones appear to have experienced severe Pleistocene droughts, but there is no indication of mass extinction, whereas there is indication of rapid species re-expansion during the early Holocene. For example, riparian forests may have been refuges for conserving mesic plant diversity in Central America (Meave and Kellman 1994). The floristic attributes of the riparian forest are similar (52 species/ha) to those in the upland forests, while stem density is higher and biomass lower, thus increasing the potential to maintain species richness. Moreover, riparian trees in a savanna matrix are often younger because more mature trees are removed by wind and fire (Kellman and Tackaberry 1993). Frequent removal of mature trees reduces the rate of competitive exclusion from the community while enhancing the potential for greater numbers of coexisting species. However, extremely diverse riparian forests occur in other tropical rainforests such as Amazonia (>225 species/ha) and southeast Asia (283 species/ha) where the role of riparian forests as refugia may have been more limited (Dumont et al. 1990, Gómez Peralta 2000).

Factors Controlling Species Richness

Earlier we discussed several environmental factors influencing the ability of plants to colonize and survive in riparian zones. However, there are only two investigations that actually test biodiversity theory for riparian vegetation under naturally occurring conditions. One investigation related flood frequency, site productivity, and spatial complexity to plant species richness in coastal Alaska (see Figure 4.18; Pollock et al. 1998). At the 1,000-m^2 scale, their model was able to predict 78 percent of the variation in species richness, whereas at the 1-m^2 scale, the model predicted only 36 percent. In general, they found support for Huston's (1979, 1994) dynamic equilibrium model with species-rich sites having low to intermediate levels of productivity, intermediate flood frequencies, and moderate to high microtopography. The other investigation is by Everson and Boucher (1998), who found a weak positive relationship between tree species richness and topographic complexity (standard deviation of slope) along the Potomac River in West Virginia and Maryland. Interestingly, they also found a negative relationship between species richness and the width of the riparian zone, but this seems related to the occurrence of unique geologic features (cliffs) and may not be applicable to many other riparian situations.

Additionally, it is suspected that diversity patterns, and controls on those patterns, will be different for riparian trees as opposed to herbaceous vegetation. While most theoretical discussions of species diversity emphasize the importance of habitat productivity and disturbance regimes, many other factors (e.g., species pools, plant litter accumulations, plant morphology) are known to be important, especially for

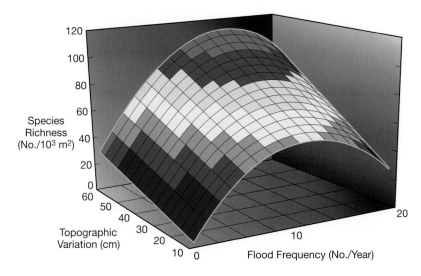

Figure 4.18 Three-dimensional representation showing relations among species richness, flood frequency, and topographic variation for riparian areas in southeast Alaska. Modified from Pollock et al. 1998.

herbaceous understory vegetation (Grace 1999). The available literature suggests an inverse relation between total community biomass and the density of herbaceous species, but there are many other factors involved (e.g., gradients in species pools, time lags, soil effects). Nevertheless, even though the relation between disturbance and species richness is complex, it is generally consistent with the intermediate disturbance hypothesis.

Faunal Diversity

Animal diversity in riparian zones probably exceeds that for plants, especially for invertebrates, while the percentage of vertebrates from the regional pool using the riparian zone is nearly equal to that for plants. In this section we first examine the diversity of soil organisms and then address the diversity of animals generally found above the soil surface.

Diversity of Soil Organisms

Soil organisms are not just inhabitants of the soil; they are part of the soil. The soil biota heavily influences soil properties such as hydrology, aeration, and gaseous composition—all of which are essential for primary production and for decomposition of organic materials. A primary component of the soil community is microorganisms. It is the highly diverse bacteria and fungi assemblages that regulate nearly all processes associated with decomposition and nutrient cycling. At any one time, most microbes are dormant but quickly become active when suitable metabolic substrates are available under suitable environmental conditions (Lavelle 1997). Several studies have illustrated the great diversity of microbes in riparian soils by using sole carbon source utilization methods (e.g., Martin et al. 1999, Clinton et al. 2002). The microbial diversity is a response to the patchy riparian environment as well as to the diversity of plant roots and their exudates.

Overall, riparian soils appear to have an impressive diversity (Brussaard et al. 1997). There may be as many as 10,000 species of ectomycorrhizal fungi, >3,000 species of bacteria, >5,000 species of nematodes, and tens of thousands of species of mesofauna (mites, collembolans) and macrofauna (ants, termites, earthworms). This is a rich area for future study because so little is known, despite its importance for the functioning of riparia everywhere.

Aboveground Fauna Diversity

Aboveground faunal diversity is better known but it is far from being completely understood. Reasonably comprehensive studies exist for some groups, especially birds, whereas knowledge about other groups is nearly nonexistent (e.g., most invertebrates). Flooding certainly affects invertebrate community dynamics—distribution, abundance, and diversity—with species richness often increasing with elevation above the river even though flood duration will affect groups differentially (Uetz et al. 1979). Among the invertebrates, spiders (Arachnida) and beetles (Coleoptera; e.g., Gerken et al. 1991) are the better-investigated groups.

Spiders in particular are especially diverse in riparian zones. Along the River Lech in Austria, 205 species from 17 families have been recorded (Steinberger 1996). The individual spider communities are distinct and can be differentiated according to distance from the waterline, vegetation, sediment grain size, moisture, and physical patch size. In the United Kingdom and southern France, 147 species of spiders from 17 families are associated with riparian zones (Bell et al. 1999). Like the Austrian situation, elevation from the river and flood frequency and duration (affecting moisture) are the most important variables affecting diversity.

In many areas, a large proportion of vertebrates utilizes the riparian zone, but it is not always apparent whether individual groups are more or less diverse there. For example, in western Oregon and Washington, 87 percent of the 414 resident species of amphibians, reptile, birds, and mammals use riparian zones or wetlands (Oakley et al. 1985), but only 10 percent use specialized habitat within riparian zones. More is known about those species depending on the streams for foraging or reproduction than about species using riparian zones only occasionally. Certainly, amphibians are more diverse in riparian zones because of their reproductive needs. For reptiles, birds, and mammals, it depends on the local environment and specific life history needs.

The best information available is for birds; this is because conservation of riparian vegetation has, in part, emphasized providing habitat for a locally diverse avifauna. In the drier western United States, bird diversity at the site scale (alpha) is generally high relative to the number of potential species (Knopf and Samson 1994). Between habitat, diversity (beta) varies across a catchment, with riparian assemblages differing most from upland assemblages at the highest and lowest elevations. This pattern is attributed to enhanced avian movements within the riparian corridors. The corridors for bird movements, in turn, facilitate faunal mixing on a broader scale, increasing regional diversity (gamma) within landscapes.

This pattern is repeated for riparian zones throughout the drier regions of Australia and southern Africa (Knopf and Samson 1994, Robertson et al. 1998, Shelly 2000) but not in more mesic regions where there is little differentiation between riparian and upland bird communities (Murray and Stauffer 1995, Whitaker and Montevecchi

1997). Nevertheless, certain bird species in mesic regions show strong fidelity to riparian areas (Murray and Stauffer 1995), especially during nesting periods (Dècamps et al. 1987, Lock and Naiman 1998). In the Amazon Basin, a substantial portion (15 percent) of the nonaquatic avifauna is restricted to habitats created by rivers (e.g., beaches, sandbar scrub, river-edge forest, varzea forest, transitional forest, and water edge; Remsen and Parker 1983). As many as 169 bird species in the lowland neotrophics may have evolved in Amazonian river-created habitats, with 99 of these spreading secondarily to other areas. Neither the Zaire nor the Mississippi basin avifaunas show such a high percentage of species related to river-created habitats—possibly because they lack the huge seasonal water-level fluctuations experienced by the Amazon.

Mammal diversity in riparian zones appears to be great, but this may be deceptive. Mammals are highly mobile, and many, such as moose and elk in winter, make frequent use of riparian zones for feeding and cover. Certainly bats feeding on emerging stream insects are abundant at times, and other medium-sized predators (badgers, foxes, cats) come to feed and drink (Sanchez et al. 1996, Virgos 2001). Browsers and grazers, as well as rodents and shrews, are often abundant, but most can be found in the uplands too (Kelsey and West 1998). For many small mammals, their presence in riparian areas depends on the vegetative cover. In Arizona, small mammals respond to the amount of herbaceous vegetation rather than to the presence of larger trees (Anderson and Nelson 1999). In Washington, the presence of large woody debris in riparian areas increases the species richness of small mammals (Steel et al. 1999). However, in Oregon, the species richness of small mammals does not differ between riparian and upland habitats among different seral plant stages (Gomez and Anthony 1998).

Are riparian zones more diverse than the surrounding uplands? The answer is that it depends on the environmental setting and the taxon being considered (e.g., Sabo et al. 2005). Ecological theory would suggest that, in general, riparian areas are highly diverse because of their mosaic structure, presence of refugia, broad ranges of environmental settings, and species assemblages. However, most biotic inventories are shockingly incomplete, making this question difficult to answer with certainty.

Biotic Functions of Riparia

5

Overview

- This chapter explores water use and fluxes, nutrient dynamics, production ecology of vascular plants, decomposition processes, strategies for information flow (i.e., mechanical and chemical communication), and the maintenance of riparian microclimates and associated moisture gradients. These topics are the foundation for understanding and appreciating the roles of riparian organisms in the larger catchment as well as the diversity of functions displayed by riparia.

- Water for riparian vegetation comes from diverse sources. Typically, the vegetation is a mixed assemblage of obligate and facultative phreatophytes—deeply rooted plants that obtain water directly from the stream or from groundwater in the alluvial aquifer. The reliability of groundwater encourages riparian trees to develop roots predominantly in the capillary fringe and saturated zone rather than throughout the soil profile, especially if precipitation during the growing season is unreliable. Annual evapotranspiration (ET) by riparian vegetation can be impressive, especially in semiarid catchments.

- Vegetative growth and production integrate environmental conditions, disturbance regimes, and successional dynamics. Basal area increase (BAI = growth) and stem production (P) for dominant riparian tree species respond to physical conditions and are expressed in their temporal patterns. The highest stem production rates are from temperate coastal rain forests and from semitropical regions. Annual litterfall, as a component of production, is influenced by forest type, successional stage, stream size, site productivity, and latitude. The highest litterfall rates are associated with tropical equatorial, tropical montane, and warm temperate forests. Information on mortality of riparian vegetation from disease, senescence, herbivory, and catastrophic disturbances is limited, having been mostly derived from the perspective of large woody debris recruitment to streams rather than from an understanding of the production ecology of riparian forests.

- Litter quality and local abiotic conditions exert strong influences on decay rates of plant litter and on nutrient immobilization potentials. Decay rates generally decrease with increasing biochemical complexity (e.g., more lignins) and increase with increasing nitrogen content. At least three distinct phases are recognized in the dynamics of decomposing plant litter: (1) a brief leaching period during which 10 to 25 percent of the litter's initial biomass can be lost, (2) a period of net immobilization during which the absolute amount of nitrogen in the decomposing litter increases, and (3) a period of net release or net mineralization of nitrogen.

- Although empirical evidence is scant, riparia may act either as conduits, filters, or breaks in the flow of genetic, chemical, and vocal information. Seven important attributes of riparian corridors operate at the landscape scale to affect information fluxes: connectivity, resistance, narrows, porosity, influence fields, spatial orientation, and distance.

- Riparian forests exert strong controls on local microclimates. Limited data suggest strong differences between upland and riparian microclimates, especially for temperature and humidity but not for solar radiation or wind velocity.

Purpose

Riparia possess an impressive array of functional attributes that are comparable to their great diversity in structure and pattern. Functional adaptations allow riparian organisms to survive in virtually every climate where liquid freshwater is available for at least part of the year, and under highly unstable and unpredictable environmental conditions. The adaptations are manifested in how water and nutrients are acquired and used, how organisms metabolically allocate resources to growth and reproduction, how dead biomass is recycled, how information (e.g., genes, chemicals, and sounds) is communicated and received, and how riparian vegetation modifies the local climate.

Interest in the functional aspects of organisms began centuries ago, but since around 1970 there has been a surging interest in organisms surviving in apparently stressful environments—and a common perception is that riparian zones are one of those environments. Certainly, riparian organisms have ways to meet the physiological demands of floods and drought, to deal with the physical aspects of sediment scour and deposition, to capture highly mobile nutrients, to decide on whether to grow further or reproduce, to determine the best means of communication, as well as to cope with the myriad of other day-to-day challenges. Perhaps the riparian zone is not especially stressful? And perhaps the organisms are preadapted for living well under unstable and sometimes unpredictable conditions?

This chapter explores these two questions as we discuss water use and fluxes, nutrient dynamics, production ecology of vascular plants, decomposition dynamics of dead material, strategies for information flow, and the modification of microclimate gradients by riparian plants. The overall goals are to provide a foundation for understanding and discovering the functional roles of riparian organisms in the larger catchment, as well as to develop an appreciation of the wonderful diversity of functional attributes displayed by riparian communities.

Water Use and Flux

Water used by riparian organisms comes from several sources via numerous flow paths. They range from the apparently obvious—stream water—to precipitation, to shallow water moving laterally and relatively slowly downslope through soils, to deeper and often older groundwater (Ziemer and Lisle 1998). Subsurface flow accounts for nearly all water delivered to stream channels from undisturbed vegetated hillslopes and, in the process, passes close to the roots of riparian vegetation. As discussed in Chapter 2, precipitation infiltrating the soil surface travels through the soil as either shallow subsurface flow (soil water) or deeper seepage that replenishes groundwater. Under natural conditions, groundwater storage is relatively stable, sustaining stream flows during dry periods. Water is transmitted within the soil along two different flow paths of importance to riparian vegetation: through the soil matrix (micropores) and through macropores or zones of preferential flow, including root holes, soil cracks, animal burrows, soil pipes, and paleochannels. Subsurface flow velocities vary widely. Flow in micropores is very slow (10^{-7}–10^{-6} m/hr), while flow in soil pipes and sandy soils (10^{-1}–10^{2} m/hr) can be as rapid as unchannelized overland flow.

Riparian vegetation is responsive to the various water sources and is characteristically a mixed assemblage of obligate and facultative phreatophytes—deeply rooted woody plants that obtain water directly from the stream or from groundwater in the alluvial aquifer. Obligate phreatophytes send roots into or below the capillary fringe to use shallow groundwater, while facultative phreatophytes can also survive in upland environments where groundwater is not available. In either case, extreme spatial and temporal availability of water, especially in arid and semiarid regions, places severe constraints on the ability of riparian plants to meet evapotranspiration (ET) requirements during key periods of the growing season (Snyder and Williams 2000). Additionally, some plants transfer water from deep soil layers to overlying dry soils through their roots (e.g., hydraulic lift, Richards and Cadwell 1987) by physiologically initiating a vertical lift of water beyond that attributable to capillary movement. Likewise, soil water can be actively transferred downward from the surface to deeper soil horizons by plant roots in addition to the influence of gravity. This reverse phenomenon allows a hydraulic redistribution, which maintains root viability, facilitates root growth in dry soils, and modifies resource availability (Burgess et al. 1998). Both hydraulic lift and redistribution have not been demonstrated in woody riparian plants, but it is conceivable that, during dry conditions, it may occur.

Surprisingly, as we will see below, some riparian trees may not always use stream water for ET. Likewise, not all tree species use only groundwater for ET, as the term *phreatophyte* implies. Use of growing season precipitation can vary considerably among different woody species as well as spatially across the floodplain in relation to topographic position (Bush et al. 1992, Thorburn and Walker 1994). Characterization of the conditions that promote use of various water sources and identification of those species most likely to use these water sources are necessary to fully understand and appreciate the complex, dynamic relationships between riparian plants and their water sources.

Water sequestered by vegetation is not returned instantaneously to the atmosphere; much of it is stored in roots and stems with the time of storage depending on the species. The herbaceous layer generally has the highest water content in terms of percentage of water in the plants, ranging from 80 to 98 percent with small (~2 to 5 percent) daily and seasonal variations (Tabacchi et al. 2000). Riparian shrubs generally have a lesser percentage of water stored (55 to 75 percent), while trees are often slightly less (~54 to 67 percent). However, in terms of absolute volume of water stored, most is found in trees because of their total biomass.

The rates of annual ET by riparian vegetation can be an impressive component of many semiarid basins' groundwater budgets—comparable in magnitude to mountain front recharge and surface water discharge. Daily ET is often correlated with solar radiation and the mean air temperature, but the relationship varies seasonally in response to leaf phenology. Most (~90 percent) ET is associated with trees, with the herb and shrub layers contributing very little (Tabacchi et al. 2000). Along the Rio Grande River in New Mexico, ET can exceed $110 \, cm \, yr^{-1}$ in dense or mature saltcedar (*Tamarix*) and cottonwood (*Populus*) stands, with less dense stands still having ET rates $>74 \, cm \, yr^{-1}$ (Dahm et al. 2002). Along the San Pedro River in Arizona, cottonwood ET approaches $127 \, cm \, yr^{-1}$, closely followed by willow at $120 \, cm \, yr^{-1}$ (see Table 5.1). Mesquite ET, depending on stand density, ranges from 49 to $76 \, cm \, yr^{-1}$ (Scott et al. 2000). When integrated over large riparian areas, ET can account for 20 to 33 percent of the total estimated water depletions along the Rio Grande and other rivers.

The long-term reliability of groundwater may encourage riparian trees to develop roots predominantly in the capillary fringe and saturated zone rather than throughout the soil profile, especially if precipitation during the growing season is unreliable (Ehleringer and Dawson 1992). Conversely, plants that maintain roots in many soil layers, or that can rapidly deploy roots into moisture-rich patches in the soil, may respond opportunistically to precipitation. The use of the stable isotopes of oxygen ($\delta^{18}O$) and hydrogen (δD) has provided evidence for both modes of root system function. In Australia, riparian *Eucalyptus* spp. uses various combinations of groundwater, rainfall-derived shallow soil water, and stream water (Mensforth et al. 1994,

Table 5.1 Riparian Vegetation Categories and Associated Water Use in the Sierra Vista Sub-Catchment of the San Pedro River Basin

Category	Area (ha)	Evapotranspiration (m³/day)	Evapotranspiration (cm/yr)
Cottonwood	541	18,840	127
Dense mesquite	630	13,045	76
Medium-dense mesquite	1,164	15,495	49
Willow	17	558	120
Dense grass	34	889	95
Total	*2,386*	*48,827*	*75*[a]

[a]Prorated by proportional area.
Modified from Scott et al. 2000.

Thorburn and Walker 1994, Dawson and Pate 1996). In California, trees along perennial, montane streams take up water from upper soil layers early in the growing season, then use primarily groundwater when the soil dries (Smith et al. 1991). In western Arizona, *Populus fremontii* and *Salix gooddingii* use groundwater throughout the entire growing season on perennial and ephemeral streams, regardless of depth to groundwater, although responses of the trees to precipitation have not been assessed (Bush et al. 1992). Similarly, mature boxelder (*Acer negundo*) in northern Utah use only groundwater and do not always appear to use perennial stream water or shallow soil water (Dawson and Ehleringer 1991; see Sidebar 5.1). This latter observation, however, may be a response to a unique geomorphic setting where female boxelder growing adjacent to the stream used stream water early in the season, while male boxelder grew upslope in association with groundwater seeps. In contrast, boxelder did use soil water from precipitation at ephemeral and perennial stream sites in Arizona (Kolb et al. 1997). In South Africa, the natural abundance of stable isotopes is not sufficiently distinct for the purposes of quantifying the source(s) of xylem water in riparian trees. Nevertheless, qualitative results suggest that some tree species in the

---

Sidebar 5.1 Gender-Specific Ecophysiology, Growth, and Habitat Distribution in Riparian Boxelder

Throughout the semiarid intermountain west of the United States, boxelder dominates riparian habitats. Boxelder is a deciduous, dioecious tree (see Figure SB5.1) that exhibits significant habitat-specific sex-ratio biases. Although the overall sex ratio (male/female) does not deviate significantly from 1, the sex ratio is significantly male biased (1.62) in drought-prone habitats, whereas it is significantly female biased (0.65) in moist, streamside habitats. It was determined that the spatial segregation of the sexes was maintained by differences in gender-specific physiological characteristics (Dawson and Ehleringer 1993).

Leaf-level investigation showed that under both field and controlled-environment conditions, males and females differed significantly in a number of ecological and physiological traits related to water use. Males maintained lower stomatal conductance to water vapor (g), transpiration (E), net carbon assimilation (photosynthesis; A), leaf internal CO_2 concentration (c_i), and carbon isotope discrimination (Δ; an index of time-integrated c_i and water-use efficiency). This in turn meant that males also had higher instantaneous (A/g) and long-term water-use efficiency (inferred from Δ) than females. Furthermore, male trees exhibited greater stomatal sensitivity to declining soil water content and increasing leaf-to-air vapor pressure gradients, measures of evaporative demand. Higher rates of carbon fixation in female trees were correlated with higher g, higher leaf nitrogen concentrations, and greater stomatal densities. Females growing in both wet and dry habitats had higher vegetative shoot growth rates compared with reproductive shoots, whereas for males, growth rates of the two shoot types did not differ. In streamside habitats, female trees exhibited significantly greater vegetative shoot growth when compared to male trees. In contrast, males showed slightly greater vegetative and much greater reproductive shoot growth in nonstreamside habitats. Regardless of habitat or growing conditions, females allocated proportionally more aboveground biomass to reproduction than did males.

Recently, the aforementioned findings have been extended on a whole-plant and more temporally integrated scale through an analysis of the variation in the carbon isotope ratio of tree rings in male and female trees exposed to different levels of water stress in a common garden (Ward et al. 2002). They found that under wet conditions, female trees had higher growth rates and were less conservative in their water use than males—a finding consistent with earlier work. However, in dry, stressful years, male and female trees had similar growth (ring widths) and carbon isotope ratios, indicating similar integrated gender-specific responses when water stressed. These data pointed out the importance of gathering long-term (tree ring) and short-term (physiology) data when studying riparian tree responses to fluctuations in water availability.

The study of Dawson and Ehleringer (1993) shows that gender-specific physiological traits—some of which are directly related to water use—can help explain the maintenance of habitat-specific sex ratio biases in boxelder along a soil moisture gradient, and that the combination of gender-specific physiology, growth, and allocation differences contributes to spatial patterns in the size and age structure of male and female plants within the population. Then, the work of Ward and her colleagues (2002) nicely demonstrates that as precipitation patterns change and their impacts are seen on riparian zones, tree species such as boxelder that are very sen-

Figure SB5.1 Riparian boxelder (*Acer negundo*).

sitive to water stress can serve as indicators of riparian zone impacts. Assessments of changes in the tree ring widths and C-isotope composition of the rings serve as important sources of information for not only assessing changes in the hydrological status (wet versus dry years) of riparian zones, but perhaps for parameterizing predictive models aimed at understanding how riparian zone trees will respond to future changes.

Todd E. Dawson, University of California, Berkeley
Joy K. Ward, University of Kansas, Lawrence

semiarid Kruger National Park do not use river water in either the wet or the dry season (Innis 2003). Rather, the trees appear to use precipitation-derived soil water moving from the uplands and groundwater.

Water transpiration by riparian vegetation from groundwater and unsaturated soil layers may depend on the interaction between site conditions and species assemblage. In southeastern Arizona, trees along the San Pedro River have an opportunity to use several water sources depending on their location relative to the channel (Snyder and Williams 2000). *Salix gooddingii*, an obligate wetland species, does not take up water from the upper soil layers during the summer rainy period but instead uses only groundwater, even at ephemeral sites where depth to groundwater exceeds 4 m. *Populus fremontii*, a facultative wetland species and a dominant "phreatophyte" in semiarid systems, also uses mainly groundwater once it matures. However, at ephemeral stream sites, it can derive 26 to 33 percent of its transpired water from upper soil layers during the summer rainy season. *Prosopis velutina*, a facultative upland species, derives a greater fraction of its transpired water from the upper soil layers at ephemeral stream sites than at perennial stream sites where the groundwater depth is <2 m. Nevertheless, *Populus* transpires a greater *quantity* of water from the upper soil layers at ephemeral sites than at perennial sites because of the timing of availability and the site-specific root deployment strategies.

There may be ecological advantages in using soil water and groundwater rather than stream water (Thorburn and Walker 1994). First, groundwater is a more reliable source of water for riparian trees. Adaptations for its use aid the trees' survival during droughts and give them an advantage over shallow-rooted species in dry seasons. Second, the level of stress imposed on the trees by relying on soil water and groundwater may, in the long term, be less than that imposed by committing resources to utilize irregularly available surface water. Finally, the use of surface soil-water benefits the trees by aiding in the acquisition of nutrients. Nutrient concentrations are generally higher in surface soils than the stream water, and uptake of soil water is required to access critical soil nutrients. As we will see later in this chapter, nutrient retention is a major ecosystem function and ecological service provided by riparian zones. Overall, it is becoming clear that riparian trees encompass a wide spectrum of functional types that respond uniquely to spatial and temporal variation in the distribution of available water in the rooting zone. Likewise, they can use substantial amounts of precipitation to meet ET requirements even when they are living above shallow groundwater.

Nutrient Fluxes

Riparian zones are fundamental components of regional and global biogeochemical cycles, acting to regulate fluxes and as sites for elemental transformations. Generally, riparia form a particularly dynamic link between upland communities and nearby river ecosystems. Nutrients derived from uplands are accumulated in riparian zones before being either transferred downstream to estuarine systems or back to the atmosphere. These fluxes, however, are highly dynamic as the nutrients participate in a variety of chemical and biological interactions and transformations (Sterner and Elser 2002). The interactions are dominated by sequential processes of biological uptake, remineralization, and microbially mediated transformations. We address the basic processes in this chapter and discuss the ecological implications of the processes in Chapter 6.

Overview of Cycles and Processes

Several elements are of key importance to the structural and physiological requirements of riparian biota, but only nitrogen (N) has been studied extensively. Phosphorus (P), sulfur (S), and many other elements are important but, unfortunately, less is known about their roles in riparian systems despite being intricately woven into the ecological balance of aquatic and riparian systems. Whereas climate and geology produce the soils and moisture regimes supporting and shaping riparian biota, the riparian community largely regulates fluxes of N and P from atmospheric and geologic sources. Biotic processes exert primary controls on the natural distributions of riparian N and P, but other abiotic processes, such as adsorption to the surfaces of soil minerals, also serve to retain nutrients. As we will see in Chapter 6, the efficiency of nutrient retention by the riparian vegetation is normally strong; however, there are bidirectional fluxes as a small percentage is continually transferred to streams and also from streams to riparia.

N and P occur in a variety of chemical forms, which are molecularly transformed as cycling reactions proceed and the oxidation-reduction environment is altered (McClain et al. 1998). In the case of nitrogen, biotic assimilation is almost always accompanied by a decrease in the oxidation state of the element. The main exception is the assimilation of ammonium (NH_4^+), which involves no redox reactions. Phosphorus occurs in only one oxidation state (+5), so redox reactions do not directly influence its cycling.

N and P availability depends on the molecular form. Nitrogen is available to plants primarily in the form of ammonium (NH_4^+) and nitrate (NO_3^-, but only after it has been microbially converted to NH_4^+). There are blue-green bacteria (e.g., *Nostoc, Anabena*) as well as free-living bacteria (e.g., *Clostridium, Bacillus*) associated with wet, anaerobic riparian soils that are capable of nitrogen fixation, utilizing gaseous nitrogen (N_2) and converting it to NH_4^+. Additionally, certain plants containing symbiotic nitrogen-fixing bacteria (*Frankia, Rhizobium*) in their roots are also able to utilize gaseous nitrogen and convert it to NH_4^+. Alder (*Alnus*) is one riparian genus that is particularly important in this regard (Rhoades et al. 2001). Phosphorus is available to plants in the form of orthophosphate (PO_4^{3-}), which is the only inorganic form where phosphorus occurs in appreciable amounts.

Once sequestered by plants and animals, each element serves vital metabolic and structural functions. Nitrogen is a component of all amino acids and therefore proteins. In this form it contributes to many internal functions, including structural support, movement, and defense against foreign substances. As a component of enzymes, it catalyzes chemical reactions in cells. Nitrogen also occurs in alkaloids and urea [$CO(NH_2)_2$]. Organic forms of phosphorus include such important molecules as nucleic acids (RNA and DNA), adenosine triphosphate (ATP), and phospholipids (components of cell membranes). As a component of bones and teeth, phosphorus occurs with calcium (Ca) as the mineral apatite [$(Ca_5[PO_4]_3F)$]. Sulfur is less abundant

han nitrogen and phosphorus in living organisms, but nevertheless it serves essential
functions as a component of two amino acids (cysteine and methionine) and several
other less abundant molecules. Occurrence in living tissue represents the intersection
point in the cycles of nitrogen, phosphorus, and sulfur. In the organic form, these elements co-occur in more or less constant proportions, roughly 35:2:1 in plants
McClain et al. 1998).

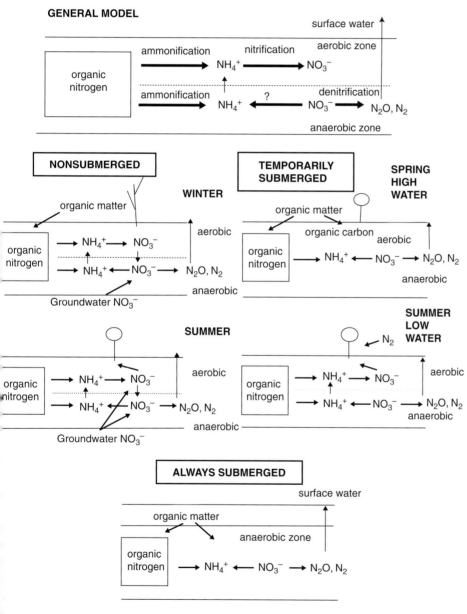

igure 5.1 A combined hydrological and biogeochemical model of nitrogen cycling and transport in an
gricultural riparian zone in France. From Pinay and Décamps (1988).

During decomposition, N emerges in the form of ammonia (NH_3) and NH_4^+. Although some of the NH_3 may volatilize in the normally slightly acidic soil water, most rapidly reacts with free protons (H^+) to produce NH_4^+. At this point, remineralized nitrogen is available for utilization by organisms. However, various other energy-producing reactions may occur prior to N assimilation (see Figure 5.1). In anaerobic environments, NH_4^+ remains the stable form of dissolved inorganic nitrogen, but in aerobic environments, its concentrations are generally low. In well-oxygenated situations, natural concentrations of NH_4^+ are generally less than 1 μM/L, but in oxygen-depleted hyporheic waters, concentrations may exceed 20 μM/L. Another factor contributing to low ambient concentrations of NH_4^+ is its adsorption to cation exchange sites on sediments and adjacent soils. However, at Little Lost Man Creek, California, concentrations of exchangeable NH_4^+ reach 1,150 μM/g soil 18 m inland of the channel (Triska et al. 1994).

Under aerobic conditions, NH_4^+ is oxidized to NO_3^- in a two-step *nitrification* process. Bacteria of the genera *Nitrosomonas* and *Nitrobacter* carry out the reactions in tandem. There are several possible intermediate compounds, the most important of which is nitrous oxide (N_2O), a greenhouse gas that may be lost from the system by gas exchange. Nitrate is relatively unreactive with organic or mineral soil surfaces. Thus, if not assimilated by riparian plants, it moves freely downslope until taken up by the biota or transformed via redox reactions. If NO_3^- is transported into an anoxic environment, such as occurs in saturated soils or some micro-zone within a particle of decaying organic matter, it may be reduced via *denitrification* to N_2. The gases NO (nitric oxide) and N_2O are also by-products of denitrification and are generally lost from the system but play important roles in atmospheric warming and ozone depletion (see Chapter 7). Biological denitrification, discovered in 1886, has been studied extensively for over 50 years, but a global estimate of how much N is denitrified, when and in what locations—including riparia—remains uncertain.

Phosphorus released from decomposing organic matter reenters the riparian system as PO_4^{3-}. Natural concentrations of PO_4^{3-} are generally low in nonvolcanic regions. At neutral or acidic pH, PO_4^{3-} is generally bonded to one or two hydrogen (H^+) atoms in the forms of phosphoric acids, HPO_4^{2-} and $H_2PO_4^-$. This PO_4^{3-} is immediately available for reutilization, and the idealized *cycle* is complete. However, like N, several reactions may occur prior to assimilation. Concentrations remain low in many natural systems due to strong *adsorptive* reactions with iron (FeO_x) and aluminum oxides (AlO_x) and clay minerals. These adsorption reactions, along with *co-precipitation* reactions with Fe III, Ca, and Al, make PO_4^{3-} biologically unavailable to organisms. Due to adsorption, co-precipitation, and biotic assimilation, particulate forms of the element generally dominate total phosphorus concentrations in riparian corridors. This is especially true where phosphate minerals such as apatite occur.

Many biogeochemical characteristics of riparian zones require that the soils be anaerobic or of low oxidation/reduction potential (redox, Eh) at least part of the year. The belowground processes that result in low Eh are composed of a series of biogeochemical reactions, which occur in a defined order based on energy yield (Correll 1997). These reactions transfer electrons from organic matter, released from the riparian vegetation, to various terminal electron acceptors. The availability of terminal electron acceptors determines which level in the series will dominate belowground processes at any one time and place. Some of the more commonly important reac-

Table 5.2 Chemoautotrophic and Terminal Electron-Accepting
Catabolic Processes Occurring within Riparian Zones

Chemoautotrophic Process	Kcal/Equivalent
Methane Oxidation: $CH_4 \rightarrow CO_2$	−9.1
Sulphide Oxidation: $S^{-2} \rightarrow SO_4^{-2}$	−23.8
Iron Oxidation: $Fe^{+2} \rightarrow Fe^{+3}$	−21.0
Nitrification: $NH_4^+ \rightarrow NO_3^-$	−10.3

Terminal Electron-Accepting Process	Kcal/Equivalent
Aerobic Respiration: $CH_2O \rightarrow CO_2$	−29.9
Denitrification: $NO_3^- \rightarrow N_2$	−28.4
Iron Reduction: $Fe^{+3} \rightarrow Fe^{+2}$	−7.2
Sulphate Reduction: $SO_4^{-2} \rightarrow S^{-2}$	−5.9
Methanogenesis: $CO_2 \rightarrow CH_4$	−5.6

Energy yields for each biogeochemical process are given along with general
chemical reactions. Solution pH is assumed to be 7 and CH_2O is used as an
"average" organic substance.
Adapted from Dahm et al 1998.

ions are denitrification, manganate ion reduction, ferric iron reduction, sulfate reduction, and methanogenesis. The sequence of electron acceptors proceeds in order NO_3^-, N_2O, Mn^{4+}, Fe^{3+}, SO_4^{2-}, and CO_2 as a result of thermodynamic considerations, and none can proceed in the presence of molecular oxygen (Stumm and Morgan 1996, Hedin et al. 1998; see Table 5.2). Once processes such as aerobic respiration or sulfide and ammonium ion oxidation have consumed oxygen, denitrification can proceed. Once nitrate is reduced or if none occurs at a site, manganate ion reduction can proceed, and so on. Collectively, these processes are fundamental for understanding why riparian zones are considered as control points for nutrient fluxes—which are discussed further in Chapter 6.

Production Ecology

Vegetative growth and production reflect environmental conditions, disturbance regimes, and successional dynamics, and, when expressed as net primary productivity (NPP), allow comparison across systems with different species assemblages, structure, and disturbance histories (Clark et al. 2001a,b). However, the same arguments that motivate production studies also limit their interpretation. Fundamental differences among systems need to be considered when interpreting variability in production rates. Although the literature on productivity of upland forest is abundant, there is little consistency in measurement methods and in definitions of what constitutes "reportable" forest productivity. As a consequence, there is significant concern about comparing production rates calculated from different measurement methods as well as from studies limited in time and space (Clark et al. 2001a,b).

Growth and Metabolism of Riparian Trees

Clark et al. (2001a) provide a thorough theoretical and practical overview of NPP for forests (see Figure 5.2). NPP represents the amount of organic matter accumulated

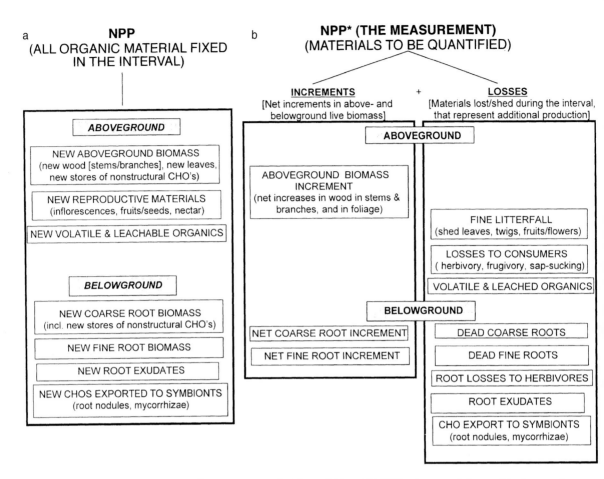

a
NPP
(ALL ORGANIC MATERIAL FIXED
IN THE INTERVAL)

b
NPP* (THE MEASUREMENT)
(MATERIALS TO BE QUANTIFIED)

INCREMENTS
[Net increments in above- and
belowground live biomass]

+

LOSSES
[Materials lost/shed during the interval,
that represent additional production]

ABOVEGROUND

NEW ABOVEGROUND BIOMASS
(new wood [stems/branches], new leaves,
new stores of nonstructural CHO's)

NEW REPRODUCTIVE MATERIALS
(inflorescences, fruits/seeds, nectar)

NEW VOLATILE & LEACHABLE ORGANICS

BELOWGROUND

NEW COARSE ROOT BIOMASS
(incl. new stores of nonstructural CHO's)

NEW FINE ROOT BIOMASS

NEW ROOT EXUDATES

NEW CHOS EXPORTED TO SYMBIONTS
(root nodules, mycorrhizae)

ABOVEGROUND

ABOVEGROUND BIOMASS
INCREMENT
(net increases in wood in stems &
branches, and in foliage)

FINE LITTERFALL
(shed leaves, twigs, fruits/flowers)

LOSSES TO CONSUMERS
(herbivory, frugivory, sap-sucking)

VOLATILE & LEACHED ORGANICS

BELOWGROUND

NET COARSE ROOT INCREMENT

NET FINE ROOT INCREMENT

DEAD COARSE ROOTS

DEAD FINE ROOTS

ROOT LOSSES TO HERBIVORES

ROOT EXUDATES

CHO EXPORT TO SYMBIONTS
(root nodules, mycorrhizae)

Figure 5.2 The components of (a) forest NPP and (b) NPP*—the sum of all materials that together represent (1) the amount of new organic matter that is retained by live plants at the end of the interval, and (2) the amount of organic matter that was both produced and lost by the plants during the same interval. CHO = carbohydrates. From Clark et al. (2001a).

over a fixed period of time and after respiration is subtracted from total carbon fixation. In practice, few of the NPP components shown in Figure 5.2 are ever measured. Most frequently, measurements are restricted to litterfall and aboveground biomass growth, with their sum considered equivalent to aboveground NPP. Belowground components are usually not measured or are estimated as some theoretical proportion of aboveground values. As a result, NPP estimates for the same forest type are quite variable depending on the vegetative components included in NPP, as well as the estimation methods used.

Stem production is the most commonly measured component of NPP. Stem production usually refers to annual increases in stem biomass or stem volume. Biomass is related to stem diameter and height using either regression equations taken from the literature or from regression equations developed by destructive sampling (Waring and Schlesinger 1985; Clark et al. 2001a,b; Balian and Naiman 2005). Stem diameter increments are surveyed by sequential measurements of bole diameter (including

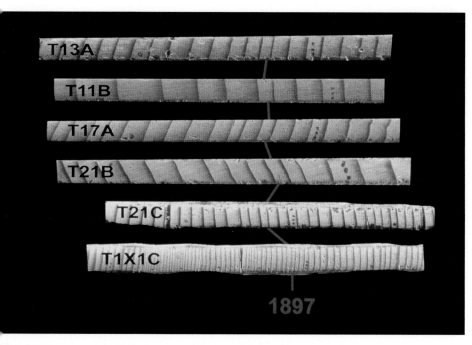

Figure 5.3 Increment cores from riparian trees as well as from dead wood in the river channel can be used to estimate age, growth rate, and time since death. (Photo: T. Hyatt.)

bark) at the beginning and end of the growing season, or by ring width measurements on increment cores or stem cross-sections (see Figure 5.3). Time series data from permanent plots (sequential measurements) require long periods of repeated measurements; consequently, most studies addressing long-term production rely on ring width measurements, which can be obtained from tree cores. The most commonly used time periods are 2, 5, or 10 years for current stem production (Clark et al. 2001a,b), with the measurement methods depending on the research theme. Both approaches are assumed to give comparable estimates of current stem production, but caution is warranted depending on the species and the environmental setting (Balian 2001).

Timing of Growth and Rates of Net Primary Production

Growth and production by riparian vegetation depend on large-scale environmental factors such as climate, elevation, and geomorphology, and on small-scale factors controlling disturbance regimes and resource availability. These factors are particularly variable in larger floodplains where interactions between river and riparian systems are less constrained than in steep upstream valleys and along small headwater streams (see Sidebar 5.2). Most information on forest productivity comes from uplands, which is somewhat surprising considering the ecological value of riparian forests and their importance in the regulatory aspects of resource management. Available observations on riparian forest productivity focus on specific species, primarily hardwoods and shrubs (e.g., Campbell and Franklin 1979), trees on large floodplains (e.g., Piedade et al. 2001, Harner and Stanford 2003), or commercially valuable species like Douglas-fir (*Pseudotsuga heterophylla*; Means et al. 1996).

Riparian zones in large floodplains and in steep narrow valleys do not have the same species composition and successional dynamics; therefore, their productivity patterns also differ. Most production studies have examined riparian plants along small to midsized channels (<10 m wide), which limit the lateral extent of riparian vegetation. Species composition in upstream riparian areas tends to be dominated by upland species, while less attention is focused on large downstream floodplains that are more diverse in physical conditions and species composition, and have a greater abundance of hardwoods (e.g., Pabst and Spies 1999, Nierenberg and Hibbs 2000, Piedade et al. 2001).

Basal area growth (BAI; m^2/ha/year) and bole wood production (P; Mg/ha/year) for dominant riparian tree species differ by successional stage and physical template. This is because each successional stage and template has a characteristic disturbance regime, soil composition, subsurface flow pattern, and nutrient composition and availability (Piedade et al. 2001, Balian and Naiman 2005). In the várzea floodplains of the Amazon basin, wood production increases as succession proceeds. Two-year-old monospecific stands of *Salix humboltiana* add 1.5 Mg/ha/yr, while 12-year-old *Cecropia latiloba* produce 8.1 Mg/ha/yr and 80-year-old *Pseudobombax munguba* produce 7.2 Mg/ha/yr. The rates are similar to those measured for riparian trees in temperate rain forests. The Queets River, Washington, has three major templates: active floodplain, paleo-floodplain, and low terrace, all with contrasting tree assemblages—e.g., either pure stands or mixtures of red alder (*Alnus rubra*), Sitka spruce (*Picea sitchensis*), big leaf maple (*Acer macrophyllum*), western hemlock (*Tsuga heterophylla*), black cottonwood (*Populus trichocarpa*), vine maple (*Acer circinatum*), and willow (*Salix* spp.). Total BAI is not significantly different among the physical templates, but the mean values tend to decrease from the active floodplain (x = 2.84 m^2/ha/year), to the paleo-floodplain (1.92 m^2/ha/year), to the low terrace (1.42 m^2/ha/year). In contrast, stem production (P) is significantly higher on the oldest sites—the low terrace (x = 10.3 Mg/ha/year)—than on the younger paleo-floodplain (6.5 Mg/ha/year) or on the active floodplain (3.2 Mg/ha/year).

Depending on the physical template, individual species make different contributions to BAI and P. For the Amazonian floodplain forest, it is well documented that young stands are dominated by trees with life history characteristics typical of pioneering organisms; that is, short life expectancy, fast growth, and low wood density (*Salix humboldiana*, *Cecropia* spp., *Pseudobombax munguba*), whereas the oldest stands are dominated by trees with opposing life history features (*Piranhea trifoliate*, *Manilkara* sp., *Tabebuia barbata*). Unfortunately, it is not yet possible to partition the total plant growth and production among species for the várzea floodplains as it is for the temperate rainforest. There, willow has a larger contribution (70 percent) than red alder to BAI on the active floodplain because of its dominance in stem density, but P is more equally distributed between willow and red alder (56 and 49 percent, respectively; see Table 5.3). The largest proportion of BAI and P is represented by red alder (58 percent of total BAI and 80 percent of total P) on the paleo-floodplain and by Sitka spruce (56 percent of BAI and 70 percent of P) or the low terrace.

Sidebar 5.2 Sexual Reproduction, Establishment and Growth of Cottonwoods

All five species of North American cottonwoods are dioecious, with individual trees being either male or female (Braatne et al. 1996). The flowers are clustered in catkins, which tend to be borne in the tree crown. Male catkins are typically smaller and reddish-purple, whereas female catkins are significantly larger and greenish in appearance. Males typically initiate flowering before females, and both sexes flower ~1 to 2 weeks prior to leaf initiation in the early spring, thus coinciding with springtime peaks in river flows. The seeds are small (~0.3 to 0.6 mg), and females produce a large and dependable crop annually, often exceeding 25 million seeds per tree. Even though seed viability is short (1 to 2 weeks), germination is rapid, and late spring and early summer commonly find seedlings in large numbers along point bars as well as on other moist substrates within alluvial floodplains. Seedling densities can reach >4000/m². The growth, development, and mortality of seedlings are intimately tied to the relative abundance of light and soil moisture and will vary in relation to the microtopographic position within the floodplain (see Figure SB5.2). Cottonwood seeds have little endosperm; therefore, full sunlight is critical for photosynthesis by cotyledons and juvenile leaves to sustain growth and development. Likewise, the soil must remain moist throughout the early stages of plant establishment (1 to 2 weeks), whereas subsequent water table patterns will influence survivorship and mortality throughout the remainder of the first growing season. If the rate of water level decline exceeds the rate of root growth (4 to 6 mm/d for most species but up to 12 mm/d in *P. trichocarpa*), significant mortality will occur. By the end of the growing season, juvenile plants can tap groundwater at depths of 75 to >150 cm. Vulnerability to drought persists until sapling roots reach alluvial water tables at depths of >2 m. Interactions between fluvial processes and cottonwood reproduction and establishment are complex, with apparently minor changes in the fluvial regime having important long-term consequences for cottonwood demography.

1 Seed Dispersal and Germination

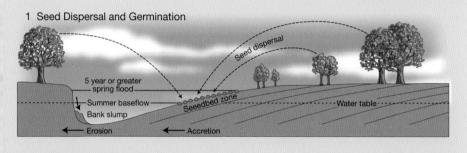

2 Established Saplings (5-10 years later)
with new seedling cohort

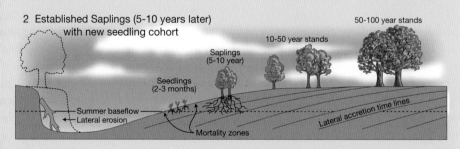

Figure SB5.2 Patterns of seed dispersal, germination, and establishment in relation to microtopographic position and river stage in an alluvial valley. From Braatne et al. (1996).

Growth and stem production tend to have contrasting temporal patterns in the temperate rainforest. Along the Queets River, total BAI decreases with plot age, whereas production increases over time (see Figure 5.4; Balian and Naiman 2005). The highest BAI (up to 3.2 m²/ha/year) is observed for willow in the active floodplain plots. However, the highest production rate is for Sitka spruce (up to 13.9 Mg/ha/year) on the oldest surfaces (100 to 230 years old) of the low terrace. Temporal changes in pro-

Table 5.3 Stem Growth (BAI; m^2/ha/year) and Productivity (P; Mg/ha/year) Estimates for Each Riparian Tree Species on the Three Biophysical Templates of the Queets River, Washington

Tree Species	BAI (m^2/ha/year)	P (Mg/ha/year)
Active Floodplain		
Willow	2.01 ± 0.40[a]	1.80 ± 0.50[a]
Red alder	0.82 ± 0.35[b]	1.36 ± 0.54[a]
Sitka spruce	<0.01[c]	<0.01[b]
Black cottonwood	<0.01[c]	<0.01[b]
Total Average	**2.84 ± 0.50[x]**	**3.17 ± 0.63[x]**
Paleo-Floodplain		
Willow	0.80 ± 0.44[a]	1.28 ± 0.71[a]
Red alder	1.13 ± 0.13[a]	5.18 ± 1.04[b]
Sitka spruce	<0.01[b]	<0.01[c]
Vine maple	<0.01[b]	<0.01[b]
Total Average	**1.92 ± 0.37[x]**	**6.46 ± 0.49[y]**
Low Terrace		
Willow	<0.01[a]	<0.01[a]
Red alder	0.19 ± 0.07b[c]	1.38 ± 0.51[b]
Sitka spruce	0.80 ± 0.11[d]	7.27 ± 1.5[c]
Vine maple	0.02 ± 0.01[c]	0.09 ± 0.07[d]
Big leaf maple	0.35 ± 0.12[b]	0.72 ± 0.26[b]
Black cottonwood	0.04 ± 0.02[c]	0.46 ± 0.20[b]
Western hemlock	0.02 ± 0.01[c]	0.42 ± 0.20[b]
Total Average	**1.42 ± 0.13[x]**	**10.34 ± 1.63[z]**

[a,b,c,d]: For each variable, values followed by different letters within a physical template are significantly different ($p < 0.05$). [x,y,z]: Values followed by different letters are significantly different among physical templates ($p < 0.05$).
Derived from Balian and Naiman (2005).

duction closely follow changes in vegetation composition. The initial successional stage is a fast-growing community of small willow attaining maximal productivity at ~10 years, but total production remains low. This willow stage is replaced in the next 10 years by red alder that reach maximal productivity around 40 years, with both high P (up to 7.9 Mg/ha/year) and high BAI (up to 1.4 m^2/ha/year). Finally, Sitka spruce out-compete red alder with a fast and continuous increase from 40 to 140 years, reaching a plateau after 150 years (up to 13.9 Mg/ha/year). The other species show little change in production with increasing tree age.

Other than on the Queets River, wood production rates >10 Mg/ha/year have been found only in Florida or Louisiana swamps colonized by large bald cypress (*Taxidium distichuum*) or in mixed hardwood forests in North Carolina. Brinson (1990) reports an average P of 6.9 Mg/ha/year for freshwater riparian forests. The estimate of stem production in the Queets floodplain equals or exceeds this average in all stands older than 30 years. However, *total* net primary production on the Amazonian várzea floodplains, which includes the annual production of fine litter and dead wood in addition to plant growth, ranges from 23.8 to 33.6 Mg/ha/yr in 40- to 80-year-old stands. Comparable data are not available for riparia at other locales.

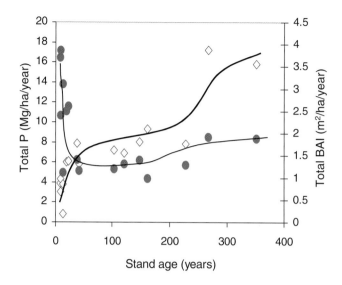

Figure 5.4 Total BAI (●) and total P (◇) changes over time for all tree species in individual plots of differing ages; trend lines are added (modified from Balian and Naiman 2005).

Many studies have demonstrated the importance of water availability to riparian tree growth (e.g., Stromberg and Patten 1995). However, not all riparian trees, nor all populations of a single species, show a response in growth rate to water availability. The strength of the relationship varies with climate, hydrogeology, and tree density, among other factors (Piedade et al. 2001, Stromberg 2001). In mesic regions, where surface flow is relatively high and constant from year to year, growth is less frequently linked to flow but can increase in response to overbank flooding. In dry regions, growth may be augmented in gaining river reaches where springs or regional aquifers supply a fairly constant source of water from year to year (Harner and Stanford 2003). In wetter areas, several other factors, such as nutrient availability and sediment characteristics, can influence growth by riparian trees. In the Amazonian várzea floodplains, a strong relationship exists between length of flood inundation, community assemblages, and plant production (Piedade et al. 2001). Herbaceous plants and trees in the floodplain—and associated production rates—are distributed according to present and past flood inundations, with the mechanisms for exclusion mainly related to differences in species-specific tolerances to low soil oxygen concentrations that may last for 200 to 300 days per year when the floodplain is saturated.

Comparing productivity estimates between riparian and upland forests is made difficult by variability in measurement methods (litterfall collections, diameter measurements, increment cores) and by different tree components included in the estimates (leaves, branches, stems). Zavitkovski and Stevens (1972) report stem production of 10.6 Mg/ha/year for red alder, 55 percent higher than estimates of stem production at 28- to 40-year-old red alder sites along the Queets River (5.8 to 7.9 Mg/ha/year). Stem production for older mature forests (between 100 and 630 years old) ranges from 3.5 to 11.4 Mg/ha/year, with an average of 7.0 Mg/ha/year (Balian and Naiman 2005). The highest stem productivity (11.4 Mg/ha/year) was estimated as half of the total NPP (22.8 Mg/ha/year) for a 121-year-old western hemlock stand (which included stem, branch, and foliage production; Grier and Logan 1977). The estimate of total stem P (10.3 ± 1.6 Mg/ha/year) on the Queets' low terrace plots is in the same range as upland estimates, while the contribution of riparian Sitka spruce to total production in plots older than 150 years is similar to the maximum production of western hemlock in upland forests.

Temporal models of growth and productivity are instructive for understanding the dynamic nature of riparian forests. In early successional stages, stem production is driven by the high number of stems but remains low due to their small size. In contrast, basal area growth is maximal immediately after the initial colonization stage because of high stem density. For example, willow dominates the colonization stage in density and basal area on the Queets River but is sometimes associated with red alder (see Figure 5.5). A few years later, red alder becomes dominant. Stem production mainly results from an increasing red alder biomass—until the canopy closes. A third stage is initiated by the replacement of red alder by Sitka spruce and the rapid increase in biomass of young Sitka spruce. Total stem production by the woody community shows a continuous increase as trees age and vegetative

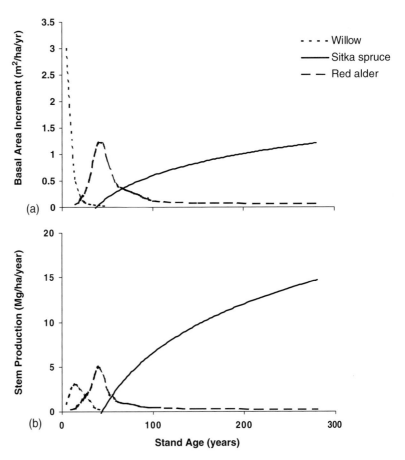

Figure 5.5 Conceptual models of (a) basal area growth (BAI) and (b) stem production (P) over time for three key riparian tree species in the Queets River floodplain. The three tree species are willow (- - -), red alder (– – –) and Sitka spruce (——). Data from Balian and Naiman (2005).

communities switch from pioneer hardwood to late seral coniferous species. Stem basal area increment exhibits an opposing pattern as it slows over time. Sitka spruce maintains high productivity in the floodplain even after 300 years and grows larger, delaying the replacement phase by western hemlock that is better expressed in uplands.

Litterfall

The amount of plant litter produced annually from riparian vegetation is influenced by forest type, successional stage, stream size, site productivity, and latitude. Many of these factors, of course, are interrelated. There have been several attempts to identify factors influencing litterfall patterns across broad spatial and ecological scales (e.g., Benfield 1997, Xiong and Nilsson 1997). Litter may be especially important in riparian corridors because of the volumes produced and its relatively fast decomposition, which have direct effects on nutrient retention and cycling and on benthic faunal assemblages (Wallace et al. 1999).

Generally, the highest litterfall rates are associated with tropical equatorial forests (10.7 Mg/ha/yr) followed by tropical montane (6.3 Mg/ha/yr) and warm temperate forests (5.0 Mg/ha/yr; see Table 5.4). The lowest litterfall rates are associated with alpine, arctic, and cool temperate forests (1.0 to 3.4 Mg/ha/yr). Young successional forests have rates that are only half of those from mature forests. For example, at sites dominated by young alder and willow (<50 years old) on the Queets River, annual litter fall averaged 2.8 Mg/ha/yr as compared to 6.0 to 7.6 Mg/ha/yr on ~100- to ~350-year-old floodplain and terrace sites (T. O'Keefe and R. J. Naiman, unpublished data). However, some caution is warranted since litter traps normally are not designed to capture materials growing close to the ground, as is common in early seral stages. Currently, there is little published information on the magnitude of litter production by riparian vegetation other than woody species, especially trees. It is expected that litter production in early seral stages may be higher than expected since these vegetative communities are growing actively and generally are poorly characterized for litter production.

There is an inverse relationship between riparian litter production and latitude, similar to the pattern found in upland forests (Benfield 1997, Xiong and Nilsson 1997). Litter production rates are greater near the equator than near the poles. Most riparian forests, deciduous as well as coniferous, exceed global

Table 5.4 Summary of litterfall in Major World Biomes

Biome	Average Litterfall (Mg/ha/yr)	Standard Deviation	Max	Min	n
Alpine and arctic	0.98	0.33	1.50	0.60	6
Cool temperate	3.41	1.17	6.90	1.00	159
Warm temperate	4.98	1.49	8.10	2.40	38
Equatorial	10.72	3.00	15.30	5.50	18
Montane tropical	6.27	1.76	11.00	3.10	22

Sample size (n) is number of studies.
From data in Bray and Gorham 1964, Vitousek 1984, Vitousek and Stanford 1986, Veneklaas 1991.

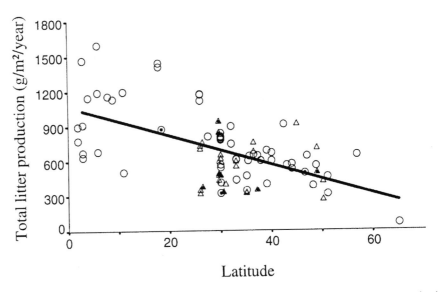

Figure 5.6 The relationship between total litter production and latitude is described by the equation Y = 1075.1 − 12.6X (r²adj = 0.25, P < 0.01, n = 87), where Y is the total litter production and X is latitude (in degrees N or S). Data for direct input of CPOM (course particular organic matter) to headwater (first and second order) streams are shown. Riparian broadleaf and needleleaf combined (solid line); riparian broadleaf (open circle); riparian needleleaf (open triangle); stillwater habitat on floodplain (dotted circle and dotted triangle). (Adapted from Xiong et al. 1997.)

averages for upland litter production, although there are exceptions (see Figure 5.6).

Litter production appears to be related to the total net production of aboveground biomass with litter averaging 47 percent of the annual NPP in riparian forests (Brinson 1990, Piedade et al. 2001), which is slightly higher than upland forests. On average, leaf litter production accounts for 72 percent and 80 percent of the total litter production in deciduous and coniferous riparian forests, respectively. It remains to be seen, however, whether litter production is consistently related to total NPP, both in riparia and in uplands.

We have already discussed the importance of water level fluctuations and soil conditions to the riparian forest and how they influence nutrient cycling and plant productivity. Not surprisingly, these same factors affect litterfall (Brinson 1990, Xiong and Nilsson 1997, Piedade et al. 2001). Riparian forests flooded infrequently produce nearly twice as much litter (~5.5 Mg/ha) as frequently or nearly permanently flooded forests (Conner et al. 1981, Cuffney 1988). Further, leaf litter production is highest in the transition area between the riverbank and the uplands in the intermediate flooding zone. The pattern is not quite so clear in the Amazonian várzea floodplains, where litter production varies from 7.8 Mg/ha/yr in a 12-year-old *Cecropia latiloba* stand to 13.6 Mg/ha/yr in a ~60-year-old mixed forest (Piedade et al. 2001). Although the different studies have site-specific variations in local topography, community composition, and successional stage, collectively they support the concept that the magnitude and timing of flooding, soil water content, and soil fertility largely control litter production as well as NPP in the riparian zone.

Mortality Rates

Mortality in woody riparian vegetation results from disease, senescence, herbivory, and catastrophic disturbances such as high wind or undercutting of stream banks. Most knowledge of riparian tree mortality has come from the perspective of large woody debris recruitment to streams rather than from the perspective of understanding the production dynamics of riparian forests. Therefore, little tangible information is available on causes of tree mortality by species, age, or site. Acquisition of this information over the next decade will be fundamental to understanding basic processes governing the dynamics of riparian forests. Even though little is known of the relative rates contributed by each cause, it is known that mortality rates are species specific owing to susceptibility by their age, life cycle, chemical makeup, and spatial position along the stream channel. Studies of plant demography show exponential decreases in willow, alder, and other woody plant densities once shading and availabilities of water or nutrients become issues (Oliver and Larson 1996). Those same species, depending on local environmental conditions, are also subject to wind and flood, contributing substantial amounts of large woody debris to channels via blow-down, bank erosion, and hillslope failure (Bilby and Bisson 1998, Naiman et al. 2002a).

These points are illustrated by examining mortality rates as a function of tree species and successional stage. In the Pacific Northwest of the United States, red alder is a common early successional riparian species. It has a relatively short life span, beginning to senesce ~60 years after stand establishment, although significant mortality does occur earlier (Grette 1985). Shade-tolerant conifers, such as western red cedar or western hemlock in the alder understory, often occupy the sites in great numbers but suffer high mortality rates due to stem suppression. The rate of mortality among the conifer species is dependent on the density of seedlings established beneath the alder overstory. There is evidence that suppression during successional development continues to play a strong role in mortality in stands up to 300 years old in the western Cascade Mountains of Washington (Rot et al. 2000). In stands greater than 300 years old, mortality of larger trees from disease and wind-throw become the dominant drivers of mortality.

Relatively rare, severe disturbances, including windstorms, fires, or floods, contribute to widespread mortality (Harmon et al. 1986). Avalanches, landslides, and debris torrents remove woody vegetation from hillslopes and headwater riparian zones, depositing the large woody debris and associated sediment in downstream channels. Severe windstorms can result in the episodic death of a large number of trees with the wind strength and direction relative to the channel, soil moisture, tree species, and a number of other interrelated factors being important. Likewise, fire varies as a function of aspect, elevation, and other factors, recurring in many western forests in the United States at intervals ranging from decades to over 1,000 years (Benda et al. 1998). In some riparia, fire frequency and severity has been less than in adjacent uplands, but in other areas, fires burn riparia with comparable frequency and severity. Fortunately, many riparian plants possess adaptations to fluvial-based disturbances that also facilitate survival and reestablishment following fires, thus contributing to the rapid recovery of many streamside communities (Dwire and Kaufman 2003). Finally, very severe floods also kill large numbers of trees through

Figure 5.7 In alluvial valleys, the undermining of riparian trees by the meandering river is a major cause of tree mortality, and an important source of sediment and large woody debris to river channels. (Photo: J.J. Latterell; South Fork Hoh River, Washington.)

accelerated bank-cutting and transport of wood stored on the floodplain into the channel (Keller and Swanson 1979). This mechanism of mortality tends to be particularly prevalent in large, laterally migrating channels with extensive floodplains (see Figure 5.7).

The relative importance of mortality mechanisms varies by stream size and watershed characteristics. In gentle terrain, where landslides or avalanches are rare, trees growing along the channel network generally die from disease, senescence, or herbivory (e.g., Johnston and Naiman 1990, Johnston et al. 1993). In unstable landscapes, however, landslides and resultant debris torrents are significant factors. From 10 percent to over 50 percent of the wood in fish-bearing streams of the Oregon Coast

Range in the United States is generated by landslides that initiate debris torrents in low-order stream channels. These torrents can extend far downstream, actively influencing geomorphology and successional properties of floodplain reaches (Naiman et al. 2002a). The relative importance of mortality factors also varies with valley form. Wind-throw is the primary mechanism of mortality along tightly constrained channels with erosion-resistant banks (Swanson et al. 1982). In unconstrained stream reaches, undercutting of trees by bank erosion becomes a more important cause of mortality (see Figure 5.7). In unconstrained channel reaches in southeast Alaska, undercutting of the stream bank produces over 40 percent of the wood in channels (Murphy and Koski 1989).

Root Production

Root production, in terms of biomass increase per unit area, seldom has been measured and remains the most poorly understood portion of riparian plants. It is suspected that root production rates are quite high considering biomasses encountered and the needs to access water, buttress against flood flow and high wind, acquire

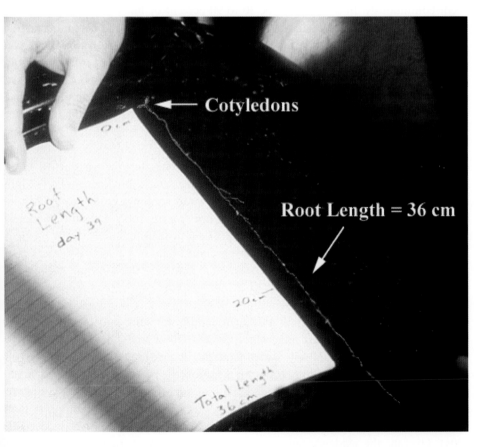

Figure 5.8 The roots of cottonwood seedlings grow quickly after seed germination, especially as water availability declines. The photo depicts a 39-day-old seedling with a 40-cm root. Note the tiny cotyledons at the top of the photo. (Photo: R. Stettler.)

resources from often nutrient-poor environments, and successfully compete against other plants in high-density communities. Roots of some riparian plants are known to grow rapidly in length after germination. Cottonwood seedling put nearly all their energy into increasing root length immediately after germination to follow a rapidly declining water level in the spring (Braatne et al. 1996, Pregitzer and Friend 1996). It is not uncommon for seedlings with minuscule leaves to have roots nearly 40 cm long only 30 days after germination (see Figure 5.8). In one study of cottonwood hybrids, there was an estimated 296,000 km of roots per hectare to a depth of 3.2 m, a value far above most plant communities except for grasslands (Heilman et al. 1996). European grey alder (*Alnus incana*) shows a pattern similar to cottonwood with fast root growth in well-drained sandy loam soil when the water level is declining 1 to 3 cm/day (Hughes et al. 1997). In contrast, the slowest growth rates for alder roots are when soils are saturated or when sandy or gravelly soils experience water level declines exceeding 1 cm/day. In cottonwood, one encounters thick root mats in surface soils for accessing nutrients and deeper roots for accessing moisture and for providing stability but, if water availability is altered, it will alter root growth along with most other parameters associated with growth and mortality (reviewed by Rood et al. 2003).

The formation of new roots, death of old roots, and decomposition of dead roots occur continuously. In upland forests, 30 to 90 percent of fine roots (<1 mm diameter) are lost and replaced annually in relation to the end and the beginning of the growing season. The turnover of fine roots can represent several thousand kg/ha (Fogel 1985) and over half the roots growing and dying are in the upper 20 cm of soil. In cottonwood, preliminary data suggest the average lifetime of roots to be between 15 and 30 weeks (Pregitzer and Friend 1996). Likewise, the turnover time of fine roots and associated mycorrhizae in Sitka spruce forests can be as rapid as 16 weeks, with slightly larger roots (1 to 5 mm) having a turnover time of nearly 4 years. High rates of fine root production generally occur prior to canopy expansion. This can be seen in cottonwood-willow communities along the Yellowstone River in Montana (Boggs and Weaver 1994). The average mass of fine roots (<1 mm diameter) increases from 2.2 Mg/ha for cottonwood seedlings to 8.7 to 9.3 Mg/ha in more mature stands. The average mass for coarse roots (>1 mm diameter) increased from 3.4 Mg/ha to 94.3 Mg/ha, peaking in mature stands. Root biomass in cottonwood often exceeds that of aboveground biomass. Collectively, roots may provide two to five times the amount of organic matter to the soil than is provided by leaf and branch litter.

Roots vary among species in their ability to grow in different soil media. Hemlock roots often grow in rotting wood, alder can endure wet soils, and tupelo and bald cypress can tolerate extended anaerobic periods (Oliver and Larson 1996). Many species cannot live continuously under anaerobic conditions and, as discussed in Chapter 4, develop adventitious roots for aeration. In riparian zones experiencing long periods of inundation, the flooded roots of some species partially die only to regrow downward as the water table recedes. Other species endure a temporary high water table for different lengths of time—days to many months (Piedade et al. 2001). Overall, the evidence suggests that the dynamics of riparian roots are of enormous ecological importance, but the empirical data needed to provide more than just glimpses into a few specific situations are elusive.

Decomposition Dynamics

Decomposition processes are fundamental to understanding nutrient fluxes in riparian zones, relationships between nutrient cycling and plant productivity, and formation of riparian soils. Decomposition processes are dependent on initial litter quality, exogenous nutrient supply, temperature, oxygen tension (i.e., aerobic versus anaerobic conditions), and the nutrient immobilization potentials of plant materials (Melillo et al. 1984).

Principles of Decomposition Dynamics

Two major groups of factors influence the rate of decay of plant litter: those of litter quality and those of the abiotic environment. Indexes of litter quality include concentrations of nutrients and various types of organic compounds in the plants. Among the major environmental factors influencing decay rates are ambient nutrient availability, temperature, and oxygen tension.

Litter Quality

In general, decay rates decrease with increasing biochemical complexity (e.g., more lignins) and increase with increasing nutrient content, particularly nitrogen (Kaushik and Hynes 1971). There is a close relationship between the initial nitrogen concentration of plant litter and its decay rate. For example, the decomposition of organic matter (OM) from plants grown under different fertilization regimes shows a direct linear relationship between decay rates and the initial nitrogen concentration of OM (Marinucci et al. 1983). Likewise, either lignin alone or lignin in combination with nitrogen is correlated with decay rate of wood (Melillo et al. 1983). When decay rates are plotted as a function of initial lignin concentration for wood with widely varying lignin and nitrogen contents, the relationship is described by an inverse linear function. Phrased more simply, the higher the initial lignin concentration of a material, the lower its decay rate.

While the simple relationships between initial litter quality and decay rate often have site-specific predictive power, they cannot answer questions about cause-and-effect relations between initial litter quality factors and decay rate because the underlying mechanisms are often unclear. Consider, for example, that even though lignin is resistant to microbial attack, lignin-rich materials also are often rich in compounds that inhibit decomposition (e.g., certain polyphenols) and are poor in compounds preferred by microbes (e.g., soluble carbohydrates and proteins). Given the "covariance" complications, the observed inverse relationship of decay rate to lignin concentration probably represents a complex casual relationship involving concentrations of soluble organic fractions, lignocellulose, total and available nutrients, and inhibitory compounds (Heal et al. 1981).

The relationship between enzyme activity and substrate disappearance assists in establishing the cause-and-effect relations between initial litter quality and decay rate. Breakdown of many macromolecular constituents of plant litter is extracellular and is catalyzed by enzymes. If the hydrolysis of some litter component, such as cellulose, limits the rate of decomposition, then the amount of enzyme affecting that hydrolysis should be proportional to the overall rate of decomposition (as measured by CO_2 evolution) and to the rate of substrate (in this case, cellulose) disappearance. There is growing evidence documenting the relationship between extracellular enzyme activity and CO_2 evolution (Linkins and Neal 1982, Suberkropp 1998). What has often been implied but not demonstrated is the relationship between enzyme activity and substrate disappearance.

Exogenous Nutrient Supply

Supplementary nitrogen added to natural materials often stimulates the rate at which they decompose. A number of laboratory experiments on decomposition of plant materials in microcosms (e.g., Kaushik and Hynes 1971) have shown increased rates of plant litter decomposition with the addition of inorganic nitrogen. However, when nitrogen is available in excess of microbial demand, the addition of inorganic phosphorus can also stimulate decay rate. The decay rate of red oak leaves (*Querus robur*) and wood from various species of trees is increased by phosphorus additions (Elwood et al. 1981, Melillo et al. 1984). For example, a high phosphorus treatment (0.025 mg P/L) dramatically increases the decay rate of alder wood relative to a low phosphorus treatment (0.005 mg P/L) when the N supply is inherently high, as is the supply of other nutrients. In contrast, decay rates in spruce wood are not much affected by altering phosphorus supply. The high initial lignin content of spruce (~25 percent) causes a slow decomposition rate not easily accelerated by additions of phosphorus, whereas the decay rate of alder, a material with a low initial lignin content (~13 percent), is easily accelerated by phosphorus additions. In other words, spruce wood decay may be more carbon limited than alder wood decay. Thus the potential to alter decay rate by changing exogenous nutrient supply may be limited by the carbon chemistry of the OM. This suggests that the further a material is decayed, the less likely there will be an increase in decay rate in response to an increase in exogenous nutrient supply.

Temperature

Temperature has long been recognized as one of the most important environmental conditions in determining how rapidly OM decays. Decay is more rapid at high temperatures than at low ones. Even at low temperatures, however, appreciable decomposition may occur. The temperature profile of cellulase activity associated with the decomposition of leaf litter indicates that at 0°C cellulase retains ~30 percent of the activity displayed at 25°C (Sinsabaugh et al. 1981). Cellulase activity at low temperatures, combined with a lack of catabolic enzyme activity within microbes at low temperatures, may explain the increase in labile DOC in water during the winter months without a corresponding increase in CO_2 production.

Oxygen Tension

Traditional wisdom suggests that plant materials decay more slowly under anaerobic conditions than under aerobic conditions (Alexander 1977). However, this may not always be true. Rice straw under laboratory conditions will initially (i.e., the first 2 weeks) decay faster under aerobic conditions but faster under anaerobic conditions over the next six months (Acharya 1935). Nevertheless, molecular oxygen is required for degradation of some organic compounds such as aromatic hydrocarbons and, apparently, lignin (Fenchel and Blackburn 1979). Thus, for litter rich in these compounds, or in well-decomposed litter, a lack of oxygen will reduce or halt the decay rate, and the litter will persist in anaerobic soils for long periods. Additionally, the organisms and biochemical processes involved in decay will be quite different in aerobic and anaerobic situations.

Nutrient Dynamics During Decay

At least three distinct phases are recognized in the nitrogen dynamics of decomposing plant litter. First, there is a brief period of leaching during which 10 to 25 percent of the litter's initial nitrogen mass can be lost. The leaching phase is followed by a period of net immobilization during which the absolute amount of nitrogen in the decomposing litter increases. Finally, there is a net release or net mineralization of nitrogen from the decomposing litter (Melillo et al. 1984). This three-phase model also can be used to describe phosphorus dynamics during OM decay.

The absolute increases of nitrogen and phosphorus observed during the immobilization phase of decay require the addition of these nutrients to the decomposing OM from the surrounding environment. Decomposing plant litter in the immobilization phase therefore reduces the amount of nitrogen and phosphorus available to support riparian plant growth or for flux to the adjacent stream, and therefore may be an important process influencing water quality and site productivity. Agriculturists and foresters have long recognized the potential for competition for nitrogen between decomposers and primary producers. For example, Viljoen and Fred (1924) observed long ago that the incorporation of woody materials into agricultural soils reduced crop (oat) growth rate. They concluded, and rightly so, that nitrogen immobilization by microbes was the mechanism responsible for the reduction.

Some early work was done toward quantification of relationships between initial litter quality and maximum nitrogen immobilization potential, but no quantifiable relationships were found (Richards and Norman 1931). Later Aber and Melillo (1981) and Melillo et al. (1983) were able to show that maximum nitrogen immobilization potential could be predicted from initial litter quality. The implication is that changing plant assemblages—and thereby litter chemistry—will have strong system-scale consequences.

Factors Controlling Immobilization of Nitrogen

Initial Litter Quality

Most observed patterns are consistent with general principles of microbial ecology, which can be stated as follows: (1) substrates broken down more easily result in a higher degree of nitrogen immobilization than substrates difficult to decompose, such as lignified tissues (Fenchel and Blackburn 1979); and (2) nitrogen-poor detritus immobilizes more nitrogen than nitrogen-rich detritus (Hutchinson and Richards 1921).

Exogenous Nutrient Supply

The slope of the inverse linear function describing the relationship between percent organic matter (OM) remaining and percent N in the remaining tissue is sensitive to

151

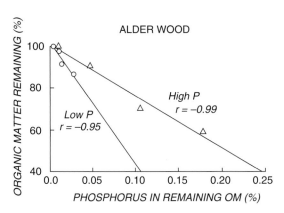

Figure 5.9 Alder wood shavings decomposing in a laboratory setting show different responses to high (O, 50x) and low (Δ, 1x) exogenous supplies of phosphorus. Alder decay is P-limited because of inherently high N concentrations in the wood tissue; therefore, the addition of P accelerates decay rate as well as the amount of P immobilized from the surrounding environment. From Melillo et al. 1984.

large changes in exogenous N supply. For a given litter material under high levels of available N, the slopes will be shallow, whereas under low levels of available N, the slope will be steep. Likewise, it is also known that a similar response occurs with P. When the percent OM remaining is plotted against the percent P in the remaining alder wood tissue decomposing under a low P supply, it exhibits a steep slope over twice that of alder wood decomposing under a high P supply (see Figure 5.9). While laboratory studies show that large changes in exogenous N supply can alter the N immobilization potential of a given litter material, initial litter quality exerts a more important control on N immobilization in field studies. For example, decaying birch wood of uniform initial quality exhibits the same immobilization potential in a variety of environmental situations (Melillo et al. 1984).

Anaerobic Decay

Under anaerobic conditions, the relationship between percent OM remaining and percent N in the remaining material conforms to the inverse linear model.

Temperature

The slope of the inverse linear function describing the relationship between percent OM remaining and percent N in the remaining tissue does not appear to be sensitive to temperature changes alone. The slopes of the inverse linear regression model are the same regardless of temperature for elm leaves (*Ulmus* spp.) decaying at 10°C and 22°C (Kaushik and Hynes 1971). This suggests that the maximum net nitrogen immobilization potential in the leaves is not influenced by temperature. However, the time it takes to reach maximum N immobilization (N_{max}) is longer at 10°C than at 22°C.

Mechanisms of Nitrogen Immobilization

Two mechanisms of nitrogen immobilization are possible: accumulation of nitrogen in microbial biomass and accumulation of nitrogen in by-products of microbial activity.

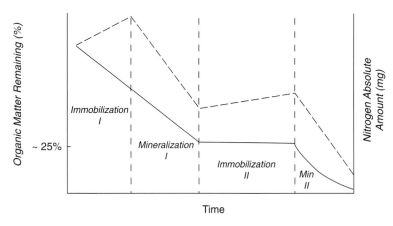

Figure 5.10 A generalized model showing the phases of nitrogen immobilization and mineralization as organic litter decays. The dashed line represents changes in the absolute amount of nitrogen during each of the four phases. The solid line represents the percentage of original litter material remaining as a function of the nitrogen concentration in the remaining litter material. From Melillo et al. (1984).

Nitrogen Accumulation in Microbial Biomass

At one time, the absolute increase in detrital nitrogen was thought to be due to the accumulation of protein nitrogen associated with microbes colonizing the detritus (Tenny and Waksman 1929). However, it is now known that nitrogen in microbial biomass is a relatively small fraction of the total amount of nitrogen in detritus. During beech leaf litter decomposition, the nitrogen in microbial biomass (largely protein) accounts for 1 to 4 percent of the total nitrogen (Iversen 1973). However, as much as 30 percent of the nitrogen content of aged (3 weeks) vascular plant detritus can be in the form of nonprotein compounds. It is thought that these compounds might be decay-resistant, nitrogen-containing, organic compounds such as chitin or humic acids (Odum et al. 1979). Collectively, these as well as other studies on the decay of marine plants suggest that the accumulation of N in microbial biomass is not the primary mechanism of N immobilization (Suberkropp 1998).

Nitrogen Accumulation in By-Products of Microbial Activity

Nitrogen accumulation in decaying plant debris follows a two-phase mechanism (Melillo et al. 1984). First, after initial leaching of soluble materials from fresh detritus, microbial exoenzymes depolymerize the detritus substrate producing reactive carbohydrates, phenolics, small peptides, and amino acids. During this period, microbial growth is rapid, with microbes converting substrate nitrogen and exogenous nitrogen into microbial biomass and exuded products of microbial activity. As microbes die, these nitrogen compounds become part of the detrital nitrogen pool that undergoes enzymatic depolymerization.

The reactive carbohydrates and phenolics condense with components of the detrital nitrogen pool to produce nitrogen-enriched compounds. This is a period of net nitrogen accretion and is referred to as "Immobilization I" in Figure 5.10. During this period, there is a continual loss of litter mass. The magnitude of net nitrogen accumulation appears to be related to the amount of water-soluble compounds in the detritus, particularly the water-soluble polyphenols. During "Immobilization I," there is often an absolute increase in the mass of the "lignin" fraction of the decaying litter.

Net nitrogen release occurs during the next phase of decay, referred to as "Mineralization I" in Figure 5.10. Cellulose degradation dominates this phase, and cellulase activity is high. Although there is a net loss of nitrogen from the detritus-microbe complex, the depolymerization–recondensation cycle continues. Exoenzymes and other nitrogenous materials exuded by microbes become incorporated in the condensates. As decay proceeds through "Mineralization I," the residual nitrogen-containing compounds become progressively more recalcitrant.

When between 10 and 25 percent of the original detritus biomass remains, further mass loss stops for a prolonged period. During this time, there are absolute increases in both the total nitrogen and the "lignin" fraction of the detritus, and lignin-shielded cellulose becomes the dominant carbon source for the microbes. Lignin must be degraded to make the cellulose available. The products of the lignin degradation (i.e., phenolics) recondense with both the reactive carbohydrates from cellulose breakdown and with nitrogenous compounds, including exoenzymes. This recondensation produces additional nitrogen-rich recalcitrant compounds. This phase is referred to as "Immobilization II" in Figure 5.10. Again, the depolymerization-recondensation cycle continues, and the nitrogen-rich condensates become progressively more recalcitrant, and humic compounds are formed. Finally, degradation of some of these recalcitrant humic compounds is achieved as energy-rich simple carbon compounds are added to the detritus. Mass loss is evident and nitrogen is released during the "Mineralization II" phase in Figure 5.10.

While this model describes nitrogen dynamics of litter decomposing in forest systems, it also may be appropriate for describing nitrogen dynamics in freshwater and marine systems. The model is consistent with many of the classical concepts of the biogeochemistry of the formation of humic substances and nitrogen accumulation in humus of soils (Stevenson 1982) and is entirely appropriate for understanding mechanisms contributing to nutrient retention in riparian zones.

Decomposition of Riparian Litter

The large amount of plant litter returned to riparian soils annually is processed via the principles outlined above. The seasonal flooding of riparian zones, or at least seasonal fluctuations in soil moisture, provides a hydrologic connection that influences a variety of functional responses. Moisture is necessary for decomposition to proceed efficiently, with insufficient moisture leading to decreased rates of decomposition, declining nutrient immobilization, inhibition of plant growth, and occasionally increased fire frequency.

Three factors suggest that moisture availability plays a key role in regulating both the amount and timing of leaf decomposition, especially in arid riparian environments. Along the Rio Grande River in New Mexico, over-winter biomass loss varies among years, with the greatest decomposition occurring during wet years; rapid biomass loss coincides with flooding; and biomass loss during flooding is greatest following a dry winter (Ellis et al. 1999). Leaves at experimental flood sites lose 30 to 50 percent of their initial mass during flooding, whereas leaves at reference sites (no flooding) lose only 1 to 5 percent during the same period. Early summer flooding thus increases the breakdown of leaves and consequently the release of essential nutrients at a time critical for riparian plant growth, which may increase primary production. At sites remaining dry throughout the early growing season, and especially during years with low winter precipitation, nutrients remain bound in dead leaf tissue. Several other studies (e.g., Bell et al. 1978, Peterson and Rolfe 1982, Shure et al. 1986) corroborate the effects of moisture availability on riparian leaf decomposition and the subsequent water-driven movements of nitrate through the soils (Bechtold et al. 2003).

Collectively, the principles of decomposition outlined above are fundamental to understanding—and managing—the nutrient characteristics and the plant productivity of riparian zones. Decomposition processes provide the biochemical basis for nutrient retention and cycling that are so important for efficient system functioning. There is still much to be learned, however. For example, the role of the soil fauna in the decomposition process remains poorly understood for riparian zones, especially in areas repeatedly subjected to strong variations in environmental conditions, whether they are floods, drought, erosion, or deposition.

Information Fluxes

There are three primary types of information by which organisms communicate: genetic, chemical (other than genetic), and vocal (i.e., transmission of sound). Genetic information concerns the transmission of propagules, whether in the form of protected DNA, bacteriophages, seeds, eggs, sperm, or plant fragments up, down, or laterally through the riparian corridor. Chemical information concerns communication between organisms using biologically produced compounds, such as pheromones. Vocal information concerns the propagation of sound waves at all frequencies for communication between individuals or among groups. Sound waves, of course, are subject to absorption, amplification, modification, and filtration by a variety of physical structures in the riparian system. All forms of information are vital to the efficient and continued functioning of riparian communities, and all are affected by the orientation, density, and dimension of riparian structures.

In theory, riparia can act either as conduits, filters, or breaks in the flow of information and organisms (Forman and Godron 1986). Normally one thinks of riparian zones as being conduits for the easy movement of plants and animals, which is not always true. The process is much more complicated, and the quantification of the processes is a major scientific challenge. Riparian zones, depending on their physical structure, appear to modify fluxes of organisms as well as their means of communication, whether they are visual, auditory, or chemical. Sounds are effectively absorbed by vegetation and soils, it is difficult to see (and be seen) in dense vegetation, and controlling the directional movement of chemicals is problematic in the interior of forests.

Nearly all of what is known about these important attributes is either theoretical or has been investigated for uplands.

There are seven important attributes of riparian corridors affecting fluxes of information: connectivity, resistance, narrows, porosity, influence fields, spatial orientation, and distance (see Table 5.5). Collectively, these are important landscape–scale parameters that have been best examined by landscape ecologists and geographers. Riparian zones, as networks in the larger landscape, unfortunately have not received the same attention as transportation or human communication networks, but many of the same principles may be operative (Forman and Godron 1986). Landscape ecologists view networks as a collection of nodes and corridors, each with specific but interconnected functions. Nodes, such as riparian floodplains at tributary junctions, are sources, sinks, or modifiers of information. Corridors, on the other hand, are routes connecting the nodes, and a network is a series of corridors connecting the nodes. These are important principles in understanding the roles played by riparia in regulating information fluxes throughout the larger catchment. Much of the theory and mathematics already has been developed for geographic questions but has yet to be applied to riparian systems, a highly fertile topic for future investigations.

Microclimate

Microclimate is the suite of climatic conditions measured in localized areas near the earth's surface. These environmental variables—which include temperature, light, wind speed, and moisture—provide meaningful indicators for habitat selection and other ecological activities. In seminal studies, Shirley (1929, 1945) emphasized microclimate as a determinant of ecological patterns in both plant and animal communi-

Table 5.5 Seven Important Attributes of Riparian Corridors that Affect Fluxes of Information

Attribute	Description and Characteristics
Connectivity	The relationship between movement and number and kinds of barriers. Riparian zones with high connectivity have few or no barriers for the information being transmitted.
Resistance	Results from two boundary characteristics (crossing frequency and boundary discreteness) and two landscape characteristics (hospitableness and length).
Narrows	Unusual width or height constrictions in the riparian network. Can either accelerate or slow information flux depending on means of transport.
Porosity	Homogeneity and heterogeneity of internal patch types forming the riparian matrix affect fluxes. Percolation of information depends on internal structure relative to the kind of information being transferred.
Influence fields	Area under the influence of a particular portion (or patch) of the riparian zone. Intensity of influence varies with distance (e.g., genetic material or disease).
Spatial orientation	Riparian shape and position relative to the direction and means of movement have profound influences on information flux.
Time-distance	Basically the route taken by a particular type of information; not all information flux is linear, and longer routes can be more effective than shorter routes.

Adapted from Forman and Godron 1986.

ies and a driver of such processes as growth and mortality of organisms. The importance of microclimate in influencing ecological processes such as plant regeneration and growth, soil respiration, nutrient cycling, and wildlife habitat selection has become an essential component of current ecological research (Chen et al. 1999).

Each component of the microclimatic environment exhibits unique spatial and temporal responses to changes in riparian structural elements. Further, relationships between microclimate and biological processes are complex and often nonlinear. These features are easy to visualize when one considers that temperature, solar radiation, and humidity affect plant growth by influencing physiological processes such as photosynthesis, respiration, seed germination, mortality, and enzyme activity. Therefore, it follows that ecosystem processes such as decomposition, nutrient cycling, succession, and productivity are partially dependent on microclimatic variables too. Many animals are also adapted to specific microclimatic conditions. Wind speed, air temperature, humidity, and solar radiation can influence migration and dispersal of flying insects. Soil microbe activity is affected by soil temperature and moisture. In addition, most fish have specific thermal ranges in which they are able to survive and reproduce, suggesting that changes in variables that affect stream temperature, such as solar radiation, cause changes in habitat suitability.

It is well known that riparian forests exert strong controls on the microclimate of streams (Meehan 1991, Naiman 1992, Maridet et al. 1998). The amount and quality of solar radiation reaching streams are determined by forest vegetation height, forest canopy density, stream channel width, and channel orientation in relation to the sun's path. Light is important to streams because of its influence on primary production by aquatic plants and on the behavior of organisms. The amount of solar radiation also affects water temperatures. Stream water temperatures are highly correlated with riparian soil temperatures but change as water flows downstream. The water temperature regime is an important factor for the vitality of streams because of its controlling influence on the metabolism, phenology, and activity of stream organisms.

Surprisingly, there is only one comprehensive study of the microclimate within riparian forests, even though riparian zones have particularly complex microclimates related to changes in vegetative structure and processes from the water's edge to the uplands. The limited data, however, suggest strong differences between upland and riparian microclimates.

The data show microclimatic gradients in the riparian forest for air, soil, and surface temperatures, and relative humidity, but not for short-wave solar radiation or wind velocity (Brosofske et al. 1997). Microclimatic data collected from the active channel to the uplands of a small (<2 m wide) mountain stream in Washington suggest that the stream environment (active channel) differs strongly from the riparian and upland forest environments. The stream significantly influenced air temperatures in the riparian forest for up to 60 m on either side of the channel in summer, either through direct cooling or by supplying water for evaporative cooling by vegetation. Over a 12-day period, the average air temperatures for the riparian zone, adjacent forest, and an upland clearcut were 19°C, 21°C, and 25°C, respectively. Gradual changes in soil temperature and relative humidity from the stream to the upland were also detected. In contrast, there was no difference for wind velocities.

Microclimate plays a critical role in plant regeneration, growth, and distribution in upland situations, and it is strongly expected that the same is true for riparian zones

(Brosofske et al. 1997). Upland researchers have found strong relationships between the distribution of some vegetation associations and various microclimatic factors such as soil moisture, air temperature, and humidity. Others have proposed that internal water budgets, which result from rates of both absorption of water from the soil and transpiration (controlled by local temperature, light, and wind), are probably more important to plant growth than absorption alone. Others have postulated that the high productivity and diversity of plants near streams might be partially accounted for by the ideal combination of microclimate and moisture conditions. Overall, the ecological consequences of many of these fundamentally important microclimatic gradients and processes remain to be discovered and quantified for riparian environments.

Conclusions

The specific functions described in this chapter are the building blocks underpinning and shaping the broader integrative characteristics one associates with riparian systems, such as nutrient buffering and food web dynamics—characteristics that are essential for long-term ecological vitality. It is the integrative characteristics that the public and the management agencies normally associate with riparian goods and services (Chapter 6), and it is these characteristics that are so easily altered by natural and human-generated environmental changes (Chapter 7). Inevitably, one needs to comprehend the building blocks in order to understand the complexities and subtleties involved in the more integrative processes, to fathom how easily environmental change can alter them, to understand how difficult they are to effectively restore once altered, and to grasp the enormous challenge faced by resource managers.

Biophysical Connectivity and Riparian Functions

6

Overview

- The biophysical complexity and the distinctive ecological functions of riparia depend on high levels of spatial and temporal connectivity between riparian habitat patches as well as between riparia and adjacent riverine and upland systems. This chapter examines the vectors connecting riparian components and the role of system connectivity in underpinning riparian functions.

- Riparia are composed of resource patches linked by flows of water, nutrients, and energy. It is the linkages that shape the attributes of individual patches in riparian corridors. Biogeochemical cycles and food webs integrate the interactions of organisms with resource patches and with hydrologic flow paths within riparia as well as between adjoining systems.

- The hydraulic connectivity of riparian zones with streams and uplands, coupled with enhanced internal biogeochemical processing and plant uptake, make riparian zones effective buffers against high levels of dissolved nutrients from uplands and streams, while geomorphology and plant structure make them effective at trapping sediments.

- The capacity of riparian zones to retain dissolved and particulate nutrients like N, P, Ca, and Mg is controlled by hydrologic characteristics (e.g., water table depth, water residence time and degree of contact between soil and groundwater) and by biotic processes (e.g., plant uptake and denitrification). The relative influence of these factors varies depending on soil characteristics, nutrient input rates, and vegetative assemblages.

- Stream water exchanges continually with the interstitial waters of bed and bank sediments in a mosaic of surface–subsurface exchange patches that are both sources and sinks of nutrients. The degree of influence on stream nutrients depends on the volume and extent of hyporheic zones, the fraction of river flow diverted through them, the residence time of individual flow paths, and the specific biological and biogeochemical processes operating along flow paths.

- Riparian food webs are connected to those of adjoining aquatic and terrestrial environments, with the integrity of both depending on the flow of energy and nutrients between them. Energy and nutrients derived from riparian leaf litter forms the base of food webs in many streams, and higher consumers, primarily fish, may depend on riparian insects seasonally. Riparian food webs, although to a lesser degree, also may be based on aquatic primary production, and aquatic insects and fish are important energy sources for a number of riparian-based consumers.

- Animals strongly influence the structure and function of riparia through their movements, feeding strategies, and modifications of the physical environment. Large animals alter hydrologic and geomorphic characteristics of riparia, causing fundamental changes in energy and nutrient cycles and in plant assemblages and structure. The net effect of animal influences is generally an increase in heterogeneity among riparian patches and even the creation of new patch types.

- Effective riparian management requires maintenance of connectivity within riparia, and between riparian patches and adjoining systems.

Purpose

The biophysical complexity and distinctive ecological functions of riparia described in proceeding chapters depend on a high level of connectivity between riparia and adjacent riverine and upland systems. Riparian communities are dependent, to differing degrees, on inputs of water, sediments, energy, and nutrients from both uplands and adjoining riverine systems. Diversity in the structure of riparian vegetation as well as associated processes are strongly influenced by the grazing activities of herbivores that move between aquatic, riparian, and upland ecosystems. Riparian predators feeding on aquatic prey are another important vector for transporting energy and nutrients to riparian communities. At the same time, many stream communities depend on energy and nutrient inputs from riparian vegetation and organisms. Energy and nutrient subsidies from riparian zones enhance aquatic productivity, and a portion of this productivity is returned to riparia through hydrologic flow paths and biotic feedbacks. Riparia are almost certainly influenced by high internal connectivity as well, resulting in exchanges of resources between riparian habitats. However, to date there has been little research into these latter connections and their roles in maintaining riparian biodiversity and ecological functions.

This chapter examines the role of connectivity with adjoining ecosystems in determining riparian ecosystem structure and function. It also considers the role of riparian-aquatic exchanges in maintaining the water quality of streams and rivers and in subsidizing aquatic food webs. The three-dimensional textural heterogeneity of riparian soils, vegetation, fauna, moisture, organic matter, and nutrients form a mosaic of discrete resource patches and unique biogeochemical environments distributed along river corridors from headwaters to sea. The enhanced connectivity of this riparian mosaic with adjoining ecosystems derives from its position in valley bottoms and its elongated form, which leads to a high ratio of edge length to core area. Gravity transports water and other materials from uplands to and through riparia, and animals concentrate in riparian zones seeking access to water and food.

The importance of connectivity in riparia is clearly reflected in biogeochemical cycles and food webs. Biogeochemical cycles are shaped by complex flow paths of surface water and groundwater (Amoros and Bornette 2002). As we learned in Chapter 2, water flow reaches all corners of riparia, either as surface flow or groundwater flow, or both. In stark contrast to the unidirectional flows of water in upland systems, flows in river corridors are spatially complex and variable with time (Bencala 1993). In addition to impacting moisture regimes, these hydrologic flows supply or retain organic matter, oxygen, nitrogen, phosphorus, and micronutrients. Flows within food webs are driven by the exchange of resources between ecosystems and the wide variety of feeding strategies utilized by consumers. Because riparia are characterized by steep resource gradients at their aquatic and upland boundaries, energy exchanges across ecotones are often fundamental in shaping riparian structure and function.

Additionally, the movement and activities of animals are important components of both food webs and biogeochemical cycles. In one sense, animals simply act as another vector connecting patches within riparia and adjacent ecosystems. Animals are drawn through riparia from surrounding uplands to access surface waters, to forage on riparian plants, or to hunt fauna. Aquatic and semiaquatic animals also move from the riverine environment into riparia either as part of their life cycle or as carcasses deposited by predators such as bears and birds. In these movements, animals are important in the redistribution of organic matter and nutrients. Animals are also important agents of disturbance and system restructuring. Dam building by beaver (*Castor canadensis* and *C. fiber*) and herbivory by moose (*Alces alces*) are illustrative examples of these influences (Naiman et al. 1988, Pastor et al. 1988, Kielland et al. 1997).

We begin by reviewing the main concepts of landscape ecology, which are used as a synthesizing framework for understanding riparia from a catchment perspective. We then illustrate the importance of connectivity in riparian systems by integrating examples of biogeochemical cycles and riparian food webs. Special attention is devoted to the role of animals, both as transport vectors and as agents of disturbance and system restructuring. We conclude our treatment of basic riparian ecology by recognizing that effective management requires maintaining connectivity between riparian zones and adjoining systems.

Patch Dynamics and a Landscape Perspective of Catchments

Riparia extend across the landscape in corridors tracing valley bottoms and stream channels. *Corridor*, along with *patch* and *matrix*, are precise terms in the field of landscape ecology describing three types of spatial elements that compose the landscape mosaic (Forman and Gordon 1981). The matrix is the dominant background ecosystem or land use, such as forest or savanna in undisturbed settings, or agriculture or silviculture in managed landscapes. Patches and corridors are relatively homogeneous, smaller areas that differ from the surrounding matrix and differ from each other in the manner their names suggest. Patches are polygonal features that may be rectangular, elongated, or quite convoluted in shape (Forman 1995). Corridors are strips that may act as conduits for movement across landscapes or barriers to perpendicu lar flows. These landscape elements are linked by flows of water, nutrients, and energy and it is the nature of these linkages that shapes the expression of spatial element and determines the dynamics of the mosaic as a whole. These basic concepts of land scape ecology also may be translated to finer scales in a hierarchical fashion, in tha new patches and corridors may be defined within smaller scale elements.

As preceding chapters have shown, riparia are heterogeneous systems. When viewec at the scale of an individual stream or river corridor, riparia form mosaics composec of finer-scale patches and may be among the most complex, or patchy, systems on Earth. These patches may be defined from a riverine perspective as a hierarchy o habitat types (see Figure 2.4; Frissell et al. 1986) or as physical process domain (Montgomery 1999). From a purely riparian perspective, however, patches are mos easily and effectively distinguished by cover type (vegetation type or barren), whicl is easily perceived and effectively integrates subsurface heterogeneity (see Figure 6.1) The subdivision of riparia into patches implies some level of order governed by organ izing forces and sustained by dynamic interactions between patches (Pickett and Whit 1985). Individual riparian patches may be characterized as relatively homogeneou areas conforming to one type of biotic assemblage or, in a related fashion, to funda mental geomorphic units like channel and point bars, levees, or backswamps (se Chapter 2). Patches also may be characterized from the perspective of individua organisms or even the ecological problem to be addressed (Pringle et al. 1988). Th

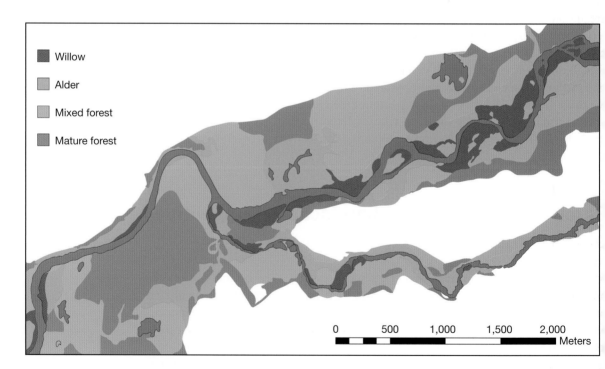

Figure 6.1 The migrating river channel is responsible for the establishment of a heterogeneous assemblage of vegetative patches within the riparian zone. The patches shown represent young willow, mid-age alder, young spruce, and mature spruce-dominated forests along the Queets River, Washington (T. C. O'Keefe and R. J. Naiman, unpublished data).

organism-oriented characterization of patches requires that patch boundaries be perceivable by the organisms under study and thus directly relevant to their sustaining ecological interactions (Farina and Belgrano 2004).

Patch dynamics, and the view of streams and river systems as mosaics of interacting patches, has proven useful in understanding community organization and fundamental ecological processes in fluvial corridors. For example, the field of community ecology views competition as the fundamental force in structuring communities. The simplest view is one of competitive exclusion from unvarying trophic niches, but more complex views acknowledge nonequilibrium or stochastic forces as well. Stream corridors are inherently nonequilibrium systems because of regular, and sometimes extreme, variability in energy conditions driven by changing hydrologic regimes. Habitats, or patches, are continually redistributed in space and time because of disturbances linked to seasonal variability (including complete drying in some systems), individual storms and floods, or a host of other processes. Recovery of community structure in such dynamic systems may be founder-controlled, where the first organisms to recolonize habitats determine future structure (Yodzis 1986), or they may be relict-controlled, where more resistant species not removed by disturbances become dominant and control future structure (Townsend 1989). For plants, recolonization immediately following high flow disturbances will be by propagules of local flora surviving the disturbance in patches that provided refuge, such as in sediments, backwaters, floodplains, or other low-energy environments. Recolonization may also come from similar upstream or downstream patches that were less impacted by the disturbance event. Disturbance-driven, nonequilibrium processes such as these promote greater biodiversity because species undergo periods of recruitment at different times, setting up interactive successional sequences (Townsend 1989). It is the flow of water, propagules, nutrients, energy, and animals among riparian patches and among riparia and adjoining ecosystems that provides the cohesion for riparian structure and function. We examined hydrologic flows in Chapter 2. In the remainder of this chapter we will examine flows of nutrients, energy, and the actions of animals as fundamental agents shaping the structure and function of riparian ecosystems.

Nutrient Flows

In the wider science and resource management communities, riparian zones are best known for their ability to intercept and retain excess particulate and dissolved nutrients flowing from surrounding uplands, thereby protecting the quality of surface waters. Recognition of this service stimulated a great deal of research into riparian nutrient dynamics, which continues today. Detailed and prolonged research, not surprisingly, has produced a far more complex picture of the nature of nutrient fluxes within and through riparian systems. Riparia continue to be viewed as areas regulating nutrient flows in catchments, but instead of being viewed as simple nutrient filters against upland contamination, they are now known to be far more hydrologically interconnected. Additionally, recent research has shown that riparia regulate longitudinal as well as lateral flows across the landscape and act as both nutrient sources and sinks (see Sidebar 6.1).

Riparian Zones as Buffers Against Nutrient Pollution from Upland Runoff

The role of riparian zones as nutrient filters for water flowing from agricultural catchments to streams is well documented (Karr and Schlosser 1978, Lowrance et al. 1984, Peterjohn and Correll 1984) and has led to major government-supported programs in North America and Europe to conserve and restore riparian buffers. Riparian vegetation was recognized as an effective buffer against upland sediment flows by the mid 1960s (Haupt and Kidd 1965, Wilson 1967), and by the mid 1970s to early 1980s there was growing recognition of their effectiveness in retaining nutrients from agricultural and feedlot runoff waters (Gilliam et al. 1974, Gambrell et al. 1975, Young et al. 1980, Lowrance et al. 1984). The filtering capability of riparian zones is due to their position in the landscape and to their geomorphic, hydrologic, and biotic processes. Because riparian zones lie at the interface of terrestrial and aquatic ecosystems, virtually all surface and shallow subsurface runoff in catchments must pass through them in order to reach the stream channel.

Geomorphically, riparian zones often have lower gradients than surrounding uplands and this, plus the baffling effect of riparian vegetation, dissipates the kinetic energy of surface flows during storms and causes entrained sediments to be deposited. This is an especially effective mechanism for retaining phosphorus and other chemical pollutants associated with sediment particles. The capacity of riparian zones to retain dissolved nutrients like N, P, Ca, and Mg is controlled by hydrologic characteristics (e.g., water table depth, water residence time, and degree of contact between soil and groundwater) and by biotic processes (e.g., plant uptake and denitrification, Sabater et al. 2003). The relative influence of these factors depends on soil characteristics, nutrient input rates, and vegetation type.

Elucidating the volume and pathway of water moving through riparia is fundamental for understanding nutrient removal and retention. If local groundwater passes beneath the rooting zone or if extensive piping occurs, the roots of riparian vegetation cannot access the nutrients (see Figure 6.2). Riparian vegetation along small streams normally has good access to water moving downslope from the uplands, and many studies have shown that in those situations, riparian vegetation buffers streams from diffuse, nonpoint-sources of N and P (Haycock et al. 1997). However, hydrologic pathways are often more complex along larger rivers, especially where deposits of coarse alluvium are extensive. For example, mixing of nitrate-rich groundwater with nitrate-poor river water can account for most of the change in nitrate concentration along groundwater flow paths, even though there are important biological controls occurring (Pinay et al. 1998). Further, because there are good flows of subsurface river water, the deeper zones remain oxygenated and little or no denitrification occurs.

Riparian zones are known to be especially effective at protecting surface waters from nitrate runoff. Early studies documented total nitrogen retention by riparian zones ranging from 67 to 89 percent of total upslope inputs (Jacobs and Gilliam 1983, Lowrance et al. 1984, Peterjohn and Correll 1984). Later, a plethora of studies showed nitrate concentrations decreasing in groundwater as it moved laterally through riparian zones (Cirmo and McDonnell 1997, Correll 1997). In a large-scale study of N budgets in 16 catchments covering 250,000 km^2 of the northeastern United States

(a) Shallow confining water

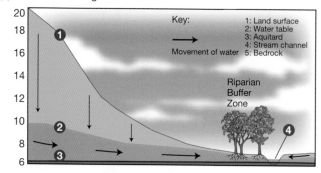

(b) Bedrock overlaid with coarse gravel

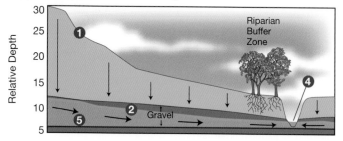

(c) Deeply incised channel and a sand aquifer

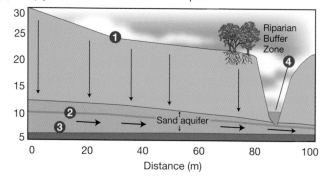

Figure 6.2 Depending on the local stream geomorphology, riparian vegetation will have varying degrees of interaction with subsurface water moving toward the channel. In (a) and (b), the plant roots have opportunities to take advantage of the subsurface water and the nutrients contained therein, whereas in (c), the lateral flow of water and nutrients takes place far below the rooting zone. (From Correll 1997.)

iver exports of N accounted for only 25 percent of the load from the catchments Boyer et al. 2002, van Breemen et al. 2002), implying that most of the N was retained n some manner. Although the effectiveness of riparian zones for retaining N is vident, specific mechanisms responsible for the widely documented retention of itrate have proven difficult to isolate.

Candidate mechanisms preventing nitrogen flowing through riparia from entering treams include denitrification, assimilation, and retention by the vegetation, as well s uptake by biota followed by storage in (nonliving) organic matter (see Figure 6.3). Denitrification is invoked most often as the primary mechanism of nitrate retention nd the most important given that N is removed permanently from the system and

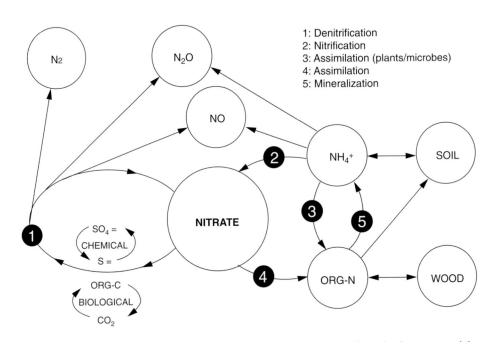

Figure 6.3 There are several mechanisms by which nitrate can be removed from riparian zones, and the key pathways are summarized here (from Correll 1997).

returned to the atmosphere as N_2 and N_2O. However, the extreme spatial and temporal variability of denitrification rates in riparian zones makes it difficult to determine accurate fluxes and to extrapolate these to wider areas (Pinay and Naiman 1991, McClain et al. 2003). By difference, denitrification was found to account for 51 percent of N losses from catchments of the northeastern United States (van Breemen et al. 2002). In individual site studies, denitrification rates of 1 to 295 kg N/ha/yr have been recorded; the fastest rates occur at the riparian–stream boundary, where nitrate-enriched water enters organic surface soil (Naiman and Décamps 1997). It had long been suspected that, in addition to nitrate, saturated conditions, available carbon (Hedin et al. 1998, Hill et al. 2000), topography, and soil grain size (i.e., water-logging potential) are pertinent environmental factors for denitrification. Indeed, there is a significant relationship between denitrification and soil texture, with the highest rates occurring in soils with a high silt and clay content. Below a threshold of 65 percent silt and clay, it is difficult to detect any significant denitrification; above that threshold, denitrification increases linearly (Pinay et al. 2000).

Field studies of nitrate mass balance show that nitrate is effectively taken up (assimilated) by plants from groundwater moving in soil and subsoil at depths of several meters at all times in temperate climates (Correll 1997). However, studies find that most uptake is in the top few centimeters of riparian soil. Conditions in the deeper subsoil—lower temperatures, low pH, and patchy low-to-high concentrations of dissolved organic matter—lead some to conclude that vegetative assimilation is the primary mechanism for nitrate removal. Plant uptake results in a short-term accumulation of nutrients in nonwoody biomass and a long-term accumulation in woody biomass. Riparian forests are especially important sites for biotic accumulations of nutrients because transpiration may be quite high, increasing the mass flow of nutri ent solutes toward root systems, and because morphological and physiological adaptations of many flood-tolerant species facilitate nutrient uptake under low-oxygen conditions. In some species, such as water tupelo (*Nyssa aquatica*), saturated conditions enhance nutrient uptake and growth (Hosner et al. 1965). Examining assimilation by the vegetation, and recycling to the forest floor as litter, may be important in discovering the primary mechanism for nutrient retention. The rate of assimilation in Peterjohn and Correll's (1984) classic study of riparia as nutrient filters along the Rhode River, Maryland, was 77 kg N/ha/yr, while litterfall plus throughfall was 66 kg N/ha/yr. Thus, assimilation by the forest could be the primary mechanism of nitrate removal from groundwater during the growing season.

While vegetative uptake may be important in explaining nitrate removal, nitrate also is removed in winter when vegetation is thought to be dormant. Further, budgets of nitrogen accumulation in woody biomass usually account for only a minor amount (<30 percent) of annual nitrate removal. Collectively, it appears that assimilation and storage of nitrogen in woody biomass are important mechanisms but not the primary mechanisms, especially in winter. If this is true, what are the other possible mechanisms or pathways? If plant assimilation cannot explain nitrate removal during the winter, another pathway may be nitrate removal from the subsoil by chemical rather than biological denitrification (Correll 1997). Strong reducing agents such as iron sulfides may react with nitrate to produce N_2 and sulfate (SO_4^{-2}). Certainly, some riparian zones reduce sulfate to sulfide (S^{-2}), and at times, sulfide is oxidized back to sulfate and released. If chemical denitrification does take place it would still depend on adequate supplies of organic matter in the subsoil to maintain a low redox potential (Eh). If this is the primary mechanism at such sites, the system's soils will eventually become nitrogen saturated—and there is some evidence that this may occur—with significant consequences for stream water quality. However, if nitrogen saturation occurs, phosphorus may become the limiting factor for tree growth, making vegetation an effective phosphorus sink (Aber et al. 1989).

Nitrogen removal efficiency by riparian buffers is mainly positive across a wide range of climate conditions, nitrate inputs, soil characteristics, and vegetation types. In Europe, nitrate removal rates generally range between 5 and 30 percent m^{-1} (Sabater et al. 2003). Average nitrogen removal rates are similar for herbaceous (4.4 percent m^{-1}) and forested (4.2 percent m^{-1}) sites. Nitrogen removal efficiencies are not affected by climate, at least for the range encountered in 14 sites shared between seven European countries in the Sabater et al. study, or by season. When nitrate inputs are low, there is a wider range of removal efficiencies. However, when nitrate inputs are greater than $5 \, mg \, N \, L^{-1}$ or there is a strong hydraulic gradient, nitrate removal efficiency declines rapidly. The data strongly suggest that the nitrate removal efficiency is closely tied to the nitrate load and the hydraulic gradient.

The current consensus that most riparian zones effectively remove nitrate from subsurface water is based largely on studies where groundwater inputs are restricted to shallow subsurface flow paths by impermeable layers that force maximum interaction with riparian soils and vegetation (Hill 1996). Limited research suggests that there is less effective nitrate retention in riparian areas connected to large upland aquifers where riparian hydrology is often dominated by surface transport or groundwater transport below rooting zones. Additionally, it has been discovered that considerable nitrogen can be released from riparia to streams as dissolved organic nitrogen

(McDowell et al. 1992, Hedin et al. 1995). The importance of this pathway, and the forms and subsequent fates of the organic nitrogen, are unknown. Collectively, the evidence assembled so far raises doubts about all riparian zones being efficient filters for nutrients Sidebar 6.1—some may be leaky while others release nutrients after they have been transformed to organic matter. This topic is highly important but in need of additional investigation.

Riparian Zones as Buffers Against High In-Stream Nutrient Levels

In much the same way that riparian zones intercept and retain nutrients from upland, overland, and subsurface runoff, they may also intercept and retain nutrients flowing in adjacent streams. The same mechanisms of N and P removal are at work, including biotic uptake, physical adsorption, and microbial denitrification, and the effectiveness of riparian buffering depends on the characteristics of riparian soils, vegetation, and connectivity between the river and riparian zone along surface–subsurface exchange pathways. Water in streams exchanges continually with the interstitial waters of bed and bank sediments in what may be viewed as a mosaic of surface–subsurface exchange patches (Malard et al. 2002) (see Chapter 2). Large-scale patterns of downwelling and upwelling depend on characteristics such as valley width and the depth and grain size of alluvial deposits. At the scale of individual stream reaches (~100 m), exchange flows are controlled by bed topography and sinuosity; at the habitat scale (~1 to 10 m), exchanges are controlled by fine-scale variability in hydraulic gradients (brought about by microtopography) and substrate grain size (Bencala 1993, Harvey and Bencala 1993). Sites of downwelling commonly develop at the upstream end of geomorphic features such as point bars, channel bars, and riffles, and the rate of downwelling depends on the pressure gradient into the subsurface and the hydraulic conductivity of bed materials. Hydrologists refer to this subsurface component of river flow as *transient storage* and recognize its importance in slowing the downstream movement of surface waters (Harvey and Wagner 2000, Fernald et al. 2001), while ecologists recognize its importance as habitat for stream biota (Williams 1984, Stanford and Ward 1988). This zone of mixed surface and subsurface waters, or hyporheic zone, may extend from a few centimeters to a few kilometers from the channel margin, and the volume of surface water moving along subsurface flow paths can be equal to or greater than that moving in the channel (Stanford and Ward 1993, Jones and Holmes 1996).

Hyporheic flow paths act as both sources and sinks of nutrients to streams, and the degree of hyporheic influence on river nutrient levels depends on the extent of hyporheic zones, the fraction of river flow diverted through them, the residence time of individual flow paths, and the specific biological and biogeochemical processes operating along flow paths (Findlay 1995, Edwards 1998, Hill et al. 1998). In general, streams with greater hyporheic exchange tend to retain and process nutrients more efficiently (Valett et al. 1996). Because the amount of surface–hyporheic water exchange relative to channel volume decreases exponentially with increasing channel size, the efficiency of nutrient removal linked to hyporheic exchanges tends to decrease with increasing stream size. The most important sites of nutrient retention therefore lie in the low-order stream networks of any river basin. For example, first-order

Sidebar 6.1 Are Riparian Zones Sources or Sinks of Nitrogen to Streams?

A large body of evidence collected over the past 20 years has demonstrated that riparian zones strongly influence the delivery of nitrogen from terrestrial to aquatic ecosystems. Most studies focused on riparian zones as NO_3^- sinks due to denitrification, which is generally recognized as the dominant riparian process altering N fluxes to streams. "Hot spots" for denitrification occur where organic carbon and NO_3^- are available to bacteria in anaerobic environments (Hedin et al. 1998, McClain et al. 2003). Temperature, pH, organic carbon content, sediment grain size, and geomorphology are among the factors affecting denitrification rates *in situ*, highlighting the challenges in predicting denitrification rates over relatively short spatial and temporal scales. Although symptoms of riparian N saturation (i.e., N availability in excess of biotic demand) have been reported, there is currently no evidence that chronic N loading reduces the potential for riparian zones to function as nitrogen sinks. Seasonal or human-induced lowering of the water table such as occurs in urban areas may, however, cause a disconnection between carbon-rich shallow soils and the zone of anoxia, reducing the denitrification potential in riparian soils (Groffman et al. 2002). It is therefore critical for researchers to continue pursuing the linkages between hydrology and riparian N transformations and losses. In particular, quantifying groundwater flow pathways is important to a catchment-scale examination of the role of riparian zones in regulating stream N losses.

In some cases, riparian zones can be significant sources of inorganic N (largely ammonium) to streams. In anaerobic soils, nitrification (microbial conversion of NH_4^+ to NO_3^-) is reduced, resulting in the buildup of NH_4^+ released during organic matter decomposition. Dissimilatory reduction of nitrate to ammonium (DNRA) may also produce NH_4^+ in riparian zones, as has been recently shown to occur in upland soils (Silver et al. 2001) and coastal sediments (Tobias et al. 2001). Riparian soils may therefore serve as a conduit delivering ammonium-rich drainage water to stream banks, where coupled nitrification-denitrification occurs to limit delivery to the stream. This coupled nitrification-denitrification appears to be particularly important in some tropical catchments, where large losses of ammonium-N occur within a meter of the stream channel, and denitrification is the most plausible pathway for net N loss (see Figure SB6.1; McDowell et al. 1992, McClain et al. 1994, Chestnut and McDowell 2000). Nitrification at the aquatic–riparian interface without coupled denitrification could also result in the riparian zone being a source of NO_3^- to streams, but this phenomenon is typically not reported in the literature.

Although research on riparian zones has largely focused on inorganic N, recent evidence suggests that riparian zones may also play an important role in regulating organic N delivery to streams. Several studies in temperate biomes have supported the notion of riparian wetlands as dissolved organic N (DON) sources to streams (Fölster 2000, McHale et al. 2000, Pellerin et al. 2004). Leaching and decomposition of organic matter in the riparian zone provides a ready source of DON, and if it is relatively resistant to microbial decomposition it might be transported through the riparian zone and delivered to the stream channel. Anoxia in riparian soils might reduce DON retention due to reductive dissolution of the Fe-oxides known to bind DOC in aerobic mineral soils.

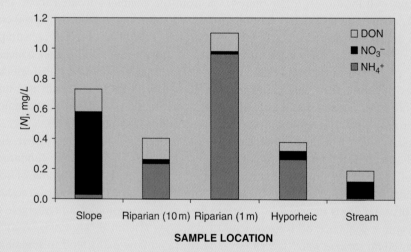

Figure SB6.1 NO_3^--N, NH_4^+-N, and DON-N concentrations in the Rio Icacos basin, Puerto Rico. Upland data (slope) are from McDowell et al. (1992). Other data are from Chestnut and McDowell (2000). Riparian 1m and 10m reflect the distance of the well from the stream channel.

Net production of DON in riparian zones is not universal, as in some sites DON is lost or passes unaltered through the riparian zone. Recent work in upland soils suggests that a novel pathway of NO_3^- loss, conversion of NO_3^- to DON, might play a role in riparian N dynamics. Rapid abiotic transformation of NO_3^- to soluble organic N was recently demonstrated to occur in laboratory studies of carbon-rich soils (Dail et al. 2001).

We suggest the following research foci to better assess the potential significance of riparian zones in the landscape: (1) additional focus on the role of DON and DNRA in riparian N dynamics, as not all nitrate loss may be due to denitrification, the only transformation that permanently removes N from riparian zones; (2) more rigorous and integrative basin-scale analysis of the quantitative significance of riparian N transformation (the scaling of plots to watersheds); (3) examination of N transformations in urban and suburban riparian zones, which are poorly studied and often have severe hydrologic modifications; and (4) document the effectiveness of management strategies (e.g., vegetation management, addition of organic substrates) to enhance N sinks.

W. H. McDowell and B. A. Pellerin, University of New Hampshire, Durham

N retention rates in streams of the Mississippi River basin average 0.45 day^{-1}, while the rate of removal in the mainstem river is two orders of magnitude less, or 0.005 day^{-1} (Alexander et al. 2000).

One of the first studies to describe and quantify the complexity of hyporheic processes on in-stream nitrogen levels was conducted along a 327-m reach of Little Lost Man Creek in northern California (Triska et al. 1989). This creek has a steep gradient (0.066 m/m) and a highly permeable bed composed of sediments ranging from sand to boulders. Consequently, it experiences a large volume of surface–subsurface exchange. Injections of the conservative tracer chloride revealed the volume of hyporheic water to exceed that of surface water, and stream water was found to compose 47 percent of water in wells 10 m from the stream channel. NO_3^- was co-injected with chloride, and retention (or release) was calculated as the difference between chloride and NO_3^- after correcting for dilution. Nitrate retention along the reach was found to be variable depending on the position of the sampling well relative to the stream channel and preferred subsurface flow paths (see Figure 6.4). Well locations 2 to 4 m from the channel appeared to be sites of NO_3^- production, whereas well locations greater than 4.3 m from the channel appeared to be sites of NO_3^- consumption. Nitrate retention also varied according to biogeochemical conditions, reflecting the redox status of groundwater. Locations of NO_3^- consumption tended to be those with smaller percentages of stream water and higher concentrations of NH_4^+, suggesting reducing conditions that would favor denitrification processes. Subsequent measurements and laboratory experiments confirmed anoxic conditions at these sites and estimated that denitrification accounted for as much as 30 percent of NO_3^- retention in the stream corridor (Triska et al. 1990).

Research across a range of stream types and climate regimes has further illuminated specific reactions occurring along subsurface flow paths and the fundamental role of redox conditions in determining whether a flow path will be a sink or source for nutrients (Malard et al. 2002). Redox conditions tend to vary with the residence time of water along individual subsurface flow paths as a result of a predictable thermodynamic sequence of biogeochemical reactions (Hedin et al. 1998). Assuming downwelling surface water is well oxygenated, rapid flow paths tend to remain oxygenated, while slower flow paths are likely to become anoxic until mixing with oxygenated flow paths either in the subsurface or upon upwelling. Rapid, well-oxygenated flow paths tend to be sources of inorganic N, and NO_3^- in particular, as organic

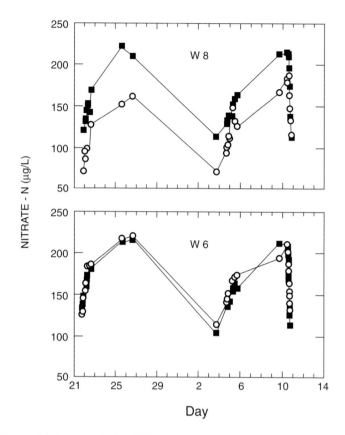

Figure 6.4 Measured (○) versus calculated (■) nitrate-nitrogen at two well locations in Little Lost Man Creek, California, during a nitrate addition experiment. Nitrate concentrations in Well 8 were below predictions, indicating nitrate uptake or conversion along the flow path feeding the well. Conversely, nitrate concentrations at Well 6 suggest nitrate is actually generated along the flow path. Well 8 is located only 1 m from the channel margin, whereas Well 6 is located 3.5 m from the channel (from Triska et al. 1989).

nitrogen is mineralized and nitrifying bacteria (*Nitrosomonas* and *Nitrobacter*) convert NH_4^+ to NO_3^- (see Figure 6.5). In streams dominated by rapid flow paths, the hyporheic zone may act primarily as a source of NO_3^- to surface waters (Jones and Holmes 1996). This same sequence of reactions occurs along slower flow paths, but given sufficient time, the available oxygen will be consumed, and *Thiobacillus denitrificans* and other bacteria will begin to denitrify NO_3^-, at which point the flow path becomes a sink for N from surface water. Anoxic flow paths, by contrast, tend to be sources of orthophospate (PO_4^{3-}) to surface waters because of the dissolution of iron and aluminum oxides that contain PO_4^{3-} (Carlyle and Hill 2001).

Mixing of surface and subsurface waters is especially important in hyporheic zones because surface and subsurface waters often differ in physicochemical characteristics, and the gradients established in mixing zones stimulate nutrient-consuming biological and biogeochemical reactions. Recognizing this, hyporheic zones can be subdivided into two subzones; (1) a *surface subzone* (consisting of >98 percent stream water) in which hyporheic water is chemically similar to infiltrating stream water, and (2) an *interactive subzone* characterized by strong gradients in key chemical parameters such

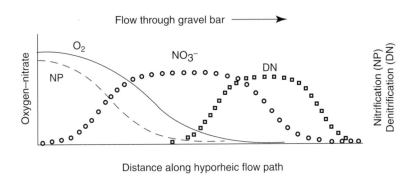

Figure 6.5 Trends in oxygen (O_2), nitrate (NO_3^-), nitrification (NP), and denitrification (DN) along a hyporheic flow path through a gravel bar. In the presence of organic nitrogen and O_2 there will be high rates of nitrification and NO_3^- production near the point of downwelling. When all O_2 has been consumed, the microbial community will switch to denitrification for their energy needs and NO_3^- will be consumed (from Malard et al. 2002).

as NH_4^+, dissolved organic carbon (DOC), NO_3^-, and O_2. In the model, the interactive subzone is the key zone of mixing between groundwater, which contributes DOC to the denitrification reaction, and stream water, which contributes NO_3^- and O_2. Subsequent studies have both confirmed and further elaborated this model for stream corridors of different sizes and in different climate regimes (Hill et al. 1998, Dent and Grimm 1999, Peterson et al. 2001).

Particle-Nutrient Considerations

Riparian vegetation also facilitates the removal of suspended sediments, along with their nutrient contents, from overland flow, whether from the uplands or from the adjacent river. Sediments and sediment-bound pollutants carried in surface runoff are effectively deposited in mature riparian forests and in streamside grasses. Sediment trapping is facilitated by sheet flow runoff, which allows deposition of sediment particles and prevents channelized erosion of accumulated sediments. Riparian areas remove 80 to 90 percent of the sediments leaving agricultural fields in North Carolina (Daniels and Gilliam 1997). Sediment deposition may be substantial in the long term, with coarse material deposited within a few meters of the field–forest boundary, and finer material deposited further into the riparian forest (Lowrance et al. 1986). Grassy areas are especially effective as they often transform channelized flows into expanded shallow flows, which are more likely to deposit sediment. However, the performance of grassy vegetation seems to be highly variable and to be of short duration when several floods occur within a limited period. For example, sediment trapping efficiency may decrease from 90 percent in a first rainfall event to 5 percent in a sixth rainfall event (Correll 1997; see also Chapter 10).

Removal of fine sediment from runoff by riparia occurs as a consequence of the interactive processes of deposition and erosion, infiltration, and dilution. This is important because fine sediments carry higher concentrations of labile nutrients and adsorbed pollutants. In forested catchments, with relatively low nutrient concentrations, fine sediments in riparian zones can be sources or sinks for nutrients, depending on how oxidation-reduction conditions affect absorption/desorption to fine particles. For example, the riparian zone of a small deciduous forest stream in eastern Tennessee was a net source of inorganic phosphorus when dissolved oxygen concentrations in riparian groundwater were low, but a sink when dissolved oxygen concentrations were high (Mulholland 1992).

In contrast to nitrogen, phosphorus adheres strongly to particles, and considerable movement is normally associated with sediment flux; most studies of phosphorus removal have focused on this pathway. Reductions of 50 to 85 percent in total phosphorus are observed, with the greatest removal occurring in the first few meters of the riparian zone (e.g., Peterjohn and Correll 1984). Results for soluble phosphorus in surface runoff are less consistent due to variability in water flow (i.e., sheet flow, channel flow, or percolation). In general, significant amounts of phosphorus may first accumulate in riparian zones but then be transported to aquatic ecosystems in a different form via shallow groundwater flow, possibly as a result of increased decomposition of organic matter.

Only limited research has been conducted on subsurface P transport in riparian zones (Carlyle and Hill 2001). However, the same environmental conditions leading to redox gradients in soils that are responsible for nitrogen removal by denitrification also mediate desorption and release of phosphorus (Hillbricht-Ilkowska et al. 1995). Although riparian zones may act as effective physical traps (sinks) for incoming particulate phosphorus, they may enrich runoff waters in available soluble phosphorus (Dillaha et al. 1989). There are, however, well-organized patterns of groundwater soluble-reactive-phosphorus (SRP) in riparian zones that reflect the interaction of hydrologic flow paths and environments of contrasting redox conditions, especially the influence of redox conditions on microbial reduction of Fe^{3+} to soluble Fe^{2+} (Carlyle and Hill 2001). Finally, unlike nitrate, which can be converted to N gases by denitrification, phosphorus removal by soil retention and biotic uptake results in accumulation within the system. Consequently, the long-term performance of riparian zones receiving high P inputs remains unclear.

One final consideration is that plant cover also influences the efficiency of riparian zones in filtering nutrients and pesticides. A riparian zone vegetated with poplar is more effective for winter nitrate retention than one vegetated with grass (Haycock and Pinay 1993). Some trees are better than others in filtering nitrate: *Populus x canadensis* effectively removes nitrate from saturated soils with a subsequent accumulation of nitrogen in root biomass (Naiman and Décamps 1997). Roots of alder, willow, and poplar seem to favor colonization by proteolytic and ammonifying microorganisms and, particularly for alder roots, to inhibit nitrifying microorganisms. It follows that changing plant cover may affect water quality. In a set-aside riparian zone in New Zealand, 12 years after retirement from grazing, dominant vegetation returned to native tussock (*Poa cita*)—leading to a zone likely to be a sink for sediment-bound nutrients and dissolved nitrogen but a source for dissolved phosphorus.

Energy Flows and Food Webs

Energy flows between riparia and adjoining ecosystems in multiple dynamic food webs. Like river systems, food webs are hierarchical networks that adapt to constantly changing energy regimes (Power and Dietrich 2002). Food webs are linkages between consumers and resources, such as between predators and their prey. The length of a given chain (or energy path) is determined by the number of consumers along the chain and depends on a number of interacting constraints, including history of community organization, resource availability, types of predator–prey interactions, disturbance, and ecosystem size (Post 2002). *Food chains* become *food webs* through the multiple linkages introduced by omnivorous consumers. New energy (i.e., organic matter) is added to a given riparian patch through primary production or transport from adjoining patches, and energy flows up the hierarchy of trophic levels from primary producers to higher-level consumers. Flows up food webs are regulated by controls exerted by changes in the population size and dynamics of consumers (Paine 1980). Patches of primary producers are heterogeneously distributed across and along riparian corridors but are generally fixed in space over short time scales. This basic level of food webs is spatially dynamic over longer time periods, however, because disturbances regularly restructure riparian communities (Naiman et al. 1998b). Energy begins to flow between patches to higher trophic levels through foraging by herbivores and hunting by predators.

Energy Flows Between Riparia and Adjoining Aquatic Systems

Riparian food webs are connected to those of adjoining aquatic environments, and the integrity of each, to differing degrees, depends on the flow of energy and materials between them (Polis et al. 1997). Flows across aquatic-riparian boundaries include physical and biotic vectors. Gravity, wind, and water are the main physical vectors, while mobile prey and consumers are the main biotic vectors. Habitat edges, or ecotones, are well known to exhibit strong gradients for physical parameters, resource availability, and species richness, but they are also sites of enhanced species interactions and community dynamics (Fagan et al. 1999).

By far the most thoroughly investigated and best understood connection between riparian and stream food webs is via the transfer of riparian plant litter to streams (Webster and Meyer 1997). Riparian organic matter inputs represent allochthonous sources of energy as opposed to the autochthonous organic matter contributed by aquatic primary producers. In low-order streams beneath closed-canopy riparian forests, the influx of carbon from riparian plant sources, both surface and subsurface, may amount to 80 to 95 percent of total organic carbon influx to streams (Conners and Naiman 1984). Riparian litter fluxes per area of stream surface in the eastern United States range from 40 to 700 g C m^{-2} yr^{-1} (Webster et al. 1995). Area-normalized fluxes are highest in low-order streams and decrease exponentially as streams widen and the canopy opens. For example, in the Matamek River catchment of eastern Quebec, litter fluxes to first- and second-order streams range from 100 to 250 g C m^{-2} yr^{-1}, whereas in a sixth-order stream, the flux was less than 10 g C m^{-2} yr^{-1} (Conners and Naiman 1984). In temperate climate zones, riparian litter fluxes are strongly seasonal, peaking during autumn leaf fall. Although the area-normalized flux of riparian litter to river systems decreases downstream, the total amount of riparian litter input to the river continues to rise and varies as a function of channel morphology and riparian forest structure and composition. Conners and Naiman (1984) calculated that total inputs of allochthonous organic matter increase exponentially with stream order, approaching 500 g C m^{-1} yr^{-1} in fifth- and sixth-order streams. And in the definitive example of downstream effects, leaves of riparian trees were found to be a major source for the more than 36 million metric tons of carbon carried by the mainstem Amazon River annually (Devol and Hedges 2001). That carbon not directly traceable to leaves appeared to come from soil organic matter, which for the mainstem Amazon is largely derived from erosion of riparian and flood plain soils (Dunne et al. 1998).

Once in the stream, riparian organic matter is decomposed by a variety of specially adapted microbial and invertebrate fauna (see Chapter 5). When litter (mainly leaves and needles) first enters streams, there is a brief period (a few days) of rapid leaching in which 25 percent or more of the initial dry weight can be lost (Giller and Malmqvist 1998). Biotic decomposition is initiated by hyphomycete fungi that break up the litter's structural integrity by secreting enzymes to hydrolyze cellulose, pectin, chitin, and other difficult-to-digest compounds (Suberkropp and Klug 1976). Fungal community composition is closely tied to the riparian forest, and fungal species richness has been positively correlated with riparian tree richness (Fabre 1996). With time, fungi give way to bacteria as the dominant microorganism in the decay process. Decomposition rates are driven by substrate quality, stream nutrient concentrations, redox conditions, and temperature (Suberkropp and Chauvet 1995). Fungal decomposition alone can fragment leaves into flakes of finer particulate organic matter within weeks (Gessner and Chauvet 1994). This fragmentation process is critical to energy dispersion in streams and rivers because finer fragments tend to be more mobile and to therefore fuel metabolism in downstream river sections. Aquatic ecologists classify fragments into coarse (>1 mm) and fine (<1 mm but >0.5 μm) particulate organic matter (CPOM and FPOM) fractions, which tend to vary consistently in quality and abundance along river networks.

Microbially colonized litter is said to be *conditioned* and generally has higher nutrient concentrations than noncolonized litter. Conditioned litter is therefore the preferred choice of macroinvertebrate consumers that make up the next link in aquatic food webs (Irons et al. 1988, Suberkropp 1998). Benthic macroinvertebrates have evolved a number of strategies to capitalize on the energy of riparian litter, to the extent that species can be generally classified into functional feeding groups (Hershey and Lamberti 1998). The most important groups in transferring riparian-derived energy up to higher consumers are *shredders*, which shred and consume litter material, and *collectors*, which simply consume litter particles they collect in the water column or on the streambed. Shredders include caddisflies (Tricoptera), stoneflies (Plecoptera), nonbiting midges (Diptera), and certain families of beetles (Coleoptera). Collectors include these groups plus a wide range of other aquatic animals, including shrimp (Malacostraca) and worms (Oligochaeta and Turbellaria). In small streams, shredders are by far the dominant group, constituting up to 50 percent of the entire macroinvertebrate community, whereas collectors become the dominant group in larger rivers (Hawkins and Sedell 1981, Hershey and Lamberti 1998).

Sharp changes in the supply of wood and leaf litter to streams (e.g., by removal of riparian forests) may cause significant changes in the abundance and biomass of macroinvertebrate functional feeding groups, and decreases in aquatic insect abundance may have direct feedbacks to riparian food webs (discussed later in this section). Wallace et al. (1999) investigated the ecosystem-scale consequences of altering riparian litter fluxes to an experimental stream in the Coweeta catchment of western North Carolina. By suspending a canopy above the stream channel and erecting fences along the stream margin, they reduced the combined vertical and lateral inputs of litter to the stream by 94 percent over a 4-year period. In the fourth year, they also removed small woody debris from the stream. Decreasing litter influxes lead to a 50 percent decrease in organic matter standing crop, from approximately 2,200 g m^{-2} to 1,100 g m^{-2}, in the stream. The invertebrate community responded with an 80 percent decrease in both abundance (individuals m^{-2}) and biomass (g m^{-2}), and total secondary production declined to only 22 percent of pretreatment values. Functional feeding groups responded differentially during the experiment; shedders, gatherers, total primary consumers, and predators declined significantly, while scrapers and filterers did not.

Riparian arthropods are also important energy sources to stream consumers. Arthropods fall into streams from overhanging foliage by accident, and the flux is proportional to arthropod abundance in the canopy (Nakano and Murakami 2001). Arthropods may also wash into streams and rivers during overland flow events. The normalized flux (per square meter of channel area) is higher in smaller streams flowing beneath a closed riparian canopy but even in larger streams and rivers the flux may remain substantial at the channel margins. Once in the aquatic system, riparian arthropods are consumed by drift foraging fish and may constitute a major proportion of their diet. In a detailed study of the annual resource budget of fish in a northern Japanese stream, riparian arthropods accounted for 46 percent of the diet of rainbow trout (*Oncorhynchus mykiss*), 51 percent for white-spotted char (*Salvelinus malma*), and 57 percent for masu salmon (*Oncorhynchus masou*). Terrestrial arthropods have also been found as a significant component of the stomach contents of redbreast sunfish (*Lepomis auritus*) and bluegill (*Lepomis macrochirus*) in a Virginia stream (Cloe and Garman 1996). In temperate regions, riparian arthropod fluxes to streams are greater during warm months (Mason and MacDonald 1982, Cloe and Garman 1996), whereas in tropical regions, increased fluxes have been linked to the rainy season when high arthropod productivity coincides with frequent washing of riparian surfaces by floods and precipitation (Angermeier and Karr 1983). Consumption of terrestrial prey by aquatic consumers is viewed as an energy subsidy to aquatic food webs, and the energy derived from riparian arthropods sometimes even exceeds that available from aquatic arthropods (Cloe and Garman 1996).

Riparian-derived arthropods are higher-quality food than riparian litter and are directly available to top consumers such as fish. Experimental evidence shows that withholding this energy input from streams has consequences that reverberate through aquatic food webs and ultimately upset the basic composition of the stream community. Nakano et al. (1999) conducted a manipulative experiment in which they blocked the influx of riparian arthropods to a stream on the northernmost island of Japan by building a 50-m-long greenhouse-type enclosure over the channel. There were four experimental treatments: reduced riparian arthropod input with fish present, natural

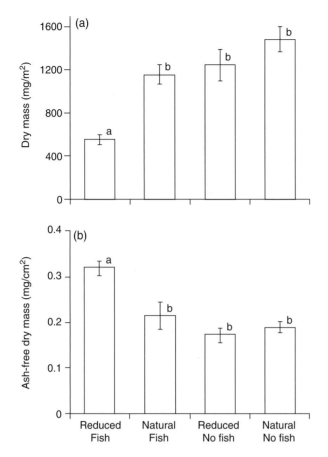

Figure 6.6 Biomass (mean ± 1 SE) for (a) aquatic herbivorous arthropods and (b) periphyton for four experimental treatments: reduced riparian arthropod input with fish present, natural riparian arthropod input with fish present, reduced riparian arthropod input with fish absent, and natural riparian arthropod input with fish absent. Significant differences are indicated by the letters above the bars (from Nakano et al. 1999).

riparian arthropod input with fish present, reduced riparian arthropod input without fish present, and natural riparian arthropod input without fish present. Fish were initially removed from the experimental reach, and then Dolly Varden (*Salvelinus malma*), one of three common fish species inhabiting the stream, were reintroduced to two treatments to assess their impact on macroinvertebrate communities. At 14 and 28 days after initiation of the experiment, investigators monitored the biomass of aquatic arthropods, periphyton, and the stomach contents of fish. The consequences of removing riparian arthropods from the diet of Dolly Varden were dramatic. These fish, which primarily had been drift foragers consuming riparian arthropods and drifting aquatic arthropods, shifted to active foraging and significantly reduced the biomass of benthic aquatic herbivorous arthropods (see Figure 6.6a). The reduction in benthic herbivorous arthropods led to a concomitant increase in periphyton biomass (see Figure 6.6b) and thus a fundamental shift in the stream's community structure and composition. Although the loss of riparian arthropods to stream ecosystems is minor compared to the energy flux in leaf litter, the shift in predator dynamics reverberated through the system, significantly impacting community-based functions.

Energy flows in both directions across the terrestrial-aquatic interface, and riparian food webs are also subsidized by aquatic resources (Jackson and Fisher 1986, Collier et al. 2002, Sabo and Power 2002b). Aquatic insects are an important energy source to a variety of riparian arthropods, and this energy subsidy is passed to higher trophic levels by the lizards, bats, shrews, and birds that consume riparian arthropods. The reliance of riparian arthropods on aquatic prey is greatest where high productivity gradients exist across the aquatic–riparian interface. Riparian arthropods inhabiting resource-scarce habitats such as exposed gravel bars and desert riparian environments appear to rely almost exclusively on aquatic prey (Jackson and Fisher 1986, Sanzone et al. 2003). For example, aquatic insects composed 80 to 100 percent of the diet of certain staphylinid and carabid beetles and about 50 percent of the diet of lycosid spiders inhabiting gravel bars of the Tagliamento River in northeast Italy (Paetzold et al. 2005). The more important aquatic insects to riparian food webs are those that emerge on land (like many stoneflies) as opposed to those that emerge from the water surface (e.g., mayflies). Interestingly, detritivorous aquatic insects—those that consume largely leaf litter—made up the largest proportion of the riparian arthropod diet along the Tagliamento River, representing an important energy feedback to riparian food webs. Aquatic insects thereby transform the energy transported to streams as leaf litter into higher-quality food that is returned to riparian zones (Paetzold et al. 2005).

Reciprocal energy subsidies such as these are especially important over the course of the year in temperate regions due to strong seasonal variability in the emergence and abundance of different insects. For example, aquatic arthropod abundance peaks

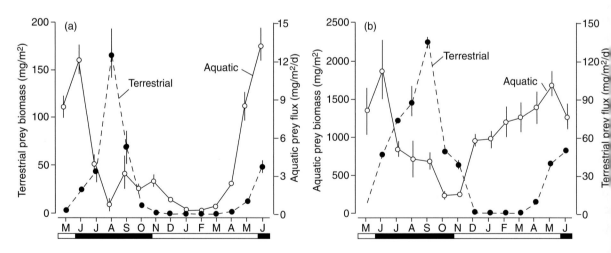

Figure 6.7 Contrasts in seasonal fluxes of prey invertebrates between a riparian forest and stream in northern Japan demonstrating the exchange of reciprocal energy subsidies. (a) Aquatic prey fluxes to the riparian forest are high in times of low terrestrial prey biomass, whereas (b) riparian prey fluxes to stream are higher during periods of low aquatic prey biomass. Both biomass ($P < 0.01$) and flux ($P < 0.01$) differed significantly among months (black and white circles represent mean values for riparian and aquatic prey, respectively). Black and white portions of horizontal bars at the bottom of figures indicate leafing and defoliation periods, respectively (from Nakano and Murakami 2001).

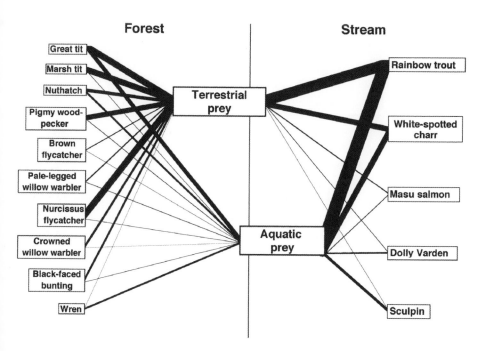

Figure 6.8 Food web linkages between a riparian forest and a stream in northern Japan. The relative contributions of terrestrial and aquatic prey to the total annual prey consumption of each species are represented by line thickness (from Nakano and Murakami 2001).

following "leaf-out" of riparian forests in spring and defoliation in autumn (see Figure 6.7), whereas riparian arthropod abundance peaks during the summer when forest productivity is maximal (Nakano and Murakami 2001). Western fence lizards (*Sceloporus occidentalis*) living on the banks of the South Fork Eel River in California rely on the energy subsidy provided by aquatic arthropods, and when aquatic arthropod availability was experimentally reduced there was a concomitant decrease in lizard abundance and more intense predation of terrestrial arthropods (Sabo and Power 2002a). The reciprocal linkages between terrestrial and aquatic food webs are also well illustrated by the bird community of a northern Japanese stream (see Figure 6.8). Ten species of riparian birds were found to rely on a diet of both riparian and aquatic arthropods. Migratory riparian birds rely heavily on aquatic arthropods during May and June, with aquatic arthropods accounting for nearly 90 percent of the consumption of the brown flycatcher (*Muscicapa latirostris*) and nearly 80 percent of the consumption of the pale-legged willow warbler (*Phylloscopus tenellipes*). Over the course of the entire year, aquatic arthropods accounted for 39 percent of the total annual energy budget of the great tit (*Parus major*) and 32 percent of that of the nuthatch (*Sitta europea*).

Large Animal Connections

Animals strongly influence the structure and function of riparian systems through their movements, feeding strategies, and physical modifications of the environment.

a
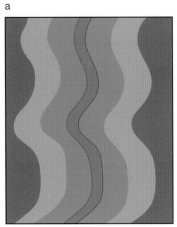

b
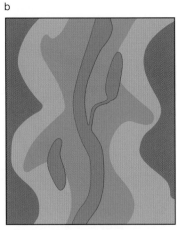

c

Low Disturbance Few Animals   Moderate Hydrologic Disturbance   Hydrologic and Animal Disturbances

Figure 6.9 Hypothetical riparian patch structure in relation to human alteration of both hydrologic regime and large animal movements. (a) Plant structural complexity is minimal under highly regulated conditions. (b) Complexity improves under natural flow regimes. (c) Maximal complexity is achieved, however, only by including the movements and feeding activities of large animals (from Naiman and Rogers 1997).

We have discussed the fundamental role of microbes and invertebrates in regulating nutrient and energy flows, but large animals also influence these flows by consuming and redistributing energy and nutrients within riparia as well as across adjacent system boundaries. More importantly, large animals may alter the hydrologic and geomorphic characteristics of riparia, causing fundamental changes in energy and nutrient cycles and altering plant community composition and structure. The net effect of animal influences is generally an increase in the spatio-temporal heterogeneity among riparian patches and even the creation of new patch types (see Figure 6.9; Schreiner et al. 1996).

The Functional Grouping of Large Animal Interactions

The influences of large animals are best understood in terms of functional groups (Naiman and Rogers 1997). This concept groups animals according to the specific changes that their activities cause relative to habitat-level structure and function in river corridors. Animals that pond water, dig holes, trample plants, or move materials cause fundamental geomorphic changes. For example, hippopotamus (*Hippopotamus amphibious*) increase the ponding of water in African stream networks by creating and maintaining deep pools and forming trails between channels and adjacent terrestrial feeding areas (Allanson et al. 1990, Rogers 1997). Burrowing mice, voles, and shrews are important agents of soil mixing in North American stream corridors, creating patches of soils with distinct nutrient compositions. Similar processes in upland systems are known to alter the composition and productivity of plant communities (Huntly and Inouye 1988), and a similar alteration may be expected in riparian systems. Earth movers like warthogs (*Phacochoerus aethiopicus*) strongly influence

Figure 6.10 New beaver pond and associated wetlands in Voyageurs National Park, Minnesota. Note dead trees that may remain as standing snags for 10 to 12 years (Photo: R. J. Naiman).

the microtopography of riparian zones by cutting ruts in search of grass rhizomes and tubers (Rickard 1993).

Beaver (*Castor canadensis* and *C. fiber*) profoundly influence the short- and long-term structure and function of riparia of drainage networks in the forests of northern latitudes by cutting wood and building dams (see Figure 6.10; Naiman et al. 1988). Beaver build dams in first- to fourth-order streams and in side channels and floodplains of larger rivers (Johnston and Naiman 1990a). Dams retain water and sediments, forming ponds that flood adjacent riparian forests and kill many riparian trees. Anoxic conditions generally develop in pond sediments, which slows decomposition and promotes the accumulation of organic C and nutrients. In catchments where beaver are abundant, there may be 2 to 16 dams per kilometer of stream length, and each dam may retain between 2,000 and 6,500 m^3 of sediment (Naiman et al. 1986, 1994). Ponds are eventually abandoned as they fill with sediment or as local food resources are depleted (Fryxell 2001) and, once abandoned, dams fail and ponds drain to produce nutrient-rich wetland meadows.

In river corridors with beaver, the cyclic pattern of pond creation and abandonment produces a shifting mosaic of patches containing diverse hydrologic, morphologic, and chemical characteristics (Naiman et al. 1988). Some abandoned ponds may be rapidly recolonized by riparian plants and recover to pre-ponded conditions in a few years to decades, whereas others may develop distinct and stable wetland features that persist for decades or centuries. In southeastern Alaska, these persistent wetlands associated with beaver ponds exhibit plant species richness exceeding that for other wetland types. Enhanced species diversity in these pond-associated wetlands correlates with flood frequency and fine-scale spatial variability in how flood frequency is distributed across the wetlands (Pollock et al. 1998). In the Adirondacks of New York, riparian meadows linked to beaver activities do not have greater biodiversity than adjacent forest patches, but they do support species assemblages that are very different from the forested patches (Wright et al. 2002). Only 17 percent of 125 species surveyed in the riparian corridor are found in both beaver meadows and riparian forest, and meadows contributed a minimum of 25 percent of the total herbaceous plant species richness of the riparian zones. Thus, although not being points of enhanced biodiversity, beaver-affected areas enhanced the overall biodiversity of the riparian corridor.

Animals browsing riparian and aquatic vegetation constitute another functional group that strongly influences riparian community structure, soil development, and propagule dispersal. Animals that browse selectively keep preferred plant species from dominating the plant assemblage and thereby provide an advantage to species not browsed. For example, moose (*Alces alces*) prefer willow (*Salix*) and poplar (*Populus*), thus giving a competitive advantage to white spruce (*Picea glauca*), which is not browsed (Pastor et al. 1988, 1993). Willow and poplar leaves are enriched in N relative to spruce, so by preferentially browsing these species, moose not only modify the vegetation structure but also alter litter quality, leading to other long-term effects on riparian nutrient status, forest composition, and soil attributes (McInnes et al. 1992). This shift in litter quality is partially compensated for by the flux of nutrients in moose feces. On an annual basis, moose consume up to 6 Mt of riparian and aquatic biomass and return approximately 60 percent of that to the soil. Provided the moose remains in the riparian zone, nutrients returned to the soil will again be available to riparian plants and animals. By contrast, hippopotamus consume ~135 kg of riparian grasses daily, but much of its defecation occurs in the river. One hippopotamus may therefore transfer approximately 9 Mt of riparian-derived organic material to adjacent aquatic systems annually (Heeg and Breen 1982). Through their browsing, hippopotamus also produce areas of "mowed" grass, which stand out as distinct patches in the matrix of tall riparian grasses and shrubs (see Figure 6.11). The influence of browsing animals on riparian vegetation also depends on the presence or absence of top predators. Wolves (*Canis lupus*) were extirpated in Yellowstone National Park during the mid-1920s, allowing elk to browse riparian plant communities unmolested for approximately 70 years (until wolves were reintroduced in the mid-1990s). During this period, increased grazing by elk virtually eliminated the recruitment of cottonwood trees (*Populus* spp.), and only with the reintroduction of wolves has recruitment begun again (Beschta 2003, Ripple and Beschta 2003).

Beaver are an example of animals that fit into multiple functional groups. In addition to creating important geomorphic change in fluvial corridors, beaver also exert strong biotic influences through their foraging practices (Pollock et al. 1994). Beaver cut trees to feed on bark (especially the soft cambium beneath the bark) and have been reported to fell as much as 1.4 Mt biomass ha^{-1} yr^{-1} (Johnston and Naiman 1990b). Like other herbivores, beaver forage selectively, preferring trees with soft, brittle bark, including trembling aspen (Populus tremuloides), willow (*Salix*), alder (*Alnus*), maple (*Acer*), and ash (*Faxinus*). Felled trees may be used in dam construction or they may be left in place, adding structural complexity to riparian zones. As

Figure 6.11 Pool used extensively by hippopotamus along the Sabie River in Kruger National Park, South Africa. Areas grazed by hippos (lawns) are outlined and a web of animal-generated paths is shown criss-crossing the riparian zones (Photo: R. J. Naiman).

with moose, beaver foraging significantly impacts the composition and successional dynamics of riparian vegetation (Barnes and Dibble 1988).

The actions of all large herbivores influence the belowground components of riparian systems as well, with consequent effects on interactions that determine long-term ecosystem function (Wardle and Bardgett 2004). Herbivore foraging changes the quantity and quality of resources delivered to belowground communities and, over longer time scales, it also changes the successional trajectory of riparian plant species. Defoliation of plants leads to a reallocation of carbon and nutrients belowground and the stimulation of belowground microbial communities (Bokhari and Singh 1974, Seastedt et al. 1988). Increased microbial activity in the vicinity of grazed plants may then enhance nutrient immobilization in the subsurface and eventually lead to increased nutrient stocks (Hamilton and Frank 2001). Successional changes brought by grazers may greatly impact the belowground community (Wardle and Bardgett 2004). Early successional vegetation generally has the highest-quality litter and therefore supports higher belowground production. Thus, grazing activities that stimulate the growth of early successional species, and therefore retard succession, help to maintain higher belowground productivity.

Pacific Salmon Influences on Riparian Ecosystems

A remarkable example of the consequences of animal-mediated nutrient and energy flows in riparia is the migration of salmon (*Oncorhynchus* spp.) from the North Pacific

Ocean to spawning areas in fresh water. Migrating Pacific salmon transport marine-derived carbon and nutrients upstream and, upon death after spawning, hydrologic and animal pathways distribute these elements throughout aquatic and riparian systems. In an important system-scale feedback, fertilization of riparian plant communities with marine-derived nutrients enhances the growth of some riparian plants, positively influencing salmon over the longer term by supplying stream organisms with an increased supply of nutritious litter and by improving salmon habitat via an influx of large woody debris (Naiman et al. 2002b).

Historically, spawning salmon represented a flux of nearly 7,000 Mt of nitrogen and more than 800 Mt of phosphorus to river corridors in California, Idaho, Oregon, and Washington (Gresh et al. 2000). Although declining populations have reduced fluxes by >90 percent during the past century salmon still are an important source of nutrients to many river and riparian systems of Canada, Alaska, Russia, and Japan. The timing and duration of marine-derived nutrient input to stream corridors depends on the number and species of salmon returning to spawn. For example, in one Washington stream used by three species of salmon, marine-derived nutrients are input over roughly half the year as sockeye (*Oncorhynchus nerka*) arrive in midsummer, chinook (*O. tshawytscha*) arrive in early autumn, and coho (*O. kisutch*) arrive in November and December (Bilby et al. 1996). Salmon carcasses and eggs are consumed directly by other fish in the stream (e.g., juvenile salmon, trout, sculpin), by riparian invertebrates (e.g., flies) and by riparian predators (e.g., bears, otters, birds). Soluble organic matter and nutrients released from carcasses are also taken up by algae and vascular plants (Schuldt and Hershey 1995, Johnston et al. 1997) or assimilated by heterotrophic organisms (Piorkowski 1995, Schuldt and Hershey 1995). At times of low stream and riparian productivity, such as November and December, chemical sorption of dissolved material may be the primary mechanism for retaining dissolved marine-derived energy and nutrients for later incorporation into the trophic system (Bilby et al. 1996).

Fish-eating predators are key to riparian transfers because they generally remove whole carcasses from the water and consume them onshore in the riparian zone (Ben-David et al. 1998, Hilderbrand et al. 1999). Predators and scavengers have been reported to remove as much as 40 percent of salmon carcasses from streams of the Olympic Peninsula in Washington State, and most removed carcasses were consumed and deposited in the riparian zone (Cederholm et al. 1989). In a mass-balance model of nitrogen in the riparian zone of Lynx Creek, Alaska, Helfield and Naiman (2005) calculated that bears alone removed ~35 percent of salmon from the stream, and the nitrogen associated with salmon carcasses accounted for ~15 percent of the total nitrogen input to the riparian zone. During years of maximal salmon numbers, marine-derived nitrogen accounted for 25 percent of total nitrogen input to the riparian zone.

Marine-derived nutrients can be traced through stream and riparian food webs using stable isotopes because salmon tissue is highly enriched in the heavy stable isotopes of nitrogen (^{15}N), carbon (^{13}C), and sulfur (^{34}S). Nitrogen isotopes are the most commonly used of these and provide compelling evidence for the importance of marine-derived nutrients in riparian food webs. Using this technique, marine-derived nitrogen was found to supply more than 90 percent of the diet of brown bear (*Ursus*

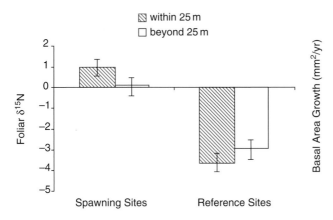

Figure 6.12 Mean (±1 SE) foliar δ^{15}N in riparian Sitka spruce at spawning and reference sites on Chichagof Island in southeast Alaska. Foliage located adjacent to spawning sites is significantly more enriched with ^{15}N derived from decaying salmon carcasses (from Helfield and Naiman 2001).

arctos) in Alaska and, historically, as much as 90 percent of the diet of grizzly bear (*Ursus arctos horribilis*) in the Columbia River valley (Hilderbrand et al. 1999). Foliar δ^{15}N levels have also been used to determine that, in the Wood River Lakes System of Alaska, marine-derived nitrogen accounts for 24 percent of total nitrogen in spruce (*Picea*), 24 percent in poplar (*Populus*), and 20 percent in willow (*Salix*) near salmon spawning streams (Helfield and Naiman 2002). By comparison, N-fixing alder (*Alnus*) contained <1 percent marine-derived nitrogen. Uptake of marine-derived nutrients into vegetation along salmon streams has been confirmed in Washington State, British Columbia, and Alaska, and although abundance of marine-derived nutrients decrease with distance from streams (see Figure 6.12), significant quantities have been detected in plants growing as much as 200 m from streams (Helfield and Naiman 2001).

Salmon-derived nutrients are transferred to riparian vegetation by a number of very different pathways. Predators and scavengers (including insects, such as flies) transfer and distribute marine-derived nutrients to the surface of riparian zones in the form of partially eaten carcasses and in feces and urine. In Lynx Creek, Alaska, partially eaten carcasses were much more important than feces and urine, accounting on average for 99 percent of the marine-derived nitrogen flux associated with bears (Helfield and Naiman 2005). The nitrogen distributed in feces and urine was important too because it was much more widely distributed; partially eaten carcasses were generally discarded very close to the stream margins. Marine-derived nutrients are also carried into riparia from streams by hyporheic flows (O'Keefe and Edwards 2002). In the mass-balance of N for Lynx Creek, hyporheic flows accounted for 3.5 percent of the total N input to the riparian zone during years with normal salmon runs and as much as 14.5 percent of total N during years of maximal salmon numbers (Helfield and Naiman 2005).

Marine-derived nutrients can significantly increase riparian nutrient stocks, which may enhance the productivity of riparian vegetation and even change community composition. Vegetation growing along salmon spawning streams in southwest Alaska was found to have higher foliar N contents, greater basal area growth rates, and higher densities of stems than vegetation growing along reference streams without spawning salmon (Helfield and Naiman 2001, 2002; Bartz and Naiman 2005). On average, foliar N content in white spruce (*Picea glauca*) was 40 percent greater along salmon streams in comparison with reference streams. Sitka spruce (*Picea sitchenisis*) along salmon streams grew three times faster than those living along streams without salmon. Sitka spruce with access to marine-derived nitrogen reached a diameter at breast height of 50 cm in ~86 years, whereas trees growing along reference streams with no salmon required more than 300 years grow to this size (Helfield and Naiman 2001). Stem densities of overstory willows (*Salix* spp.) were 500 percent higher along salmon streams compared to reference sites (Bartz and Naiman 2005). Observed changes in community composition include a reduction in the abundance of N-fixing species (i.e., *Alnus* spp.), suggesting that the competitive advantage enjoyed by such species may be reduced by marine-derived nitrogen inputs (Helfield and Naiman 2002).

After being taken up into vegetation, marine-derived nutrients may cycle through riparian systems multiple times via litterfall, mineralization, and reuptake. Marine-derived nutrients also accumulate in riparian soils, allowing for longer-term storage and subsidies to riparian species year-round. Herbivores are also important for the rapid and potentially long-distance redistribution of marine-derived nutrients. In this process, marine-derived nutrients will be distributed further and further from stream and river margins.

Conclusions

Riparia are highly complex physical and biological systems. Their complexity and distinctive ecological functions are maintained through strong spatial and temporal biophysical connectivity with adjacent riverine and upland systems. Water, sediments, and nutrients enter riparia from adjacent uplands and streams, mixing and reacting along dynamic surface and subsurface flow paths. Under normal flow conditions, riparia retain a significant portion of these materials and generally return chemically purer water to streams and rivers. At the same time, riparia are important sources of energy to both upland and aquatic systems in the form of plant and insect tissue. Many stream food webs fundamentally depend on these resources and many upland animals depend on them as important subsidies to their diets. At the same time, riparian communities benefit from the enhanced productivity of adjoining ecosystems (especially aquatic systems) through physical and biotic feedbacks that return a portion of that productivity to riparia in the form of organic matter and nutrients.

This and preceding chapters have detailed our understanding of the unique ecological functions of riparia and how these functions are linked to dynamic biophysical processes and interactions across multiple spatial and temporal scales. We recognize that maintaining these interactions and the connectivity driving them is a fundamental requirement for maintaining healthy riparia and the many services they provide; effective management requires maintaining connectivity, both in the timing and extent of flows as well as in the movements and types of animals. Management to achieve these outcomes is challenging at best. Most riparian management to date as reduced connectivity, simplified riparian spatial complexity, and eliminated or reduced the movements of large animals. In the following chapters we examine environmental changes and management activities in riparia, seeking a new model that will enable maintenance of riparia and their unique services, both for ourselves and for other organisms that depend on them.

Disturbance and Agents of Change

7

Overview

Globally, environmental alterations to riparia are generally pervasive and severe. This chapter examines (1) major avenues and ecological consequences of human-mediated change, (2) status and trends associated with major changes, (3) ecological principles associated with responses to changes in flow regimes, and (4) the environmental consequences of significant changes expected to occur in the near future, especially from climate and land use.

Anthropogenic disturbance is a human-mediated event or activity virtually unknown in natural systems in terms of type, frequency, intensity, duration, spatial extent, or predictability. Many contemporary disturbances are far outside the norm for historical events. Additionally, the response times of woody vegetation are long and, therefore, changes may not be immediately discernible.

A "distress syndrome" characterizes stressed systems. The symptoms are reduced biodiversity, altered productivity, increased disease prevalence, reduced efficiency of nutrient cycling, increased dominance of exotic species, and increases in smaller, short-lived opportunistic species. In contrast to natural perturbations, anthropogenic stress is not a revitalizing agent but a debilitating one. Multiple anthropogenic stresses added to a natural disturbance regime can either proliferate or reduce the number of successional pathways in riparian vegetation, often truncate successional processes and, depending on the type of alteration, increase the presense of terrestrial species in the riparian zone.

Massive hydrologic alterations—to ensure water and power for agricultural, industrial, and domestic purposes or for flood protection—have changed riparian characteristics throughout the world. The main responses of riparia to flow regulation depend on the type of regulation—dam characteristics and operations, dikes, and diversions—and the local geology and climate.

Flow regulation affects the ecological integrity of riparia by disrupting sediment transport as well as by reducing flood peaks, flooding frequency, and channel-forming flows. Riparian zones undergo a stabilization process, which undermines the natural processes, facilitates invasion by upland and exotic species, and alter the standing stocks and fluxes of energy and nutrients in the reconfigured system.

- Projected changes in climate and land use place additional pressures on already stressed riparian ecosystems. Although riparia are generally viewed as resilient and able to maintain healthy and self-sustaining conditions, rapid climate and land use changes impose new environmental regimes that may exceed limits of resilience. Significant changes in temperature, precipitation, nutrients, and invasive plants are reflected in riparian characteristics.

Purpose

Human attitudes toward riparia have changed enormously over the centuries—and this, in addition to the broader environmental alterations, has resulted in pervasive changes. At first, riparia were rather fearful places, made of inextricable and impassable thickets, deemed as unhealthy because of the wetlands that they included. Riparia also were transformed and exploited as fertile fields and as pastures. Riparian vegetation was removed because it impeded towing of boats by ropes, but it was also planted with flexible willows that were cut regularly for local uses. After the navigation era many cities turned away from their rivers, neglecting riversides. During the 19th century, construction of dams and river embankments marginalized riparia. Only during last quarter of the 20th century were riparia again recognized as being worthy of conservation or rehabilitation. Overall, the net result of human activities over many centuries has been one of imposing large and lasting changes to riparian environments.

Human exploitation of the natural benefits of riparia often takes place without an understanding of how these systems maintain their vitality (Gleick 1993, Naiman et al. 1995, Naiman and Turner 2000). We now know that alterations to disturbance regimes, and spatial patterning and uses associated with the terrestrial landscape, have important long-term consequences. Further, changes to land cover and use, and how societies view and value the land and its resources, can no longer be divorced from effective riparian management (Naiman et al. 1998a,c; Dale et al. 2000). Today, with ever-increasing demands being made on fresh water and riparia by an exponentially increasing human population, a basic understanding of trends in resource use, of the ecological consequences of system alterations, and of human perceptions and attitudes toward riparian resources is essential for formulating sound management and policy decisions (NRC 1998, Postel 1998).

This chapter forms a segue between previous chapters on basic ecological principles and the three chapters to follow on riparian management, conservation, and rehabilitation. In contrast to the previous chapter on biophysical connectivity, which addressed interactive pathways and functions, this chapter focuses on "severed" interactive pathways and functions (see Figure 7.1). Four topics related to human-mediated changes of riparian systems are addressed:

1. We reexamine the term *disturbance*, identifying major avenues and ecological consequences of human-mediated changes to riparia.

Figure 7.1 The strip of riparian vegetation along a second-order stream in the agricultural landscape of Brittany, France, is symbolic of changes to riparia everywhere. During the 1980s and 1990s, re-allotment programs in France contributed to the channelization of small streams such as this one, reducing the extent of flooded and waterlogged areas in the original floodplain. (Photo: Gilles Pinay.)

2. We outline the status and trends in major avenues of riparian change.
3. We discuss ecological principles associated with disturbance, illustrating them with a discussion of the responses of riparia to changes in flow regimes.
4. We examine major changes expected to occur in the next two to three decades, especially by climate and land use.

However, before beginning, it is instructive to examine the ultimate drivers of human-mediated changes so as to better appreciate the variety of changes taking place, now and into the future.

Major Categories of Change

The four ultimate drivers of human-mediated change affecting all ecological systems, including riparian ecosystems, are human demography, resource use, technology development, and social organization (Naiman and Turner 2000). Collectively these result in one or more types of changes: physical restructuring of river and riparian systems, introduction of exotic species, discharge of toxic substances, or over-harvesting of resources. Ecological systems generally lack the capacity to completely adapt to these stresses and thereby maintain fully normal functions and structures. Thus, stress results in a process of degradation, which is commonly marked by signs such as less biodiversity, reduced primary and secondary production, and lowered capacity to recover to an original state (i.e., resilience; Rapport and Whitford 1999).

Human Demography

Projected changes in the world's population are well known (Cohen 1995). The world's population likely will increase by ~50 percent in the next 50 years or so, about 90 percent of the increase will be in developing countries, and as much as 60 percent of the population may reside in urban centers.

The population expansion will not be evenly experienced across the globe. Using the United States as a regional example, the pattern of population growth will be stronger in most of the western and southern states, while the national population is expected to increase by 27 percent from 275 million in the year 2000 to 350 million by year 2030 (U.S. Department of Commerce 1997). All states will have more people, but in some states the increases will be significant. California, Texas and Florida will gain >6 million persons (8 percent), while 12 states (all in the West with one exception) will experience >4 percent increase in population size. Thus, although the per capita rate of population growth in the United States is relatively low compared to many other countries, the number of individuals to be added to the population in coming decades is substantial. This is especially noteworthy because all states except Alaska already experience severe water-based environmental challenges.

Resource Use

The distribution of major land uses reflects a complex pattern of historical conversion of native lands, especially forests and grasslands, to human-dominated land (Turner et al. 1990). Again using the United States as a regional example, the area occupied by forests continues to decline, even though large-scale conversion to agriculture has diminished. This is mainly a result of urban and suburban expansion (Turner et al. 1998). In effect, land use change represents an enormous uncontrolled experiment in the ways human-mediated changes influence the movement of water, nutrients, and sediments from land to freshwaters through riparian zones.

Today, <5 percent of the original forests remains in the lower 48 states in the United States, croplands now occupy about 16 percent of the land base, nearly 50 percent of

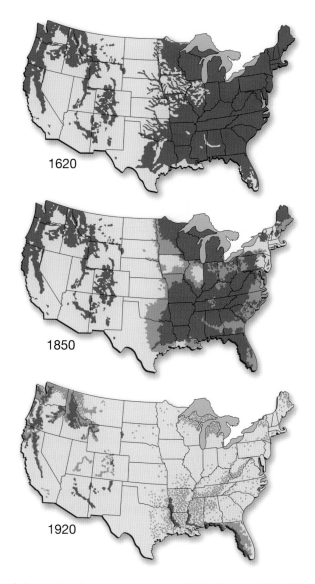

Figure 7.2 Area of virgin mature forest in the contiguous United States in 1620, 1850, and 1920. It is suspected that modification of riparian forests closely mirrored the removal of upland forests as human expansion progressed. Note that this figure does not depict forests that have regrown after the initial harvest (from Meyer 1995).

the wetlands and 70 percent of the riparian forests have been converted to other uses, nearly half the land is cultivated or grazed by livestock, and about 3 percent of the land is in urban settings (see Figure 7.2; Turner et al. 1998). Although the percentage of area in urban setting seems low, urban centers have a disproportionate "footprint" on the environment. Overall, changes in drainage and erosion accompanying all land use changes have had, and will continue to have, substantial effects on riparia in the United States as well as throughout the world (NRC 1992, 1998; Naiman et al. 1995).

Direct changes to rivers and riparia have been equally severe. Over a decade ago, the Nationwide Rivers Inventory estimated a total of 5.2 million kilometers of

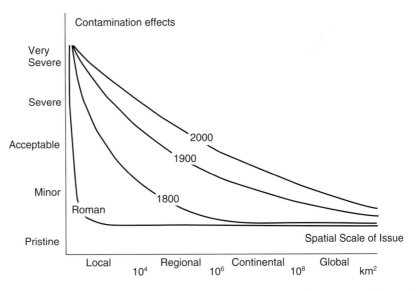

Figure 7.3 Schematic evolution of the contamination of continental aquatic systems at various scales from Roman time to the present. From Meybeck 2001.

streams in the contiguous United States, but only 2 percent (~10,000 km) have sufficiently high-quality features to be considered relatively natural rivers and thus worthy of federal protection (Benke 1990). The situation has not improved. In North America (north of Mexico), in Europe, and in the republics of the former Soviet Union, 77 percent of the runoff from the 139 largest rivers is strongly or moderately affected by fragmentation of the river channel by flow regulation, interbasin diversion, and irrigation (Dynesius and Nilsson 1994). As a result of these and numerous other changes related to exploitation, exotic introductions, pollution, and erosion, there is a general malaise in riparian systems to the point that their vitality and spatial extent are seriously compromised (see Figure 7.3; Swift 1984, Décamps et al. 1988, Petts 1989).

Water consumption is a basic symptom of wasteful resource use. In the United States, water consumption has doubled in the last 40 years, and over the next few decades this consumption is expected to increase dramatically. The scenario is similar elsewhere, especially for agricultural regions where about 1,000 T of water are required to produce 1 T of grain. Globally, the volume of irrigation water annually available to crops (as soil moisture) would need to increase by 2,050 km over current demand to meet agricultural needs in the year 2025, the equivalent of the annual runoff of 24 Nile Rivers or 110 Colorado Rivers (Postel 1998, Jackson et al. 2001). The consequences for riparia are obvious.

The United States alone has built more than two million dams and ponds to support household and agricultural consumption. Reservoirs have become a significant component of the nation's hydrologic cycle because they have the capacity to store an amount of water equal to three years' annual runoff from the nation's landscapes (Graf 1993). Globally, the scenario is similar, with the total number of large dams representing a 700 percent increase in the standing stock of river water and the average age of water retained by dams spanning <1 day to several years (Vörösmarty et al. 1997). This aging leads to significant changes in net water balance, flow regime, deoxgenation of surface waters, and sediment transport (see Sidebar 7.1). All in all, the nature and severity of water constraints remain ill-defined, largely because of national inadequacies in governmental coordination, data collection and management, and effective application of knowledge, thereby hampering the development of ecologically appropriate water and agricultural strategies (NRC 1998, Postel 1998, GWSP 2004).

Sidebar 7.1 Fragmentation of the World's Rivers by Dams and the Ecological Consequences for Riparian Zones

According to the World Commission on Dams (WCD 2000), there are between 45,000 and 48,000 large dams and an estimated 800,000 small dams in the world. A large dam is defined as one measuring 15 m or more from foundation to crest. Dams are built on rivers for many different reasons, such as hydropower, flood control, and freshwater supply, resulting in fragmentation of the river channels and regulation of their flow. Rivers are further regulated by irrigation schemes, dikes, channelization, and removal of floodplain forests.

Dynesius and Nilsson (1994) and Revenga et al. (2000) reviewed the 227 largest river systems in the Americas, Africa, Europe, the republics of the former Soviet Union, China, and in mainland Southeast Asia and found that 37 percent are strongly affected by fragmentation and altered flows, 23 percent are moderately affected, and 40 percent are unaffected. These rivers all have a virgin mean annual discharge (VMAD) ≥ 350 m^3/s. Strongly affected systems include those with less than one quarter of their main channel without dams, most of those with dams on major tributaries, as well as rivers whose annual flow patterns have changed substantially. Unaffected rivers are those without dams in the main river channel and, if tributaries have been dammed, river discharge has declined or been contained in reservoirs by <2 percent of the VMAD. In all, strongly or moderately fragmented systems account for nearly 90 percent of the total water volume flowing through the rivers analyzed. Approximately 70 rivers of this size or larger remain to be assessed before the world list becomes complete. In addition, many dams are planned and under construction, especially in developing countries (McCully 2001).

The global situation for dams is well illustrated by the 21 largest rivers in the world, which are situated on five of the six major continents and encompass all dominating biogeographic provinces: all are strongly or moderately regulated (see Table SB7-1). Chang Jiang is the world's largest strongly regulated river. The largest free-flowing river in the world is the Yukon in North America.

The extensive damming of the world's rivers has degraded ecological qualities of riparian zones (Nilsson and Svedmark 2002). Strongly regulated rivers, especially those regulated for hydroelectric power, face elimination of waterfalls and rapids, including their riparian areas, in reaches where water is transferred underground. Even along dammed and regulated reaches where the channel remains water-filled, riparian zones may become completely displaced because of altered water levels. The new riparian zones, and those that remain in place after regulation, develop modified plant communities because of changes in flow and water-level regimes. Such communities are often species-poor compared to those along free-flowing rivers.

Riparian responses differ according to the hydrologic changes. Prolonged inundation alters or eliminates plant communities because of anoxic conditions in soil. Increased variation of flow and water levels increases scour and washout of plants, organic matter, and inorganic matter, thus reducing the area of potential plant habitat. In some reaches (e.g., reservoirs in boreal regions), the shoreline may eventually develop into desert-like habitat. Stabilized flow and water levels result in reduced plant recruitment and plant growth rates, in reductions of plant cover where water availability becomes poor, and in reduced diversity because of fewer disturbances. Loss of seasonal flood peaks may lead to succession toward wooded wetlands and invasion of exotic plants. In some rivers, exotics comprise 20 to 30 percent of the entire riparian flora. Some exotics are just added to the existing community, whereas others (e.g., *Tamarix* spp. in the United States) can give rise to substantial habitat modification and loss of natural species.

Free-flowing rivers are effective in dispersing propagules of riparian plant species, and plant communities show successive change downstream. Dams disrupt longitudinal pathways, making species migration and dispersal difficult, thus causing fragmentation of riparian plant communities. Such obstructions to dispersal may have far-reaching consequences for the ability of rivers to transfer genetic material across landscapes.

Table SB7-1 Data on Channel Fragmentation and Flow Regulation for the 24 Largest River Systems in the World

River System	Continent	Discharge MAD[1] (m³/s)	Fragmentation[2]		Flow Regulation[3] Reservoir Capacity (percent of MAD)	Impact Class[4]
			Main Channel	Tributaries		
Amazonas-Orinoco	S. America	200,000	0	2	3	MR
Congo (Zaire)	Africa	41,000	1	2	0.2	MR
Chang Jiang (Yangtze)	Asia	29,460	≥2	2	12	SR
Ganges-Brahmaputra	Asia	22,102	1	1	8	MR
Rio de la Plata	S. America	21,000	3	2	28	SR
Yenisey	Asia	20,000	≥2	2	29	SR
Mississippi	N. America	18,400	3	2	16	SR
Lena	Asia	16,900	0	1	4	MR
Mekong	Asia	15,900	2	2	3	MR
Ob	Asia	12,800	2	2	8	MR
Irrawaddy (Ayeyarwadi)	Asia	11,953	0	2	1	MR
Amur	Asia	10,900	0	2	10	MR
St. Lawrence	N. America	10,800	3	2	11	SR
Zhu Jiang (Pearl)	Asia	10,700	3	2	31	SR
Mackenzie	N. America	9,910	1	1	12	MR
Volga	Europe	8,050	4	2	38	SR
Magdalena	S. America	7,500	1	2	1	MR
Columbia	N. America	7,500	4	2	24	SR
Zambezi	Africa	7,070	2	2	30	SR
Indus	Asia	6,564	3	2	13	SR
Danube	Europe	6,450	3	2	5	SR
Yukon	N. America	6,370	0	2	<1	FF
Niger	Africa	6,100	3	2	15	SR
Fly	Australasia	6,000	0	0	0	FF

[1]MAD = Mean Annual Discharge (m³/s). In rivers that have lost water and where old flow records are available, the virgin MAD is given.

[2]Fragmentation is ranked into five classes describing the longest main-channel segment without dams in relation to the entire main channel (0 = 100 percent; 1 = 75 to 99 percent; 2 = 50 to 74 percent; 3 = 25 to 49 percent; and 4 = 0 to 24 percent). For the tributaries, fragmentation is described by three classes (0 = no dams; 1 = dams only in the catchment of minor tributaries; and 2 = dams also in the catchment of the largest tributary).

[3]Flow regulation is calculated as the percentage of MAD that is retained in reservoirs. Gross capacity is the total water volume that can be retained by a dam, including the bottom water that cannot be released through the lowest outlet. Live storage is the gross capacity excluding this bottom water, and these are the values used whenever available. One half of the gross capacity is used when live storage data are unavailable. Because many small and/or recent dams are undocumented, actual flow regulation may be higher than shown.

[4]SR = Strongly Regulated, MR = Moderately Regulated, FF = Free-Flowing.

Christer Nilsson, Umeå University, Umeå, Sweden and Mid Sweden University, Sundsvall, Sweden
Elisabet Carlborg and Catherine A. Reidy, Umeå University, Umeå, Sweden

Technology Development

History is replete with examples of how new technologies profoundly affect societal use of the landscape (Diamond 1999). Inventions such as powerful pumps, labor-saving machinery, and herbicides and pesticides have fundamentally altered agriculture and forestry. Meanwhile, construction of highways and river locks has altered forever the transportation and use of essential goods, and medical advances have reshaped the age structure and size of human populations (Dale et al. 2000). New technologies continue to emerge, and some will have strong influences on the distribution of the human population and on land use. Collectively, these technological "innovations" have severe ecological consequences for riparia as human influences expand into formerly natural areas.

As one looks to the future, there are numerous technologies that will alter land and water use in fundamentally significant but as yet unspecified ways. For example, advances in telecommunications are changing the ways business is conducted and thereby influencing patterns of human settlement. Emerging technologies targeted at improving resource production (e.g., food and fiber) and water efficiency are making more intensive use of increasingly scarce land and water resources. Consider managed forests—these are being cut on shorter rotations, plantation areas are increasing, and additional uses are being found for fiber that is not currently utilized—with concomitant effects on environmental quality.

Social Organization

A comprehensive discussion of this complex subject is beyond the scope of this book but will be addressed more fully in Chapters 8 and 9. However, a brief introduction is needed to understand why ecological systems have been changed in the past and will change in the future.

The variety of human attitudes, traditions, values, and perceptions are embodied in our cultures and institutions, which ultimately shape the character of freshwater systems. There are at least 22 federal agencies in the United States, and scores of state and local agencies, which have responsibilities for the hydrologic cycle, often with dramatically different perspectives and goals (NRC 1998)—and it is similar in many other countries. Environmental conditions directly influence human perceptions about freshwater availability and resources, and those perceptions determine much of the policy related to environmental regulation. In turn, policies directly influence the future environmental state. Unfortunately, attitudes and traditions, as well as institutional missions, are highly resistant to change. Effectively resolving most ecological issues means finding better ways to communicate between people and organizations (Naiman et al. 1998c, NRC 1998). We return to this subject later in the book.

Riparian Disturbances

Disturbance is a complicated subject, especially so when human alterations are added to the suite of natural disturbances. One needs to carefully examine the nature of anthropogenic disturbance as well as understand the disturbance history of riparian sites.

Defining Anthropogenic Disturbance

As discussed in previous chapters, research on riparian disturbance mostly concentrates on flow-related events, either water or sediment. However, defining, quantifying, and separating natural from anthropogenic disturbances are other matters altogether. The subject is complex because riparian systems are intrinsically unstable systems subjected to the vagaries of the environment, which means having disturbance regimes that vary in frequency, intensity, duration, spatial extent, and predictability. An *anthropogenic* disturbance implies that humans have an important influence on the event in question.

Defining disturbance is a constant cause of consternation for ecologists. Many prefer the perspective where disturbance is a mortality-causing event (for only some taxa perhaps) or an event that alters resources (Connell and Sousa 1983, White and Pickett 1995). Others define a disturbance as an event falling outside a "predictable" range (e.g., 95 percent CI; Resh et al. 1988) but others (e.g., Poff 1992a) do not share that specific perspective. Rather, they distinguish between an ecological and an evolutionary perspective on disturbance. In any of these cases, disturbance acts as an essential structuring agent that sustains the ecological integrity of ecosystems—the lack of (natural) disturbance is itself a "disturbance" and a common feature of many human-dominated systems.

Here we define *anthropogenic disturbance* as a human-mediated event or activity that is virtually unknown in natural systems in terms of type, frequency, intensity, duration, spatial extent, or predictability over the last century (<5 percent probability of occurring under historic conditions). This "definition" illustrates the difficulty of crafting a perfect description, and should be considered more of a guideline in separating natural from human-mediated disturbances. Nevertheless, it should convey the basic concept that many disturbances experienced today are far outside the norm of events experienced by riparia for millennia.

Attempts to conceptualize disturbance have led to a better appreciation of the complexity of this topic. Three generic temporal patterns are recognized—pulse, press, and ramp—and they are common to the full suite of disturbance types (Lake 2000). *Pulses* are short-term and sharply delineated disturbances (e.g., floods); *presses* may arise sharply but then reach a constant level that is maintained (e.g., sedimentation after hillslope failures); and *ramps* occur when the strength of disturbance steadily increases over time or reaches an asymptote (e.g., drought, or sedimentation as a catchment is progressively cleared of vegetation; see Figure 7.4). Further complexity is introduced when one examines the possible biotic responses to the temporal trends. Pulse disturbances may show a pulse biotic response, a press biotic response, or even a ramp biotic response. The same is true for both press and ramp disturbances, and the responses vary further according to the type of anthropogenic alteration. All in all, trying to formulate the perfect definition is an impossible task. Rather, it is probably more informative to construct conceptual models and provide case studies as illustrations of human-mediated environmental change.

Understanding History: Basic Concepts and Approaches

Chapter 2 considered the common types of *natural* disturbances, or, perhaps more properly stated, the characteristics of natural hydrogeomorphic processes in shaping

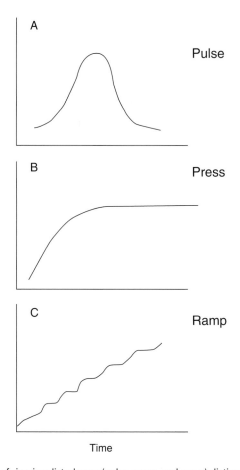

A   Pulse

B   Press

C   Ramp

Time

Figure 7.4   Three types of riparian disturbance (pulse, press, and ramp) distinguished by temporal trends in the strength of the disturbing force. The disturbing force or activity is generated by significant deviations from normal conditions. From Lake 2000.

riparian characteristics. We learned that interactions between climate and geology determine discharge patterns and sediment bed loads. The resulting discharge and sediment regimes, acting in concert with position along the riparian corridor and the available species pool, shaped the biophysical characteristics of riparia in historic landscapes.

Today the ultimate determinants of riparian characteristics are the same. However, the frequency and intensity of natural geomorphic processes, as well as the climate, have been altered by human actions and, in some cases, new types of disturbances have been added and other disturbances have been suppressed. Indeed it is a very different world than the one that originally shaped the biotic characteristics of natural riparian assemblages. In order to understand existing riparian patterns and processes, one must consider how past human-mediated disturbances shaped contemporary characteristics (i.e., *legacies* and *lag-times*), as well as understand how several distinct types of disturbances acted in concert to produce the contemporary situation (i.e., *cumulative effects*). Additionally, looking to the future is necessary for successful system rehabilitation since one must consider the consequences of changing climate and land uses on environmental regimes and species pools.

Legacies and Lag-Times

J. J. Magnuson (1990) eloquently articulated a concept he termed "the invisible present" in order to understand mechanisms shaping extant patterns. This concept is especially relevant for riparia where processes acting over decades are hidden and reside in current patterns. The invisible present corresponds to the temporal scale of acid deposition, the invasion of nonnative organisms, the introduction of synthetic chemicals, CO_2-induced climate warming, and deforestation. In the absence of a factual temporal context, serious misjudgments occur not only in attempts to understand and predict change in riparia but also in attempts to manage riparia. Processes acting in the past leave biophysical legacies whose biotic expression may take place immediately or sometime into the future (i.e., lag-time). The effects of legacies and lag-times on riparia have been well documented. Examples include the Trémolières et al. (1998) description of how river management history altered the community structure and nutrient status of the Rhine River's alluvial forest, the Harding et al. (1998) examination of past land use, and the Décamps et al. (1988) and Johnson (1994) discussions of human influences on riparian succession, to name just a few.

Legacies are perhaps easier to conceptualize than lag-times. In most cases, a legacy is a physical template that has been changed by a human action and that remains as a feature of the environment. The physical template can be a permanently altered flow regime, changes to the sediment grain size or soil structure, modification of nutrient availability, or some other physical change. The biotic response to that physical legacy is the extant riparian community or ecological process. Lag-time refers to how quickly the biotic community responds to the initial physical change. In many cases, the response will be almost immediate, as in the re-colonization of a gravel bar by willow after a scouring flood. However, some lag-times may take decades to centuries. Often this is a result of either remnants of the original community remaining behind or the time needed for the new riparian community to be fully expressed at mature successional stages.

Cumulative Effects

Environmental changes are often affected by more than one human-mediated activity. Cumulative effects are not new types of impacts but rather a new way of examining the multiple impacts that ecosystems have confronted (Reid 1998). Technical issues that complicate analysis of cumulative effects include the large spatial and temporal scales involved, the wide variety of processes and interactions that influence cumulative effects, and the lengthy lag-times that often separate a human activity and the ecological response to that activity.

The concept behind cumulative effects is simple, but the quantitative evaluation is technically difficult. Consider, for example, what might radically change the vegetative composition of a riparian community in a catchment undergoing resource extraction and urban development. What are the relative contributions of the various types of disturbances to the altered community assemblage from new flow regimes, increased sediment loads, atmospheric deposition of pollutants, modification of local climate, invasion of nonnative plants, and so on? Additionally, are the responses to each type of change linear, or not? Are there thresholds, adaptations, feedbacks, amplifications, synergisms, or other nonlinear biotic responses that cannot be foreseen? Fortunately, approaches are being developed to address these complicated issues and are the subject of much contemporary investigation at the landscape scale. Perhaps the two best-known and most effective approaches are *Watershed Analysis* and *Ecosystem Analysis* (summarized in Naiman et al. 1998a, Reid 1998).

Collectively, legacies, lag-times, cumulative effects, biotic resistance and resilience, landscape position, and other factors make the forecasting of human-mediated change an inexact science. This remains as one of the grand challenges for all of ecological science, not just for riparian ecology. However, riparian ecology has much to contribute in this regard as an integrator of catchment processes.

Historical Examples of Riparian Alterations

The natural relationship between floodplains and the main channels of many large rivers in North America and Europe no longer exists, and it is rapidly disappearing on other continents. The influence of the rivers on floodplain riparia has been reduced by the removal of large woody debris from the channel, the construction of dikes, the draining of floodplains for agriculture and urbanization, and flood control by upstream dams. The combined effects alter the extent, composition, and nutrient capital of riparian communities. Two examples capture the range of changes that many riparian zones have been subjected to over the last few centuries: One is for Europe with a long history of human-modification of rivers and the other is for the western United States with a relatively short history of land use change.

European rivers have been harnessed and managed for numerous uses for nearly a thousand years. Some of the first modifications were for waterpower and navigation throughout Flanders, Germany, France, Italy, and England in the 12th and 13th centuries (see Table 7.1). Flood control, channelization, and land reclamation followed by *ca.* 1500 AD, and the construction of water supply dams followed by *ca.* 1600 AD. Concomitant impacts included artisan fisheries and severe pollution. Collectively, these and other modifications to rivers and uplands greatly modified riparian communities (Décamps et al. 1988, Petts 1989). This is especially well illustrated for the upper Rhône River in France where the chronosequence of changes since *ca.* 1750 AD have resulted in a river corridor that is almost unrecognizable from its original condition (see Figure 7.5).

A classic example from the western United States is provided by the Willamette River, Oregon, one of the older settled regions of the American West (see Figure 7.6). The pre-European settlement river (*ca.* 1805 AD) had banks 1.5 to 2.6 m above the low water line, floodplains 1.6 to 3.2 km wide, and dense woodlands of Douglas fir, cottonwood, red alder, and willow. Between 1870 and 1950 AD, over 65,000 dead trees (i.e., snags, large woody debris), 30 to 60 m long and 0.5 to 2.0 m diameter, were removed from the river (550 snags/km). Today there are only 2 to 3 snags/km in place. Additionally, revetments were constructed to confine the river to one channel. The net result was that the >250 km of shoreline described in 1854 AD was reduced to 64 km by 1983 AD. By 1970, the gallery forest had been reduced to a narrow, discontinuous ribbon of vegetation immediately adjacent to the river and its major tributaries. This was a common occurrence in the American West since "stream improvement" has been well financed by the United States government since 1870. In the lower 1,600 km of the Mississippi River, >800,000 large snags (mean diameter .7 m) were removed over a 50-year period. The situation was the same for many other

Table 7.1 Development of River Regulation in Europe

Year	Historical Sequence	Significant Developments	Other Impacts
1250	Weirs for water power	Stanches widespread in Flanders, Germany, France, Italy, and England	
	River improvements for navigation	1398 First summit canal (River Stecknitz)	Artisanal fishery
		1400 Bertola designed channelization of River Adda	
		1497 Leonardo designed pound-lock with mitre-gates	
		Verona (River Adige) and Florence (River Arno) established river authorities	
1500			
	Flood-control and land reclamation	Dredging using endless chain technology developed by 1561	Artisanal fishery
		Pound locks widespread	
		Small rivers channelized (e.g., Yevre and Havel)	
		1550 Lupicini designed flood-defences for River Po	
	Science of regulating rivers established	1577–1643 Castelli (Founder of modern hydraulics)	
		1594 Alicante dam (41 m-high masonry)	
		1692 Completion of Languedoc canal	1616 R. Thames pollution problems
		Guglielmini (1697) and Baratteri (1699) scientific approach to river regulation	
1750			
		Large rivers channelized (e.g., River Oder)	Commercial and artisanal fishery
		River Guadalquivir: length to Seville reduced by 50 km (40 percent)	
		Earth bank water-supply dams spread in headwater catchments	Pollution
1850			
	Extensive floodplain reclamation	Major rivers channelized (e.g., Alsatian Rhine and Alpine Rhône)	R. Thames severely polluted
		1845 River Tisza (Theisz) shortened by 340 km, 12.5×10^6 ha reclaimed	
		1849 River Danube 4×10^6 ha floodplain reclaimed along 230 km reach	
		Masonary headwater supply dams (50 m high) common	
	Water supply dams spread	1898 Hydroelectric power dam at Rheinfelden	Overfishing
1900			
	Hydroelectricity dams	1937 First $1,000 \times 10^6$ m^3 reservoir: Ivankovo, R. Volga ($1,120 \times 10^6$ m^3)	
		1941 First $25,000 \times 10^6$ m^3 reservoir: Rybinsk, R. Volga ($25,400 \times 10^6$ m^3)	Severe pollution widespread
		1950 First 150 m high dam: Noce-Aldigo, Italy	
	River regulation dams	1955 First $50,000 \times 10^6$ m^3 reservoir: V. I. Lenin dam, R. Volga ($58,000 \times 10^6$ m^3)	
		1957 First 200 m high dam: Mauvoisin dam, Switzerland (237 m high)	
		1961 First 250 m high dam: Vaiout dam, Italy (262 m high)	
	Impounded rivers	1962 Grand Dixence 285 m high dam, Switzerland	Conservation

From Petts 1989.

rivers as the United States improved river navigation and opened up riparian land for agriculture (Maser and Sedell 1994).

Certainly, a major consequence of large woody debris removal from river channel has been a major loss of heterogeneity in the river corridor (Naiman et al. 2005). As discussed in previous chapters, the presence of large woody debris shapes riparian successional processes as well as the formation of riverine islands. Acting alone, or synergistically, large woody debris jams cultivate new landforms, reinforce existing landforms, and transform or reconfigure existing landforms within stream reaches. Differentiation in form and dispersion of large woody debris jams strongly influences the availability of resources (e.g., habitat quantity and quality) for aquatic organisms and riparian vegetation, and subsequently riparian system function (Pettit et al. 2005).

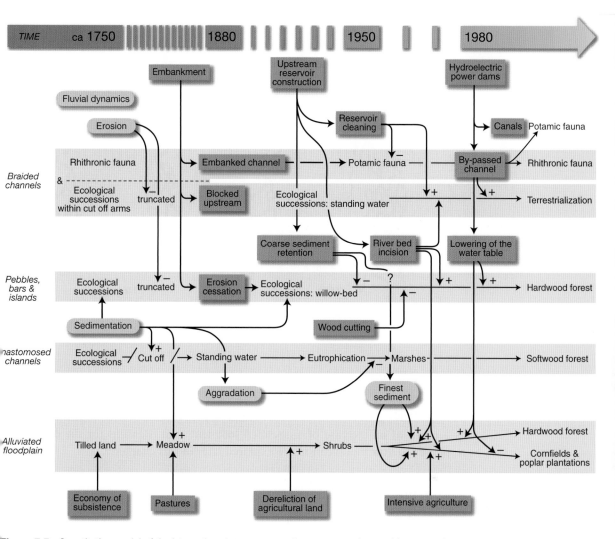

Figure 7.5 Quantitative model of the interactions between natural processes and several human actions responsible for the ecological changes within a braided-anastomosed sector of the French Upper Rhône from 1750 to 1988 AD. From Roux et al. 1989.

In contrast to prevailing perceptions, a few rivers have seen riparian zones expanded by human modifications. For example, riparian areas associated with the Arkansas and South Platte rivers in eastern Colorado have increased dramatically since *ca.* 1830 AD. Most riparian expansion in the South Platte River occurred between the 1930s and the 1950s as 50 to 90 percent of the active channel became wooded (Johnson 1994). Earlier, they were relatively straight, wide, braided, and intermittent streams. Expansion was a result of percolation of irrigation water into and through valley alluvium, which raised water tables, and flow regulation that produced more uniform flow in the rivers. As a result, new floodplains formed, and bank and floodplain vegetation became denser as if a climatic change to more humid conditions had occurred (Nadler and Schumm 1981). The rivers became narrower and more sinuous because of perennial stream flow, abstraction of sediment with irrigation water, and a decrease in discharge during drought. However, the changes were not uniform among reaches.

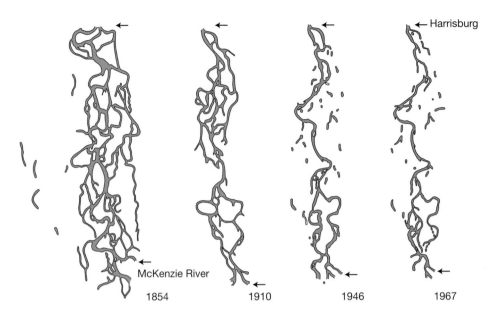

← Harrisburg

McKenzie River

1854   1910   1946   1967

Figure 7.6 The Willamette River, Oregon, from the McKenzie River confluence to Harrisburg, showing the reduction of multiple channels and loss of shoreline 1854–1967. From Sedell and Froggatt 1984.

The South Platte River and a reach of the Arkansas River narrowed and developed a single thalweg. In contrast, another reach of the Arkansas River began to meander as a result of an increase of suspended sediment load. Reasons for the differential responses to flow modification are explored later in this chapter.

Pervasive Human-Mediated Changes

The transformation of the Earth by human action, especially over the last three centuries, has been remarkable (Turner et al. 1990). It is well accepted that land use change is the leading cause of habitat loss and fragmentation for aquatic as well as terrestrial systems. The total land used by humans has increased dramatically to the point that nearly all habitable land and most of the available freshwater is dedicated to human use (Richards 1990, Postel et al. 1996). The transformation of rivers and riparia, as discussed earlier, has been at the forefront of this human-mediated change. However, there are several human-mediated changes that are increasingly ubiquitous, transcending the catchment scale and having strong influences on riparian characteristics. These are land use change and climate change, which not only have local effects but also link continents through atmospheric transport of nutrients and other materials.

Changes in land use alter the fluxes of greenhouse gases (e.g., water vapor, carbon dioxide, methane, nitrous oxide), therefore making it difficult to cleanly separate land use from climate changes. The most important climate change is in the mobility of water. With every 1 C increase in air temperature, the atmosphere can hold 6 percent more water vapor. What does this mean for riparia other than the obvious physiological effects of increasing temperature? There is an increasing consensus that substantial warming of the Earth's climate would produce more clouds and rain on average but varying drastically from one area to the next; more violent precipitation events locally and regionally with aggravated risk of flooding; and severe drying of soils in many locations, including a potential 10 to 30 percent decline in summer soil moisture over much of North America (Houghton et al. 2001). The consequences for riparia are numerous and severe. We return to this topic at the end of this chapter.

Disturbance Ecology: Responses to Stress

In native riparian systems, changes caused by extreme but natural perturbations are seldom more than a temporary setback and, over the long term, have a beneficial effect on total system functioning. The immediate transformation sets the stage for recovery, which allows the system to adapt to changing environments (Rapport and Whitford 1999). In contrast to natural perturbations, anthropogenic stress is not a revitalizing agent but a debilitating one. Stressed riparia may not fully recover, and further degradation may follow.

There are remarkable similarities in the responses of different ecosystems to anthropogenic stress, and riparia fit existing models very well. Stressed systems are characterized by a "distress syndrome" indicated not only by reduced biodiversity and altered productivity but also by increased disease prevalence, reduced efficiency of nutrient cycling, increased dominance of exotic species, and smaller, short-lived opportunistic species. Indeed, there is an emerging research effort aimed at gaining a broad perspective of the functional traits of plants in response to extreme disturbances (e.g., McIntyre et al. 1999, Rood et al. 2003a).

Riparia are subjected to four broad types of anthropogenic stress: flow regulation, pollution, climate change, and land use (see Table 7.2). Each type of stress has common, as well as unique, effects on riparia. Flow regulation, which includes direct alteration of the flow regime by dams, water abstraction, channelization, or levee construction, either lowers the water table, alters the flooding pattern, or isolates riparia from the river. Pollution either adds excess nutrients or toxic materials, such as by acid rain. Depending on the type of pollutant and its application regime (concentration, timing in relation to growth cycle, frequency of application, and so forth), it can either increase or decrease productivity but certainly changes community assemblages. Climate acts through the precipitation and temperature regimes. The response of riparia to climate stresses is location specific because of regional differences in climate changes acting in concert with the local species pool. This is probably the most difficult environmental change to forecast because of the great heterogeneity expected in the newly emerging precipitation and temperature regimes. Land use alters vegetative cover, encourages the invasion of nonnative species, and results in a never-ending array of new management prescriptions as societal conflicts proliferate. The net effects are equally difficult to predict because of strong feedbacks to flow regulation, pollution, and climate. In essence, land use change is an experiment set into motion without adequate means to monitor the changes, formulate solutions, or accurately predict the outcomes.

Table 7.2 Summary of the Major Types of Anthropogenic Environmental Change and Their Principal Effects on Riparia

Environmental Change	Principal Effects on Riparia
Flow Regulation	
Flow Regime	Alters community composition and successional processes; loss of life history cues
Dams	Lotic to lentic; inundation above dam; altered flow, nutrient, sediment, and temperature regimes below dam
Withdrawals	Lowers water table; alters flow regime; decreases alluvial aquifer recharge; system simplification
Channelization and Dredging	Lowers water table; desiccates riparian forest causing terrestrialization and change in community composition; possible decline in biodiversity
Levees	Isolates river from floodplain, thereby reducing hydraulic connectivity laterally and vertically. Constrains channel migration; alters riparian successional trajectories
Pollution	
Nutrients	Increases productivity; shifts community composition toward tolerant species; organic loading leads to redox changes
Toxic Materials and Acid Rain	Decreases productivity; declining biodiversity; system simplification; shifts community composition toward few tolerant species
Climate	
Precipitation	Modifies entire flow regime, groundwater–surface water exchanges, and channel morphology and stability; loss of life history cues
Temperature	Changes spatial patterns and phenology of riparian species
Land Use	
Vegetative Cover	Modifies albedo and feedbacks to climate; changes local microclimate and successional trajectories
Invasive Species	Produces introgression and hybridization; increased competition for space and resources; may reduce biodiversity
Resource Management	Usually alters successional trajectories and community composition

Generalized responses to anthropogenic stress depend on stress type and on landscape position. As discussed in Chapter 6, interactive pathways and their individual strengths are variable; for example, depending on whether the riparian zone is high in the mountains or low in the valley. Changes in landscape position and dominance of interactive pathways are concomitant with alterations in species pools, nutrient regimes and availability, and other biophysical characteristics. Consider atmospheric deposition of nitrogen, either as a nutrient or as increased acid precipitation. It generally has stronger effects at higher altitudes where nutrients are more limiting or where the pH-buffering capacity is naturally low.

One important consequence of having multiple anthropogenic stresses added to natural disturbance regimes is a decline in the number of vegetative successional pathways. Under natural conditions, there is a relatively well-defined number of possible successional pathways, depending on the specific region. However, the increasing number of stressors decreases the number of possible successional pathways. Along France's Garonne River, modern agriculture and urbanization have isolated the river from its floodplain, fragmented the forest cover, and substituted species in the riparian zone. As a consequence—and as we saw in Chapter 4—there are now several "new" successional pathways and "new" mature communities for the vegetation on the alluvial terraces replacing the natural successional avenues (see Figure 4.12). When fluvial disturbance from natural flooding is intact, a diversity of riparian assemblages and seral stages is sustained. When flooding is artificially suppressed, rejuvenation does not occur and the successional trajectory becomes unidirectional. In the absence of fluvial dynamics and periodic inundation, species formerly restricted to the terrace or the uplands are able to invade the historically dynamic riparian zone and competitively exclude true riparian species. The greatest number of successional trajectories normally occurs in a natural system, which has a disturbance (flood) gradient ranging from highly disturbed within the active floodplain to completely undisturbed by fluvial processes on the hillslope. The overall effect of flow regulation on floodplains is to impose equilibrium conditions on nonequilibrium communities. In the case of the Garonne River, human-mediated changes resulted in a buildup of silt and clay, allowed terrestrial species to persist, added nonnative species, and acidified soils—which collectively form a sharp contrast to the historic successional pathways. Identifying, characterizing, and predicting when a specific successional pathway will occur is becoming increasingly challenging as new human-mediated disturbance regimes appear.

Some human-mediated changes cascade throughout the broader ecological system in ways that are often unpredictable but with severe consequences for riparia. A pervasive example is the proliferation of reactive nitrogen (Nr), whose inputs to terrestrial systems have more than doubled worldwide (Galloway and Cowling 2002). As the new Nr cascades through various environmental reservoirs, it contributes to a wide variety of environmental changes—one of which is N saturation in riparia (as well as other parts of the catchment). Increased Nr alters net primary production and nutrient cycling, interacts with elevated carbon dioxide, causes significant reductions in biological diversity, influences rates of nitrogen transfer via hydrologic pathways, and drives acidification and eutrophication in aquatic ecosystems (Matson et al. 2002). Altered N loading may eventually decrease the resistance and resilience of riparia via pH-mediated cation losses, nutrient imbalances, or increased plant susceptibility to stresses such as frost, ozone, or insect attack, depending on the ecoregion and the nutrient composition of soils. The complexity of this one issue—nitrogen—is enormous, underscoring the unforeseen ways by which human actions cascade throughout the environment.

Ecological Consequences of Flow Regulation

Massive hydrologic alterations—to ensure water and energy for agricultural, industrial, and domestic purposes or for flood protection—have changed riparian characteristics throughout the world. An estimated two-thirds of the fresh water flowing to the oceans is obstructed by ~45,000 large dams >15 m high (GWSP 2004), at least 800,000 slightly smaller dams, and literally millions of "minor" diversions such as artificial ponds and roof catchments (e.g., more than 2.5 million dams <8 m high exist in the United States; Postel and Richter 1997). The world's dams alone store 10,000 km^3 of water; seven times more than the total volume of water in all rivers and equivalent to a 10 cm layer spread over the world's dry land surface. In the United States (excluding Alaska), at least 90 percent of the total discharge of rivers is strongly altered, mainly by damming and water abstraction (Jackson et al. 2001). Additionally, long stretches of many rivers are further constrained by artificial levees and dikes. More than 500,000 km of waterways have been altered for navigation worldwide, and more than 63,000 km of canals have been constructed. Globally, the net result is severe regulation of water flow in most catchments, with equally pervasive alterations to riparia (Nilsson and Berggren 2000).

The main responses of riparia to flow regulation depend on the type of regulation—dam characteristics, dikes, and diversion—and the local geology and climate. Dams can have both upstream and downstream effects by altering fluxes of nutrients and migrations of organisms (Pringle 1997). Dikes and bank stabilization isolate rivers from their floodplains. Diversions either dewater rivers, thereby isolating floodplains, or they add water to rivers (e.g., interbasin transfers), thereby modifying natural moisture regimes. The responses of riparia to flow regulation are conditioned by local fluvial processes, which are determined by the geologic and climatic factors governing flow variability and sediment load.

Forecasting how flow regulation will change a particular riparian community requires quantifying and integrating the multiple and cumulative influences of the water regime on the existing vegetation—a complex task. Nonetheless, there are well-documented cases of the effects of flow regulation on riparian communities, and a body of general theory and ecological principles that can be applied. First we explore current theory addressing responses and recovery of riparia to flow regulation.

Theory

The principle body of theory addressing flow regulation by dams is the serial discontinuity concept (SDC), which views impoundments as major disruptors of longitudinal gradients along rivers (Ward and Stanford 1983). According to the SDC, dams result in upstream–downstream shifts in biotic and abiotic patterns and processes. The direction and extent of displacement depend on environmental variables of interest and are a function of dams' spatial position along the river. As originally formulated, the SDC did not consider interactions between rivers and their floodplains. Later, a floodplain perspective was added using a three-reach characterization: a constrained headwater reach, a mid-catchment braided reach, and a lower-catchment meandering reach (see Figure 7.7).

The constrained headwater reach has characteristics similar to those described in the original SDC, but the braided and meandering reaches, with their extensive riparian zones, provide a theoretical perspective on the consequences of flow regulation for riparia. The theory predicts that flow regulation by dams (or any flow-modifying human activity) will have ecological consequences that can be characterized by the magnitude of change from natural conditions and the downstream distance of ecological recovery to natural conditions appropriate for that section of the river corridor. For example, the relative strength of connectivity between natural rivers and their riparian zones varies along the fluvial corridor. A dam placed in the constrained head-

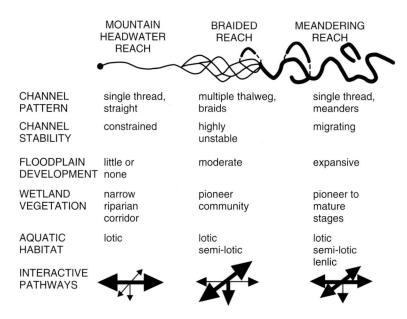

Figure 7.7 Three-reach model river-riparian system and some general features that distinguish reaches. Arrows indicate the relatively strengths of interactions along longitudinal (horizontal arrows), vertical (vertical arrows), and lateral (oblique arrows). From Ward and Stanford 1995.

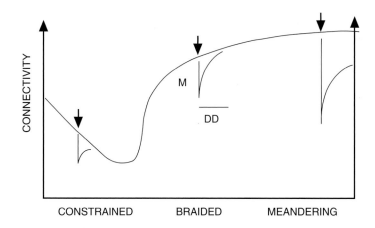

Figure 7.8 The thick line shows an idealized downstream pattern of ecological connectivity between the main river channel and its riparian zone. Arrows represent the placement of a dam. M = magnitude of displacement from natural conditions by the impoundment. DD = downstream distance required for connectivity to return to natural conditions. From Ward and Stanford 1995.

waters has a different ecological effect from one placed in the meandering section of the lower river in both magnitude of ecological adjustments and downstream distance required for recovery of "natural" ecological conditions (see Figure 7.8). In general, this perspective provides a conceptual model allowing one to envision how flow regulation affects the biophysical characteristics of riparia. Unfortunately, like any pervasive human modification, the model becomes highly complex when flow regulation devices are spread throughout the catchment, when additional human modifications are present (i.e., cumulative effects), or when numerous biophysical responses are considered simultaneously.

Extent of Flow Regulation

Flow regulation is now global in scope. Humans actively manage a large—over 30 percent—and increasing fraction of the world's runoff in inhabited regions, and this fraction is likely to intensify to nearly 70 percent in two decades (Postel 2000, Vörösmarty and Sahagian 2000). Management of the world's rivers has resulted in profound changes to the dynamics of the terrestrial water cycle (GWSP 2004).

The extent of dam construction and associated flow regulation on a single catchment can be massive (Rosenberg et al. 2000). For example, there are 19 large dams on the main channel of the 2,000-km-long Columbia River in the United States and Canada; only 70 km (<4 percent) of the river remains free flowing. The Columbia catchment as a whole contains 194 large dams. Almost 200 reservoirs occupy the Danube River catchment. Eleven large hydropower stations and 200 small and large reservoirs (inundating 26,000 km^2 of land) have been built on the Volga-Kama River catchment, and more than 130 reservoirs have been built on the River Don catchment (inundating 5,500 km^2). The list is seemingly endless.

Examples of flow alterations on the extent, duration, and frequency of floodplain inundation are pervasive. After closure of the Aswan high dam, the Nile River showed a reduced annual discharge, truncated annual floods, higher base flow rates, and a several month shift in the timing of the flood peak. The maximum-to-minimum discharge ratio decreased from $12:1$ to $2:1$, with far-reaching consequences on floodplain inundation (see Figure 7.9). The Senegal River, however, showed a gradual decrease in peak, average, and low water discharge, primarily as a result of increasing water abstraction for irrigation. During the dry season, the Senegal River now frequently ceases to flow. The Danube River (downstream of Vienna) has a relatively unaltered hydrology with frequent "flood" and "flow" pulses. However, the floodplain is disconnected from the main channel by artificial levees, which drastically reduce the duration and frequency of floodplain inundation (Tockner et al. 2000). The predictable historic pattern of regular flood peaks of the Columbia River is now reduced to a chaotic pattern responding to human-generated needs for energy (Stanford et al. 1996).

Currently, about 3,800 km^3 of water is withdrawn annually worldwide, primarily for agriculture (Revenga et al. 2000). Hydrological alterations by water withdrawal can be extreme in some systems; for example, less than 1 percent of the natural flow of the Colorado River reaches the mouth. The Murray River in Australia now discharges only 36 percent of its natural flow into the sea, flood duration on the fringing floodplains has decreased from two months to a matter of days, and the timing of floods has shifted from spring to late summer. Similar fates have affected the Nile, Ganges, Amu Dar'ya, Syr Dar'ya, and Huang He (Yellow) rivers, and over-pumping of groundwater plagues the central United States, California's central valley, China's northern plains, and major portions of India (Postel 1997, 2000).

An interesting index of flow regulation is shoreline length, the interface between terrestrial and aquatic elements of the floodplain. In alluvial systems, shoreline length

(a) Nile (Aswan)

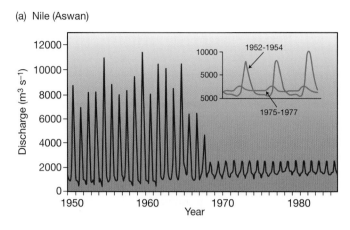

(b) Senegal (Bakel)

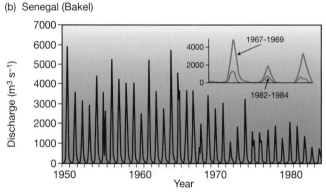

(c) Danube (Vienna)

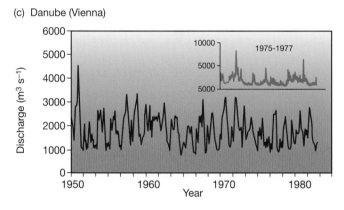

Figure 7.9 Discharge patterns of selected rivers: (a) Nile River below the Aswan dam, (b) Senegal River at Bakel, and (c) Danube River downstream of Vienna. Modified from Tockner and Stanford 2002.

can be up to 25 km per river km and remains high throughout the annual cycle, except during major flood events. In flow-regulated rivers, however, shoreline length drops to about 2 km per river km (Arscott et al. 2002, Tockner and Stanford 2002). Reduction in shoreline length not only affects habitat availability of already endangered riparian communities but also impedes the exchange of matter and organisms between rivers and their riparian zones.

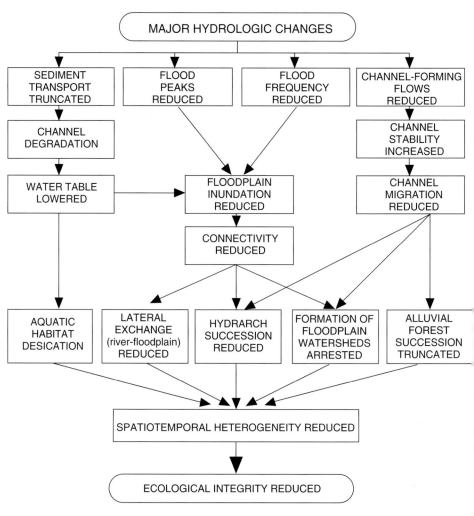

Figure 7.10 Some ecological implications of major hydrological changes induced by flow regulation on downstream river–floodplain systems. From Ward and Stanford 1995.

Effects on Native Species and Processes

There are multiple mechanisms by which flow regulation affects the ecological integrity of riparia. Flow regulation alters sediment transport and reduces flood peaks, flooding frequency, and channel forming flows (see Figure 7.10). Ultimately, these modifications affect riparia by lowering water tables, reducing lateral fluxes of water and materials, accelerating and modifying successional processes, and stopping the formation of new habitats (Ward and Stanford 1995). Essentially the riparian zone undergoes a "terrestrialization" process that undermines the natural ecological vitality.

Because flow regulation typically reduces the frequency and extent of floodplain inundation and lowers the water table, there are major ramifications for riparia. Productive pioneer species tend to be replaced by less productive upland species that are able to invade the floodplain under artificially enhanced conditions of environmental stability (Décamps 1984). In essence, the overall effect of flow regulation on floodplains is to impose equilibrium conditions on nonequilibrium communities. It should not be surprising that communities developing after flow regulation are poorly adapted to the infrequent but catastrophic flows resulting from "flood control" measures.

In earlier chapters we learned that recruitment, establishment, and survival of many riparian tree species (e.g., *Populus*, *Salix*) are tightly coupled to flow regimes—as is true for many other riparia taxa. Plant reproductive strategies and floods are linked. However, downstream of dams and diversions, floods are reduced and may not coincide with seed dispersal. Reduced rates of channel migration, coupled with reduced flooding and sediment transport, slow the formation of new bare substrates required as seedbeds. Lowered water tables from channel degradation (e.g., incision, arroyo formation) may induce drought stress, killing seedlings and old trees as well as reducing the growth rates of surviving trees (Rood et al. 2003a,b). The pioneering forest is adversely affected, being gradually replaced by a late successional community dominated by species that, under natural conditions, occur in the nearby uplands.

This is well illustrated in semiarid South Africa where riparian communities are often subjected to multiyear droughts (du Toit et al. 2003). Normal wet periods are characterized by strong plant production, more litter of better quality, relatively light herbivory with good-quality feces added to the soils, and a general increase in N availability (see Figure 7.11). The generally excellent upland savanna conditions mean that herbivory is spread across the landscape rather than being concentrated in the riparian zones. However, as drought severity becomes more acute, savanna conditions begin to deteriorate, there is a decrease in plant biomass and production, and palatable plant species are over-grazed in the riparian corridors. The consequences are that litter quality and N availability are reduced, the incidence of stand replacing fires in the riparian zone is more frequent, and terrestrial plant species begin to invade the river corridor. The net result is a proliferation of successional pathways depending on fire severity, subsequent rainfall, sediment dynamics, and herbivory.

Contrasting responses of the riparian vegetation to flow regulation can occur too. In the Platte River, Colorado–Nebraska, flow regulation was accompanied by expansion of riparian woodlands (Johnson 1994). River regulation transformed the Platte River from a wide braided system largely devoid of woody vegetation to a single thread channel with a dense riparian corridor. This contrasts with nearby meandering rivers, such as the Missouri, where flow regulation greatly reduces riparian vegetation. The expansion of *Populus* and *Salix* in braided rivers is positively correlated with low flow periods rather than peak flow events. Flow regulation, by reducing spring flows, provides suitable conditions for seedling establishment across much of the braided channel. In general, the long-term response throughout many riparian systems is a decrease in pioneer species and an increase in shade-tolerant trees, shrubs, and herbs that can reproduce in the absence of physical disturbance.

Specific responses to flow alterations largely depend on interactions between species-specific life history traits and the new environmental conditions. For example, along rivers in the arid interior of the western United States, flood control has been associated with the spread of the nonnative shrub saltcedar (*Tamarix* spp.). Like its native counterparts, cottonwood and willow, saltcedar produces abundant wind- and water-dispersed seeds capable of becoming established immediately on bare, moist substrates. Compared with cottonwood and willow, adult saltcedar is more tolerant

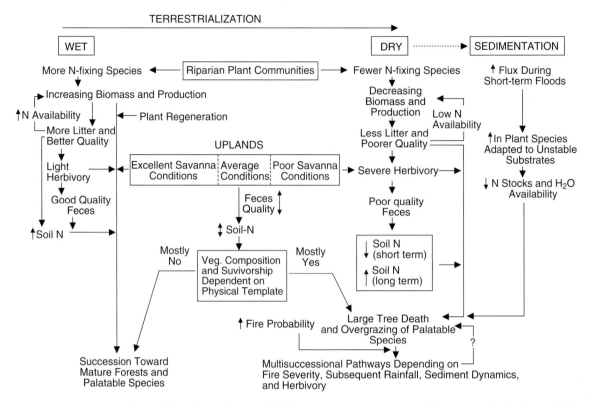

Figure 7.11 A conceptual model of the environmental responses of riparian corridors in the semiarid region of South Africa to increasing drought conditions. In effect, the riparian system becomes more and more terrestrial as species and processes normally associated with the savanna uplands move closer to the river.

of high salinity, fire, and drought but less tolerant of flood disturbance (Friedman and Auble 2000). Whereas viable seeds of cottonwood and willow are present for only 1 to 2 months in early summer, saltcedar releases seed throughout the summer. Flood control and diversions foster the spread of saltcedar and stifle cottonwood and willow through reduction of moisture (also allowing the accumulation of salts on the floodplain), decreasing physical disturbance, accumulating flammable leaf litter, and delaying peak flows beyond the seed viability period.

Flow regulation also fragments the continuity of plant communities and reduces plant diversity. In Sweden and elsewhere, dams obstruct the dispersal of plant propagules and cause shifts in downstream community structure (Jansson et al. 2000a,b). Riparian species richness along Sweden's regulated rivers is lower for most plant groups, and in no case higher than in free-flowing rivers. Reasons for this are that dams retain species with poor floating capacity, causing them to become unevenly distributed along the riparian corridor and, in cold northern climates, many nonnative species fail to become established.

Alterations in flow are suspected to not only affect plant communities but also the associated fauna. One of the best examples is from eastern Colorado. Almost 90 percent of the 82 breeding bird species seen each spring were not present in *ca.* 1900 AD (Knopf 1986). The new species use forests that developed during channel-narrowing along rivers such as the South Platte as well as trees planted in shelterbelts. At the same time, flow-related channel narrowing degraded the habitat for birds such as whooping crane (*Grus americana*) and the least tern (*Sterna albifrons*) that seek wide braided channels. In general, however, comparative studies on mammals and birds in regulated and unregulated river systems are almost entirely lacking. Most existing data are in unpublished reports evaluating the environmental impacts of dam construction or generated by limited personal observations (Nilsson and Dynesius 1994). Nevertheless, at least locally, the most destructive effects of flow regulation appear to be the inundation of valley bottoms or the dewatering of streams. In these cases, the food base, communication routes, and feeding areas disappear for native species.

Alterations to Energy and Nutrient Budgets

Alterations to the riparian community structure by flow regulation are reflected in the standing stocks and fluxes of energy and nutrients in the reconfigured system. In Europe, there are two classic studies of riparian functional responses to flow regulation. The first took place along the lower reaches of the rivers Dyje and Morava (Czech Republic) when flow was regulated to prevent flood damage to crops and other structures (Penka et al. 1991). The second took place along the Rhine River (France/Germany) in three areas: one receiving regular floods, one not flooded for 30 years, and one not flooded for ~130 years (Trémolières et al. 1998).

The Czech study is remarkable for its longevity (1970 to ~1985), complexity, and detailed synthesis. Numerous researchers examined dozens of biophysical riparian responses to water regulation, summarizing the results in a comprehensive book (Penka et al. 1991). Some of the key findings related to flow regulation were (1) soil moisture content remained almost optimal for plants in the lowest lying areas but declined to the wilting point for plants at the higher floodplain elevations, (2) input of nutrients from flood waters to the soils ceased, (3) organic matter decomposition increased as decomposition processes were fundamentally altered by having more oxygen in soils, (4) primary production and physiological responses by the plants were species-specific, and (5) there were predictable changes in the community structure of the shrub and herb communities as soils became dryer. Various species of riparian animals also responded, depending on their individual life history characteristics, to ecological conditions and the structure of the plant community. Not surprisingly, the pathways and fluxes of water, energy, and nutrients were altered forever by flow regulation.

The basic patterns seen in the Czech study are generally repeated in the Rhine investigations (Trémolières et al. 1998). The community structure and species diversity of three stands (i.e., flooded, unflooded 30 yr, and unflooded 130 yr) were substantially changed by flow regulation. The unflooded sites responded with increases in plant biodiversity and tree density. Similarly, there were ~3 times more mycorrhizal species in the unflooded sites. However, the production of autumn leaf litter biomass remained about the same (~4 Mg/ha/yr for all sites).

As a consequence of the collective environmental changes instigated by flow regulation of the Rhine River, the construction of partial nutrient budgets reflects changes

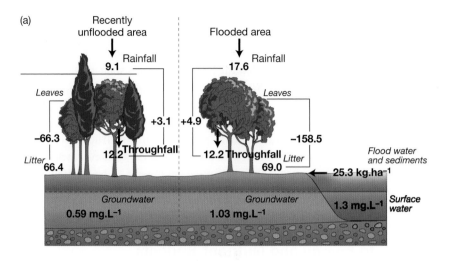

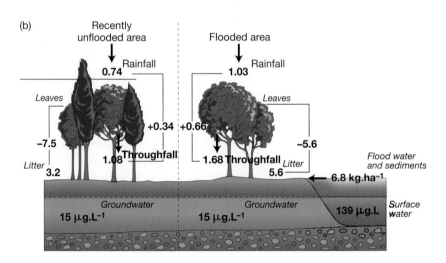

Figure 7.12 Nitrogen (a) and phosphorus (b) budgets in two riparian stands along the Rhine River. One stand remained naturally flooded while flooding was prevented in the other for 30 years. Inputs are expressed in kg/ha/yr and outputs in concentration, mg/L for N and μg/L for P. From Trémolières et al. 1998.

in community structure and ecological processes (see Figure 7.12). Nitrogen and phosphorus budgets differ mainly because of community composition and the presence or absence of nutrients from regular floods. However, nutrient contents in soils, mature leaves, and leaf litter generally differ significantly in their N and P concentrations but not in their potassium (K), calcium (Ca), or magnesium (Mg) concentrations between flooded and unflooded sites (see Table 7.3). Nitrogen concentrations of mature leaves from the unflooded sites are ~63 percent of concentrations in leaves from the flooded site. However, soil N concentrations are ~18 percent greater at the unflooded sites. Similarly, P concentrations in mature leaves are slightly lower (by ~10 to 20 percent) at the unflooded sites, whereas P concentrations in soils are only ~58 percent of those at the flooded site. Surprisingly, only one significant change (soil Ca) was observed

Table 7.3 Nutrient Levels in Soil, Mature Leaves (August) and Leaf Litter for the Two Vegetation Units (Flooded Hardwood and Unflooded Hardwood for 30 Years)

	Flooded	Unflooded for 30 Years	Unflooded for 130 Years
Nutrient level in the mature leaves ($kg\,ha^{-1}\,yr^{-1}$)			
N	227.5 ± 58.7	134.7 ± 33.5*	151.9 ± 28.6*
P	11.2 ± 2.7	10.7 ± 1.4	8.3 ± 1.4*
K	106.7 ± 20.8		105.6 ± 32.7
Mg	26.2 ± 9.0		25.7 ± 14.7
Ca	198.7 ± 51.2		211.6 ± 52.6
Nutrient level in the leaf litter ($kg\,ha^{-1}\,yr^{-1}$)			
N	69.0 ± 15.2	66.4 ± 14.9	68.5 ± 14.2
P	5.6 ± 1.7	3.2 ± 1.0*	3.6 ± 1.2*
K	53.0 ± 15.6	62.2 ± 9.9	55.1 ± 23.0
Mg	14.7 ± 6.8	8.6 ± 1.1	13.3 ± 6.8
Ca	172.8 ± 43.9		181.3 ± 48.7
Nutrient level in the upper soil layer, 10 cm ($kg\,ha^{-1}$)			
N	58.8 ± 11.0	71.3 ± 6.2*	67.8 ± 7.4*
P	36.3 ± 15.2	24.5 ± 7.2*	17.3 ± 7.2*
K	97.8 ± 16.3	142.5 ± 11.6*	124.2 ± 57.3
Mg	169.7 ± 27.7	192.5 ± 21.9	150.0 ± 18.0
Ca	8958 ± 1264	8709 ± 383	6971 ± 109*

Data expressed in $kg\,ha^{-1}\,yr^{-1}$. The species analyzed are the most frequent species of the hardwood forest, i.e., *Fraxinus excelsior, Quercus robur, Ulmus minor, Acer campestre, Carpinus betulus, Corylus avellana, Cornus sanguinea, Clematis vitalba, Hedera helix.*
*= significant at the 0.05 level.
Trémolières et al. 1998.

for K, Ca, and Mg between the flooding treatments, even after 130 years of flow regulation.

The Czech and Rhine studies point to the importance of soil processes in shaping ecosystem characteristics after flow regulation. Soil conditions—as related to moisture regimes—governed the changes in plant succession, the disappearance of pioneer stages, and the appearance of new communities and their attendant flux processes. There are, of course, lag-times involved because of the persistence of many plants for decades after the initiation of flow regulation. Changes in vegetation are most noticeable after a century because of continued lowering of the water table, which causes the disappearance of hydrophilous plants, and because of changes in soil nutrients. The decline in availability of N and P is rapid (<30 yr), and is related to the severed connectivity between the river and its floodplain and to changes in plant productivity. In light of these system-scale shifts in structure and nutrient fluxes, equally long lag-times may be expected for riparian restoration and rehabilitation (see Chapter 10).

Basic Ecological Principles

The alteration of flow regimes is the most serious contemporary threat to the ecological sustainability of rivers and their associated floodplains (Naiman et al. 2002c, Nilsson and Svedmark 2002). While the obvious and often irreversible impacts of large impoundments are well recognized, there is a growing awareness of the pivotal role of the flow regime as a key "driver" of the ecological integrity of riparian zones. There is still much to learn about the ecological significance of individual flow events and sequences of events, and descriptive science can take us only so far in unraveling these linkages. General advice on environmental flows might be regarded as a series of largely untested hypotheses about how flows affect organisms and how riparia function in relation to flow regimes (Bunn and Arthington 2002).

There is a small set of overarching ecological principles that, if used to guide river management, may ameliorate many of the difficult flow regulation issues facing resource managers and policy makers. The principles, which are deceptively simple, are distilled from decades of research in river ecology and, most importantly, on several spatially broad overviews of river and riparian characteristics (e.g., Puckridge et al. 1998, Richter and Richter 2000). The basic principles are:

- The flow regime determines the successional evolution of riparian plant communities and ecological processes.
- The river serves as a pathway for redistribution of organic and inorganic material that influences plant communities along rivers.
- Every river has a characteristic flow regime and an associated riparian community.
- Riparian zones are topographically unique in occupying nearly the lowest position in the landscape, thereby integrating catchment-scale processes.

The major challenge for riparian management is to place water resource development within the context of these fundamental ecological principles in order to maintain ecological vitality (i.e., goods and services) for the long term.

Despite growing recognition of the relationships between riparian characteristics and flow regimes, ecologists still struggle to predict and quantify biotic responses to altered regimes. One obvious difficulty is the inability to distinguish the direct effects of modified flow regimes from effects associated with other changes that often accompany water resource development. One often encounters river systems affected by multiple stressors, making it nearly impossible to definitively separate the effects of altered flow regimes from those of the myriad of other factors and interactions associated with climate and land use changes.

Consequences of Global Climate and Land Use Changes

Climate Change

The global climate regime is responsive to the release of greenhouse gases. Today, the Earth's climate is vastly different from even a few thousand years ago, when ice sheets covered much of the Northern Hemisphere. Although the Earth's climate continues to change, climatic change in the distant past was driven by natural causes, such as variations in the Earth's tilt of axis or the carbon dioxide content of the atmosphere. Modern changes in the Earth's climate, however, are being driven increasingly by human actions (Schneider and Root 1998). Humans are indirectly altering natural flows of energy by enhancing the natural capacity of the atmosphere to trap radiant heat near the Earth's surface—the greenhouse effect. Average global surface temperatures are projected to increase by 1.5 to 5.8°C by 2100 AD (Houghton et al. 2001), but increases may be higher in North America (Wigley 1999). The seemingly small projected increase in temperature is expected to result in broad ecological changes, with strong implications for riparian communities.

The enormous emphasis placed on greenhouse warming by the international community is a quest to understand global change (e.g., Houghton et al. 2001, McCarthy et al. 2001, Metz et al. 2001). This emphasis is clearly justified from the standpoint of water resources owing to the evidence that climate and weather systems have departed from the historic patterns to which ecological systems and human civilizations have become adapted (Lettenmaier et al. 1999). Yet, the several additional sources of human intervention on the global water cycle discussed earlier—land cover change, urbanization, industrial development, and water resources management—have hydrological impacts beyond the greenhouse effect alone (Rosenberg et al. 2000, GWSP 2004, Kabat et al. 2004). In fact, the cumulative impact of these factors, *considered over the full global domain*, is likely to surpass the effect of recent and anticipated climate change over decadal time scales (Meybeck and Vörösmarty 2004).

Documentation of changes in runoff is geographically limited but collectively shows that runoff patterns are being altered, with attendant consequences for riparian systems. In the United States, there have been strong increases in monthly mean stream flow during the cool season months (November–April) at nearly half of 1,000 stations between 1948 and 1988 (Lettenmaier et al. 1994). The largest changes occurred in the North Central states, including the Upper and Middle Mississippi basins and the western Great Lakes. In general, the conterminous United States is becoming wetter (Lins and Slack 1999). Perhaps most alarming, high stream flow (>90th percentile) has increased in the eastern United States since about 1940, and this increase is correlated with increasingly heavy and extreme precipitation during the same period (Karl and Knight 1998, Groisman et al. 2001). However, trends in Canada have the opposite pattern (Zhang et al. 2001). Annual mean stream flow generally has decreased since 1947, with significant decreases detected in the southern part of the country. Also, Canada is not experiencing the more extreme hydrologic events seen in the eastern United States.

Outside North America, there are fewer studies for direct comparison. However, international trends in "great floods," defined as discharges exceeding the 100-year return interval in river basins with drainage areas exceeding 200,000 km^2, are compelling. The focus on extreme floods reduces the influence of nonclimatic factors like land use change and water management. In 29 river basins distributed throughout the globe, 21 exceedances of the 100-year return interval occurred over the periods of record for all sites, of which 16 were in the second half of the data set (Milly et al. 2002). This proportion is statistically significant, suggesting that the risk of "great floods," and therefore the catastrophic replacement of riparian communities, is increasing.

Ecological responses also are expected to be uneven and will depend on the geographic region. For example, even though the world is undergoing a steady increase in temperature because of the greenhouse gases, various regions have differential capacities to absorb and hold the energy. Further, the existing general circulation models cannot resolve local or regional details of how the resulting weather will affect biotic communities—nor the importance of how regional topography, water bodies, vegetation, or heterogeneous soils will modify local warming processes. The fundamental fact emerging for riparian zones is that the ability to accurately forecast the effects of climate change will be formidable. Synergistic interactions, thresholds, buffering, and other system-scale properties, especially when combined with habitat fragmentation, present daunting challenges to those trying to develop reliable forecasts—but that does not mean that they cannot be achieved. It just means that the task is difficult, requiring focus, innovation, and patience.

Projected changes in climate will place additional pressures on already-stressed river and riparian systems. Although riparia are generally viewed as resilient and able to maintain a healthy and self-sustaining condition despite large year-to-year variations in hydrologic and temperature conditions, rapid climate change imposes new environmental regimes that may exceed biotic limits. Even though riparia have historically experienced elevated temperatures similar to those projected for the next 100 years, the projected rate of change falls outside the natural range of variation and is therefore unprecedented (Peters 1989). Moreover, the extent of landscape fragmentation, within which riparian systems are embedded, is historically unprecedented (Dale 1997). Riparian systems are increasingly isolated and disconnected, making adjustments to rapid climate change through animal and plant dispersal problematic. Thus, climate change clearly represents a significant threat—one that will interact in complex ways with existing human-caused stresses (Carpenter et al. 1992, Firth and Fisher 1992, Décamps 1993).

The ecological consequences of an altered climate largely depend on the rate and magnitude of change by two critical environmental drivers: temperature and water availability. As we have seen, these factors regulate many riparian processes, both directly and indirectly (see Sidebar 7.2). Even though average temperatures will increase across North America and Europe, the increase will be greater at higher latitudes (Wigley 1999). Some continental areas will become wetter and some drier, and the timing and quantity of precipitation will change, thereby altering runoff patterns. Although the precise geography of the regional shifts is uncertain because of

---

Sidebar 7.2 Possible Effects of Changing Climate on Riparian Zones

Climate changes can have far-reaching impacts on catchment-scale hydrologic processes, influencing the quantity of runoff draining to the river, the preferential hydrologic pathways followed by the runoff, and thus the frequency and magnitude of river floods, low flows, and riparian soil and groundwater regimes. These hydrologic changes in turn influence the sediment sources that can be tapped by the runoff, the caliber and quantity of sediment delivered to the river system, and the sediment transport regime of the river (Steiger and Gurnell 2003). In this way, climate change can have major impacts on fluvial processes controlling the character of riparian zones and so can change riparian morphology, sedimentary structure, and disturbance characteristics. Further, since the requirements for seed germination and growth of different riparian plants are highly sensitive to microclimate, inundation duration and frequency, and soil water regime, variations in climate can have far reaching impacts on vegetative composition.

Gurnell and Petts (2002) proposed a simple model for river island formation, growth, and destruction along the Tagliamento River, Italy, which considers the impact of differences in dominant reproductive strategies of riparian trees subject to the same river flow regime. In the Alpine climate of the upper reaches of the river, the main riparian tree species is *Alnus incana*. This species does not reproduce very freely by vegetative reproduction and so colonization of gravel bars is relatively slow and is largely achieved from seedlings (vegetation trajectory A in Figure SB7.2). In the Mediterranean climate of the lower reaches of the river, the main riparian tree

---

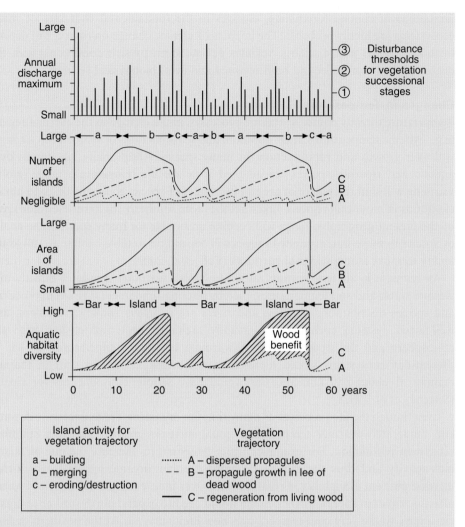

Figure SB7.1 Temporal variations in the significance of three vegetation recruitment and growth trajectories for the number and areal extent of riverine islands and for aquatic habitat diversity. A model based on observations along the Tagliamento River, Italy. The vegetation successional stages in the top graph represent transitions from (1) seedlings and saplings to pioneer shrubs; (2) pioneer shrubs to maturing riparian trees; (3) maturing riparian trees to fully mature riparian woodland. From Gurnell and Petts 2002.

species, *Populus nigra*, reproduces freely from both vegetation fragments and seeds. In particular, vigorous sprouting from uprooted, flood-deposited trees results in rapid growth rates. A substantial and robust vegetation cover can develop quickly, which is capable of protecting bar surfaces from moderate floods and inducing high sedimentation rates to support island development (vegetation trajectory C in Figure SB7.2). The production and deposition of large amounts of dead wood protect delicate seedlings, giving an intermediate vegetation growth trajectory (vegetation trajectory B in Figure SB7.2). There is little overlap in the two riparian tree species along the course of the Tagliamento, reflecting their different climatic requirements. Therefore, the large differences in the extent and persistence of islands under the three trajectories depicted in Figure SB7.2 could reflect changes at a single site, if a change in climate occurred that was sufficient to induce vegetation species change or to induce a change in the vigor of colonizing plants.

Angela M. Gurnell, King's College London and Geoffrey E. Petts, University of Birmingham, UK limitations in climate forecasting, changes in the fundamental character of many ecosystems are highly probable. The impacts of climate change on riparia will depend on how thermal and discharge regimes deviate from present conditions, with the degree of deviation reflecting both regional and local biophysical settings.

Changes in Temperature Regimes

Temperature directly controls vital life processes, and a change in the thermal regime (e.g., extreme temperatures, their duration, and seasonal rates of temperature change) will alter growth and reproduction for many species. Additionally, since individual aquatic and riparian species are adapted to a specific temperature range, global warming will shift the potential geographic ranges of species to the north, or to higher elevations in mountain regions. Likewise, the more southerly (or lower-elevation) part of the present geographic range will become unsuitable for many species. The ability of species to move into expanded ranges will depend on suitable habitat as well as the ability to move along dispersal corridors (Poff et al. 2002).

An increase in air temperature caused by global warming translates directly into warmer water and soil temperatures, thereby altering fundamental riparian processes and species distributions. Streams and rivers are relatively shallow, turbulent, and well-mixed systems, meaning they readily exchange heat and oxygen with the atmosphere and the riparian zone. A warming of water temperatures by 4°C in present-day ecosystems would represent a northward latitudinal shift in thermal regimes of about 680 km, and this would have serious consequences for riparia (Sweeney et al. 1992).

Even though the life processes of many riparian organisms are temperature dependent, moderately warmer temperatures could increase growth rates and stimulate ecosystem production. For example, higher temperatures increase the rate of microbial activity and thus the rate of decomposition of organic material, which may increase nutrient availability in riparian soils. Over time, even groundwater will warm, affecting riparian species further. Some would need to expand northward or to higher elevations, where very cold temperatures now limit survival. At the northern or high elevation end of current distributional ranges, conditions should improve for cold-adapted species because of the favorable warming. In northern, circum-polar areas, winter water temperatures are likely to increase by several degrees Celsius, eliminating extensive ice cover and permafrost, and allowing invasion of cool-adapted species (Poff et al. 2002).

Changes in Precipitation and Runoff Regimes

Regional patterns of precipitation and runoff dictate how discharge varies seasonally, affecting the amount of available riparian habitat, soil moisture regimes, and what kinds of species can flourish. Yet, because climate varies from region to region, the hydrologic regime also varies among regions and among catchments. Nevertheless, a change in regional climate that alters the extant hydrologic regime can greatly modify habitat suitability for many species and cause significant ecological change (even if the thermal regimes remain unchanged).

Prophetically, a modified seasonal pattern of runoff in response to climate change will alter riparian composition and system productivity—and the timing and magnitude of flooding are central in this process. In ecological terms, flood peaks are critcal events (Poff et al. 1997, Meyer et al. 1999). Streams in the northern United States, in Canada, and in much of the mountainous western United States have snowmelt runoff regimes with a predictable period of high flow in late spring followed by a predictable period of late summer, fall, and winter base flow. By contrast, in the deserts of the American Southwest, winter flows are variable; summer discharge is usually low because of lack of rainfall, but intense monsoon-driven thunderstorms can create flash flooding. In the eastern United States, precipitation is generally high enough to sustain year-round flow in streams. Thus, knowledge of particular regional changes in discharge regimes brought about by climate change is critical to anticipating ecological impacts. Some projections are reasonably well supported by data, whereas others need to be considered as scenarios evaluating the range of ecological concerns related to changing climate (Poff et al. 2002).

Early Snowmelt Runoff

A major consequence of global warming will be a shift from spring peak flows to late winter peaks in snowmelt-dominated regions (Frederick and Gleick 1999). Many of the life history characteristics (e.g., reproductive strategies) of both aquatic and riparian species have evolved to avoid or take advantage of predictable high spring flows. For example, successful reproduction by cottonwood trees depends on snowmelt that creates high spring flows inundating floodplain habitat (Rood and Mahoney 1990, Auble et al. 1994; also see Chapter 5). A warmer climate that reduces the magnitude of spring flows could have broad impacts.

Another significant consequence of shifting from snow to rain at high elevations or northern basins is the reduction of discharge in late summer. This is expected even if winter precipitation increases in northern latitudes because excess precipitation will not be stored as snow, which provides a source of runoff to sustain late summer baseflow in arid highlands (see Frederick and Gleick 1999). Lower summer discharge translates into a "terrestrialization" process for riparia. Further, less water in the stream channel means less water flowing into streamside groundwater tables, which are important for sustaining riparian tree communities (Stromberg et al. 1996, Scott et al. 1999). As a consequence, riparian communities are likely to experience conspicuous changes in species composition and productivity.

Increased Precipitation Variability

Some climate change models predict possible increases in the intensity of rainfall on fewer rain days (Kattenberg et al. 1996), and this already has been observed in the United States over the last century (Karl et al. 1996). The implications for rivers and riparia are significant. For example, in a historical reconstruction of flood histories for upper Mississippi River tributaries over the last 7,000 years, small shifts in temperature (1 to 2°C) and precipitation (10 to 20 percent) caused sudden changes in flood magnitude and frequency (Knox 1993). Because the transport and storage of nutrients and pollutants depend on flow, an increase in flood magnitude or frequency would result in more silt and pollutants entering streams and rivers. In fact, the peak daily loads of the herbicides Atrazine, Alachlor, Cyanazine, and Metolachlor attained 7,110 kg/day, 720 kg/day, 3,130 kg/day, and 2,250 kg/day, respectively, near the mouth of the Mississippi River during the 1993 flood (Goolsby et al. 1993). Peak daily loads of an additional 16 herbicides used in the catchment were not measured. Annual applications of Atrazine, Alachlor, Cyanazine, and Metolachlor in the basin ranged from ~8,000 to 21,000 Mg from 1987 to 1989 (Gianessi and Puffer 1991)—many of which come in contact with riparia. Not surprisingly, the corresponding flood-related degradation in water quality could lead to a loss of sensitive riparian species. Further, because floods often rearrange the structure of riparian sediments, they indirectly affect physiological and reproductive processes through impacts on moisture regimes. The literature suggests that a substantial increase in flood frequency would cause major shifts in species composition and system processes.

Reduced Discharge

Even if flooding increases in magnitude and frequency, earlier snowmelt and higher temperatures could still result in lower summer discharge in many areas. In addition, some areas could become generally drier and thereby become particularly stressful for river and riparian systems.

Prolonged dry periods selectively eliminate biota requiring wetter conditions. In humid regions characterized by greater annual precipitation, such as the eastern United States, periods of no flow are less common. Accordingly, increasing the duration of no-flow periods would represent a large deviation from normal conditions, and thus be ecologically damaging. One previous analysis suggested that a 10 percent decline in annual runoff in eastern U.S. streams having little groundwater input could result in nearly half the streams ceasing to flow in some years (Poff 1992b).

Many aquatic communities in large rivers are partially dependent on riparian floodplains, either for nursery habitat for fish or for seasonal export of nutrients from floodplain wetlands to the river (Sparks et al. 1990, Bayley 1995). If these floodplains become disconnected from the main rivers because of reduced discharge, it is obvious that aquatic productivity and diversity would decline.

Can Riparia Adapt to Climate Change?

In general, the ability of ecosystems to adapt to climate change is limited. Expected rates of climate change are probably too great to allow adaptation through natural genetic selection. Moreover, many types of habitat will be diminished or lost entirely. Not all species needing to disperse northward or to higher elevations will be able to move to hospitable locations. Further, most high-quality riparian habitats are now spatially isolated, making successful dispersal even more difficult and local extinctions easier. Many surprises also are expected as migrating species come into contact with new species for the first time, particularly nonnative species.

Overall, there is strong scientific consensus that riparian systems are especially vulnerable to projected climate change. However, the amount of ecosystem alteration directly attributable to climate change will be difficult to ascertain because other powerful agents of environmental change (e.g., human activities, nonnative species) continue to independently and interactively degrade riparia (Carpenter et al. 1992, Firth and Fisher 1992, Poff et al. 2002). One critical uncertainty in projecting future aquatic ecosystem response to a changing climate is how humans will interact with changing river and riparian conditions. Human activities have changed many aquatic and riparian ecosystems by diversion, groundwater pumping, and the building of dikes, levees nd reservoirs, all of which have modified natural processes and increased system vulnerability to the additional stresses associated with climate change. Cumulatively, these alterations fragment the aquatic landscape, making dispersal between ecosystems more difficult. We examine these changes further in the next section on land use.

Land Use Change

In little more than a few decades, the fragmentation of large land areas into smaller parcels has become an environmental issue of worldwide proportions. Yet fragmentation is but a phase in the broader sequence of transforming land by natural or human causes from one type to another (Forman 1995). The phrase *land use change* has several meanings: here we use it to include changes in both the vegetative cover and how the land is used (Turner et al. 1998). *Land cover* refers to the habitat or vegetation type present, such as forest, agriculture, and grassland. *Land cover change* describes differences in the area occupied by cover types through time. *Land use* is usually defined more strictly and refers to the way by which humans employ the land and its resources. Land use change encompasses all those ways in which human uses of the land have varied through time (see Sidebar 7.3).

Land use change has pervasive consequences for riparia, and a comprehensive discussion is well beyond the scope of this book (but see the treatments by Turner et al. 1990 and Forman 1995). Here we address the causes and consequences of the ones deemed to be most important: temperature regimes, nutrient enrichment, and invasive plants. Terrestrialization, which is equally important, was covered earlier in this chapter.

Temperature Regimes

Temperature, a critical driver of biotic processes, is strongly affected by land use change. Riparia influence stream temperatures, while the temperature regimes of streams also influence air and soil temperatures within the riparian zone. Tempera-

---

Sidebar 7.3 Historical Changes in Poplar Populations Along the Garonne River, France

Riparian woodlands have been drastically reduced in surface area and genetic diversity along most European rivers as a result of human activity, as exemplified by popular (*Populus* spp.) along the Garonne River, France (see Figure SB7.3). Riparian woodlands occupied almost 1,200 ha in 1810 AD over a 42-km stretch. Most were destroyed in two distinct periods: a first wave around 1830 AD (disappearance of 598 ha) and a second wave around 1970 AD (disappearance of 395 ha). Only 221 ha remain (19 percent of the woodland existing in 1810 AD), which corresponds to a shift of the natural riparian woodland bordering the river from a width of 277 m in 1810 AD to 53 m today.

Poplar plantations mostly replaced the natural riparian woodland during the first wave of destruction. However, patterns of erosion and sedimentation remained reversible, and the width of the riparian woodland fluctuated between 135 m and 168 m. During the second wave of riparian woodland loss, intensification of river management activities and gravel mining resulted in a radical breakdown of natural river dynamics, causing a vertical channel incision of about 20 m. Since then, the river has followed a stabilized and increasingly incised channel at low flows. When erosion and sedimentation remained reversible, the sinuosity of channel fluctuated between 1.30 and 1.38 and the length of the side channels between 54 and 58 km. Today, the sinuosity is stabilized at 1.30, and 27 km of side channels have disappeared.

Genetic diversity was equally affected by the human-mediated changes. Black poplar (*Populus nigra* L.), a pioneer tree species growing along the banks of rivers in Europe, is still self-regenerating along the Garonne,

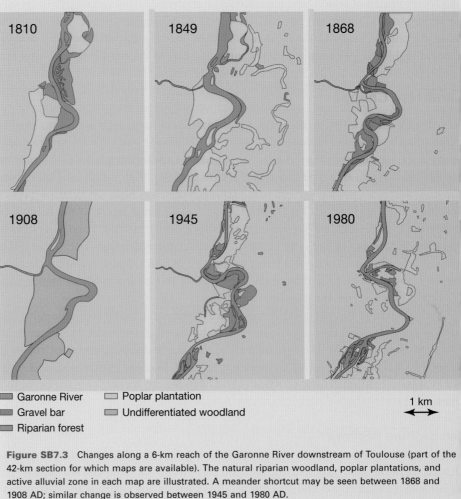

Garonne River   Poplar plantation
Gravel bar   Undifferentiated woodland
Riparian forest

1 km

Figure SB7.3   Changes along a 6-km reach of the Garonne River downstream of Toulouse (part of the 42-km section for which maps are available). The natural riparian woodland, poplar plantations, and active alluvial zone in each map are illustrated. A meander shortcut may be seen between 1868 and 1908 AD; similar change is observed between 1945 and 1980 AD.

although its potential recruitment areas represent only 38 percent of what they were in 1945. However, *Populus nigra* is threatened by widespread planting of poplar clones, which have the potential to weaken native genetic stocks. Wide-scale planting of introduced cultivars is occurring with the dominance by only a small number of clonal types such as *Populus nigra* L. cv. "italica," *P. nigra* L. cv. "de Garonne," *P. deltoïdes* Marsch ssp *angulata* cv. "carolin," and, in the 20th century, Euramerican clones (i.e., hybrids of *P. nigra* and *P. deltoïdes*) such as Robusta, Regenerata, I 45/51, and I 214. In 1810 AD, half of the floodplain woodlands were already planted. Today, plantations occupy 91 percent. It is therefore debatable whether the genetic potential of the Garonne's black poplar populations have declined through mixing with stands of poplar cultivars or because of the historic changes. Particularly, radical changes in hydrologic conditions may have favored genotypes better adapted to drier conditions and nitrogen-enriched water.

Etienne Muller, Centre National de la Recherche Scientifique (C.N.R.S.) and Université Paul-Sabatier, Toulouse, France

ture regulates metabolism, growth, solubility of gases, sequencing of life history traits as well as many other biophysical processes and interactions.

Variations in water temperature are the result of multiple factors that interact with one another. Direct solar radiation on the water surface is a dominant source of hea

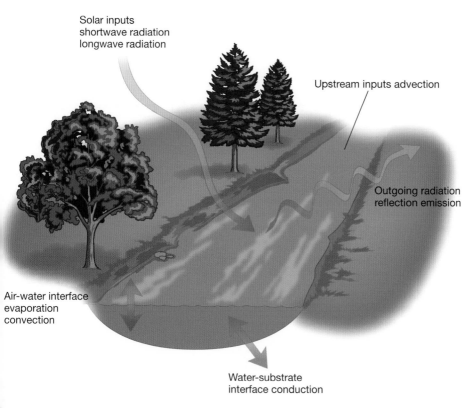

Solar inputs
shortwave radiation
longwave radiation

Upstream inputs advection

Outgoing radiation
reflection emission

Air-water interface
evaporation
convection

Water-substrate
interface conduction

Figure 7.13 Factors and mechanisms influencing stream temperatures. From Johnson and Jones 2000.

for streams, but other sources and fluxes of energy also contribute to water temperature at any given point (see Figure 7.13). These include energy exchange by conduction between stream water and the stream substrate, evaporation and sensible heat exchange with the atmosphere, and energy contributed by advection of water from deep groundwater sources and from upstream.

Forest harvest in riparian areas can increase stream temperatures in summer and cool them in winter, with the magnitude varying among sites and regions. It is widely assumed that changes to stream water temperatures also alter the temperature regimes of riparian soils and roots. Sites where only overstory vegetation is removed generally have smaller changes than sites where the understory is also removed. Responses among sites also vary by the amount of stream surface exposed by harvesting. Unfortunately, most studies of forest harvest effects are limited to a few post-harvest years or the research design does not allow characterization of the preharvest temperature regime. A notable exception is the analysis by Johnson and Jones (2000) of streams in the H.J. Andrews Experimental Forest in Oregon. After forest harvesting, maximum stream temperatures increased by 7°C and occurred earlier in the summer, and June diel variations increased by 6°C. Stream temperatures gradually returned to preharvest regimes after 15 years as the forest recovered, better regulating short-wave radiation and heat conduction from terrestrial soils. No studies have been conducted in winter, but it is strongly suspected that riparian forest removal affects ice formation, especially frazzle and anchor ice. This is because the lack of an overhead canopy

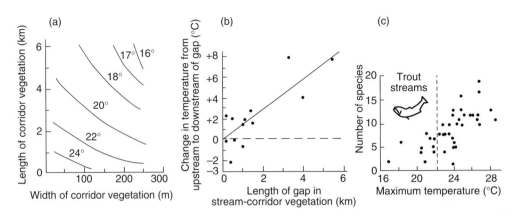

Figure 7.14 Riparian vegetation, water temperature, and fish populations. (a) Effect of upstream vegetation corridors of varying length and width on predicted stream temperature (degrees Celsius). (b) Change in stream temperature according to the length of the gap in riparian vegetation. (c) Fish species richness as related to stream temperature. Of many variables studied, only weekly maximum water temperature clearly distinguished between trout and nontrout streams; streams with temperatures >22°C had marginal or no trout populations. All graphs based on weekly maximum temperature for pool-riffle portions of 40 streams in southern Ontario, Canada. From Forman 1995, based on Barton et al. 1985.

in an open reach of stream will promote anchor ice formation as a result of radian heat loss from the substrate on cold clear nights in early winter. A canopy cover, especially a coniferous one, acts like an insulator to reflect radiation back to the stream and, thereby, reduces the likelihood of supercooling within the stream, which is necessary for initializing anchor ice formation (Gridley and Prowse 1993).

Does a riparian corridor need to be continuous in order to provide a reasonable temperature regime? And what is the effect of gap size in the corridor on temperatures? It is well known that land use changes, causing increased stream temperature in regions with cold-water fisheries, have been linked to fish mortality, prevalence of disease, and changes in interspecific competition (e.g., Meehan 1991). Only one study however, has addressed the riparian continuity questions (Barton et al. 1985). They found that in southern Ontario trout streams, the proportion of banks covered by vegetation is positively correlated to water temperatures required by trout (see Figure 7.14). As the proportion of vegetated banks increases upstream, water temperatures concentrations of fine particles, and variability in water flow decrease, further benefiting trout. Of these variables, trout are most sensitive to the weekly maximum water temperature. An impressive 56 percent of the variation in temperature could be explained by the proportion of forested stream bank within 2.5 km upstream of a site Short gaps in a riparian corridor probably have little effect on water temperature However, in Ontario, riparian gaps >1 km significantly increase stream temperature (see Figure 7.14), with consequences for the entire stream community and, presumably, for the riparian community too.

Nutrient Enrichment

Human activities greatly affect the amounts and ratios of nutrients being cycled between the living world and the soil, water, and atmosphere (Melillo et al. 2003) Perhaps the best-known case is for nitrogen (N). Humans have already doubled th rate of nitrogen entering the land-based N cycle, and that rate continues to climb (Galloway and Cowling 2002, Vitousek et al. 2002). Alteration of the N cycle is important because added N has myriad effects on nontarget ecosystems. Enrichment with N often affects the species composition, productivity, dynamics, and diversity of recipient ecosystems. Increased plant growth in response to moderate N additions has been widely documented. The growth of many animals—indeed the health of humanity—is constrained more by protein supply than by energy, and decomposition and other microbial processes can be shaped substantially by the supply of available N.

There are two primary mechanisms for the production of reactive nitrogen (Nr). Since the scientific discovery of the Haber-Bosch process in 1913, which allowed the mass production of NH_3 from N_2 gas, there has been virtually an unlimited supply of Nr to grow food (and make explosives). Today, more than half of the food eaten by the world's peoples is produced using nitrogen fertilizer from the Haber-Bosch process (Galloway and Cowling 2002). Energy production also creates Nr, although in contrast to food production, it is produced by accident as a consequence of fossil fuel combustion. The sources of Nr in energy production are twofold: (1) conversion of atmospheric N_2 to NO_x and (2) conversion of fossil organic N in the fuel to NO_x. Both sources result in increased Nr—one from the creation of new Nr and one from the release of sequestered Nr.

The net results of these human-mediated developments are that N deposition on land from direct application and from the atmosphere are increasing exponentially, and that the ecological consequences are severe. In addition to reactive nitrogen's well-known contribution to acid rain, it also contributes to a decoupling of carbon and nitrogen cycles (Asner et al. 1997) and to a saturation of an ecosystem's capacity to effectively assimilate and utilize the added nitrogen (Hanson et al. 1994). Additionally, there are fundamental shifts in the chemical stoichiometry of available elements needed for growth, survival, and other ecosystem processes (NRC 2000, Sterner and Elsner 2002). One poorly appreciated aspect of the worldwide alteration in elemental stoichiometry is that, proportionally, the global P cycle has changed substantially more than the N cycle as materials rich in P are mined and redistributed across the landscape, mostly in fertilizer (Sterner and Elsner 2002).

Collectively, alterations to the global N and P cycles—and the ecological consequences to uplands—are suspected to strongly affect riparia because of their spatial position at the land–water interface. Eventually, most of the added nutrients will make contact with riparia. Changes in basic chemical stoichiometry and system saturation will be manifested in decreasing retention capacity for one or more elements by riparia, with attendant downstream effects. In fact, this is already documented for most of the developed world with N export rates to the coastal zone continuing to rise, accelerating increases in harmful algal blooms, coastal hypoxia, and fish kills (NRC 2000).

What about effects of the added nutrients on the riparian zones themselves? One would expect that species composition and productivity of floodplain organisms would be influenced by the quality of in-flowing water. Detailed site-specific data on this topic for riparian zones are sparse; however, there are a few examples. Artificially high nutrient inputs led to an impoverishment of floodplain communities in the lower Rhine River (van den Brink et al. 1996). Other riparian areas have become major sources of nutrients and pesticides, especially when intensively cultivated. In many Southeast Asian rivers, huge amounts of pollutants are discharged from cultivated riparian zones into the main river during seasonal floods (e.g., Dudgeon 2000).

In developing countries, an estimated 90 percent of wastewater is discharged directly into rivers and streams without treatment, and in many parts of the world, rivers are so polluted that their water is unfit even for industrial uses (Johnson et al. 2001). In China, 80 percent of the 50,000 km of major rivers are too polluted to sustain fisheries, and in 5 percent fish have been completely eliminated. Despite major treatment efforts, pollution is still a major problem in the northern hemisphere. For example, 75 percent of the water in the Vistula, Poland's largest river, is unsuitable even for industrial use (Oleksyn and Reich 1994). High nutrient concentrations are also major obstacles to restoring floodplains along many rivers such as the Danube and the Rhine (Buijse et al. 2002). What are the possible consequences to riparian organisms in such environments?

Nutrient saturation, especially N saturation, has a number of damaging consequences for riparian organisms. These impacts became apparent in Europe in the late 1970s when scientists observed significant increases in nitrate concentrations in lakes and streams. Additionally, as ammonium builds up in riparian soils, it is increasingly converted to nitrate by microbial action (see Chapter 5), a process that releases hydrogen ions, acidifying the soil. The buildup of nitrate enhances emissions of nitrous oxides and also encourages leaching of highly water-soluble nitrate into streams and groundwater. As these negatively charged nitrates seep away, they carry with them positively charged alkaline minerals such as calcium (Ca^{2+}), magnesium (Mg^{2+}), and potassium (K^+). Thus, the human modification of the nitrogen cycle not only increases N losses from the riparian soils, it also accelerates the loss of Ca and other nutrients that are vital for plant growth. Plants growing in soils replete with N but starved of Ca, Mg, or K develop nutrient imbalances in their roots and leaves, resulting in reduced photosynthetic rates and efficiencies, stunted growth, and increased mortality. Unfortunately, specific studies of riparian vegetation in response to increased nutrient pollution are few.

Invasive Species

The Earth's biota is being rapidly homogenized as human activities introduce numerous species outside their natural ranges. Species invasion is probably the second most important cause, behind habitat loss, of the overall decline in biodiversity and the changing composition of riparian communities. The combination of land transformation, altered hydrology, and numerous deliberate and accidental species introductions has perpetuated widespread invasions. Mechanisms for introduction can be direct, such as planting nonnative species in gardens, or they can be subtle, such as propagules being moved passively by automobiles. The normal grime from a single car can transport a hundred species and carry thousands of seeds (Schmidt 1989).

Many riparian systems are already dominated by alien species. The high percentage of exotic plants and animals observed in floodplains, as compared to neighboring uplands, demonstrates the vulnerability of riparian zones to invasion (DeFerrari and Naiman 1994, Pyšek and Prach 1994). In fact, the same factors supporting high diversity in riparian habitats (transport of propagules, flooding disturbance, water availability) may also increase their susceptibility to invasion by exotic species. Along

France's Adour River, 198 riparian plant species disappeared between 1989 and 1999, and in 1999, 153 species were new (Tabacchi and Planty-Tabacchi 2000). Today, invaders account for one-quarter of the total species richness of 1,558 species and locally can constitute up to 40 percent of all species. Despite great differences in climate, species richness, and land use history, the proportion of invasive species along the Adour River is similar to rivers along the Pacific Coast of the United States and to South African rivers. About one-quarter of the riparian plant species along the Willamette River (Oregon); the Hoh and Dungeness rivers (Washington); the Santa Ana and Santa Margarita rivers (California); and the Sabie, Olifants, Letaba, and Crocodile rivers (Kruger National Park, South Africa) are nonnative (Planty-Tabacchi et al. 1996, Hood and Naiman 2000). In general, frequently disturbed sites near the active channel contain higher percentages of nonnative species, and the local cover by nonnative plants may be virtually complete.

Theoretically, there are two alternative hypotheses about the invasibility of plant communities by nonnative species. The first hypothesis dates to Charles Darwin and suggests that the high plant species richness should reduce invasibility because communities composed of many strongly interacting species are thought to limit the invasion possibilities of similar species (Case 1990). The second hypothesis suggests that highly diverse plant communities are intrinsically unstable, with some species dropping in and out routinely, and native species dropping out might be replaced by nonnative species (see Stohlgren et al. 1998). Under the second hypothesis, nonnative plant species might invade and coexist with high numbers of native plant species as long as basic resources were not limiting. Proximity to source areas for propagules also might be important. Which hypothesis is correct? A comparison of whole river systems for the north temperate zone shows no relationship between species richness and the proportion of exotic species (Hood and Naiman 2000). However, on a smaller patch scale, contrary to the predictions of the first hypothesis, species richness is positively correlated with the percentage of nonnative species in the riparian zones of France, Washington, Oregon, the central United States, and other locations (DeFerari and Naiman 1994, Planty-Tabacchi et al. 1996, Stohlgren et al. 1998). Species richness may be important in reducing the invasibility of undisturbed communities (however, see the discussion in Stohlgren et al. 1999 for uplands) but not in communities experiencing frequent disturbance, such as in riparian zones where abiotic factors (including soil fertility) often dominate biotic interactions.

Invasions by trees can be severe in riparia, both ecologically and economically. In the southwestern United States, as a result of widespread human-induced changes in hydrology and land use, native cottonwood–willow stands are being replaced by nonnative woody species such as Russian olive (*Eleagnus angustifolia*) and tamarisk (*Tamarix* spp.; also known as saltcedar). The superior drought tolerance of tamarisk, relative to native phreatophytes and its ability to produce high-density stands and high leaf area indices, are major reasons why it is favored when flow is reduced (Cleverly et al. 1997). Based on a cost–benefit calculation, the presence of tamarisk in the western United States will cost an estimated US$7 billion to US$16 billion in lost ecosystem functions within the next decades (*ca.* US$15,600 to US$24,600 per ha; Zavaleta 2000).

Can one predict the ecological outcome of a single species colonizing a riparian zone? The answer is probably not. After all, only a very small percentage of the available species pool has the life history characteristics and physiological tolerances needed for successful colonization. Those that do colonize either remain rather inconspicuous in the community or they spread quickly throughout the riparian corridor in a few years—and these latter species are noticed. All colonization patterns are characterized by large variance and exceptions, therefore precluding confident predictions of the outcome of any particular introduction. Developing accurate predictions remains a formidable challenge for ecology (Clark et al. 2001) and will be discussed further in Chapter 11.

Conclusions

The pervasive environmental changes being experienced by riparia are receiving careful scrutiny in many parts of the world. They are the subjects of management debates and strategies, widespread conservation practices, and costly rehabilitation efforts. We now turn our attention to those important topics, which collectively have captured the attention of scientists, decision makers, and the public.

Management

8

Overview

Riparian management is based on understanding fundamental ecological and related social processes. This chapter concentrates on the management of riparian resources and the ecological services they provide, and acts as a prelude to the two following chapters that examine conservation and restoration.

Riparian management is a recent and evolving concern requiring an economic evaluation, a social and cultural perspective, a suite of suitable institutions, and the collection and dissemination of information. Whether socioeconomic systems affect the health and integrity of riparian systems has consequences for many resources and ecological services, including bank stability, fish production, water quality, scenic amenities, and recreation.

An understanding of human perceptions of riparia is crucial to planning, designing, and implementing sustainable management. Riparian management represents cultural challenges as well as the long-term retention of distinctive management-caused features and memories. Effective riparian management requires institutions that encourage people to join together to achieve common goals. Continuing education is essential for encouraging citizens to monitor socioenvironmental conditions and to develop a thoughtful and long-lasting stewardship.

Riparian management is intimately linked to catchment and river management. Catchment management provides a framework for implementing riparian management, and river management is an essential component of healthy corridors and floodplains. Inversely, an appropriate management of riparian systems may have consequences that go far beyond their limited spatial areas, propagating throughout the entire river corridor, floodplain, and catchment.

Riparian management is highly site specific. There are important community variability's in time and space, strong legacies of ancient practices, and connections between social and environmental components at multiple scales. Additionally, riparian biological communities evolve through a dynamic equilibrium between persistence and invasiveness of constituent species. These properties require one to specify what can be an adaptive, a sustainable, and a locally appropriate management approach.

- Riparian management has a strong human dimension in that it must build and maintain ecological resilience and social flexibility through participatory approaches. Communication is critical, and subjective and objective approaches should be combined when planning management strategies. Cultural, social, and historical determinants of human perception are as important as ecological rules.

- Riparian management has significant impacts on human and ecosystem health, now and into the future. Effective management requires cooperation of diverse institutions at local, regional, national, and sometimes international levels. Riparian management also encompasses broader issues as riparia are part of catchment and river systems, often revealing or expressing landscape dysfunctions as community assemblages and processes respond to new environmental conditions.

Purpose

System-scale management of riparia is the process of land and water use decision making and practices that account for the full suite of organisms and processes that characterize and comprise the system (Dale et al. 2000). The goals of riparian management are to provide self-sustaining ecological goods and services now and into the future, and to maintain important ecological linkages throughout the landscape. Successfully attaining the goals depends primarily on the maintenance of species diversity and the age composition of organisms, as well as the structural complexity of riparian soils, plant cover and associated communities. In practice, riparian management zones often include floodplains and vary in width and extent according to ecologically significant boundaries. Riparian management is based on the best understanding of fundamental ecological and related social processes (Naiman and Bilby 1998). In the previous chapter, we saw to what extent change may affect these processes, provoking all sorts of ecological and social disruptions—many with long term consequences. This leads to the notion that managing riparia is managing the resilience of systems—systems that may be considered to be socioecological entities (Walker et al. 2002). Such management should be adaptive and cooperative, recognize the need for social communication, and consider both local and catchment scales as well as short- and long-term processes. Effective management integrates conservation, management, and restoration—the CMR trio—in all projects (van Andel and Aronson 2005).

Although management, conservation, and restoration are closely allied, their viewpoints and goals differ. Therefore, we examine their individual implications for riparian resilience in three separate chapters. The focus of this chapter is on the management of resources and ecological services provided by riparia—a recent and evolving concern that is firmly rooted in river and catchment management but highly site specific in its adaptiveness, its sustainability, and its appropriateness. Riparian management has a strong human dimension, and this appears as an emerging issue. In contrast, Chapter 9 examines riparian conservation in the context of the current biodiversity crisis with a view toward preventing species extinctions and the maintenance of biophysical functions. Finally, Chapter 10 considers restoration, which is the process of assisting riparia toward recovery to a functional condition.

Riparian Management: A Recent and Evolving Concern

Early civilizations in a large part of the developing world probably did not pay much attention to riparia as natural systems. Two of their primary concerns were obtaining drinking water and providing irrigation. Later, the development of urban centers led to increasing water demand and to construction of aqueducts and delivery systems. Natural waterways were modified for transportation of goods and people, and artificial ones were constructed when needed. The river's potential energy was also used to grind grains—thanks to the invention of waterwheels—further modifying rivers and shorelines. Basically, water rather than riparia was the focus of early civilizations (see Table 8.1), and this focus eventually resulted in pervasive modifications of natural riparian systems. The extent of the modifications may be imagined from the following excerpt (cited by Cech 2003) from the Justinian Code (533 AD), an illustration of the "Riparian Doctrine" that provided the framework for water allocation throughout the Roman Empire:

> The public use of the banks of a river is part of the law of nations, just as is that of the river itself. All persons therefore are as much at liberty to bring their vessels to the bank, to fasten ropes to the trees growing there, and to place any part of their cargo there, as to navigate the river itself. But the banks of a river are the property of those whose land they adjoin; and consequently the trees growing on them are also the property to the same persons.

It may be added that dams to control floods in drainage basins were constructed as early as the first three centuries AD in Tripolitania, Libya (see Figure 8.1). One of the dams erected to protect a city from silt was 7 m high, 133 m long, and 7.25 m thick at its base (Vita-Finzi 1969).

Riparian management is a recent concern for those responsible for the integrity and protection of natural resources. Riparia have been long considered as areas to be drained and destroyed, or to be used as disposal areas. Even today one can find riparia and islands covered with garbage. Consider the Mississippi River, where since 1997

Table 8.1 Selected Ancient Water Development Events

Year BC	Event	Present Location
3200	King Scorpion proclaims "Day of Breaking the River"[1]	Egypt
2280	Yu the Great constructs various waterworks	China
1000	Quanats[2] constructed	Middle East
560	Cheng State Irrigation Canal completed	China
500	Water harvesting developed	Middle East
500	Dikes and levees constructed for flood control	China
312	First Roman aqueduct built	Italy
100	Waterwheels used to grind grain	Greece

The initial cutting of ground for a new canal.
Underground water delivery system consisting of a long tunnel with vertical shafts.
After Cech 2003.

235

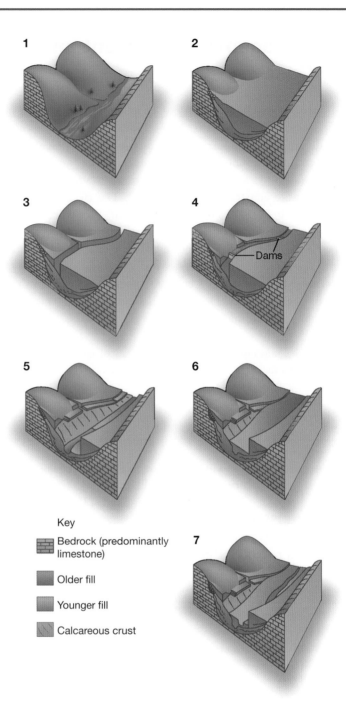

Figure 8.1 History of a Tripolitanian wadi in Libya: (1) original valley; (2) deposition of older fill, with calcareous enrichment and crust formation in the early stages (last glaciation); (3) the wadi cut down into the fill as far as the first major crust, which it breaches in a few places (post-Neolithic and pre-Roman times); (4) dams are erected to accumulate silt; the wadi continues to shift over the crust toward the axes of the valleys originally cut in bedrock (Roman period); (5) the crust is breached at many dam sites; drainage density is increased by the development of gullies (late glacial period); (6) the younger fill is deposited to a maximum depth of 10 m, with a consequent reduction in drainage density (Middle Ages); and (7) both fills are eroded; the wadi flattens its longitudinal profile and gullies once again increase drainage density (present day). From Vita-Finzi (1969).

Chad Pregracke has been cleaning more than 1,600 km of shoreline and camping on the islands of this braided system (Cech 2003). Pregracke found these areas filled with old household appliances (e.g., refrigerators and TV sets) and other items and decided to do something about it (Kilen 2000). This is far from an isolated example: rivers and riparian areas are widely used as depositories for old bicycles, tires, car batteries, and other items that are difficult to dispose of properly, making it necessary to organize beautification and restoration projects (see www.livinglandsandwaters.org).

In fact, riparia have been greatly modified by many means and in many ways since the origin of human civilizations on the riverbanks of most continents 2,000 to 7,000 years ago. The origin of these modifications—and the neglect of riverbanks—is related to managing river flows and can be illustrated with examples from many parts of the world. In Europe, primeval flooded forests covered about 200,000 ha on both sides of the Rhine in the 18th century, whereas only 15,000 ha remained in 1970 after river channelization, and only 400 ha in France and 12,000 ha in Germany in 1990 (Schnitzler 1994). Fortunately, in the last two or three decades, some appreciation has emerged of the strong and intimate links between terrestrial and aquatic systems that shape the character and productivity of riparia (Carbiener and Schnitzler 1990), as well as the natural resources and environmental services provided by intact riparian systems.

It is important to acknowledge that riparia have not always been perceived as important aspects of the landscape, as they are today in many countries. Human perceptions of riparia change in time and space in relation to social and cultural biases. It is against the backdrop of broader environmental alterations—such as climate and land use change—that the shift in human perceptions must be addressed if one is to understand how sustainable uses of river riparia have been maintained in historical times, and may be maintained in the future within different cultures. Addressing the diversity of human perceptions is paramount for sustaining river riparia through management. Additionally, the scientific perception of riparia is only part of the human perception—and it too has changed profoundly during the last century. Nancy E. Langston (1998), a forest and environmental historian at the University of Wisconsin, says it well: "perceptions of the natural world are refracted through a set of cultural lenses" that "profoundly shape the practice of science and the ways science is used to transform the Earth." All sciences are affected by the culture and ideologies of their practitioners. Recognizing this in ecology is to admit that there are cultural biases in our scientific and personal models of nature. In essence, what we see may not be all there is to see (Magnuson 1990), and societal perceptions and values underpin the management (as well as conservation and restoration) of riparia. This is reflected in the four pillars for effective riparian management: (1) economic valuation, (2) social and cultural perspective, (3) suitable institutions, and (4) information collection and dissemination.

Economic Valuation of Riparia

An economic valuation of riparia provides a basis for decision making about riparian management. For example, in northern Nigeria, the construction of large dams on the Hadejia River for intensive irrigation led to a reduction in wetlands downstream (Acreman 1997). It was immediately apparent that the economic value of water

Table 8.2 A Taxonomy of Behavioral and Motivational Sources of Economic Value

		Total Economic Value		
Use Values			Nonuse Values (Existence Value)	
Current Use Values		Future Use Value	Altruistic Motives	Stewardship Motives
		—Option Value	—Bequest Value	—Preservation Value
Consumptive Use	Nonconsumptive Use			—Intrinsic Value
—Recreational Value	—Observing Nature			
—Market Values				

After Huppert and Kantor 1998.

was many times less when used for intensive irrigation than when used for supporting fisheries, agriculture, and fuel wood production from riparian wetlands. This led, consequently, to exploring the potential for releasing water from the dams to restore the original riparian wetlands.

The taxonomy of sources of economic value (see Table 8.2) provides a means to understand personal reasons for valuing an environment. Current use values include consumptive activities that lead to deterioration of environmental quality and nonconsumptive activities such as viewing wildlife and natural aesthetics. Future use values correspond to an extension of evaluation over time—what people may be willing to pay to assure that river environments are available for future use. Nonuse value (or existence value) is the value placed, for example, on preserving a spectacular river landscape even though it may never be visited, and may reflect altruistic or stewardship motives. Such a taxonomy explains how socioeconomic systems might affect the health and integrity of riparian ecosystems—with real consequences for bank stability, fish production, water supplies, scenic amenities, recreation, and other goods and services.

However, economic values represent just one of several important considerations when deciding riparian management strategies. First, it is pragmatically difficult to place values on many biotic and hydrologic processes in riparian systems, such as their role in buffering floods and nonpoint source pollution or in maintaining biodiversity. Second, some riparian forests may be considered as having a value in themselves, beyond any human economic use. As landscapes, riparia are cultural expressions as well as nature and tend to become part of the scenery. There are cultural differences in space and time in the way human groups embrace nature and landscape myths— the endurance of which is sometimes surprising (Schama 1995). Large rivers such as the Nile, Danube, Rhine, and Mississippi, among many others, are known for having created regional identities linking rivers and their riparian landscapes. These regional identities are reflected in riparian management strategies and practices.

Social and Cultural Perspective

Riparia can be thought of as landscapes and, like every landscape, have a natural and a cultural identity. Therefore, their management requires a social and cultural perspective. Viewing riparian areas as landscapes is not a new idea among geographers and ecologists. G. P. Malanson (1993) developed this approach in his book *Riparian Landscapes*, R. T. T. Forman (1995) devoted an entire chapter on stream and river corridors in his comprehensive book *Land Mosaics*, and several ecological papers about riparian systems refer to a landscape perspective (e.g., Lee et al. 1992, Décamps 1996, Ward 1998). It must be emphasized, however, that as landscapes, riparia include much more than patches and ecological processes. Man-made social, economic, and physical structures are considered equally as landscape elements (Söderqvist et al. 2000). As a consequence, understanding landscapes is not only knowing their environmental morphology, it is also knowing the cultural, social, and historical causes of the ways they are perceived. Thus, riparia appear as "ecosymbols" similar to hedgerows, paddyfields, or mountain pastures that are ecological as well as symbolic entities, corresponding to a particular relationship between a social group and its environment (Berque et al. 1994)—and that find form and come alive at certain periods of time (Jackson 1994). In other words, riparian landscapes are relative entities where natural environmental processes and cultures interact (Nassauer 1992, 1995), and this is crucial to planning, designing, and implementing sustainable management, conservation, and restoration (Décamps 2001). Successful management cannot only consider monetary incentives and scientific facts (Burgess et al. 2000).

Continued learning also becomes an imperative prerequisite for ecological management in a world of increasing uncertainty, of changing ways of thinking and doing, and of changing beliefs about rights and responsibilities (Michael 1995). It is important that people appreciate that diverse and connected riparian landscapes are needed to maintain essential and inherent riparian functions underpinning the supply of goods and services over time. Effective riparian management acknowledges the importance of orienting people's attention and care toward diversity and connectivity—two basic properties enabling the creation and maintenance of healthy and aesthetically pleasing riparian landscapes.

Nevertheless, studies on human perceptions of riparian aesthetics are rare. One of them noted a general preference for channels without woody debris, suggesting a needed advancement in the public's appreciation of the significance of large woody debris in rivers (Gregory and Davis 1993). Knowing that large woody debris is important for riverine integrity will affect public perception of the aesthetic value of riparia (see Sidebar 8.1). This is in agreement with views on the necessary linkage between aesthetic perceptions of landscapes and ecological knowledge (Eaton 1997). In modifying the aesthetic of care, one attracts a community's sustained attention to ecological functions (Nassauer 1997) and makes people aware of the importance of change in riparia, a change independent from those who manage them.

Finally, as for land and water, ethical values anchor the sustainable management of riparia. Ethics implies fairness and integrity in the relationships between people, social groups, and the environment, and is strongly rooted in ecological thinking (Marsh 1864, Leopold 1949). It also implies that policy makers will choose the best uses of funds and other resources for the greatest public good, and every individual will act as a member of a community, which includes other people and the environment. An ethic is perhaps all the more needed in riparian management in that it applies to boundary systems having highly important effects on the equitable allocation of water within drainage basins. It has been suggested to reformulate the famous

Sidebar 8.1 Riparian Wood in Rivers: Issues and Challenges

Wood in rivers is now recognized as a key element of aquatic ecosystems, mainly in forested landscapes, which influence physical conditions and livability for human communities. It is now understood that wood, which had been removed in many temperate rivers decades ago in the United States but millennia ago in Europe, plays a positive role in river networks. Wood affects channel geometry (mainly its width and depth), velocity distribution, and grain size pattern (Gregory et al. 2003). It has direct effects on the retention of nutrients in rivers, on water quality, and on the production of invertebrates and other food resources for aquatic organisms. Wood locally creates complex habitat conditions that increase the abundance, survival, and diversity of fish assemblages.

A major contemporary issue is to preserve some wood in rivers or, in some cases, to reintroduce it. Several key questions challenge resource managers: where should wood be introduced, what structural types of wood accumulations are more effective and geomorphically stable, what designs can increase the life span and efficiency of wood accumulations, and how can riparian forests be regenerated and maintained to naturally provide inputs of large wood? A contrasting issue is the continued removal of wood. In such cases, it is important to know the consequences of wood removal and what "soft" solutions can be promoted to reduce the negative impacts.

A major social and management challenge is that wood in rivers also has some negative consequences for people living along riverbanks. Wood in a given reach increases the bed roughness, encouraging flow to move over riverbanks and onto floodplains more frequently. It also diverts flows laterally, increasing critical local shear stress and the probability of bank erosion and channel widening. This is why many countries have regulations preventing wood delivery and deposition in river channels, and why many agencies remove wood from rivers. In France, for example, landowners are required to maintain the living vegetation and to remove wood from the channel—as well to preserve the fauna and the flora. The landowner also has some "riparian" rights: water extraction, fishing, hunting, and taking gravel in limited amounts.

Leaving wood in rivers does not always cause major damage, especially where there are few buildings, roads, bridges, or other social investments. Nevertheless, there is still potential danger downstream because pieces move and become trapped where the probability of flooding and lateral erosion is much higher. Soquel Creek in California is a good example of a situation where wood originating upstream causes flooding downstream due to wood-generated obstructions at bridges (Lassettre 2003). This is a pervasive problem in Europe, especially in France and Switzerland, as well as in North America, where bridges retaining the wood are destroyed by lateral erosion and upstream water pressure on them.

Recent summaries about the dynamics and functions of wood (i.e., Gregory et al. 2003, Montgomery and Piégay 2003) illustrate the roles of wood in river ecosystems and the need to find a balance between ecosystem rehabilitation and preservation of infrastructures. Rather than "clean" rivers of wood frequently and intensively, our recommendations are to use interventions that maintain complexity to the greatest extent possible:

- Clean channels with explicit objectives about the desired health and function of the river. This requires maintenance strategies that vary in intervention frequency and intensity from one reach to another, according to the risks and values.
- Promote options to prevent flooding and erosion, such as design floodable floodplain landscapes or the construction of structures that can trap the wood in safer sections of sensitive reaches.
- Preserve the most ecologically important wood structures in order to limit the negative impacts and, in other cases, use the reintroduction of wood to enhance habitat.

Such approaches embody scientific needs and challenges—understanding what is an ecologically effective structure, what is the best structural arrangement for an accumulation, what is the most efficient density of structures within a given reach, which fish species are favored by different types and locations of wood accumulations, and how the geomorphology of the river interacts with wood to change the system-scale structural and ecological complexity. An additional challenge is understanding wood transport through a river network with the goal of preventing associated risks (Moulin and Piégay 2004). When is wood delivered from forests and floodplains, and is there a linear relation between discharge increases and the amount of wood delivered? What are the travel distances of single pieces during flood events? How does decomposition and abrasion alter the abundance and nature of wood in streams and rivers? Answers to these basic questions would permit better riparian management upstream of sensitive reaches. Further, application of this knowledge in simulation models could be a powerful tool for exploring the consequences of different management alternatives. The growing body of knowledge about the ecological roles of wood in streams and rivers offers many opportunities for scientists and resource managers to develop innovative approaches for providing effective and beneficial balances between critical ecosystem functions and diverse social and cultural needs.

Hervé Piégay, Université de Lyon 3, France
Stanley V. Gregory, Oregon State University, Corvallis, Oregon, USA

Aldo Leopold maxim of land ethics as: "A thing is right when it tends to disturb the biotic community only at normal spatial and temporal scales. It is wrong when it tends otherwise" (Callicot 1996). Although the reformulated statement is general and requires one to define what are normal spatial and temporal scales for every impacted system, such a maxim might be useful as a basis for a riparian ethic.

Suitable Management Institutions

Numerous institutions have responsibilities for stream corridors, at various levels and often with contrasting missions, and face a wide range of issues concerning resource allocations as well as management philosophies and guidelines. Indeed, an equitable allocation of resources depends on persistent and well-understood (and respected) patterns of relationships among people and among institutions. These relationships shape social life into organized patterns allowing people to join together to achieve a common goal, such as providing drinking water or managing public land (Shannon 1998). Institutions are required at national, provincial, and local levels so that all stakeholders can contribute to the decision-making process. They are also increasingly required at the global level (Acreman 1997).

In the United States, like many other countries, resource agencies span local, regional, and federal levels. Several primary federal agencies have been charged with water resources development, management, and protection for many decades—and thereby have had significant effects on riparia (see Table 8.3; Cech 2003). Soon after

Table 8.3 Selected Water Resources Agencies at the Federal Level in the United Sates

Agency	Description	URL
U.S. Army Corps of Engineers (Corps or USACE)	• The nation's oldest water resource agency, whose history can be traced to the Continental Congress in 1755 and to the placement by Congress of the Army Corps of Engineers in charge of training the nation's new federal engineers at the Military Academy at West Point in 1802. The Corps removed snags and built levees to improve river navigation on eastern rivers, in addition to building bridges, national roads, and army outposts in the West. Its current missions are flood control and navigation improvement, plus wetlands protection and environmental restorations. • Located within the U.S. Department of Defense.	http://www.usace.army.mil
Bureau of Reclamation (USBR)	• Created in 1849, primarily to develop irrigation projects to promote settlement of the arid western states. Its water storage projects (over 200 during the 20th century) produce irrigation water for agriculture, electrical power for economic development, and water supplies for urban areas. • Located within the U.S. Department of Interior.	http://www.usbr.gov
U.S. Geological Survey (USGS)	• A scientific agency created by Congress in 1879 in response to a recommendation from the National Academy of Sciences. Today, the USGS operates and maintain over 7,000 stream gauging stations (over 85 percent of the total gauging stations in the United States and its territories), and provides real-time hydrologic data to citizens; local, state, and federal agencies; universities; and others. • The USGS studies the nation's groundwater resources. It conducts research on the effects of human activities on water quality in lakes, rivers, estuaries, and aquifers; it produces various reports, maps, fact sheets, and web pages to guide resource management decisions by local, state, and federal agencies. One of its important activities is land survey and the production of topographic maps that show land and water features, elevations, and structures at various scales. • Located within the U.S. Department of Interior.	http://www.usgs.gov

Table 8.3 *Continued*

Agency	Description	URL
U.S. Fish and Wildlife Service (USFWS)	• Created in 1940 with the responsibility to conserve, protect, and enhance fish, wildlife, and plants and their habitats. The USFWS has the responsibility to administer the Endangered Species Act and thus is charged with protecting endangered species and their habitat from destruction. It is involved in numerous debates over species protection, river discharges, and minimum flow requirements, as well as habitat maintenance and protection. • Located within the U.S. Department of Interior.	http://www.fws.gov
National Park Service (NPS)	• Created by an act signed by President Woodrow Wilson in 1916, the NPS has the primary responsibility to manage the nation's national park system (more than 339,000 km^2 divided in 384 areas, many of them containing large, environmentally sensitive catchments). • Located within the U.S. Department of Interior.	http://www.nps.gov
Bureau of Land Management (BLM)	• Created in 1946 from the fusion of the General Land Office and the U.S. Grazing Service, the BLM is responsible for the management of 1.1 million km^2, representing about one-eighth of the total land area of the United States, and 40 percent of all federal lands that are owned by the American people. The Federal Land Policy and Management Act passed by Congress in 1976 reorganized management of federal lands, giving a clear mandate to the BLM. Today, land and water conservation are major management issues of the BLM, in addition to its responsibility of sustained yields from federal lands. • Located within the U.S. Department of Interior.	http://www.blm.gov
U.S. Environmental Protection Agency (EPA)	• Created in 1970 with a key position to affect change through its mission of environmental protection and its legislative authority to impose substantial monetary fines or even jail sentences for noncompliance with federal environmental laws. The EPA develops water-quality parameters for drinking water, pollution levels in rivers, and wastewater discharge. • Independent agency of the federal government.	http://www.epa.gov
Natural Resources Conservation Service (NRCS)	• Created by President Abraham Lincoln in 1862 to assist the nation's farmers. The NRCS promotes small-scale, on-farm improvements to reduce erosion and water-quality degradation and provides funds for wetlands restoration, soil erosion prevention, and wildlife habitat protection. Its role in water resource management is evolving. • Located within the U.S. Department of Agriculture.	http://www.nrcs.usda.gov
U.S. Forest Service (USFS)	• Established by Congress in 1905 but with origins going back to 1881 when Congress created the Division of Forestry within the Department of Agriculture. Its role is to manage the country's national forests, thus providing quality water and timber for the nation's benefit. This role is evolving with the increase of timber harvesting, protection of old growth forests, and the need to protect water quality and quantity. • Located within the U.S. Department of Agriculture.	http://www.fs.fed.us
Federal Energy Regulatory Commission (FERC)	• Formed in 1920 and renamed FERC (or Commission) in 1977 to regulate nonfederal hydroelectric projects, whose production represents about 56 percent of the nation's hydroelectric capacity, the remaining 44 percent being federally developed, primarily by the U.S. Army Corps of Engineers, the Bureau of Reclamation, and the Tennessee Valley Authority. Projects can consist of multiple facilities—for example, two or more dams and reservoirs—and more than one powerhouse. The relicensing of existing hydroelectric projects comprises the majority of current license applications pending at FERC. • Located within the U.S. Department of Energy.	http://www.ferc.gov.us
National Marine Fisheries Service (NMFS)	• Founded in 1871 as the U.S. Commission of Fish and Fisheries, originally to investigate fish declines in New England waters, established by the Wildlife Act of 1956 as one of the two bureaus in the U.S. Fish and Wildlife Service, and split off and placed in the Department of Commerce in 1970. The NMFS (or NOAA Fisheries) is responsible for managing most marine mammals in the United States and is extensively involved in the protection and reintroduction of endangered salmon in the Pacific Northwest. • Located within the U.S. Department of Commerce.	http://www.noaa.go/nmfs
Federal Emergency Management Agency (FEMA)	• Created in 1979 by President Jimmy Carter to merge the activities of many federal assistance programs founded since the Congressional Act of 1803 initiating response/recovery legislation in the United States. FEMA provides assistance during and after natural and human-caused disasters. It has assisted with earthquakes, hurricanes, floods, and other disasters, both natural and man-made. • An independent agency that reports directly to the president.	http://www.fema.gov

After Cech 2003.

World War II, their historic mission of water development gradually shifted to missions of environmental protection and conservation, leading to the creation of new agencies (e.g., the U.S. Environmental Protection Agency) and the expansion of missions by already existing ones (e.g., the U.S. Fish and Wildlife Service). This led to overlapping activities and sometimes conflicts due to opposing institutional missions or diverging consistencies.

At local, state, and regional scales, water-related agencies provide diverse services ranging from drinking water supply to wastewater treatment, irrigation water delivery, flood control, and water-quality protection. Most have important effects on riparia—for example through programs addressing flood control, water management, public education, and erosion control. Their relationships with federal agencies are highly variable and always evolving. Multistate water management agencies such as the Chesapeake Bay Commission formed in 1980 (www.chesapeakebay.net) and the Missouri River Basin Association formed in 1981 (www.mrba-missouri-river.com) generally address issues that concern large drainage basins.

A myriad of laws control, either directly or indirectly, human activities that affect rivers and riparia (Masonis and Bodi 1998). In the United States, some apply directly to riparia, such as federal land use laws about river corridor protection (the National Wild and Scenic Rivers Act), wetland regulation (the Clean Water Act), forest practices regulation (the National Forest Management Act and the Federal Land Planning and Management Act), and hydropower regulation (the Federal Power Act). In addition, the Endangered Species Act applies to species listed as endangered or threatened through, among other actions, the designation of critical habitats. Even though such laws possess great potential for protecting and restoring the ecological health and integrity of rivers and their riparian systems, much of this potential remains unrealized (Masonis and Bodi 1998). Among the reasons for this are that resource agencies are insufficiently funded, political will is lacking, the ecological and economic importance of healthy rivers is not understood, and the complexity of riverine systems is underestimated. At the same time, there is a global tendency to reduce the level of environmental regulation (e.g., free trade agreements, among others), which some people view as both an impediment for economic growth and an impingement on private property rights. Clearly, improvement depends on developing collaborative and consensus-driven management, as well as collecting and disseminating information about the values of riparia.

Information Collection and Dissemination

Decisions need to be based on timely, sufficient, and accurate data for resource management to be effective. Such data are lacking in many countries where water-related resources are under-appreciated and hydrologic measurements rare. It is vital to expand the collection and analysis of information about river corridors as well as to disseminate to local communities the resultant information and trends in environmental conditions. It is also vital to transfer this information not only by Internet but also through brochures, newspaper articles, radio broadcasts, and public meetings (Acreman 1997).

Continuing education is essential if citizens are to play a realistic role in monitoring socioenvironmental conditions and in developing thoughtful riparian-based stewardship. Educated citizens provide valuable information on riparian conditions over areas larger than those of just endangered sites, help ensure data continuity despite discontinuities within public agencies and large landowner organizations, and assist with conducting large-scale adaptive management experiments, making experiments possible with limited agency or landowner resources. It is also important that scientists and managers provide *knowledge* (which is different from information) about riparian processes and management techniques to citizens on a regular basis. Riparian educational efforts are developing around the world, at local or regional scales, often within water or wetland initiatives promoting appropriate use of water resources. These include the initiative promoting thoughtful use of Mediterranean wetlands (MedWet) under the umbrella of the Ramsar Convention on Wetlands, an intergovernmental treaty signed in the Iranian city of Ramsar in 1971 (http://ramsar.org).

A fundamental challenge regarding riparian management is not only to do *good* and *useful* science, but to do science that will be *used* (Naiman et al. 2002c). This requires one to formulate fundamental ecological principles in a way simple enough to be communicated to citizens, resource managers, and policy makers but comprehensive enough to capture drivers underpinning ecological integrity or health. This is probably the most important factor for maintaining the vitality (i.e., goods and services) of riparian systems. Whence the importance of management principles such as those suggested within the framework of the Scientific Committee on Water Research (SCOWAR) of the International Council for Science (ICSU) (see Table 8.4). Translating such principles into management directives requires addressing questions such as: How much water does a river or riparia need and when? How are various organisms affected by habitat fragmentation, especially in terms of their dispersal? How are different types of riparia affected by hydrologic alterations, particularly in relation to their location in the catchment? It also requires understanding cumulative impacts of water regime change at various scales and measuring ecosystem responses due to modified flow regimes and land uses.

Biological integrity and system resilience may be useful to consider in this respect. Biological integrity is defined as the capacity to support and maintain a balanced, integrated, adaptive biotic system having the full range of elements and processes expected in the natural habitat or region (Karr 1996). This definition encompasses a variety of scales, including individuals to landscapes, all elements of biodiversity (genes, species, assemblages), and all processes that maintain these elements (mutations, demography, interactions). And this definition refers to the dynamic evolutionary and biogeography processes inherent in biotic systems. Resilience to stress is the ability to return to a previous state following disturbance, the previous state being described as a reference, such as a pristine (Ward et al. 1999) or a normative (Stanford et al. 1996) state. Increasingly, "pristine" conditions appear as unattainable or even undesirable goals (for example, in the case of recurrent flooding) for conservation and restoration practices. This implies that relatively unaffected systems are useful for comparisons, but using them as models may be incompatible with the idea of humans being an integral part of riverine landscapes. And putting to use such models requires the widest discussion possible between scientists, managers, decision makers, and the general public.

Table 8.4 Overarching and More Specific Ecological Principles Giving the Context within Which to Place Water Resources Development

Overarching Principles (Naiman et al. 2002c)	• The natural flow regime shapes the evolution of aquatic biota and ecological processes. • Every river has a characteristic flow regime and an associated biotic community. • Aquatic ecosystems are topographically unique in occupying the lowest position in the landscape, thereby integrating catchment-scale processes.
More Specific Principles Riparian Dynamics (Nilsson and Svedmark 2002)	• The flow regime determines the successional evolution of riparian plant communities and ecological processes. • The riparian corridor serves as a pathway for redistribution of organic and inorganic material that influences plant communities along rivers. • The riparian system is a transition zone between land and water ecosystems and is disproportionately plant species rich when compared to surrounding ecosystems.
Biogeochemical Cycling of Nitrogen (Pinay et al. 2002)	• The mode of nitrogen delivery affects ecosystem functioning. • Increasing contact between water and soil or sediment increases nitrogen retention and processing. • Floods and droughts are natural events that strongly influence pathways of nitrogen cycling.
Biodiversity (Bunn and Arthington 2002)	• Flow is a major determinant of physical habitat in streams, which in turn is a major determinant of biotic composition. • Aquatic species have evolved life history strategies primarily in direct response to their natural flow regime. • Maintenance of natural patterns of longitudinal and lateral connectivity is essential to the viability of populations of many riverine species. • The invasion and success of exotic and introduced species in rivers is facilitated by the alteration of flow regimes.

Particularly relevant here are some key ecological concepts, italicized in the following extract from Karr (2002):

Some divergence from *integrity* may be culturally acceptable, even necessary, in some areas dedicated to support the needs of humans. Care is necessary to maintain the condition or *health* of such areas to ensure that their use by human society does not alter their ability to provide the needs of both human and nonhuman living systems over the long term. Because systems are so complex and human actions influence numerous nonlinear dynamics, the precision of human predictions about system responses is limited. *Surprise* is inevitable and *uncertainty* is high because we are relatively ignorant of the many parts of living systems, the processes that generate and maintain those parts, and their interactions.

It may be added that the assessment of riparian health should be accurate, timely, rapid, and inexpensive, as should be the assessment of river health (Boulton 1999), and that data are needed at the landscape level in order to comprehend cumulative effects and to account for the variable character of riparia—at the complex interface between land and water.

Riparian Management: A Process Linked to Catchment and to River Management

Riparia benefit from the management of catchments and rivers, particularly when management actions maintain or restore hydrologic regimes close to natural conditions. Conversely, adjacent terrestrial and aquatic systems may largely benefit from appropriate management of riparia. Thus riparian, river, and catchment management strategies are tightly linked, if not inseparable.

Riparian Benefits from Catchment Management

Ecological principles related to land use have implications for riparian conditions (see Figure 8.2). A clear understanding of goals and guidelines of catchment management is important—all the more so as emerging technologies and global environmental conditions are likely to affect land use in unpredictable ways with unpredictable consequences. Thus, the catchment scale provides a framework for implementing riparian ecosystem management and can be summarized in five broad guidelines (Naiman et al. 1998a):

1. Effective catchment management needs cooperation among citizens, industry, governmental agencies, private institutions, and academic organizations because of the complexity of information processing and the inherent socioenvironmental changes.

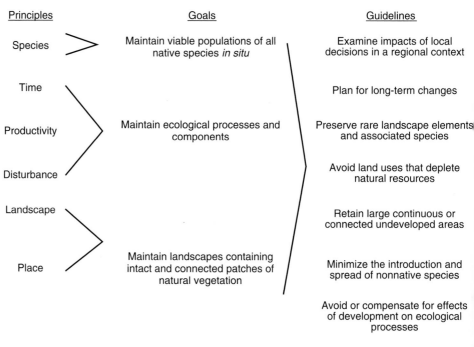

Figure 8.2 Relationship between ecological principles, land use goals, and guidelines. After Dale et al. 2000.

2. Technical solutions (e.g., fish hatcheries or waste management) to specific human-generated problems must be balanced with the maintenance of environmental components providing similar ecological services.
3. Resolution of issues depends on data-driven policy and management decisions rather than on only theoretical conceptualization or scientific perceptions.
4. The structure and behavior of the socioenvironmetal system must be regulated evenly and fairly throughout the catchment.
5. Both human activities and the structure and dynamics of the environmental components are fundamental elements of the catchment and must have *rights* to exist for the long term.

The landscape approach (or perspective) is also pertinent for the management of riparia. In the landscape approach, the spatial structure can be conceptually described in terms of matrix, patch, corridor, and mosaic (see Figure 8.3). Patches such as wetlands and lakes, and corridors such as streams, are embedded in a matrix where the natural vegetative community (e.g., prairie, forest) or land use (e.g., agriculture, urban) dominates. Moreover, the general organization of riparia at the landscape scale depends primarily on characteristics of the catchments areas, of which size is an important attribute. For example, the average slope (relief) of drainage basins decreases as catchment area increases, and drainage density (the total stream length per unit area) normally declines as catchment area increases (Kalff 2002). Therefore,

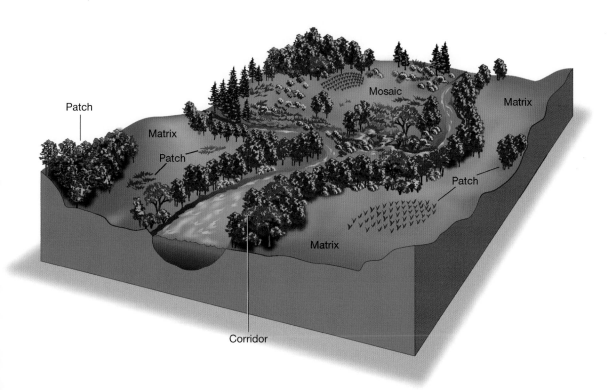

Figure 8.3 The spatial structure of landscapes can be described in terms of matrix, patch, corridor, and mosaic at various scales. After the Federal Interagency Working Group 1998.

the average lateral and longitudinal slopes of riparia tend to be higher in small catchments than in larger ones, as well as the total length per unit area. Thus, managing a riparia differs according to landscape position—for example, does it occur in the upper, middle, or lower catchment zone, is it a small or large floodplain, or does it occur at a delta or tributary junction? Effective management of riparian areas requires understanding the stream, corridor, catchment basin, and landscape as biophysically complex ecosystems interacting with neighboring ecosystems. In other words, although a riparian management plan should address a specific project in a specific physical setting, it should also consider the influencing (and influenced) larger system. Such a landscape approach has several benefits, including the establishment of a healthy, sustainable pattern of land use across the basin and a sense of stewardship for private landowners and the public.

The management of Big Spring Creek, Montana, combines agricultural, forest, and urban settings, illustrating the benefits of a landscape approach (Federal Interagency Working Group 1998). The Big Spring Creek catchment provides drinking water to the city of Lewiston and is one of Montana's best trout streams. Its management is dedicated to solving resource problems on a catchment scale, considering the cumulative effect of all actions in the landscape. A Big Spring Creek "Watershed" Partnership has been created, including the Fergus County Conservation District that is assisted by a citizen committee and funded from the Montana Department of Environmental Quality. Area landowners have been helped in implementing conservation practices such as improving riparian vegetation, treating streambank erosion, and developing water sources away from the stream for livestock. Cooperating agencies have participated in additional catchment improvements (see Figure 8.4). As a result, many positive changes have been brought about with the implementation of management practices such as fencing, off-stream water developments, and stream/riparian protection. A monitoring strategy provides feedback to measure improvements, and an initiative has emerged to manage agriculture and open space, as well as to develop recreational and environmental resources.

Cultural values, social behavior, and environmental characteristics will continue to evolve in the Big Spring Creek catchment, as they will in any catchment. Therefore, to be successful, management must be based on a common long-term vision shared by citizens, regulators, educators, and industries. Forecasts of the state of ecosystems, ecosystem services, and natural capital will greatly improve planning and decision making if uncertainties are fully specified and are contingent on explicit scenarios for climate, land use, human populations, technologies, and economic activities (Clark et al. 2001). In every case, a dialogue involving scientists, managers, and policy makers is essential to produce, evaluate, and communicate forecasts of critical riparian resources and services.

Riparian Benefits from River Management

The effect of water management on riparia has been long and widely ignored, although riparia are recognized as an essential component for the maintenance of healthy river corridors and floodplains. It is well known that rivers carry and redistribute sediment, nutrients, propagules, and organic debris, thus creating habitat for colonization by riparian species. Rivers contribute to groundwater recharge in certain

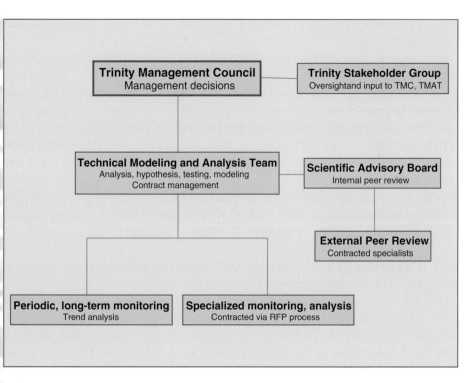

Figure 8.4 Organizational components of a successful Adaptive Environmental Assessment and Management (AEAM) program. TMC: Trinity Management Council; TMAT: Trinity Modeling and Analysis Team; RFP: Requests for Proposals. From U.S. Fish and Wildlife Service and Hoopa Valley Tribe 1999.

locations, and carry surplus water away in others. In addition, floods and droughts help create disturbance regimes that affect vegetative succession.

Flood management has long-lasting effects on riparia. As discussed in earlier chapters, floods act as both a disturbance to slow the rate of competitive exclusion and as an agent increasing spatial heterogeneity (Pollock et al. 1998, Naiman et al. 2005, Pettit and Naiman 2005). Those that do not remove physical substrate (non-destructive floods) slow the rate of competitive exclusion in riparian communities, whereas those that remove physical substrate (destructive floods) tend to increase large-scale spatial heterogeneity. The unusually high levels of diversity in riparian corridors is probably due, above all other factors, to this ability of floods to slow rates of competitive exclusion and to maintain a spatially heterogeneous environment. Indeed, much of the riparian habitat diversity is in unconstrained low gradient valleys where fluvial processes create the largest variety of landforms. In other words, diverse riparian corridors are maintained along rivers that create and destroy habitats at natural rates. Thus, an abundance of life forms are found where flood management strategies allow rivers to occasionally flood, meander, and move so as to create a dynamic mosaic of physically complex habitat patches (Salo et al. 1986, Pollock 1998, Tockner and Stanford 2002).

Management of river flow has considerable long-term effects on riparian plant communities, a topic particularly well investigated along French rivers (Carbiener 1970, Pautou 1984, Bravard et al. 1986, Décamps et al. 1988, Schnitzler et al. 1992, Trémolières et al. 1998). As just one example, the management of flows along the Rhône River in the Alps caused:

- A high mortality of willows (*Salix alba, S. triandra, S. viminalis, S. daphnoides*) and alder (*Alnus incana*);
- An increase in the shrubby willow (*Salix eleagnos*), particularly prolific on alluvial coarse sands and gravel;
- A demographic explosion of poplar (*Populus nigra*) on bulldozed areas and on sand bars in the main river;
- The establishment of common locust (*Robinia pseudoacacia*);
- High regeneration rates and expansion of several sparsely represented and recently introduced species, including *Polygonum sacchalinense, Buddleja variabilis*, and *Acer negundo*; and
- Invasion of a number of habitats by *Solidago gigantea, Urtica dioica, Impatiens glandulifera* and formation of dense thickets of these shrubs in gaps resulting from the death of willow and alder (Pautou et al. 1992).

There is a plethora of similar examples describing the unintended riparian-related consequences of flow management throughout the world.

The effect of dam management strategies is equally pervasive, with between 45,000 and 48,000 large dams (>15 m high) and over 800,000 smaller ones along rivers worldwide (McCully 1996, WCD 2000). Management strategies have created novel and relict habitats alongside of each other (Stevens et al. 2001). Novel habitats include riverine deltas formed at the upper reservoir connection to the mainstem river, lacustrine deltas formed where tributaries enter the reservoirs, and reservoir shorelines. Relict habitats are remnant floodplains existing above, between, or below reservoirs. They are likely to support the best native diversity to be found in a regulated river system, reflecting a legacy of pre-regulation channel and floodplain dynamics. Less native biodiversity may be expected in association with novel habitats and greater native biodiversity with relict habitats. However, biodiversity and spatial heterogeneity of novel habitats are likely to increase with the aging of reservoirs, whereas relict habitats demonstrate spatial homogenization and attendant decreases in biodiversity unless some recovery of flow variability and cut-and-fill alluviation occurs.

As a means of helping manage regulated systems, biodiversity ranking for habitats associated with large, regulated rivers such as the Missouri have been proposed (see Figure 8.5; Johnson 2002). Among the relict habitats, unregulated or lightly regulated

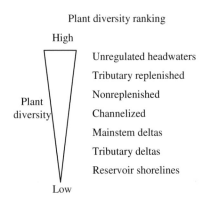

Figure 8.5 Mainstem habitats along regulated rivers ranked from the highest to the lowest plant diversity. After Johnson 2002.

reaches can be expected to retain the highest percentages of native species and communities, followed by replenished reaches and then by unreplenished reaches. Engineered channelized reaches demonstrate the greatest losses. Novel habitats all rank below relict habitats in this scheme. Mainstem deltas rank the highest as they form more quickly compared to most smaller tributary deltas in reservoirs and to reservoir shorelines. Interestingly, such a ranking is likely to change as some of the identified categories may change position with time as more species become established.

More generally, river and water regulation are at the origin of a number of long-term modifications to riparia (see Chapter 7). Numerous experiences with water regulation now suggest that it is essential for experimental floods designed for sediment management to be based on careful designs to mitigate riparian impacts, on public discussion of values and flexibility in planning, and on sound scientific information (Stevens et al. 2001). It is also essential to remember that humans are also affected by river and water regulation, sometimes with tragic consequences. Consider the Three Gorges Dam in China, the largest dam in the world, that will influence an area of ~58,000 km^2, impacting long-standing cultural and social structures. Moreover, about 1.13 million residents within the reservoir's path are being forcibly relocated to uplands in new urban environments (Wu et al. 2004).

Catchments and Rivers Benefit from Riparian Management

Riparia exert strong influences on adjacent ecological systems by modifying the flow of materials and information across the landscape. Through their significant levels of biodiversity, as well as significant rates of nutrient cycling and productivity, they play a key role in maintaining the vitality of terrestrial environments at local and catchment scales. They are also recognized as having important functions for rivers, as they provide shade, bank stability, reduced runoff of fertilizers, herbicides and pesticides, barriers to logging debris, continuous sources of wood and other terrestrial organic matter, and habitat for wildlife. Therefore, appropriate management of riparian systems may have consequences that go far beyond their limited areas and propagate throughout the entire river corridor, floodplain, and even the catchment.

Among these benefits, the effects on natural biogeochemical cycles have been particularly investigated, for example in the Pacific coastal ecoregion of the United States and Canada (McClain et al. 1998). There, rivers have been largely impacted by conversion of native forests to managed silvicultural plantations, especially on the uplands and less so in riparian zones. Logging activities, which include high-density road construction in addition to tree removal, can substantially increase suspended sediment loads (Brown and Krygier 1971, Beschta 1978) and, consequently, fluxes of particulate nitrogen, phosphorus, and sulfur. Higher amounts of sediment and dissolved nutrients to streams is an immediate consequence of tree harvest, but some effects are continued into the future with replanting and regrowth, and practices such as forest fertilization. Application of urea fertilizer in British Columbia, Canada, resulted in nitrogen peaks 20 times greater in streams without buffer strips as compared with streams with buffer strips (Perrin et al. 1984). Application of herbicides, also in British Columbia, resulted in a doubling of orthophosphate (PO_4^{3-}) concentrations for 1 to 2 years in a receiving stream, and an increase in concentrations of nitrate (NO_3^-) following large storm events (Hartman and Scrivener 1990).

Additional and more severe consequences for water quality may be found when converting forested land to agriculture or to urban and suburban uses. The destruction of natural geomorphic structures through bank engineering, reduction or elimination of hyporheic flows, and inputs of fertilizers from agriculture, sewage, and industrial effluents can substantially augment nutrient concentrations if land conversion is not properly performed (Welch et al. 1998).

In contrast, in addition to being an interesting and important aspect of catchments, riparia can serve as catalysts or focal points for ecological and socioeconomic revitalization in urban areas, benefiting cities embedded within catchments with centrally placed rivers. An example is the restoration of the Corderie Royale in Rochefort, a French arsenal from the 17th century (Lassus 1998). This historic building, situated on the River Charente where it enters the Atlantic, was originally used to make rigging for the king's ships built at nearby docks. For more than two centuries, soldiers and scientists sailed from Rochefort to all parts of North America and the West Indies. There, unknown plants, including begonia (*Begonia* sp), were returned to France to be acclimatized near the Corderie before being sold elsewhere in Europe. In 1926, work at the Corderie stopped, and trees spread into the once industrial banks of the river, hiding the building—which became invisible from the river as well as from the town. In 1992, the restoration of the Corderie symbolized the historical relationship between the former port of Rochefort and the far distant corners of the world. Today, boat passengers and riverside walkers can view the façade of the beautiful building through a screen of riparian shrubs and small trees (see Figure 8.6). On the town side of the river, the Corderie is seen from the ramparts that dominate the building along its other façade, which is lined by palm and Virginia tulip trees, evoking the past activity of the port. This new riparian urban landscape provides connections between different historical periods, which together provide an opportunity for a new future—a horticultural landscape where the past is symbolized by the begonia. Here, riparian management re-creating the *genius loci* or the "spirit of place" contributed to the revitalization of a former port.

Another example is the creation in 1999 of the Gwynns Falls Trail, a 22.5-km stream-valley trail system connecting over 30 neighborhoods in west and southwest Baltimore, Maryland. Parklands, unique urban environmental features, cultural resources, and historic landmarks may be seen along the trail (Groffman et al. 2003). Before construction, and before there was an emphasis on the lotic corridor, there was extensive degradation of the streams, poor riparian habitat, destabilized stream banks, and low water quality. This was concomitant with socioeconomic decline—~50 percent of the population was lost between 1940 and 1990 and over 60,000 houses and lots were abandoned. Restoration projects were planned to stimulate community cohesion, provide interest in improved city services, and revitalize neighborhoods. Thus, a nonprofit group working at the interface of humans and the environment initiated the stream-riparian trail after a series of rehabilitation projects. The trail now appears as a unique and valuable natural resource, which people insist be maintained. Such creations may help in reversing the negative spiral of population loss, as well as the accompanying environmental and social degradation (Groffman et al. 2003). Baltimore is an example of what is occurring in many of the world's cities that have rivers (small or large) running through them, including San Antonio (Texas) and Portland (Oregon) in the United States and St. Petersburg in Russia.

Figure 8.6 River Charente and the restored Corderie Royale of Rochefort, France. (Photo: Bernard Lassus.)

Riparian Management: A Highly Specific Process

The unique properties of riparia make their management a difficult and locally specific process (Naiman et al. 1998a,b). Balancing human needs with the remarkable variability of riparia—which is the origin of their biodiversity, productivity, and resilience—is a difficult challenge (see Figure 8.7). The challenge stems from the fact that riparian communities are the result of dynamic interactions mixing persistent and opportunistic species along river corridors. Additionally, riparia display particularly strong connections between social and environmental components at multiple scales,

253

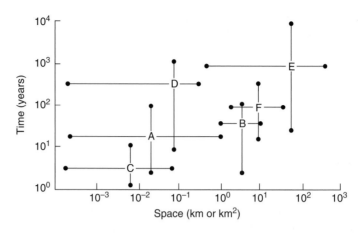

Figure 8.7 Illustration of the diversity of spatial and temporal scales influencing the creation and maintenance of riparian forests in the coastal temperate rain forest of North America. (A) Colonization surfaces created by flooding, (B) colonization surfaces created by debris flow, (C) seedling germination and establishment, (D) longevity and size of species patches, (E) persistence and movement of dead wood in channel, and (F) impact of herbivores. From Naiman et al. 1998.

and this often results in considerable uncertainty about the consequences of management actions. Finally, since human culture and institutions shape riparian characteristics, there is a strong legacy from ancient practices that not only shape present but also future conditions. Such legacies, which may be found in other ecological systems, are exacerbated in riparia. Collectively, this leads to the acute need to clearly specify what can be an adaptive, a sustainable, and an appropriate form of riparian management.

Adaptiveness

Adaptive environmental and resources management is particularly useful in riparian systems, many of which undergo rapid changes naturally. Indeed, riparia often display frequent surprises that have broad implications, from management as well as from ecological points of view. Adaptive management of riparia requires integrated and flexible policies, diverse regulations, and implies planning for learning, not simply for economic and social products (Holling 1978).

Citizen involvement and partnerships are the building blocks of "civic science" (Lee 1993), particularly when interest groups have conflicts over management goals and find it difficult to achieve consensus on common objectives. In this respect, interdisciplinary teams need to develop explicit hypotheses about how systems function as well as test hypotheses through systematic monitoring and evaluation. This allows accelerated learning through experiments, even if human understanding of nature is imperfect. Establishing and maintaining *learning for ecological management* has been identified as a major way to renew ecosystems and institutions (Michael 1995). This applies particularly to riparia, systems whose management requires considering long time scales, large decision-making units, interdependency with neighboring systems, and complexity. Predictive capabilities are limited by the dynamism of riparia, and ever-changing social values render uncertain decisions about the best end points for management efforts. Adaptive environmental assessment and management programs may allow one to face this uncertainty by learning from the outcomes of management actions, accommodating change, and improving management. The programs may also provide a process for integration of research with management, conservation, and restoration. Such a program, outlined as a 10-step process (see Figure 8.8), is relatively easy to conceptualize but very difficult to efficiently implement.

Sustainability

Sustainable riparian management is more effective when large organizations in the public sector are complemented with small, private, nonprofit or nongovernmental organizations (NGOs) designed to address diverse and complex problems. A broad bottom-up public participation is needed, as well as an extensive local knowledge, to simultaneously consider the economic, social, and ecological dimensions of riparian management (von Hagen et al. 1998). A basic rule of sustainable riparian management is accounting for human needs at local scales.

From a catchment perspective, nonprofit organizations build or at a minimum enrich the strength and diversity of public institutions. They conduct critical research, provide access to information, incite the public to conserve or restore individual rivers, and create markets for long-term community stewardship (von Hagen et al. 1998). Each of these key functions may be useful for riparian management, especially so because nonprofit organizations can make long-term commitments and adapt their structure and governance to appropriate strategies, ranging from the local to the international, and varying in terms of the issues they address. At the same time, nonprofit organizations have limitations in that they may lose innovation as operations scale up, and potentially lack incentives for efficiency and accountability. However, they contribute to effective solutions when partnerships are formed with government organizations and the private sector, and when the best attributes of each are combined (von Hagen et al. 1998).

From a landscape perspective, riparia are places where people live and are incorporated into the mental images of those places by the people. These mental images organize public perception of riparian landscapes, and thereby help determine management objectives. Therefore, one needs to understand how people perceive riparian systems in order to devise management strategies. For example, the conservation of riparian forest fragments depends on the ability to maintain structure and function in the face of external pressures over time. Such conservation demands that ecological sustainability be combined with cultural sustainability. Riparian landscapes "are more likely to be ecologically maintained in a world dominated by humans if they evoke the sustained attention of people that compel aesthetic experiences" (Nassauer 1997). Clearly, the management and vitality of riparian landscapes requires the focus of human attention. And landscape properties that invite the focus of human attention must not only be relevant but also vivid for people to take care of them (Nassauer 1992). Diversity and connectivity may be examples of such properties, helping to provide the ecological as well as cultural sustainability of riparian landscapes (Décamps 2001).

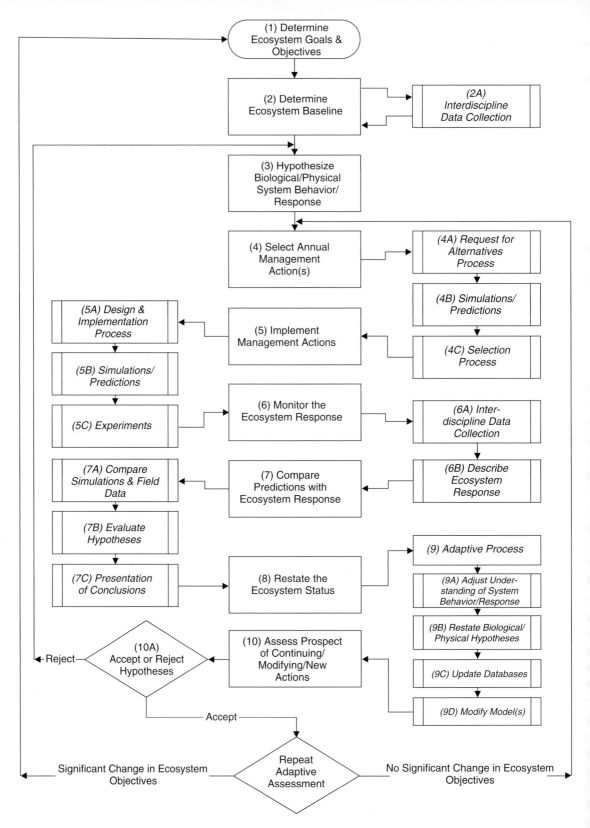

Figure 8.8 The 10-step adaptive environmental assessment and management process. From U.S. Fish and Wildlife Service and Hoopa Valley Tribe 1999.

Appropriateness

Appropriate management of riparia requires prescriptions based on local conditions, not on general notions conceived *a priori*. Communication and cooperation among landowners, regulators, and biologists is needed to allow system resilience and to avoid discontinuities or discrepancies, such as regulations respected by one landowner but not implemented by another. Two examples of the importance of local conditions are illustrated below—the guidelines for timber harvest practices and the revegetation of riverbanks.

Timber Harvest Practices

Buffer strips are often left along both sides of rivers to protect wildlife from the impacts of timber harvest in adjacent upslope areas (Kelsey and West 1998). Nevertheless, in some situations, the buffer strips may increase the probability of windthrow by suddenly exposing trees to strong winds that are no longer ameliorated by the surrounding forest (Steinblums et al. 1984). Alternative forest harvest management strategies may reduce the incidence of windthrown riparian trees, at least along first to fourth order streams. Thus, one might consider the following design: (1) a 10 m no-harvest buffer strip on each side of the stream in areas where probability of windthrow is low; (2) a 30 m buffer strip on each side where the risk of windthrow is moderate, the first 10 m from the stream being a no-harvest zone, 10 to 20 m being a 25 percent selective cut zone, and 20 to 30 m being a 60 percent selective cut zone; and (3) a 200 m diameter buffer island along headwater streams having high probabilities of windthrow, with unique riparian communities or with unstable soils (Kelsey and West 1998).

Revegetation of Riverbanks

This is an appropriate way of managing banks for protection against erosion along small streams and rivers. In North America, plants have been used to soften the impact of installing hard structures, thus protecting already damaged riparian areas (see Figure 8.9, Klapproth and Johnson 2001). Such a practice is efficient and much less expensive than creating gaps in existing levees or removing ancient levees and replacing them with new levees farther from the river. Additionally, there is no need to change existing land use over ample areas. In Europe, willows are often used because of their diversity, speed of growth, aptitude for regrowth, and branch flexibility (Lachat 1994). They are used to protect the bases of riverbanks as fascines or mattresses of branches as well as planted as trees at the top of banks or as shrubs halfway up. A good understanding of the sites and of the species is essential; for example, using species adapted to local growing conditions will have a strong possibility for successful establishment (see Table 8.5).

In addition to the importance of local situations, effective riparian management pays substantial attention to hydrologic connections between landscape elements. Riparia regulate hydrologic connectivity and, as a consequence, water-mediated transfer of matter, energy, and organisms occurs within or between elements of the hydrologic cycle at the catchment scale (Pringle 2001). Alteration of riparian hydrologic connectivity comes from flow modification by dams, groundwater extraction, and water diversion, resulting in profound effects on the biotic integrity of systems

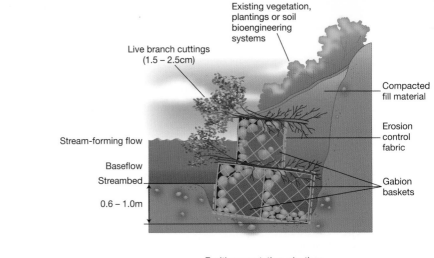

Existing vegetation,
plantings or soil
bioengineering
systems

Live branch cuttings
(1.5 – 2.5cm)

Compacted
fill material

Stream-forming flow

Erosion
control
fabric

Baseflow

Streambed

Gabion
baskets

0.6 – 1.0m

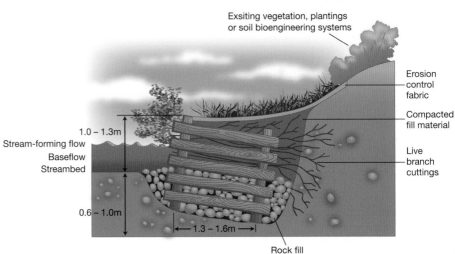

Exsiting vegetation, plantings
or soil bioengineering systems

Erosion
control
fabric

Compacted
fill material

1.0 – 1.3m

Stream-forming flow
Baseflow
Streambed

Live
branch
cuttings

0.6 – 1.0m

1.3 – 1.6m

Rock fill

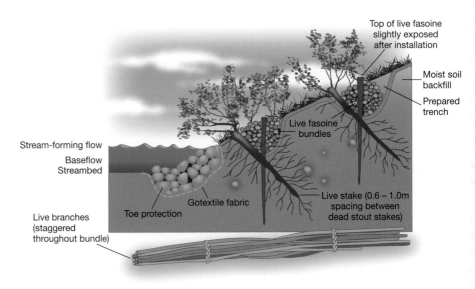

Top of live fasoine
slightly exposed
after installation

Moist soil
backfill

Prepared
trench

Live fasoine
bundles

Stream-forming flow
Baseflow
Streambed

Live stake (0.6 – 1.0m
spacing between
dead stout stakes)

Gotextile fabric

Live branches
(staggered
throughout bundle)

Toe protection

Figure 8.9 Using vegetation in stream bank stabilization. From Klapproth and Johnson 2001.

Table 8.5 Size and Growth Characteristics of Willow Species that Can Be Used for Revegetating Riverbanks Under an Altitude of 1200 m in Europe

Species	Normal Size	Bush	Shrub	Tree
Salix aurita	1–3 m	x		
Salix nigricans	1.5–5 m	x		
Salix purpurea	1–6 m	x		
Salix cinerea	3–6 m	x	(x)	
Salix atrocinerea	3–6 m	x	(x)	
Salix triandra	2–7 m	(x)	x	
Salix viminalis	2–10 m		x	
Salix daphnoides	3–15 m		x	(x)
Salix eleagnos	2–15 m	(x)	x	(x)
*Salix fragilis**	5–25 m			x
*Salix alba**	5–30 m			x

(x) = possibly used in addition to the main habitat.
* = to be planted at the top of the bank, not as weaving or fascines at the bottom.
After Lachat 1994.

downstream (terrestrial as well as aquatic) and often well outside the boundaries of a given site. This requires new and innovative partnerships between managers and scientists to predict the cumulative and interactive effects of hydrologic alterations (Pringle 2000).

Human Dimension of Riparian Management

Humans as individuals and as societies are concurrently emergent from, and parts of, the natural world. As individuals, we are biophysical entities interacting directly with ecosystems as well as persons interpreting ecosystems, giving them meaning and significance. As social beings, humans interact with one another, developing collective lives expressed as societies. These social relations generate processes within groups whose integrity depends on four interrelated perspectives: economic, political, community, and cultural (Lewis and Slider 1999; see Figure 8.10). The economic perspective views the natural world as a resource to be transformed into products that are acquired, processed, and distributed, so as to enable a society to adapt to its biophysical environment. The political perspective views the natural world as an object of control in the collective interest. For example, preserving a particular riparian area for future generations requires controlling the destiny of that area and implementing effective policies. The community perspective views the natural world as an ingredient of communal experiences and a reason for people to adhere to a norm (preserving a particular riparia), to participate in a social group, and to decide to do something, which eventually enters into the political dimension. The cultural perspective views the world as a source of knowledge, value, and experience that, in turn, plays a major role in orienting behavior with respect to the natural world.

These perspectives function simultaneously and interdependently. Together, they contribute to the integrity of social systems and provide a framework to understand

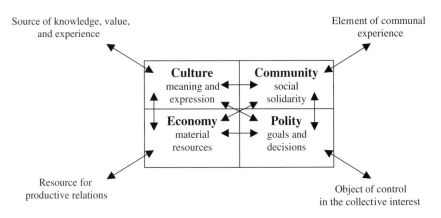

Figure 8.10 Significance of the natural world for the four dimensions of the social system. After Lewis and Slider 1999.

Table 8.6 Components of a Shared Socioenvironmental Vision for Riparian Systems

Shared Socioenvironmental Vision	

• Environmental endpoints
  — Riparian forest condition
  — Species persistence
  — Water and habitat quality
	• Long-term commitments by leadership
	
	• Empowerment of citizens

• Social endpoints
  — Literacy
  — Adaptive institutions
  — Partnerships
  — Stewardship and
    responsibility
	• Communication of visions
	
	• Education about value of vision
	
	• Active monitoring
	
	• Continued learning

After Naiman et al. 1998a.

mutual impacts between social and natural systems. Importantly, these mutual impacts have geographic dimensions that require identifying boundaries. For example the geographic dimensions might mean delineating the local riparian zone and its inhabitants, delineating inhabitants living in the catchment adjacent to the riparian zone, and delineating the regional population. Social assessments related to riparia require data about many human perspectives because the significance of riparia differs between individuals as well as groups and between cultural and economic perspectives.

Shared Socioenvironmental Visions

There is also an environmental perspective to social systems. Both human and environmental perspectives require a shared socioenvironmental vision, including individual environmental and social endpoints, to be successful (see Table 8.6). The conditions of riparian forests, species persistence, and water and habitat qualities may be among the environmental endpoints; literacy, adaptive institutions, partnerships, and stewardship and responsibility may be among the social endpoints. Shared socioenvironmental visions maintain long-term commitments through leadership, empowerment of citizens, communication, education, active monitoring, and continued learning. Of particular importance are long-term commitments by the citizens who initially provided much of the vision and leadership, and empowerment of citizens with responsibility for their future (Lee 1993).

Such shared visions greatly benefit from diversification and evolution of disciplines such as ecology and design. Ecology diversified into several new and still evolving subdisciplines during the last two decades. At the same time, the design professions acquired a sense of ecological awareness and responsibility. These two phenomena are both a cause and a consequence of an important evolution in the human perception of management (Johnson and Hill 2002). In the first phenomenon, newly distinct—although interacting and complementary—perceptions of the natural world by ecologists led to the creation of subdisciplines such as landscape ecology, biological conservation, ecological restoration, and ecosystem management (see Table 8.7). The second phenomenon is especially interesting because, whereas the ecological subdisciplines diversified, the design professions focused increasingly on ecological awareness and responsibility and adopted ecological guidelines for professional practices. Among these professions are civil engineering, landscape architecture, and urban planning (see Table 8.8). Alone, each subdiscipline has insufficient knowledge to address the combined complexities of cultural and ecological issues of land management. Their potential success depends on a capacity to build collaborative approaches to land development and management. This is particularly relevant in designing, planning, and implementing resources management as well as shaping riparian habitat conservation and restoration. On the one hand, all riparian design is ecological, as every patch in the riparian mosaic includes multiple biological and geophysical processes likely to be affected by human actions. On the other hand, every ecological management plan is cultural as it involves and affects people, distributing costs and benefits, and ultimately aiming to satisfy human needs and values.

Social-Ecological Systems

Considering riparia as social-ecological systems is another useful vision for the scientific community. Riparia are increasingly viewed by science as complex systems, changing in unpredictable ways under the evolving influences of climate, technology, and society, and characterized by sudden and turbulent periods of transition. A primary goal for the management of such systems is to maintain a capacity to face expected as well as unexpected conditions, without undesirable transformations. This can be accomplished by preserving the system's resilience to multiple changes (Walker et al. 2002).

The approach is based on the metaphor of an adaptive cycle (Holling 2001, Gunderson and Holling 2002), and is easily adapted to riparia. Accordingly, interactive natural and human systems undergo transformations that include four successive phases: (1) rapid growth and exploitation, (2) accumulation and maturity, (3) rapid

Table 8.7 Ecological Subdisciplines Seek to Use Ecological Science as a Foundation for Solving Environmental Problems

Ecological Subdiscipline	Subdiscipline Description	Society	URL
Landscape Ecology	The study of spatial variation in landscapes at a variety of scales. It includes the biophysical and societal causes and consequences of landscape heterogeneity. It is broadly interdisciplinary.	The International Association for Landscape Ecology (IALE) was established in 1982 to "develop landscape ecology as a scientific basis for analysis, planning and management of the landscapes of the world." Its core themes include the spatial patterns of structure of landscapes ranging from wilderness to cities, the relationship between pattern and process in landscapes, the relationship of human activity to landscape pattern process and change, and the effect of scale and disturbance on the landscape.	http://www.landscape-ecology.org
Conservation Biology	The field of biology that studies the dynamics of diversity, scarcity, and extinction.	The Society for Conservation Biology was formed in 1985 "to help develop the scientific and technical means for the protection, maintenance, and restoration of life on this planet—its species, its ecological and evolutionary processes, and its particular and total environment." The society encourages "communication and collaboration between conservation biology and other disciplines (including other biological and physical sciences, the behavioral and social sciences, economics, law, and philosophy) that study and advise on conservation and natural resources issues."	http://conbio.rice.edu/scb/info
Ecological Restoration	The process of assisting the recovery and management of ecological integrity. Ecological restoration includes a critical range of variability in biodiversity, ecological processes and structures, regional and historical context, and sustainable cultural practices.	The Society for Ecological Restoration was established in 1988 to "promote ecological restoration as a means of sustaining the diversity of life on Earth and reestablishing an ecologically healthy relationship between nature and culture." The society encourages the development of restoration "as a scientific and technical discipline, as a strategy for environmental conservation, as a technique for ecological research, and as a means of developing a mutually beneficial relationship between human beings and the rest of nature."	http://www.ser.org
Ecosystem Management	Management driven by explicit goals, executed by policies, protocols, and practices, and made adaptable by monitoring and research based on our best understanding of the ecological interactions and processes necessary to sustain ecosystem structure and function.	Although there are no societies dedicated to ecosystem management, land management agencies often refer to ecosystem management as a primary paradigm. The Ecological Society of America (ESA) has devoted one of its report to the topic (Christensen et al. 1996). According to the ESA report, ecosystem management focuses primarily on the sustainability of ecosystem structures and processes necessary to deliver goods and services, and incorporates the following key factors: long-term sustainability, clear operational goals, sound ecological models and understanding, complexity and interconnectedness, the dynamic character of ecosystems, attention to context and scale, humans as ecosystem components, and adaptability and accountability.	See the Ecological Society of America's home page at http://www.esa.org

After Johnson and Hill 2002.

Table 8.8 Design Professions Recognize the Need for Ecological Awareness and Responsibility

Design Profession	Profession Description	Society	URL
Civil Engineering	Develop and encourage the use of evolving technologies required to achieve an ecologically sustainable world for future generations.	The American Society of Civil Engineering (ASCE) was created as early as 1852 to include civil engineers and those in related disciplines. The society invites civil engineers to participate in interdisciplinary teams with ecologists, economists, sociologists, and professionals from other disciplines. The principles of sustainable development are used to promote "responsible, economically sound, and environmentally sustainable solutions that enhance the quality of life [and] protect and efficiently use natural resources."	http://www.asce.org
Landscape Architecture	The art and science of analysis, planning, design, management, preservation, and rehabilitation of the land.	The American Society of Landscape Architects (ASLA) was established in 1899. Its mission is "to lead, to educate, and to participate in the careful stewardship, wise planning and artful design of our cultural and natural environments." The ASLA declaration on environment and development offers a set of principles that include "how the health and well-being of people, their cultures and settlements [and] of other species and of global ecosystems are interconnected, vulnerable, and dependent on each other."	http://www.asla.org
Planning	Economic and demographic analysis, natural and cultural resource evaluation, goal-setting, and strategic planning are used as tools to offer options "so that communities and their citizens can achieve their vision of the future."	The American Institute of Planners (established in 1917) and the American Society of Planning Officials (established 1934) formed the American Planning Association, with a Planning Advisory Service in 1949, to "advance the art and science of planning and to foster the activity of planning—physical, economic, and social—at the local, regional, state and national levels." The association aims at "fostering meaningful citizen participation in planning decisions and protecting the integrity of the natural environment and the heritage of the built environment." It edited guides on "endangered species and habitat protection" and on "planning for sustainability."	http://www.planning.org

After Johnson and Hill 2002.

breakdown or release, and (4) renewal and reorganization. These cycles are adaptive; that is, they are capable of reacting when facing new difficulties and able to develop new innovations. Such cycles depend on three properties of natural and human systems: (1) their *wealth*, or the inherent potential that is available for change; (2) their degree of *connectedness* between internal controlling variables and processes, or internal controllability; and (3) their *resilience*, or adaptive capacity when facing unexpected and unpredictable disturbances.

A theoretical trajectory of such an adaptive cycle may be represented as a function of the first two properties, potential (i.e., wealth) and connectedness (see Figure 8.11).

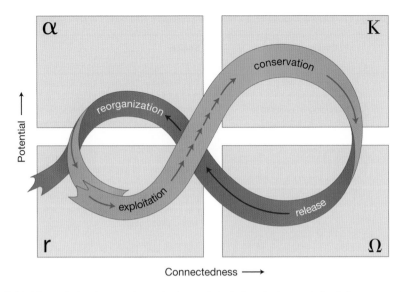

Figure 8.11 Theoretical trajectory of an adaptive cycle including the two periods of slow accumulation of resources (exploitation and conservation, r to K) and rapid creative destruction (release and reorganization, Ω to α). This representation refers to the potential that is inherent in the accumulated resources of biomass and nutrients (y axis), and to the degree of connectedness among controlling variables (x axis). In r, the potential can leak away into a less productive and less organized system. From Gunderson and Holling 2002.

First, there is a long period of accumulation and transformation of resources from the exploitation to the conservation phases. Second, there is a short period of creation of opportunities for innovation during the entry into the reorganization phases. In a riparian system, for example, plant biomass will accumulate during the first period as well as the possibility that the biomass will be used (potential), and the internal self-controlling mechanisms of the system will become more complex (connectedness). As the system loses its flexibility, vegetation and dead wood accumulate, and sudden disturbances such as floods, wind, insect outbreaks, drought, or fire can disorganize the system during a second period, thereby inducing a phase of reorganization with rapid and unpredictable changes, eventually leading to innovations for a new coming cycle. Thus, each adaptive cycle is built within two opposing and successive objectives—the first is one of growth and maturation, and the second is one of creative destruction and renewal. The accomplishment of each objective prepares the scene for its opposite. In addition to potential and to connectedness, resilience constitutes a third property of adaptive cycles. Resilience decreases during the first phase (i.e., exploitation to conservation) reaching a minimum when the mature system is the most developed, and increases during the second phase (i.e., release to reorganization) attaining a maximum when the renewing system is most innovative.

Furthermore, adaptive cycles function as nested sets in a hierarchy of spatial and temporal scales such as river catchment, floodplain, riparian system, and patches (see Figure 8.12). Each scale is semiautonomous, having its own internal interactions as well as its various phases of exploitation, accumulation, release, and reorganization. Any given adaptive cycle at one level (i.e., a riparian system) operates at its own pace

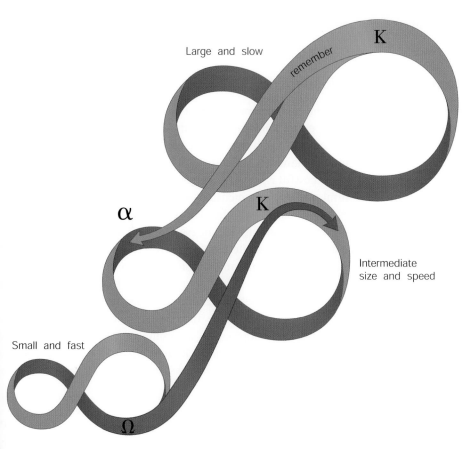

Figure 8.12 Three levels of a panarchy indicating two connections likely to create and sustain adaptive capacity: the "revolt" connection, where a critical change in one cycle can cascade up to a vulnerable stage in a larger and slower one, and the "remember" connection, where renewal in one cycle is facilitated by the potential that has been accumulated and stored in a larger, slower cycle. From Gunderson and Holling 2002.

protected by its upper slower level (i.e., a floodplain) and invigorated by its faster lower level (i.e., a patch). The hierarchy associated with adaptive cycles is termed *panarchy* by Gunderson and Holling (2002). In panarchy, different levels are not isolated, and connections may develop between them. A destruction phase of a lower level may have an impact at a higher level, creating a crisis, particularly when the higher level is at its maturity phase—that is, when resilience is low. Thus an insect outbreak in one tree may eventually affect an entire riparian system. Inversely, higher levels may impact a lower level when the latter one is reorganizing, creating a kind of memory effect or legacy. For example, a seed bank dominated by one species may shape the reorganization of a riparian patch after a flood or drought.

The adaptive cycle metaphor is a useful guide for thinking about the dynamics of social-ecological systems such as riparia. It focuses attention on changes in resilience occurring during the behavioral phases of systems. It gives a theoretical basis for riparian management, including conservation and restoration, to build and maintain ecological resilience and social flexibility, as well as participatory approaches (Holling 2001).

Communication Needs

The above visions of riparian management require an understanding of natural and cultural dimensions of the entire landscape, meaning that communication is critical among the natural sciences, among the social sciences, between the natural and the social sciences, and among managers, scientists, political leaders, landowners, and other citizens. In effect, successful riparian management is grounded in an expansive integration of the natural and social dimensions of decision making. Coupling the behavior and attitudes of humans with the reality of the laws of nature and a sense of place (see Sidebar 8.2) may provide an essential role for traditional knowledge in riparian management, as illustrated in Africa (Mathooko and Kariuki 2000) and in Amazonia (Posey 1985, McClain and Cossio 2003). In addition, the perception of riparia as landscapes implies developing a renewed management vision and adjusting concepts for planning, designing, and implementing (Langston 2004). As other landscapes, riparia are at the same time reality and the appearance of reality, and subjective and objective approaches are complementary when planning management actions. Cultural, social, and historical determinants of human perception are as important as ecological rules: worldviews and beliefs do matter (Berkes 1999).

Sidebar 8.2 Poplars: From a Popular Image of Nature to a Symbol of Productivism

Rural communities in Europe have developed familiar relationships to poplar trees for centuries, particularly along alluvial valleys where poplar is indigenous. Numerous varieties of *Populus nigra* characterize alluvial areas and have domestic uses, serving as places for daily social activities, material for household items (e.g., clogs, dishes), timber for construction, and food for cattle. Such goods and services were delivered long before engineers developed fastigiated poplars (*Populus nigra* "Italica") for linear plantations along canals and roads.

Current social relationships to poplar trees vary largely across regions. Poplar can stand as a symbol for places with particular meanings and values for inhabitants or regular visitors (see Figure SB8.1). Those meanings and values are closely linked to social practices of nature such as walking, fishing, gathering, meeting neighbors, and so forth. Poplars can also refer to popular images of nature, actually closer to representations fashioned by impressionist painters than to actual conservation.

Surveys have made clear that, through the way people talk about poplars and the areas where poplars stand, people are really talking about themselves and their personal relationships. For instance, in France's Garonne River valley, two social groups can be distinguished. One, people who settled there long ago and are well integrated in the community, view poplars as symbols of the riparian area. They call this area "les ramiers"—a vernacular term for a place that symbolizes life, freedom, and their own identity. In contrast, people who settled more recently perceive "les ramiers" as an indescribable mess of vegetation that the older settlers are keen to maintain as inaccessible to "foreigners" or even to newcomers (Le Floch 2002).

Following the emergence of the environmental consciousness in the 1970s, the poplar issue arose in social debates about the future of rural riparian areas. Except for mountainous areas, the issue affected every French region as poplar plantations increased in valleys while cattle use declined. This issue is based on criticisms by environmentalists relative to the impact of poplar cultivation on riparian environments. The critics did not exclusively refer to ecological arguments but formulated many of their objections in terms of landscape issues ("landscape closing," "banalization," or "fragmentation"). In fact, those critics express a regret of a certain way of

Figure SB8.1 Herbaceous vegetation in a poplar tree plantation along the Garonne River, France. (Photo: S. Le Floch.)

managing and inhabiting the land, over which current rural societies have less and less control. Currently, poplar plantations do not fit the existing social representation of valley landscapes being more and more rejected as a symbol of productivism (Le Floch and Terrasson 1995).

Sophie Le Floch and Daniel Terrasson, Centre d'Etudes du Machinisme Agricole du Génie Rural des Eaux et Forêts (Cemagref), Bordeaux, France

Conclusions

Riparian Management as an Emerging Issue

In many respects, the management of riparia may be viewed as an emerging environmental issue. Although not yet generally recognized, we suspect that it may have significant impact on human and/or ecosystem health as it matures and as it is more widely implemented (Munn et al. 1999). The perception of riparia by scientists, managers, and the general public has changed along with the allied issues of water availability, sustainable development, climate change, and biodiversity. Riparia are recognized as delivering irreplaceable goods and services to society and as playing a crucial role in the landscape, not only as corridors and as natural filters but also as symbolic representations of the environment by society. Nevertheless, management of riparia remains rooted in a regional or local context due to complex, controversial and uncertain dynamics that are difficult to generalize. Most often riparian issues result from an unsustainable and unintegrated management of water and/or land. As such, they are components of broader issues, but components that are more and more perceived as being crucial because they reveal dysfunctions in the landscape.

Viewing riparian management, conservation, and restoration as emerging environmental issues is to recognize that the dynamics of riparia represent a complex web of physical, biological, and social phenomena. Most importantly, these phenomena are dominated by feedbacks and synergies intimately connected with the dynamics of the most important limiting factors for development in many countries: land and water (Jasanoff et al. 1997). As such, the ecological and cultural sustainability of riparia is in itself a factor of economic development and social equity. Clearly, the true value of riparia cannot be fully accounted for through assigning monetary values (Gatto and de Leo 2000). Nevertheless, a riparian ecosystem service such as contributing to cleaner water is beneficial to society, and its loss has a cost that justifies management, conservation, and restoration decisions (Acharya 2000).

Integrated water and land management appears as a means to ensure the mutual sustainability of people and the environment (Acreman 1997), implying that to value water and the land—and to use them sustainably—is the basis for sustainable management. Under this concept, riparia (as well as water and land) are examples of common pool resources (Ostrom et al. 1999). For such resources, exploitation by one user tends to reduce resource availability for others. Empirical studies suggest that protecting institutional diversity, as related to common pool resources, may be the most important condition to prevent short-term interests from causing outcomes that are not in anyone's long-term interests. Such conclusions are of paramount importance for riparia, a resource whose successful management depends on the cooperation of numerous public institutions at local to international levels, private companies, individual landowners, and nongovernmental groups.

Conservation

9

Overview

- Effective conservation is a difficult challenge in riparian systems because many are deeply influenced by human populations. Interdisciplinary approaches integrating human and natural sciences are required to preserve an appropriate balance between the health of resources and ecosystems and the health and quality of human life. Biodiversity, ecosystem services, and hydrologic effects are relevant issues for riparian conservation.

- Principles adopted by the Conservation Convention on Biological Diversity are applicable to riparia. Among them, causes for the reduction or loss of biotic diversity should be anticipated and prevented early, and measures that could avoid additional losses of biotic diversity should not be deferred because of scientific uncertainty. Conservation actions toward riparia should be locally accepted by people, and effects on adjacent and other ecosystems should be considered, including changes likely to occur in the future.

- Conserving riparia for ecosystem or natural services is an important approach to sustaining functions, which are varied as well as complementary: trapping sediment eroding from river banks and diffuse nutrients in surface or subsurface runoff, limiting increases in summer water temperature through shading, providing wildlife habitats and facilitating movements of certain species along hydrographic networks. In addition, forage, firewood, and other specialized products are benefits of riparian conservation in many places.

- Hydrologic considerations are often advanced in support of riparian conservation. Riparian vegetation influences many processes related to surface and subsurface flow at local scales and modifies the local temperature and humidity through uptake of soil moisture, water storage, and evapotranspiration, establishing an "oasis effect."

- Site conditions have great importance in riparian conservation. The effectiveness of plant cover in protecting banks varies according to local factors—sediment type, plant species, severity and frequency of hydrologic events, and channel size. Although riparian woodlands often prevent soil erosion, there are limitations and even cases where erosion may be a positive process.

- Riparia offer aesthetic and cultural values, recreational opportunities, improved water quality, reduced sedimentation, and less flood damage, as well as income to some landowners. Societal benefits are generally recognized but not always translated into better conservation practices because attitudes vary among individuals, depending on personal beliefs. Essential factors for adoption of conservation guidelines are clearly defining problems, addressing the concerns and needs of landowners, and targeting specific areas where the greatest benefits may be expected.

- Effective riparian conservation is underpinned by two key questions linking ecology, conservation, and public policy: (1) How does one approach scientific uncertainty in ecological problems? and (2) How does one integrate ecological knowledge with the social and ethical aspects of conservation?

- Many countries have protective regulations to encourage voluntary conservation of water resources, including riparian areas. The future of riparian conservation embodies new approaches to scientific uncertainty in ecological problem solving and to integration of ecological knowledge with social and ethical issues. Riparian conservation requires iterative protocols involving scientific research and synthesis, as well as public perceptions, in an adaptive management perspective.

Purpose

Conservation biology may be seen as a multidisciplinary science addressing the current biodiversity crisis (Primack 1993). Its main goal is to understand the impacts of human activities on species, communities, and ecosystems, and to develop approaches to prevent species extinction—indeed to reinsert species into functional ecosystems. Conservation biology differs from other applied disciplines by having a more general theoretical framework and by having the conservation of biodiversity as a primary objective; economic factors are secondary. The core of conservation biology includes subdisciplines emerging from biological sciences (Soulé 1986). Nevertheless, because the biodiversity crisis results from human activities, conservation biology aims at integrating ideas and expertise from human-related sciences. This is clearly embodied in the seven principles for the conservation of living resources (see Table 9.1; Mangel et al. 1996). These principles are relevant for riparia, particularly as they focus on the inconsistency of unlimited expansion of human consumption of and demand for resources, on addressing biological diversity at various levels, and on recommending the use of the full range of knowledge and skills from natural and social sciences in promoting interactive, reciprocal, and continuous communication between all users and stakeholders.

   Historically, the idea of conservation was associated with wilderness and nature reserves, particularly with the image of the North American national parks—that is, large uninhabited areas. Thus, Yellowstone National Park, the world's first national park, was created in 1872 and established as part of the U.S. national park system "to conserve the scenery and the natural and historic objects and the wildlife herein and to provide enjoyment of the same in such manner and by such means as will leave them unimpaired for the enjoyment of future generations" (the National Park Organic

Table 9.1 Seven Principles for the Conservation of Living Resources

Principle 1	Maintenance of healthy populations of wild living resources in perpetuity is inconsistent with unlimited growth of human consumption of and demand for those resources.
Principle 2	The goal of conservation should be to secure present and future options by maintaining biological diversity at genetic, species, population, and ecosystem levels.
Principle 3	Assessment of the possible ecological and sociological effects of resource use should precede proposed use as well as proposed restriction, expansion, or ongoing use of a resource.
Principle 4	Regulating the use of living resources must be based on understanding the structure and dynamics of the ecosystem of which the resource is a part, and must consider the ecological and sociological influences that directly and indirectly affect resource use.
Principle 5	The full range of knowledge and skills from the natural and social sciences must be brought to bear on conservation problems.
Principle 6	Effective conservation requires understanding and taking account of motives, interests, and values of all users and stakeholders, but not by simply averaging their positions.
Principle 7	Effective conservation requires communication that is interactive, reciprocal, and continuous.

From Mangel et al. 1996.

Act of 1916). In such areas, despite Principle 5 in Table 9.1, human sciences have remained marginal, and conservation management remains focused on biodiversity and system functions rather than on the interactions between human populations and their environment.

Today the management of areas such as North American national parks has little in common with the management of riparia—areas where human activities are often omnipresent. Many riparian systems are deeply influenced by the strong interrelations between humans and the environment. In most cases, riparia are more *ordinary* in nature rather than *sanctuary* in nature. The term *ordinary* is used in the sense of riparia being part of our everyday landscape, whereas the term *sanctuary* denotes a system set aside for its natural features that most people only see while on vacation. In riparian systems, effective conservation is a difficult challenge because preserving an appropriate balance between the health of resources and ecosystems and the health and quality of human life requires interdisciplinary approaches integrating human and natural sciences, strong communication between scientists and managers, and a personal involvement of researchers in conservation issues. Three issues are particularly relevant for riparian conservation: biodiversity, ecosystem services, and hydrologic effects. In this chapter we discuss these issues, using them to examine how principles for the conservation of living resources may be applied to riparia—that is, systems where local conditions are important, human benefits varied, and conservation legislation still emerging. The outcomes from such a discussion are essential for establishing effective riparian conservation strategies for the long term.

Conserving Riparia for Biodiversity

The Convention on Biological Diversity was proposed at the 1992 United Nations Conference for the Environment and Development in Rio de Janeiro, Brazil.

Biological diversity was defined as the variability of living organisms from all origins, including organisms from terrestrial and aquatic ecosystems, and ecological systems that sustain them; this includes diversity within and among species, and the diversity of ecosystems and their inherent processes. Reasons for conservation are scientific, as biodiversity is the raw material on which evolution—that is, life adaptability—develops, and surprisingly little is known about the elements and processes involved or even the full suite of species and processes that one might include in a catalog of life on Earth. Reasons are also economic, as biodiversity is a reservoir for genetic variability of cultivated species, a basis for the development of biotechnologies (e.g., pharmaceuticals), and a source for ecosystem services such as the regulation of flow regimes, buffering of subsurface water quality, and control of pathogens. In addition, there are ethical reasons for conserving biodiversity, as each generation borrows the Earth from its descendants and has a duty of conservation for future generations. All these reasons apply to riparian conservation, including the recreational value of natural areas, the beauty of wild species, and the cultural and educational value of discovering knowledge about species in their environment.

Also applicable to riparia are the principles adopted by the Convention on Conservation of Biological Diversity and its approach, which includes interactions among organisms and interactions between organisms and their environment. Among these principles are that (1) it is highly important to anticipate and prevent causes in the reduction or loss of biological diversity at the beginning, and to grapple with them, and (2) the absence of scientific certainty should not be invoked as a reason to defer measures that could avoid the loss of biological diversity or the attenuation of its effects. Collectively, these become a suitable framework for conservation action in riparia as they result in management strategies that should be locally accepted, that consider real or potential effects on adjacent and other ecosystems, that apply to appropriate scales, and that admit change is unavoidable.

As an example of effects on adjacent ecosystems, we saw earlier that riparian forests affect stream productivity and the variety of aquatic plants, microorganisms, invertebrates, and fish in many ways, including seasonal inputs of leaves. As the leaves decompose, they feed a whole community of invertebrates, which in streams provide food for fish. The leaf exclusion experiment carried out in North Carolina (discussed in Chapter 6) is significant in this respect (Wallace et al. 1997, 1999). Recall that 200 m of a woodland stream were deprived of leaf litter inputs for three consecutive years. The result was unambiguous: of the 29 species of invertebrates contributing 95 percent of the stream's production, 17 species significantly declined in abundance and/or in biomass with the removal of the leaf-litter input. This decrease not only affected detritivores, which are direct users of dead leaves, it also affected predators, thus modifying the whole food chain (see Table 9.2). Conserving appropriate riparian assemblages as well as transfers across system boundaries is important for the integrity of the larger system.

Dead wood is another important source of aquatic diversity. Tree trunks and branches regularly fall into watercourses and may strand where they fall or be carried along by the current. This large woody debris is an essential component of stream channels of forested regions (Naiman et al. 2002a, Gregory et al. 2003). It fulfills many functions in lotic ecosystems, creating a wide variety of habitats, from deep pools with slow-moving water to fast-moving, well-aerated rapids—thereby

Table 9.2 Effects of Depriving a Woodland Stream of Its Leaf Litter Inputs in a 200 m Zone for Three Consecutive Years

Taxon	Abundance	Biomass	Taxon	Abundance	Biomass
Shredder			**Collector-filterer**		
Peltoperlidae	ns	*	*Diplectrona* sp	*	*
Leuctra spp	*	**	+ one taxon	ns	ns
Lepidostoma spp	***	***			
Pycnopsyche spp	*	*	**Predator**		
+ four taxa	ns	ns	*Lanthus* sp	***	***
			Cordulegaster sp	**	**
Collector-gatherer			*Beloneuria* sp	*	**
Nematoda	*	*	*Rhyacophila* spp	ns	*
Copepoda	***	***	Tanypodinae	***	**
Stenonema spp	***	***	Ceratopogonidae	***	**
Chironomidae	***	**	*Hexatoma* spp	**	**
+ four taxa	ns	ns	*Pedicia* sp	**	**
			+ three taxa	ns	ns

Probability levels: *$P < 0.05$, **$P < 0.01$, ***$P < 0.001$, and ns – no significant difference between streams. Modified from Wallace et al. 1997.

diversifying plant and animal life. Frequently, accumulations of dead leaves on submerged tree crowns become hotspots of biological activity. Dead wood is also a substrate on which fungi and bacteria can develop, attracting aquatic invertebrates, which creates a link between the microbes and detritus on which they feed and predators, including fish. Fish also make use of accumulations of dead wood for shelter, protection from predators, and reproduction. In addition to dead leaves and wood, other inputs of organic material from riparia include terrestrial invertebrates that have fallen into the water from the banks, which can be a major source of food for fish (Nakano et al. 1999). Therefore, conservation of lotic ecosystems requires intact and healthy riparia for maintaining the long-term integrity of river corridors—they are, in effect, a single ecological unit.

The importance of a particular riparia to wildlife depends on the landscape setting, the vegetation, and the species considered. In intensive agricultural landscapes, riparian-dependent wildlife can live and reproduce, and perhaps find supplementary habitat, in areas adjacent to riparia. Similarly, riparia provide reserves for maintaining wildlife in forested settings, particularly when associated with harvest of upland forests (Darveau et al. 1995). Riparia also may be important habitats in urban settings where natural areas are lacking, although activities in the immediate vicinity (e.g., traffic or industry) can impede wildlife diversity and abundance. The local vegetation affects wildlife through the availability of food, cover, and territory, as well as access to water. Diversity and complexity of habitats are important factors for maintaining viable riparian wildlife populations, as well as for maintaining suitable environmental conditions (e.g., temperature, humidity, moist and loose soils, and diverse canopy layers; Klapproth and Johnson 2000). The wildlife may vary from riparian-dependent organisms, such as beaver (*Castor canadensis* and *C. fiber*), to occasional visitors such as white-tailed deer (*Odocoileus virginianus*). Various large and small mammals, reptiles and amphibians, and birds are attracted by riparia with a diversity of sizes, forms, plant communities, and features, such as dead woody debris, wetland depressions, and nesting sites. Three biophysical continua appear to strongly influence the composition of riparian wildlife communities at both local and landscape levels: stream order, forest successional stage, and the type, frequency, duration, and severity of natural or human-induced disturbances (Kelsey and West 1998).

The utilization of riparia by birds provides support for the suitability of the ecosystemic approach to riparian conservation in agricultural and forested areas. Not surprisingly, results vary by bird species and by regions considered, and there are contrasting observations about the widths necessary to support specific birds and about differences in diversity or abundance when adjacent areas are either agricultural or forested (e.g., Croonquist and Brooks 1991, Keller et al. 1993, Darveau et al. 1995, Dickson et al. 1995, Murray and Stauffer 1995, Lock and Naiman 1998). The edge habitat that characterizes riparia may cause reproductive failure, restriction of range, and mortality (Wigley and Roberts 1997), particularly in urban settings where commercial, residential, and industrial developments attract domestic predators (e.g., house cats). Similarly, the corridor character of riparia, although important to small animals and young individuals establishing territories (Machtans et al. 1996), can promote the movement of generalist species at the expense of area-sensitive species, and thereby enhance the spread of contagious diseases, fires, predators, and exotic species. Native trees and shrubs, ground and shrub cover, high tree density, and canopy closure also appear to be important characteristics in conserving bird species richness (Miller et al. 2003).

The apparently high biological diversity associated with riparia make them key areas for conservation (Hughes 2003). As interfaces between land and water ecosystems, they provide habitats for species from the adjacent systems, in addition to true riparian species. As systems frequently disturbed by natural flow regimes, they include a mixture of differently composed plant and animal communities in various stages of succession. As systems occupying the lowest terrestrial position in the landscape, they accumulate plant propagules from entire river catchments. As a consequence, biological diversity is a primary reason for riparian conservation, a reason linked to the quality of ecosystem services provided by riparia along rivers.

Conserving Riparia for Ecosystem Services

Riparia are known to filter materials from both urban and rural catchments, with nitrogen and sediment-bound phosphorus among the most important nutrients (see Chapter 6). This natural filtering function is often cited as reason enough to justify conserving riparia. However, sustainable riparian management is based on services that are many and varied, as well as being complementary. These services concern the aquatic as well as the semiterrestrial environment (Iowa State University 1997, Federal Interagency Working Group 1998, Naiman et al. 2000, AFFA 2001).

Concerning the aquatic environment, riparia stabilize river banks, protecting soils against surface erosion and strengthen their resistance to destabilization. As a consequence, they improve water clarity by reducing the amount of sediment eroding from river banks, and prevent the silting of sensitive aquatic habitats (for example, the spawning grounds of salmon). Moreover, riparia trap diffuse nutrients in surface or subsurface runoff, thus counteracting the over-enrichment of rivers by phosphorus or nitrogen. In addition, as they reduce light penetration by shading, riparia limit increases in water temperature during the summer and reduce algal and macrophyte growth; they also maintain existing fish stocks through their effects on temperatures, on the diversity of aquatic habitats, and in supporting both aquatic and terrestrial prey species.

Concerning the semiterrestrial environment, riparia form corridors that facilitate the movements of certain species along hydrographic networks and that may prove essential to (1) the survival of fragmented populations functioning together as metapopulations, and (2) the increase both aquatic and terrestrial regional species diversity. Also, riparia complement grasslands and agriculture by providing forage, firewood, and other specialized products, they provide habitat for various species of mammals, birds, and invertebrates that may control pests in the adjoining agricultural land, and they form natural screens, protecting crops from direct winds and livestock from extreme heat and cold.

These services occur globally in a critically threatened environment (Tockner et al. 2005). As we saw in Chapter 7, many habitats have been lost or severely degraded along rivers as a consequence of dam construction, channel straightening, gravel and sand extraction, and forest clearance. In addition, new habitats created by past management practices risk being lost, for example when traditional agriculture is replaced by intensive and mechanical agriculture, or when economically less viable dams are removed (Johnson 2002).

Fortunately, engineers are increasingly aware of the importance of working with nature rather than against it, of keeping human impacts within sustainable limits, and of preserving services performed by natural systems. This requires adequate knowledge of natural landform dynamics and landform–habitat interactions (Warren and French 2001). For example, riparian conservation must consider and incorporate upslope systems (upstream and adjacent hills) that never achieve equilibrium, that lose sediments in episodic events, that inherited soils from previous climatic periods, and that have juxtaposed areas largely differing in erosion rates. Such a geomorphically dynamic perspective is a fundamental component of models addressing the conservation of riparian goods and services. For example, the Italian Tagliamento River has been proposed as a reference system for the Alps, as being "the only large morphologically intact alpine river remaining in Europe, providing insight into the natural dynamics and complexity that must have characterized alpine rivers in the pristine state" (Ward et al. 1999). Conserving these natural dynamics and complexity is useful for eventually discovering relationships among topography, sediment, and vegetation along the active zones of mountainous rivers (Gurnell et al. 2001, Tockner et al. 2003). Another fascinating example of a reference system is forest regeneration through primary succession on newly deposited riverine soils in the Upper Amazon region (Salo et al. 1986). The high between-habitat diversity that characterizes this region principally depends on preserving river dynamics that result in a mosaic of riparian forests of different ages on different soils. At a larger scale, many of the fundamental geomorphic processes of the mainstem Amazon are linked to the magnitude and variability of water and material supplied from the Andes (Dunne et al. 1998). Understanding and protecting such processes and linkages is essential for sustaining riparian goods and services.

Conserving Riparia for their Hydrologic Effects

As discussed in earlier chapters, riparian vegetation influences many processes related to surface and subsurface flow at the local scale, depending on the physical structure of the site (see Figure 9.1; Darby 1999, Bendix and Hupp 2000, Tabacchi et al. 2000). These hydrologic effects are often offered up in support of riparian conservation. For example, riparian grasses can trap more than 50 percent of sediments delivered from hillslopes when overland flow depths are less than 5 cm (Magette et al. 1989), and this may be amplified within patchy vegetation of differing heights and flexibilities (Bromley et al. 1997). In addition, dense herbaceous vegetation colonizing newly deposited sediment helps sustain high moisture levels during dry periods by shading the sediment surface. Riparian trees also play an important role in the dissipation of kinetic energy from floods. This is accomplished through the high hydraulic

Figure 9.1 Main physical impacts of riparian vegetation on hydrologic processes: (1) interaction with overbank flow by stems, branches, and leaves; (2) flow diversion by log jams; (3) change in the infiltration rate of flood waters and rainfall by litter; (4) increase of turbulence as a consequence of root exposure; (5) increase of substrate macroporosity by roots; (6) increase of the capillary fringe by fine roots; (7) stemflow; and (8) condensation of atmospheric water and interceptions of dew by leaves (after Tabacchi et al. 2000).

roughness and flow resistance imposed by trees on floodwaters. This role varies with discharge and with the relative widths of the riparian corridor and the channel. Deep-rooted trees improve drainage or infiltration by increasing substrate porosity and cap-illarity (Thorne 1990). For example, in northeastern Brazil, on some clayey soils, the Amazonian forest absorbs water from a ground depth of over 8 m during the dry season (Nepstad et al. 1994), making the root systems instrumental in upward trans-port of water from deep to shallow layers. Later, during wetter periods, the deep roots aid in the downward movement of percolating water along the root channels. Addi-tionally, riparian vegetation may attenuate the input of water to the floodplain and delay drainage from backwaters, thereby facilitating exchanges between surface and groundwater.

However, mainly as a consequence of alterations in flow regimes, invasion by pioneer trees can drastically modify the hydraulics of entire channels. Spectacular examples are provided by the invasion of the Colorado River and its tributary, the Green River, by thickets of riparian tamarisk (*Tamarix* spp; Graf 1978), and by the colonization of the Platte River, Nebraska, by willow and poplar, which increase sed-iment retention and the flow velocity in the remaining narrow channel of the river (Johnson 1994). Likewise, along the River Ouvèze, in southern France, where the mean width of riparian woods increased from 50 to 92 m downstream of Vaison la Romaine between 1947 and 1991, there was a reduction in the mean width of the surface water channel from 83 to 48 m (Piégay and Bravard 1997). This resulted in a considerable reduction in the river's capacity to carry water during major floods, as demonstrated by a catastrophic flood in 1992.

In addition, riparian vegetation stores absorbed water and releases it to the atmos-phere by evapotranspiration (see Figure 9.2; Crockford and Richardson 2000). This process may be highly variable from one species to another and, for the same species, from one site and one time to another. Measurements taken along the Morava River in the Czech Republic demonstrate this aspect (Penka et al. 1991). Between June and September, the water content of small balsam (*Impatiens parviflora*) changes with wet weights decreasing from 98 to 92 percent in the stems and from 94 to 90 percent in the leaves; for stinging nettle (*Urtica dioica*), the corresponding percentages are 88 to 85 percent in the stems and 86 to 80 percent in the leaves. During this same period, these two very common species along watercourses in Europe may increase in height from 16 to 38 cm (*Impatiens*) and from 40 to 80 cm (*Urtica*). Along the Morava River, herbs contribute 3 percent of the potential evapotranspiration, bushes 9 percent, and trees 88 percent. Daily and seasonal variations depend on local environmental con-ditions, on the plant species, and on community structure. Under arid and semiarid conditions, riparia may act to modify temperatures and humidify the surrounding air, creating an "oasis effect."

Riparian Conservation in a Management Context

Riparian conservation can be achieved in a management context, which is integrated into projects combining conservation, management, and restoration (see Chapter 8).

Figure 9.2 Main physiological impacts of riparian vegetation on hydrologic processes: (1) hydraulic lift; (2) hydraulic redistribution; (3) water storage in large roots; (4) water storage in the stems; (5) water storage in branches and leaves; and (6) evapotranspiration (after Tabacchi et al. 2000).

For example, a three-pasture rotation system can provide total annual rest for pastures along banks dominated by herbaceous riparia. Under such a system, a given pasture is grazed early during the growing season the first year, after upland grasses are seed-ripe during the second year, and totally rested from livestock grazing during the third year (Elmore 1992). Various programs have been used to encourage conservation of private lands and implement riparian buffer policies: voluntary programs (Harrington et al. 1985), economic incentives and disincentives (Malik et al. 1994), and regulation. Studies indicate that conservation of riparia in a management context requires education, technical assistance, and financial support (Lichtenberg and Lessley 1992, Johnson et al. 1997, King et al. 1997).

In addition, riparian conservation cannot be a substitute for good catchment management and land use practices. For example, the use of riparian buffer zones as natural filters must be incorporated into broader conservation plans. In addition, environmental issues are important to consider because buffer zones do not filter diffuse pollution everywhere in the hydrographic network with equal efficiency (see Figures 9.3 and 9.4). Depending on the plant assemblage, riparian buffer zones may act

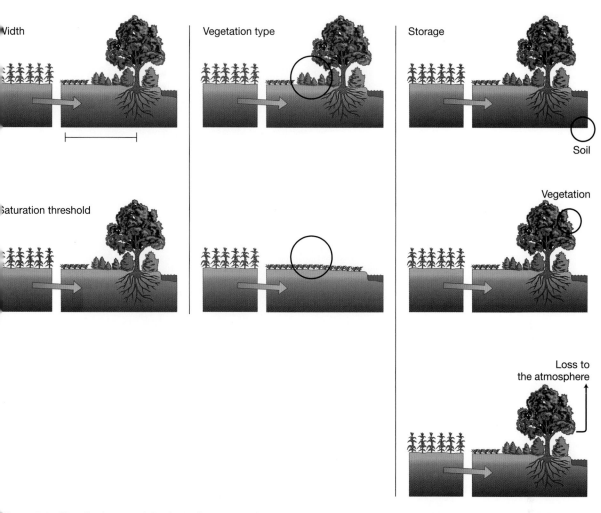

Figure 9.3 The effectiveness of riparian buffer systems depends on questions that can only be addressed at a local level. This figure illustrates some of these questions about width, saturation threahold, vegetation type, and nitrogen storage. For example, effectiveness of the buffer will differ according to the nature of the vegetation, shrubs or herbs (in circles). Similarly effectiveness will differ according to the place where nitrogen is stored, in soils or in vegetation (circles) or lost to the atmosphere (after Lowrance et al. 2000).

differently as filtration sites for dissolved nitrates as well as for phosphorus adsorbed onto sediments (see Table 9.3). Moreover, this filtering capacity, attractive though it may seem, cannot justify uninterrupted conservation throughout the entire hydrographic network. In the upper parts of catchments, streams flow through areas where much of the sediment and nutrients are being produced. These small-order streams are often primary conduits for runoff from surrounding agricultural and forestry-related activities. Further downstream, watercourses of higher order convey material originating upstream as well as receive their own influx of materials from surrounding lands. The most effective sites for filtration are associated with smaller watercourses in upstream areas, but they are needed in downstream areas too.

Riverbank erosion also illustrates the necessity for assessing environmental issues before deciding on conservation measures. The presence of riparian vegetation generally reduces the risk of erosion because (1) the plant roots provide cohesion to

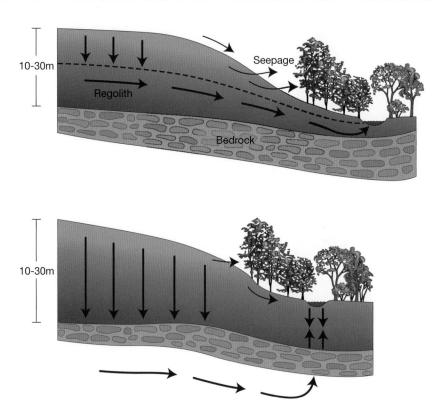

Figure 9.4 An example of a situation where components of groundwater flow are likely to bypass the riparian buffer zone (after Lowrance et al. 1995).

Table 9.3 Effects of Various Riparian Buffer Zones on the Reduction of Inputs from Surface Runoff: Evaluation Based on Studies Carried out in the Chesapeake Bay Catchment in the United States

Buffer Zone		Reduction: 100 × (Input − Output)/Input		
Width (m)	Plant Cover	Sediment (percent)	Nitrogen (percent)	Phosphorus (percent)
4.6[a]	herbs	61.0	4.0	28.5
9.2[a]	herbs	74.6	22.7	24.2
19.0[b]	trees	89.8	74.3	70.0
23.6[a]	herbs + trees[c]	96.0	75.3	78.5
28.2[a]	herbs + trees[d]	97.4	80.1	77.2

[a]inputs: sediment 7.3 mg/L, nitrogen 14.1 mg/L, phosphorus 11.3 mg/L.
[b]inputs: sediment 6.5 mg/L, nitrogen 27.6 mg/L, phosphorus 5.0 mg/L.
[c]width comprises 4.6 m of herbs plus 19 m of trees.
[d]width comprises 9.2 m of herbs plus 19 m of trees.
Modified from Lowrance et al. 1995.

riparian soils, particularly where there are many fine roots; (2) the plant roots assist in the drainage of riparian soils, thereby reducing the risk of collapse due to water-logging; and (3) plants growing at the base of the banks act as buttresses, helping protect the soils above (Poesen and Hooke 1997, AFFA 2001). However, protecting banks from erosion requires an accurate knowledge of natural conditions that vary from one site to another, as riparian plant cover is not always capable of stabilizing a bank. For example, bank steepness may prevent establishment of significant vegetation on the outside bends of meanders. Terrestrial vegetation will not protect higher banks from the risk of being undercut and collapsing. Aquatic or semiaquatic vegetation, as well as large woody debris, can increase the risk of undermining when dense stands of aquatic vegetation or large woody debris redirect the current against the banks. Further, it has been suggested that the weight of large trees on riverbanks may increase the risk of collapse (see Thorne 1990). In fact, this risk only occurs on banks that are almost vertical; in the majority of cases large riparian trees actually improve bank stability thanks to their deep root systems (Abernethy and Rutherfurd 2000). Any risk comes from trees that are shallow rooted and therefore easily thrown by strong winds, and from deep shading, which prevents herbaceous plants from growing beneath the trees. The fine networks of herbaceous roots are effective in maintaining cohesion in some riparian soils. The banks of small watercourses in Wisconsin, for example, on sandy and clayey soils, are more stable with a covering of herbs than trees (Trimble 1997). In fact, bank stability depends on a whole range of local factors: sediment type, plant species, severity and frequency of hydrologic events, and stream size.

Moreover, the conservation of riparian woodlands as a remedy for soil erosion must be placed in a wider context, preferably at the scale of an entire catchment. For example, the conservation of riparian woodland to reduce the destructive effects of soil erosion along Mediterranean rivers is inseparable from conservation practices that aim to prevent erosion in agro-sylvo-pastoral systems. Effective practices make every effort to reestablish the soils' herbaceous cover and to avoid overgrazing. For example, modern orchards, ploughed and treated with herbicides, are extremely vulnerable, whereas traditional orchards with a good herbaceous cover show much greater resistance to erosion (Décamps and Décamps 2001). In addition, erosion need not be always viewed only in a negative light (Grove and Rackham 2001). Although riparian woodlands may be considered as a remedy for soil erosion at some sites, limitations of this role in conservation (and restoration) efforts should be recognized. For example, simply planting vegetation rarely halts erosion of banks that are already heavily degraded, and a vegetative layer may reduce the channel's capacity to cope with large floods once vegetation is established on islands or on rocky shoals in the riverbed itself. Additionally, appropriate rates of bank erosion involve some positive environmental aspects and cannot be viewed only in terms of dangers: some erosion keeps the major channels open, it perpetuates the riparian mosaic, it absorbs the impact of some floods—thus counter-balancing effects downstream—and it replenishes stream substrate that is essential for life in rivers.

Finally, it must be emphasized in support of riparian management and conservation (and restoration) that riparia occupy only a relatively small proportion of agricultural land. The density of streams in mountains is ~2.5 km of stream/km²; the corresponding number in agricultural areas is ~1.5. Given these stream densities, 2.5 percent and 7.5 percent of the land would be occupied by 10-m- and 30-m-wide buffer zones in mountains, and 1.5 percent and 4.5 percent of the land would be occupied by 10-m- and 30-m-wide buffer zones in agricultural areas. Given the difficulties involved with using agricultural machinery close to watercourses, many riparian areas are difficult to farm. Further, benefits other than from conservation may be derived from buffer zones. Some benefits may be commercial (e.g., the production of wood and forage) and others noncommercial (e.g., hunting, recreation, critical habitat—both aquatic and terrestrial). Riparian conservation may be considered as resulting in a loss of net income to landowners, but in the longer turn, it is beneficial in a management context. This requires organizing communication between local examples of education, technical assistance, and finanical support that are disseminated in many countries, and remain largely unknown.

Human Benefits from Riparian Conservation

Riparia maintain ecological functions that provide many benefits to humans—whether individuals or social groups. They offer aesthetic and cultural values, recreational opportunities, improved water quality, controlled sedimentation, and less flood damage, as well as income to some agricultural landowners and artisan enterprises (Klapproth and Johnson 2001a). Many factors contribute to the aesthetic appeal of water in the landscape, such as topography, relief, form, vegetation types and arrangement, water variability and pattern, and human use and impacts (Litton 1977). The presence of rapids or a large scenic vista also increases the appeal of streams (Brown and Daniel 1991). Riparian trees provide a source of satisfaction: old, tall, large-diameter trees are enjoyed for their beauty, their shade and privacy, and the attention and care that they necessitate (Jackson 1994). Riparian areas are also rich in historical, archeological, and other cultural features (see Sidebar 9.1).

Conservation has greatly increased and diversified recreational activities in riparia and adjacent rivers during recent decades. Activities such as rafting, motorboating, hiking, biking, photography, and observing nature have expanded, along with the more traditional activities such as trapping, hunting, and fishing. Visitors to river valleys often remark that rivers and their riparia provide important sources of mental and physical refreshment (Pawelko et al. 1996). The protection of fish and wildlife habitat, water and air quality, and scenery are given as important reasons for river landscape conservation. People are also aware of protecting rivers for future generations to enjoy and use (Clonts and Malone 1990). At the same time, recreational use of riparia is a source of potential income for communities and landowners. According to a survey of Virginians (U.S. Department of the Interior and Department of Commerce 1996), 44 percent of the population participated in some form of wildlife-related recreation such as fishing (13 percent), hunting (3 percent), or wildlife-watching (37 percent), significantly contributing to the state's economy. In Maryland, waterfowl hunting alone generates US$3.5 million to local economies each year (Lynch 1997). Without proper management and conservation, all these activities can destroy the aesthetic and ecological benefits provided by riparia.

Societal benefits also come from riparia reducing the amount of sediment, nutrients, and other diffuse pollution moving toward lakes and the coastal zone (NRC 2000). This lowers costs for water treatment, water storage, and dredging (Holmes 1988). Other benefits include reducing flood damage and purifying surface water that eventually replenishes groundwater. Riparian vegetation also benefits fisheries and, in

Sidebar 9.1 The Multifunctional Landscape of the Ibrahim River, Lebanon

The Ibrahim River, the river of "love and faithfulness," lies 20 km north of Beirut. It is one of several coastal rivers flowing from sources on Mount Lebanon to the Mediterranean Sea. The river has long been recognized for its natural, historical, and religious significance as well as for the outstanding beauty of its landscape. The river and its source at Afqa were sacred to the ancient inhabitants of the region and equally sacred to the Greeks and Romans that followed. The river landscape was the scene for the mythical and tragic love story between Astarte, Greek Venus, and Adonis. The myth goes that Adonis was killed while hunting and that his lover Astarte, failing to save him, cast a spell that the blood of Adonis tint the torrential waters each spring turning them red, thus giving the river a third appellation, the River of Immortal Love. The Romans built three temples close to the source of the river at Yanuh, Aqoura, and Afqa as well as an aqueduct toward the end of the river. In centuries that followed, the pre-Greek pilgrimage road that ran along the north side to Afqa was used by caravans moving westwards to the sea up to Ottoman times. The remains of a stone bridge built during Ottoman rule in 1806 stands close to the estuary, another monument to the enduring legacy of the Ibrahim River.

The riparian floristic inventory for the Ibrahim River includes a collection of 367 plant specimens from 59 families and 147 genera (Abboud 2002). The best-represented families (*Asteraceae, Poaceae, Fabaceae, Lamiaceae,* and *Apiaceae*) contribute 38 percent of the recorded flora (Abboud 2002). There are 2 endemic plants (*Origanum ehrenbengii* and *Papaver umbonatum*), 19 riparian specific, 7 riparian generalist, 57 weeds, 13 Mediterranean invasive species, and 39 medicinal species. Seven plants are widely distributed in the catchment: *Dittrichia viscose, Cephalaria joppensis, Euphorbia peplus, Geranium purpureum, Geranium rotundifolium, Piptatherum miliaceum,* and *Rubus sanctus.*

The floristic characterization of the Ibrahim River can be explained within the historical, ecological and biological context of riparian ecosystems. A limited number of riparian-specific and general species, coupled with an exceptionally high number of weed species, are indicative of the highly modified, disturbed, and fragmented nature of the Ibrahim riparian landscape. Increasing human disturbance downstream directly affects land use, riparian vegetation, and ecosystems properties. As a result, the dense, full-canopied, seminatural vegetation characteristic of the upper reaches of the Ibrahim is increasingly degraded closer to the coast.

The Ibrahim River exemplifies the versatility of Mediterranean riparian landscapes and their enduring adaptability to a diversity of uses and management. Up to the mid-20th century, the river served agricultural and domestic uses, namely for local inhabitants. Traditional stone terraces with citrus and olive crops can be seen along the steep riverbanks and represent the traditional rural use of riparian landscapes in Lebanon. Several small villages exist within the basin with a total population of ~15,000. Agriculture, though the major occupation of the inhabitants, has more recently witnessed a shift in form and location, from traditional terrace cultivation along the middle and higher reaches of the basin toward commercial cultivation, mainly bananas, in the lower reaches and along the watercourse. Another change is recreational activities. Traditionally, the Ibrahim River, because of its exceptional landscape, was a place for water-related sports and picnicking. More recently, unregulated dumping of building debris and refuse has limited the potential for water sports. Equally disturbing are unregulated industrial activities, which include a tannery, paper mill, two quarries, and a cement factory. A hydroelectric plant was built in the 1950s and continues to maintain three dams along the river. Increasing population pressures, unsustainable industrial and agriculture practices, and the lack of wastewater treatment currently combine to undermine the integrity of the riparian system.

The Ibrahim River exemplifies the coevolution between people and natural resources in the Mediterranean region. Ecological diversity and a beautiful, heterogeneous landscape make the river at once a natural and cultural heritage in Lebanon. Eager to protect the national heritage, a Ministerial Decree declared the Ibrahim River a natural site in 1997. The implications are that a buffer zone of 500 m on both banks, and 1,500 m for quarries, be maintained. The Ministry of Environment must approve activities within the buffer. The decree is contributing to the improvement of the vitality of the riparian system, and it is hoped that eventually the river will be utilized for hiking, kayaking, and as a camping destination again.

Future planning and management can build on the diversity of natural, seminatural, and human-made ecosystems along the riverbanks, which are key to preserving the integrity of the riparian landscape. Effective local biodiversity management and conservation, however, are hindered by limited funds, manpower, and local expertise, as is so often the case. Developing cost-effective management strategies is a necessary step to ensure that needs for restoration are addressed, and with minimum costs and expert inputs. A holistic approach that evaluates riparian biological resources within the socioeconomic and cultural context has assisted in preparing a habitat assessment model that prioritizes conservation sites within the river and serves as an important tool for effective management (Abboud et al. 2004).

Jala Makhzoumi and Maya Abboud, American University of Beirut, Lebanon and Ministry of Environment, Beirut, Lebanon urban areas, it improves the quality of life by improving the environment (Groffman et al. 2003). In addition to providing services to societies, riparian conservation can yield many valuable goods, including wood products, foods, pharmaceuticals, and weaving and dying materials (Klapproth and Johnson 2001a). The leasing of hunting and fishing rights, as well as developing recreational activities, also may allow landowners to derive income.

Farmers and the forestry industry are generally concerned about soil erosion, water quality, and the environment and recognize that riparian forest buffers can provide many benefits to society at large. However, this does not always translate into adopting conservation practices, and the agricultural community—as well as other land based industries—may be resistant to the establishment of forested buffers (Klapproth and Johnson 2001b). This is not surprising, as farmers and ranchers everywhere are facing increasingly uncertain economic circumstances: they must generate personal income, meet debt obligations, and maintain future profitability. Attitudes vary among individuals, depending on personal beliefs. For example, conservation practices are more likely to be adopted by individuals who think that they have a moral obligation to maintain the land for future generations (Nowak and Korsching 1983) or who have strong views about the use of nonrenewable resources (Lynne et al. 1988). Similarly, different attitudes were observed in a cost–share program initiated in 1992 by the Maryland Department of Natural Resources to encourage landowners to install forested buffers along stream banks (Hagan 1996). Riparian landowners who were more educated, younger, and had less farm experience were those who established forested buffers through this Buffer Incentive Program; aesthetic factors and interest in fish and wildlife were important to them. Participating landowners worked generally on smaller areas and were part-time farmers. Nonparticipants were more likely to be full-time farmers, working on large areas, and deriving much or most of their income from the farm. Many full-time farmers preferred to install grass buffers instead of forested buffers.

Emergence of New Conservation Legislation

Regulatory approaches to meeting water-quality goals are gaining wider acceptance as a means to reduce nonpoint source pollution. Everywhere, more citizens are agreeing to better protection for water resources. As a result, many countries have adopted regulations to encourage communities and individuals to voluntarily protect water resources. Riparian conservation is an important part of these protective regulations.

In Europe, the Habitats Directive is a community legislative instrument establishing a common framework for the conservation of natural habitats and their fauna and flora. This directive establishes special areas for conservation, the "Natura 2000" network (EU 1992). The role of the network is to maintain and restore natural habitats and wild spaces of interest to the European Community. It involves the implementation of sustainable development—in effect, reconciling the conservation of natural habitats with economic and social demands. It also involves raising awareness of the necessity for protecting natural sites of special interest. The directive provides

Table 9.4 Riparia Included as "Special Areas of Conservation" in the European Union Habitat Directive (EU 1999)

Alluvial forests with *Alnus glutinosa* and *Fraxinus excelsior*, including riparian forests of *Fraxinus excelsior* and *Alnus glutinosa* of temperate and boreal Europe lowland and hill watercourses, riparian woods of *Alnus incanae* of montane and sub-montane rivers of the Alps and the northern Apennines, arborescent galleries of tall *Salix alba*, *S. fragilis* and *Populus nigra*, along medio-European lowland, hill, or sub-montane rivers.
Riparian mixed forests of *Quercus robur*, *Ulmus laevis*, and *Ulmus minor*, *Fraxinus excelsior* or *Fraxinus angustifolia*, along the great rivers.
Gallery forests with *Salix alba*, *Populus alba*, and related species in the Mediterranean basin.
Riparian formations on intermittent Mediterranean watercourses with *Rhododendron ponticum*, *Salix*, and others, including relict galleries.
Forests of *Platanus orientalis* and *Liquidambar orientalis*, forming galleries along watercourses, temporary rivers, and gorges in Greece and the southern Balkans.
Thermo-Mediterranean riparian galleries and thickets with *Nerium oleander* and *Tamarix* spp.

a master list of "special areas of conservation," several of which are riparian habitats (see Table 9.4; EU 1999). These include various riparian forests along small and large rivers, under contrasting climatic conditions, with particular attention to the Mediterranean region.

Allied to the riparian conservation efforts, the preservation and restoration of surface water and groundwater has been addressed in the European water framework directive (EU 2000). This is an important water policy instrument for the future development of aquatic ecosystems in member states of the European Union (see Sidebar 9.2).

In the United States, there are numerous types of legislation mandating environmental protection and conservation (Masonis and Bodi 1998). The various regulations can require the attainment of specific resource conditions (substantive laws) or consideration of environmental impacts in resource management decisions (procedural laws). Federal-level regulations govern water quality and wetland development (Clean Water Act), forestry-related activities on federal lands (National Forest Management Act, Federal Land Policy and Management Act), hydropower development (Federal Power Act), and protect biota in riverine systems (Endangered Species Act). State-level regulations establish, among many examples, water quality standards, water rights, and certifications that hydropower projects must satisfy to meet state water-quality standards. As an example of state-level regulations, there are three major pieces of legislation that give the citizens of Virginia responsibility for protecting the state's water from agricultural and forestry activities and from urban development (Klapproth and Johnson 2001b,c). These include the Chesapeake Bay Preservation Act of 1988, the Forest Water Quality Law of 1993, and the Agricultural Stewardship Act of 1996. Collectively, these regulations require citizens (and companies) to prevent pollution. State agencies, in turn, are charged with providing technical and financial assistance to help meet water-quality objectives. The citizens are joined in this effort by federal agencies and nonprofit conservation organizations (Klapproth and Johnson 2001c).

Sidebar 9.2 European Regulations Affecting River and Floodplain Management

Many regulatory instruments related to agriculture, water management, land use planning, forestry, and other policy fields create or hinder opportunities to either conserve existing floodplain forest or create new floodplain forest. Most legislation having an impact on how rivers and floodplains are managed is designed and implemented by individual countries, although there is an increasing amount of EU legislation that each member state has to comply with. Some of the most prominent are:

The EU Water Framework Directive (WFD), in force since December 2000, obliges all Member States to manage rivers on an integrated, catchment basis. A key innovation of the WFD is to set environmental objectives for achieving "good ecological status" for surface waters across whole river basin districts. The WFD makes little explicit reference to wetlands, and when it does, it suggests that only wetlands with proven links to groundwater and obvious surface water bodies are relevant to achieving "good ecological status." Nevertheless, the WFD creates important new opportunities for floodplain restoration in the context of improving the geomorphic and biotic status of rivers, as well as in providing buffer strips for nutrient retention (see Table SB9.1). Much will depend on how the WFD is interpreted and implemented by individual member states and river basin districts.

Table SB9.1 Opportunities and Obstacles to the Conservation and Restoration of Naturally Functioning Floodplains Provided by the EU Water Framework Directive (WFD) 2000

Opportunities	— Member states obliged to restore surface waters to environmentally "good status"
	— Floodplains will be important in achieving geomorphic and biotic objectives for "good status"
	— River basin management promotes integrated approaches to managing water and land use
	— Recreation and restoration of wetlands is explicitly encouraged
Obstacles	— Framework directives are not accompanied by financial instruments. Effective implementation will require the redirection of existing national and EU funding sources
	— No formal guidelines on how stakeholders can effectively participate in the process
	— Member states have a poor record of implementing existing community legislation on water
	— Timetable for implementation is long (15 to 27 years)
	— Floodplain restoration not helped by some waters being exempted from the need to achieve "good status"

Hughes 2002, adapted from Wise Use of Floodplains Project 2000.

Recent **EU Directives on Species and Habitats** require national action. The EU Birds Directive (1979 and subsequent amendments) and the EU Habitats Directive (1992 and subsequent amendments) are together producing the Natura 2000 network of protected nature conservation sites across Europe, some of which are on floodplains. The drawback is that the focus of biodiversity work is primarily on protected areas; this distracts from catchment-scale biodiversity management and can limit the scope for creating new ecosystems.

Some EU legislation and financial instruments have deleterious effects on floodplains. These are well reviewed in the Wise Use of Floodplains project document (WWF 2000) and include the Common Agricultural Policy (CAP) and EU Structural Funds. The Worldwide Fund for Nature (WWF) has identified other EU initiatives, which should have a positive effect on water management (WWF 1999), primarily the Amsterdam Treaty and the European Spatial Development Perspective (ESDP), both of which relate explicitly to water management.

Francine M. R. Hughes, Anglia Polytechnic University, Cambridge, UK and Timothy Moss, Institute for Regional Development and Structural Planning, Erkner, Germany

Riparian Conservation for the Long Term

Riparian conservation for the long term requires, on one hand, attention to ecological properties of the natural river systems, particularly reversible processes, connectivity, and thresholds (e.g., Amoros et al. 1987, Metzger and Décamps 1997, Nilsson et al. 1997, Naiman et al. 2002b,c). However, riparian conservation for the long term also requires attention to two other major questions: (1) How do we approach scientific uncertainty in ecological problem solving? and (2) How do we integrate ecological knowledge with the social and ethical aspects of conservation?

Approaching Scientific Uncertainty in Ecological Problem Solving

The capacity to precisely predict the long-term consequences of riparian conservation is limited. Riparia are basically "social-ecological systems" (Walker et al. 2002), the complexity of which makes any forecasting difficult because of the nonlinearity of change in key drivers, the reflexivity of human action in response to forecasts, and the rapidity of change. Improving the capacity to predict presupposes not only a comprehensive ecological knowledge of the functioning of riparian systems, but also (and above all) an understanding of how economic and social factors interweave with ecological and evolutionary science (Clark et al. 2001; see also Chapter 11).

Predicting how a riparian zone will react to conservation management requires consideration of certain aspects of the natural water regime—the regime that created and maintained the biodiversity and the functions. The hydrologic aspects include the frequency and the volume of floods as well as their duration, periodicity, and sequences (i.e., the way in which the high and low water events succeed each other temporally; Poff et al. 1997). Yet, in most cases, the construction of levees, channels, canals, and dams preceded the installation of flow measurement systems, which makes it difficult or even impossible to define the baseline "natural hydrological regime."

Predicting how a riparian zone will react to conservation management also requires considering the landscape context, the position along the river continuum, and local characteristics such as productivity, biodiversity, soil conditions, and nutrient levels. Past events, or legacies, also shape the trajectory followed by recently restored riparia, and species characteristics have their importance too. For example, certain species may prove to be more capable than others of naturally colonizing a riparian zone due to their capacity for dispersal or the longevity of their seeds. Similarly, genetic differences between populations of a given species can considerably affect the outcomes of reintroductions, particularly when a species is rare or endangered. Finally, if a riparian system is to be conserved primarily for its nutrient filter capacity, any enrichment by additional nutrients may lead to impoverished biodiversity by the promotion of one dominant species to the detriment of others (i.e., competitive exclusion).

In fact, highly complex interactions between plant cover, hydrologic processes, and land forms make it difficult to precisely identify the consequences of modifying the period, frequency, amplitude, or duration of floods, or the deterioration thresholds beyond which a particular riparian woodland cannot be conserved. For example, several related reasons can be responsible for a single species suddenly becoming the dominant riparian plant cover. Sometimes this can be traced to the date of the last flood, the topography of the site, or the drainage conditions with their concomitant effects on soil humidity and nutrient levels. The plant cover itself influences the very flows that were responsible for plant establishment. As an indirect consequence, hydrology is also modified, particularly by increasing the roughness of a site or by reducing the current speed and thus encouraging sediment deposition.

Integrating Ecological Knowledge with the Social and Ethical Aspects

Although ecological understanding of riparia is necessary for conservation management, it is not enough. A wider understanding and appreciation of the social and ethical aspects of conservation are essential (Ludwig et al. 2001). This is why designing a conservation strategy presupposes close collaboration with people inhabiting or using riparia in rural as well as in urban areas (Groffman et al. 2003).

Environmental and social conditions differ from one river system to the next and, therefore, it is crucial to organize opportunities for public reflection on the reasoning behind riparian conservation plans, the motives of those involved, and the perceptions of local people toward riparia. Reciprocal information plays a vital role here: many varied questions need to be examined, inevitably leading to further questions and to answers that differ from one situation to another. Flexibility is required in every case as illustrated by the management of accumulated dead wood in rivers (see Chapter 8). Removal of large woody debris may be necessary in one place to reduce the risk of flooding or erosion, whereas the maintenance of large woody debris may be necessary in another place to promote aquatic biodiversity, to preserve fishery resources, or to store water that would otherwise cause flooding further downstream. Artificial structures for large woody debris retention may be required upstream of sensitive sectors, particularly where conservation activities increase woodland regeneration along watercourses. Public understanding of these differences is needed to implement conservation practices, and this requires interdisciplinary knowledge in order to take into account human values, perceptions, behaviors, and institutional cultures (Naiman et al. 1998c, 2002c).

Conclusions

The seven principles guiding the conservation of wild, living resources given at the beginning of this chapter are applicable to riparia (Mangel et al. 1996). Likewise, this is true for mechanisms that implement the principles. Of particular importance for riparian conservation are (1) that the scientific, economic, and social aspects have been accounted for and (2) that riparia should not be modified beyond the natural boundaries of variation. An effective riparian conservation strategy is one where human and ecological systems are seen as one system with numerous feedbacks across broad scales of time and space (Folke 1996). This means paying attention to the underlying causes necessitating conservation; that is, the economic, institutional, social, and cultural factors that direct human activities (Ludwig et al. 2001). This means also paying attention to the interconnections between scientific research and public opinion, as

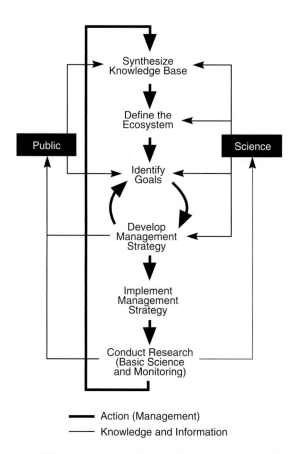

Synthesize
Knowledge Base

Define the
Ecosystem

Public

Science

Identify
Goals

Develop
Management
Strategy

Implement
Management
Strategy

Conduct Research
(Basic Science
and Monitoring)

—— Action (Management)
—— Knowledge and Information

Figure 9.5 Iterative protocol for involving scientific research, synthesis, and public opinion in adaptive ecosystem management (after Stanford and Poole 1996).

well as using iterative protocols for blending scientific research and synthesis and public opinion into an adaptive ecosystem management perspective.

There is a protocol to assist policy makers, scientists, managers, and the public to identify common objectives, to establish a body of knowledge, to design management practices, and to implement an effective strategy (Stanford and Poole 1996). The six steps of the protocol are as follows (see Figure 9.5):

1. Synthesize knowledge to root the process in facts and to help people to agree on a common pool of empirical information at the outset.
2. Define the spatial limits of the area to manage. This is not an easy task as riparia are interfaces between aquatic and terrestrial systems: their boundaries are particularly permeable to physical, chemical, and biological fluxes, and the entire catchment is the fundamental unit for management.
3. Identify goals to provide a framework within which managers and scientists can develop a management plan with public participation. The subsequent debates will point out incompatibilities and consequences of using different resource values.
4. Develop a simple, rational management strategy based on reviewed and synthesized information, taking into account uncertainties and risks. This will result in

289

a plan that is peer-reviewed, endorsed by the public, effectively conveyed to stake-holders living within the ecosystem, and eventually includes revisiting goals.
5. Implement a management strategy that meets the goals and includes associated acceptable risks and consequences.
6. Continuously reevaluate the evolving ecological and social conditions to improve management strategies and policies. This is accomplished through a redefinition of the ecosystem, a refinement of public goals, and an alteration of the conservation action, if warranted.

This protocol is iterative and adaptive over the long term to environmental change, new information, and changing societal goals. It involves citizens in a process that is founded upon empirical analyses of cause and effect. Clearly, the conservation of riparia is contributing to the emergence of a science for environmental decision making (Poff et al. 2003). This science, in addition to a solid conceptual basis, exhibits cooperative interactions among scientists, managers, and other stakeholders (Baron et al. 2002, Naiman et al. 2002c). However, riparian conservation, like other types of conservation, often suffers from the "too little, too late" syndrome (MacCall 1996): *too little* because riparia are socially invested systems that need transdisciplinary measures; *too late* because riparia have been transformed for a long time by human activities. Therefore, conservation may not be enough and, in many cases, it may be necessary to go further, adopting more proactive strategies to rehabilitate riparia, as seen in the following chapter.

Restoration

10

Overview

- Ecological restoration assists the recovery process of a system that has been degraded, damaged, or destroyed. Riparian restoration reestablishes processes affecting banks, habitat diversity, fisheries, biodiversity, and the filtration of diffuse pollution. Nevertheless, some restoration strategies may not be compatible with others, and some may not be relevant everywhere along the river corridor.

- A reasonably natural hydrologic regime is essential for restoring riparia, particularly the magnitude, frequency, and timing of peak flows to reconnect and periodically reconfigure channel and floodplain habitats. Several approaches have been developed that allow for the recovery of river flows.

- Every restoration initiative is unique in its ecological, social, and economic parameters, requiring reasonably precise plans that are crafted for local conditions. Plans are composed of several related and iterative steps, with possible adjustments during the process in light of new information.

- Setting geographic boundaries provides a spatial context for technical assessments, a sense of place to the community involved, and a framework to gather, organize, and depict information necessary for decision making. Boundaries reflect the nature of the human-induced disturbances, as well as social organizations and perceptions.

- Restoration plans include descriptions of degraded conditions and deviations from desired conditions. Desired conditions reflect a common vision consistent with the ecological goal of reestablishing riparia to a reasonable state of dynamic equilibrium. Other possible goals may be to satisfy socioeconomic desires, such as to provide specified goods and services or aesthetic amenities.

- Implementing restoration plans depends on responsible sharing of commitments by public agencies, private organizations, and individuals. Monitoring is required to assess progress and to document needed information and lessons learned. Continued evaluation of biophysical structures and processes determines whether the riparia has reestablished and will continue to maintain desired conditions.

- Assessing the ecological integrity of riparia involves formulating an integrated statement regarding current conditions and factors contributing to those conditions.

This requires information and knowledge of basic physical, chemical, and biological components and understanding how the components are formed, maintained, and interact in space and time.

- If financial resources are limited or uncertain, or if it is not acceptable to invest in highly engineered solutions requiring long-term maintenance, specific enhancements may be justified if compatible with other measures. Among these are implementing riparian silvicultural practices, reintroducing large woody debris, designing riparian buffer zones, excluding livestock, and optimizing for biodiversity.

- Successful riparian restoration establishes a planning framework that accounts for social, cultural, and political conditions—necessities for developing catchment-scale sustainability.

Purpose

Restoration of a degraded, damaged, or destroyed ecological system is the process of assisting its recovery to a former, but not necessarily pristine, state (SER and Policy Working Group 2002). Ideally the restored system will contain enough biotic and abiotic resources to continue its recovery without further assistance or subsidy, sustain itself structurally and functionally, and demonstrate resiliencies within normal ranges of environmental stress and disturbance. More specifically, a restored riparian system should have self-sustaining natural processes and linkages among terrestrial, riparian, and aquatic systems. It also should be able to provide specified natural goods and services for social benefit in a sustainable manner, particularly where populations depend on local resources for survival (Aronson et al. 1993).

Ecological restoration of riparia is not easy to achieve (Zedler 2000). It requires an adequate knowledge of original hydrologic regimes, geomorphic structures, and vegetation dynamics, as well as a deep understanding of the concomitant evolution between riparia and societal goals at the catchment scale (Postel and Richter 2003). A catchment perspective is requisite when planning riparian restoration. Only when an integrated management plan encompassing the river network and its drainage basin is agreed upon can the restoration needs of riparia be identified (see Chapter 8). An integrated management plan may lead to changes in land use practices that require an agreement on conditions for sharing future economic sustainability. In fact, successful restoration of riparian systems depends on several interrelated physical, biological, and social conditions in order to identify what should be restored—the physical habitat, a population, a community, or the entire ecological system. Factors for success include having long-term objectives, strong social commitment, good data and expertise, and initial damage that was not too catastrophic for restoration (Wissmar et al. 2003). It is also important to acknowledge that some portions of riparia will recover more rapidly than others, and to estimate the expected recovery times and the resiliencies of the subsystems to be restored.

It is widely acknowledged that the need for riparian restoration is pervasive. As discussed in Chapter 7, many riparian systems have been substantially altered by human activities. In the United States, riparia once occupied ~5 percent of the land mass (30 to 40 million ha) before extensive human settlement. Twenty years ago, riparia occupied only 10 to 14 million ha, a loss of >66 percent (Swift 1984); the total riparian area is certainly much less today. In addition, much of the remaining riparia suffer from point and nonpoint pollution, invasive exotic species devastating native populations, erosion and sedimentation from cultivation and grazing, and significantly altered hydrologic regimes (Naiman et al. 1995, Naiman and Turner 2000).

The situation in Europe is similar. It has been estimated that 88 percent of European alluvial forests have disappeared from their historic range (UNEP-World Conservation Monitoring Centre 2000), mainly through conversion to agriculture and from changes to river flows. There are also large differences among countries (Hughes and Rood 2003). Many western European countries have little or no remaining riparia, whereas eastern and central European countries have retained many impressive tracts, notably on the Danube River (Klimo and Hager 2001). However, many of the riparian forests have been converted to plantations of poplar hybrids (*Populus*) or black locust (*Robinia pseudoacacia*), and floodplains continue to be converted to agriculture. For example, in Austria, Danube floodplain forests decreased from 33,000 ha in 1930 to 8,000 ha in 2001 (Hager and Schume 2001). Similar losses are found throughout the world, calling for the development of approaches to effectively restore riparia (Tockner et al. 2005).

Most often conservation or protection, rather than restoration, is preferable. However, it is not always so simple. Theoretically, restoration addresses already degraded riparia, whereas conservation and protection address riparia perceived as intact or functional systems. In a practical sense, it may be difficult to separate conservation or protection from restoration, as is not always easy to delineate a boundary between less and more degraded systems. In reality, conservation/protection and restoration are two aspects of management strategies, requiring consensus among all resource users, including developers, agriculturists, conservationists, and landowners. Passive restoration (*laissez faire*) is sometimes advocated as it may be cost-effective to do nothing and let natural processes proceed—for example when considering the time and money needed to reestablish one species at one site. However, an active restoration process is necessary in most cases to initiate ecosystem development along a prefered trajectory, to monitor this development, and to evaluate whether autogenic processes can guide subsequent development with little or no human interference (SER and Policy Working Group 2002).

This chapter discusses how to restore riparia mainly through reestablishing "reasonably" natural hydrological regimes, developing a restoration plan, assessing sustainability of ecosystem functioning, and achieving specific enhancements. This is an emerging area of restoration science that is facing many challenges in both its theoretical and practical aspects (Wissmar and Bisson 2003). These challenges may be summarized as to preserve biodiversity, maintain or improve sustainable economic productivity, and protect or augment the supply of ecosystem goods and services—the natural capital (Aronson and van Andel 2005). All are motivated by the concern for long-term stewardship of our planet's limited resources, resilience, and health.

General Principles and Definitions

Ecological restoration is the process of assisting the recovery of an ecosystem that has been degraded, damaged, or destroyed (SER and Policy Working Group 2002). The process considers the trajectory of the ecosystem—that is, its developmental pathway, which may exhibit a range of potential expressions through time. The development pathway(s) often progresses toward a desired state of recovery that may be embodied by multiple reference sites that express various potential states within the historic range of heterogeneity. The goal is to repair the ecosystem with respect to its integrity and its health. *Integrity* is defined in terms of biodiversity—particularly species composition, community structure, and ecological functions. *Health* is defined in terms of vigor, organization, and resilience. Resilience refers to the degree to which the system expresses a capacity for learning and adaptation (Walker et al. 2002). Restoration ecologists also find three additional concepts of use:

1. *Natural capital*: Renewable and nonrenewable resources that occur independently of human action or fabrication (Costanza and Daly 1992).
2. *Emerging ecosystems*: Those that develop after social, economic, and cultural conditions change, thereby so changing the environment that new biotic assemblages colonize and persist for decades, with positive or negative social, economic, and environmental consequences (UNESCO 2003).
3. *Social-ecological systems*: Complex adaptive systems combining social, economic, and ecological factors with the managers as integral parts of the system (Walker et al. 2002).

Other terms may be used besides *restoration* to describe various activities of altering the biota and physical conditions (SER and Policy Working Group 2002). Among them, *ecological engineering* involves manipulations of natural materials, living organisms, and the physical-chemical environment to achieve specific human goals and solve technical problems. *Rehabilitation* emphasizes the reparation of ecosystem processes, productivity, and services, as opposed to preexisting biotic integrity. *Reclamation* aims at stabilization of the terrain, assurance and public safety, aesthetic improvement, and usually a return of the land to what—within the regional context—is considered to be a useful purpose. Finally, *mitigation* is an action that is intended to compensate environmental damage.

It is particularly relevant for riparia that, among other attributes, the restored ecosystem is sufficiently resilient to endure the normal periodic stress that serves to maintain the integrity of the ecosystem (SER and Policy Working Group 2002). Other useful attributes for restored riparia are that they serve as natural capital for the accrual of specified goods and services, provide habitat for rare or endangered species, maintain or enhance aesthetic amenities, and accommodate activities of social consequence. Often, riparian restoration is a part of larger development projects and programs at the catchment scale.

Several riparian functions may justify restoration: bank stability, habitat diversity, fish production, biodiversity, and buffering diffuse pollution. Riparian restoration may address several of these functions at the same time, although some types of restoration may not be compatible with others and may not be relevant everywhere along a river corridor. Moreover, when identifying a target state, it is important to recall that most riparia have been modified by human influences, leaving constraining legacies. As a consequence, restoring riparia to "pristine" states is often unrealistic, and may even be undesirable—for example, when pristine states are compared with what modern societies need in terms of goods and services, or appreciate in terms of aesthetic qualities.

Returning to More Natural Hydrologic Regimes

Reestablishing more natural hydrologic regimes, which includes the flow regime of the floodplain, is probably the most effective factor in successfully restoring riparia. Of particular importance are magnitude, frequency, duration, and timing of peak flows that reconnect and periodically reconfigure channel and floodplain habitats (Stanford et al. 1996). These ideas are nowadays recognized by scientists and practitioners, and considered as a paradigm (Poff et al. 1997). However, they may be difficult to put into practice because preregulation data are often not available and comparable pristine, unregulated rivers commonly no longer exist regionally.

Nevertheless, it is possible to estimate hydrologic attributes from historical streamflow records, cross-sections, aerial photographs, and local and scientific literature reviews. For example, 10 fundamental hydrologic characteristics of alluvial river integrity, identified as part of the restoration of the Trinity River, California (U.S. Fish and Wildlife Service and Hoopa Valley Tribe 1999), are general enough to apply to all alluvial river systems. They are:

1. The establishment of spatially complex channel morphology.
2. The predictability of variable flows and water quality.
3. The presence of a frequently mobilized channel bed surface.
4. The occurrence of periodic channel bed scour and fill.
5. The balancing of fine and coarse sediment budgets.
6. Periodic channel migration occurring by avulsion.
7. The existence of a functional floodplain.
8. The occurrence of infrequent channel-resetting floods.
9. The presence of self-sustaining diverse riparian plant communities.
10. The occurrence of a naturally fluctuating groundwater table.

Most of the characteristics are interrelated. The second characteristic is central to all biophysical processes since a varied flow regime is required to restore and maintain the overall health and productivity of a river. Together with characteristics 1, 5, and 10, characteristic 2 helps define a desired condition and quantify channel restoration goals. Characteristics 3, 4, 6, 7, 8, and 9 are process oriented and help shape adaptive management monitoring objectives related to the hydrologic regime.

The lack of hydrologic restoration at local and/or broader scales often limits the success of riparian recovery. For example, efforts initiated in the late 1980s to improve

Table 10.1 Factors Influencing Reforestation Success of Southern Forested Wetlands in the Lower Mississippi River Alluvial Valley

Phase	Factors Influencing Success	Citation Frequency
Planning	Site/species matching	9
Planting	Seed/seedling condition	11
	Seedling handling/storage	8
	Planting techniques	8
	Flooding	7
	Drought	4
	Timing of planting	3
Establishment	Drought	10
	Flooding	7
	Herbivory	4

Citation frequency includes those factors mentioned by at least three restorationists in a survey of 27 individuals.
After King and Keeland 1999.

wildlife habitat along the lower Mississippi River valley were impeded because restoration of natural hydrology on the river was generally impracticable due to drastic and widespread hydrologic alterations and socioeconomic constraints (King and Keeland 1999). Restoration of natural hydrology on lower-order tributaries may be practical, as more is known about small river restoration (Gore and Shields 1995). Similar conclusions apply to other large rivers (Gren et al. 1995, Van Dijk et al. 1995, Trémolières et al. 1998). In fact, hydrologic restoration is a complex process, the success of which depends on several factors (see Table 10.1). For example, artificial flooding may have negative consequences as in Germany where heavy metals (e.g., Cd, Zn, Co, and Cr) contaminate the sediments of the Weisse Elster and Mulde rivers (Neumeister et al. 1997). Artificial flooding is likely to increase the movement of the metals into groundwater, depending on pH, soil texture, precipitation, and climate conditions.

The need for hydrologic modifications is evident in the restoration of native species, especially when their habitats are invaded by exotics. Native species are typically more competitive than exotic ones under natural hydrologic conditions, and restoring such conditions may allow for their re-establishment. For example, along streams of the U.S. southwest, biological invasion by *Tamarix* (saltcedar) has completely transformed hydrologic regimes, elevating the banks and impeding rejuvenating floods to the floodplain. As a consequence, *Tamarix* is progressively replacing native *Populus* (cottonwood). It has been possible to increase the abundance of *Populus* at *Tamarix*-dominated sites through restoration of timed spring flood pulses, high groundwater levels and soil moisture, and exclusion of livestock grazing (Shafroth et al. 1998, Stromberg 1998). In fact, different flooding scenarios could induce a re-establishment of native cottonwood and willow (*Salix*) along *Tamarix*-dominated banks. Consider that (1) *Populus fremontii* and *Salix gooddingii* are favored when sites are moistened during spring and become dry during summer when seeds of *Tamarix* are dispersed and (2) *Tamarix ramosissima* seedling mortality is increased by post-germination summer floods of long duration or of large magnitude. However, there is a need to improve knowledge on flood flows at which seedlings of various ages and species are uprooted, on the effects of seedling height at the time of sedimentation and on growth

and survivorship rates of native and exotics under various soil moistures, temperatures, and salinities (Levine and Stromberg 2001).

A characteristic of riparia is that, in addition to seasonal flow patterns, they require
annual variations of flow patterns for the life cycle needs of many organisms. Indeed,
riparia rely on *regeneration flows* to provide the successful establishment of seedlings,
and on *maintenance flows* to allow the growth of trees once they are established.
Regeneration flows are overbank flood events that occur periodically and vary over
decadal time periods. They differ substantially from most cycles of annual flow experienced in managed rivers. Maintenance flows are variable river stage levels that are
not very different from the minimum flows established by management, and thus are
rather easy to provide throughout the year. Both types of flow are characterized by
temporal variability. Critical for successful plant regeneration is flood timing, flood
stage, and the shape of the hydrograph through the first growing season. Also critical for the growth of riparian plants surviving the initial year is the maintenance of
local water tables within the floodplain (Stromberg et al. 1996).

There are approaches for allocating river flows for riparian restoration, particularly
(1) the flushing flows and floodplain maintenance flows, (2) the floodplain recruitment
box model, (3) the multiple flow methodology, and (4) the bottom-up, top-down and
combined approaches (see Table 10.2; Hughes and Rood 2003). These approaches

**Table 10.2 Some Flow Allocation Approaches for the Rehabilitation of Riparia and Floodplain
Ecosystems**

Approach	Description
Flushing flows and floodplain maintenance flows (Kondolf 1998, Whiting 1998)	Designed to achieve instream or floodplain geomorphological objectives (flushing flows) and to maintain physical properties of a floodplain based on its dominant building processes (floodplain maintenance flows).
Recruitment box model (Mahoney and Rood 1998)	Designed specifically for initiating recruitment of floodplain tree species. Allows fine-tuning of planned releases downstream of dams to allow both well-timed flood peaks and well-tapered flood recension curves.
Multiple flow methodology (Hill et al. 1991)	Designed for measuring different flow regimes on the basis of four main flow types: instream/fisheries, channel maintenance, riparian ecosystems, and valley maintenance.
Bottom-up, top-down, and combined approaches	In bottom-up approaches—"Expert Panel Assessment Methodology" (Swales and Harris 1995), "Building Block Methodology" (King and Louw 1998), and "Downstream Response to Imposed Flow Transformation" (King et al. 2003)—a modified flow regime is constructed by adding flow components to a zero flow baseline.
	In the top-down approach—the "Benchmarking Methodology" (Arthington 1998)—the change of flow statistics before and after a regulation is linked to ecosystem impact.
	In combined bottom-up and top-down approaches—"Flow Restoration Methodology" (Arthington et al. 2000)—flow regimes are more precisely defined than either top-down or bottom-up approaches.
	These three approaches allow interannual flow variability and can be used to produce flow-related scenarios for river managers when occasional well-timed flood flows can be included in the scenarios.

After Hughes and Rood 2003.

have been developed in the semiarid southwestern United States, in Australia, and in South Africa—that is, in countries where water availability limits environmental processes and economic development.

Flushing flows maintain substrate properties of the channel, while *floodplain maintenance flows* allow for the movement and deposition of sediment making up the floodplain (Hughes and Rood 2003). Flows required to maintain a floodplain can be evaluated through conceptual models (Whiting 1998). Examples include determining the threshold for movement of sediment as bed load (floodplains built mainly from lateral or braided bar accretion) or determining minimum flows that promote movement of sediment in suspension (floodplains built through vertical accretion). Additionally, it is important to consider flow frequencies and durations because they determine the quantity of sediment deposited over time, hence floodplain maintenance. Flows maintaining a floodplain may reoccur every four or more years, such as in the Chagrin River, Ohio (Whiting 1998), or they may be more frequent in large river floodplains (e.g., Amazon). Riparia associated with large rivers need regular high flows to cause channel movement (Hughes 1997). Other positive consequences of restoring flood pulses are to replenish groundwater aquifers, to minimize fire disturbances, and to restore herbaceous biodiversity (Stromberg and Chew 2002). Appropriate flood disturbance is central for restoring biophysical interactions between main channels, backwaters, and floodplains (Gore and Shields 1995). Among many examples, Figure 10.1 illustrates a measure proposed to reconnect side channels to the main channel in the alluvial zone of the Danube River near Vienna, Austria

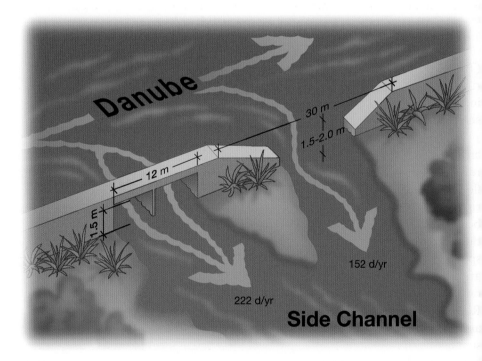

Figure 10.1 Lowering of riverside embankments and creation of openings to reconnect former side channels with the Danube River channel. Numbers indicate the duration of surface connectivity in days per year. From Tockner et al. 1998.

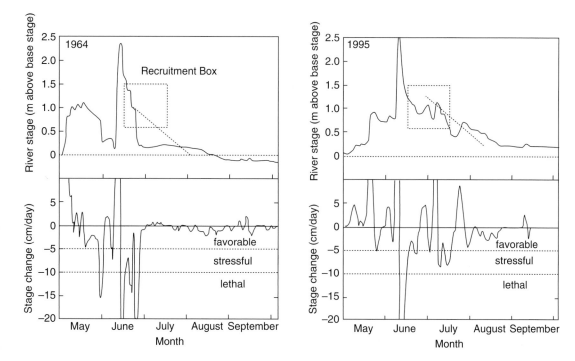

Figure 10.2 An application of the recruitment box model to floodplain forest regeneration on the lower St. Mary River in Alberta, Canada. Above: River stage hydrograph with superimposed recruitment box at the ideal time period and elevation and with ideal drawdown rates. Below: stage change rate graph (in centimeters per day). In 1964, regeneration of trees did not occur as a post-dam flood was managed for maximal cutback rather than naturalized recession; in 1995, regeneration was successfully promoted through a managed flow using the recruitment box as a guide. From Mahoney and Rood (1998).

(Tockner et al. 1998). In this example, lateral connectivity is restored as parts of the river embankment are lowered along 30-m lengths at several sites in the hope of increasing the frequency and duration of flow in side channels and the expansion of shallow waters in the floodplain.

The *floodplain recruitment box model* uses appropriately timed high flows for the creation of suitable recruitment sites in the floodplain and subsequent gradual flow recession for seedling survival (see Figure 10.2; Mahoney and Rood 1998). Indeed, flooding is necessary to create barren nursery sites for recruitment of many native riparian tree species (Rood et al. 2003). First, it is important that peak flows occur before seed dispersal to prevent scouring of established seedlings. Several studies show that dam operations may alter this timing and promote the encroachment of invasive species (Shafroth et al. 1995, Sher et al. 2000, Stevens et al. 2001). Second, seedlings are particularly vulnerable to drought stress if their root growth cannot match the rates of decline of stream stages and corresponding water table levels (Mahoney and Rood 1992, 1998; Segelquist et al. 1993). Moreover, adequate minimum water flow is also required after establishment, especially during the subsequent hot and dry summer period (Rood and Mahoney 1990). The recruitment box model may be used to prescribe flows when restoring degraded riparia. For example, it has been suggested that a decline rate in river stage of about 2.5 cm per day is a maximum after the peak flows have occurred. The actual decline rate depends on several factors, such as substrate texture, plant species, and weather conditions (Mahoney and Rood 1992, 1998; Hughes 1997; Barsoum and Hughes 1998).

The *multiple flow methodology* addresses flows needed for different geomorphic and biotic purposes. It distinguishes flows for fisheries, channel maintenance, riparian habitats, and valley maintenance. With this methodology, multiple flow recommendations have been realized for the Salmon River, Idaho, using a combination of flow analysis for fishery flows, evaluation of bankfull flows for channel maintenance flows, identification of out-of-channel riparian flows, and peak discharges (usually exceeding Q_{25}) for valley maintenance flows (Hill et al. 1991). This work has been influential in developing ideas on the provision of flows for different biophysical purposes in North America (Hughes and Rood 2003). A similar holistic conceptual approach has been used in Australia (Arthington et al. 1992).

The *bottom-up*, *top-down*, and *combined approaches* have been developed from a holistic viewpoint where all components of the riverine system are considered simultaneously. They first include an identification of water needs of in-stream and riparian systems and, second, negotiations with water resource planners concerning costs, design constraints, and competing water demands (Arthington et al. 1992). These approaches adapt themselves to the particularities of the river considered. The bottom-up approach considers flows required to maintain geomorphic processes and channel structures, and then to maintain habitats and life-cycle requirements of aquatic plants and invertebrates, fish, and riparian vegetation. Flows for wildlife and water quality are also considered, as well as nutrient and energy exchanges between the river and its floodplain, in order to identify ecologically significant river flows for riparia (Pettit et al. 2001). The top-down approach determines how a river would respond as the flow regime progressively departs from natural flow conditions, what could be the maximum acceptable departure from natural conditions, and what could be an environmentally acceptable flow regime (Brizga 1998). Bottom-up and top-down approaches are combined in the flow restoration methodology, which compares hydrologic characteristics of a river system when it is, and is not, regulated. This approach assesses the consequences of reinstating (or not) various flow characteristics on different species. Thus, the flow restoration methodology allows one to estimate which flow regimes would reinstate preregulation ecological properties in a regulated river. In Australia, such flow allocation methodologies are developed in iterative, practical processes where (1) the impacts of river regulation are assessed, (2) a preliminary qualitative and (after detailed studies) quantitative flow is recommended, (3) socioeconomic implications of selected scenarios are evaluated before (4) final recommendations and monitoring, as well as (5) feedback and (6) ongoing research (Arthington et al. 1998).

Restoration of riparia takes place over decadal time frames when varied interannual flows are provided (Hughes and Rood 2003). This may be feasible through cooperation between managers and scientists, for example, in relatively simple situations where flows are manipulated downstream of dams. However, there are more complicated situations where floodplains are densely inhabited and intensively developed (Cals et al. 1998, Pedroli et al. 2002). In these places—such as most European countries—it may be no longer possible to allow river channels to move. Moreover, significant amounts of micropollutants are present in soils along some rivers, such as the Rhine and the Meuse in the Netherlands, and reconstructing the floodplain may increase the risk of mobilizing these toxins. Here, cleaning heavily polluted sites is financially impossible and a risk assessment is a prerequisite of any restoration project (Platteeuw et al. 2001). To be successful over the longer term, such riparian restoration projects must effectively promote resilience and sustainability of biotic processes. From this perspective, using a "natural flow regime" is just a starting point. One must also develop integrated management models based on experimental flow regimes and quantified in-stream and overbank flow budgets (Stromberg and Chew 2002). Next, one needs to include flow prescriptions for natural resource values in the context of competing economic and social values (Schmidt et al. 1998, Stromberg 2001). It is increasingly apparent that carefully prepared restoration plans are essential for articulating these needs to the public and to decision makers.

Developing a Restoration Plan

A restoration plan documents processes, forms, and functions within riparia, identifies disturbances, and organizes actions—including performance measures and monitoring. It provides a framework to address critical issues, problems, and needs, as well as to organize communication with the public, interested parties and other restoration groups. It represents the common vision of the various partners. Each restoration initiative is unique in its ecological, social, and economic conditions, requiring place-based plans composed of several related steps (see Figure 10.3). These steps are iterative in an adaptive management process, with possible adjustments at any time in light of new information.

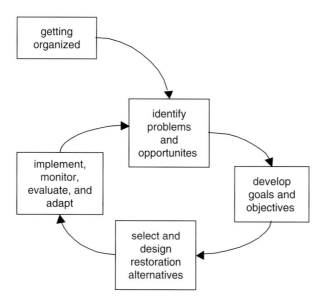

Figure 10.3 The riparian restoration plan development process. From the Federal Interagency Working Group 1998.

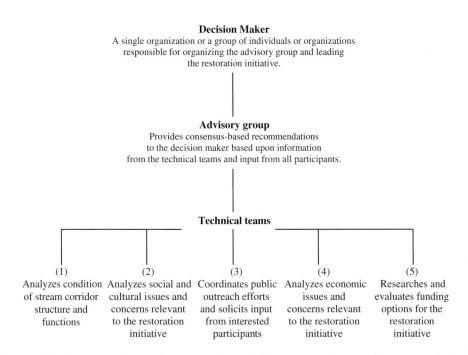

Figure 10.4 Restoration plan development requires a decision structure that streamlines communication between the decision maker, the advisory group, and the various technical teams. After the Federal Interagency Working Group 1998.

Getting Organized

Setting geographic boundaries provides a spatial context to the technical assessment, a sense of place to the community involved, and a framework in which to gather, organize, and depict information necessary for decision making. Boundaries reflect the various scales at which environmental processes influence riparia. They differ according to the nature of human-induced disturbances, particularly the spatial scales at which they operate. For example, at the reach scale, one might consider lack of shade, gaps in migratory routes, and erosion points, whereas at the catchment scale one might consider water quality and sediment loading, and at broader regional scales a reduction in biodiversity. It is necessary to account for interactions between the spatial scales and their linkages over time. Boundaries also differ in response to societal organization. This deserves consideration when forming advisory groups, technical teams, and the decision structure, as well as when identifying funding sources. It also deserves consideration when facilitating investment and information sharing among participants (see Figure 10.4).

Identifying Problems and Opportunities

Impaired stream corridor conditions regularly affect riparia (see Tables 10.3 and 10.4). Identifying problems and opportunities is the first step in formulating a rehabilitation plan. Success depends on (1) collecting and analyzing existing data, (2) defining existing riparian conditions, (3) comparing present conditions to desired or to reference conditions, (4) analyzing processes that lead to altered or impaired conditions,

Table 10.3 Common Impaired Stream Corridor Conditions

Stream aggradation or degradation
Stream-bank erosion
Increase of fine sediment in the corridor
Altered hydrology
Increased peak flow elevation
Increased bank failure
Lower water-table levels
Impaired water quality
Impaired aquatic, riparian, or terrestrial habitat
Decrease of species diversity
Loss of gene pool of native species

After the Federal Interagency Working Group 1998.

Table 10.4 Examples of Disturbances Due to Human Activities Occurring within the Stream Corridor

Displacement of aquatic and/or riparian species or habitats
Altered stream hydraulics and sediment transport capacity
Altered channel and riparian zone sedimentation dynamics
Altered surface water–groundwater exchanges
Chemical discharges and altered water quality

After the Federal Interagency Working Group 1998.

(5) determining how management practices might be affecting riparian conditions, and (6) stating problems and opportunities. Reference conditions should represent as closely as possible the desired outcome of the rehabilitation. Comparison with reaches or sites believed to be indicative of the natural potential of riparia may be especially useful. Clearly stating problems and opportunities is also critical for focusing the rehabilitation effort, developing specific objectives, and designing relevant monitoring approaches. This includes describing degraded conditions as well as deviations from desired conditions.

Defining Goals and Objectives

Defining goals and objectives is, in essence, identifying the desired future conditions. This is the common vision of all participants, consistent with the ecological goal of bringing riparia close to a state of dynamic equilibrium that is well integrated with important social, political, economic, and cultural values. It is of particular importance to include social, economic, or amenity values associated with changes from one set of riparian conditions to another. Also important is considering acceptable levels of change in the corridor. One must also consider the extent to which present conditions are susceptible to further deterioration without restoration, the extent to which restoration will achieve improved conditions, and the degree to which the restored riparia can be expected to maintain functional processes.

Goals and objectives are most often defined at the reach scale, a fundamental unit characterized by its position along the river course and its influence on stream corridor integrity. Whatever the scale chosen, constraints and issues limit the establishment of specific restoration goals and objectives. Some are technical, such as the availability of data and restoration technologies; others are nontechnical, such as financial, political, institutional, legal, social, and cultural. Sometimes causes of degradation cannot realistically be eliminated and force difficult decisions. Consider, for example, unnaturally accelerated sediment delivery from the catchment. Although this may produce lateral instability in the stream system, modifying land use may not be feasible at the catchment scale. In this case, engineering or soil-bioengineering control structures may be possible approaches to reduce channel sedimentation. Pre-disturbance conditions will not be a realistic target, and the presence of an altered substrate will persist, as well as a modified channel structure and impaired water quality. In addition, treating only one symptom of degradation may trigger other undesirable changes. For example, bank hardening in one location may interfere with critical sedimentation processes, thereby transferring instabilities to other locations.

Implementing, Monitoring, and Evaluating

Successful restoration requires the sharing of responsibilities and commitments among public agencies, private organizations, and individuals. Additionally, monitoring is fundamental to assess progress and document the needed information, the chronological and other aspects of system succession, and the lessons learned so that they may be applied in similar efforts. An example is the way the concept of "normative" condition was adapted for the restoration of the Yakima River, Washington (Snyder and Stanford 2001). The premise was that it would be possible to determine and implement nearly normal flows. That is, there would be sufficient water flowing through key river reaches to provide quality habitat for anadromous salmon (*Oncorhynchus*). The goal for this river is to achieve a shifting mosaic of ecologically functional salmon habitat along the whole corridor. An adaptive environmental assessment approach is used where the results of ongoing research modify management decisions. New knowledge generated from research is applied to specific actions, such as (1) synthesis of data relative to the Yakima River ecosystem; (2) demonstration of the range of variation in existing flow conditions; (3) construction of hydrologic and temperature models; (4) immediate implementation of a normative flow regime; (5) assessment of longitudinal and vertical hydrologic connectivity; (6) assessment of the riverine food web and salmon habitat conditions; and (7) construction of a salmon production model (SOAC 1999). As these actions have been implemented only recently, more time is needed for a realistic evaluation of the long-term results.

Evaluation is critical for protecting the restoration investment. Whether a stream corridor has reestablished and will continue to maintain desired conditions is determined through a continuous evaluation of biophysical data. A careful evaluation can ensure that future restoration experiences benefit from lessons learned in previous initiatives. An example is the floodplain reforestation undertaken by the Tennessee Valley Authority in conjunction with reservoir construction projects during the 1940s (Shear et al. 1996). Agricultural fields between reservoirs and embankments were planted with trees to reduce wave erosion during extreme high-water events. Planted species included bald cypress (*Taxodium distichum*), green ash (*Fraxinus pennsylvanica*), red maple (*Acer rubrum*), and other moisture-tolerant species. Other common bottomland forest species, such as oaks, were not included. Fifty years later, the plant communities in planted stands were compared to plant communities in similar sites on abandoned fields that had been naturally invaded and to plant communities in older stands that had never been converted into agriculture. Understory composition was similar in the three kinds of stands, whereas overstory composition of the planted stands differed from the others, as expected due to their planted origin. However, both the young stands (planted and naturally invaded) had few individuals of heavy-seeded species (i.e., oaks and hickories), contrary to the old stands. Thus, some bottomland forests can develop conditions resembling those of natural stands within 50 years, although when compared to older stands, they lack heavy-seeded species. This conclusion now supports the practice of favoring heavy-seeded species in bottomland forest restoration initiatives. It also emphasizes the critical importance of thoroughly considering plant succession for successful restoration of riparia.

Financial Incentives

Riparia are often located on private lands, frequently in steeper terrain and thus difficult to farm (see also Chapter 9). Therefore, the direct cost per unit area to farmers for riparian conservation or restoration may seem relatively low. Nevertheless, removing the land from agricultural production inevitably causes a reduction in net income, and compensation measures become an important element of successful riparian conservation or restoration. Fortunately, the cooperation of private landowners often can be secured using financial incentives (Keiter 1998). Such an approach may involve the acquisition of private lands and easements or the use of management fees paid to private landowners (Farrier 1995). In the United Kingdom, it has been suggested that the most critical factor influencing the impact on net income was the ratio of buffer zone area to catchment area: as the ratio increased, the cost of buffer zone per hectare of catchment area decreased. Estimated reduction in annual net income for four farms ranged between £323 and £680 per ha of land transformed to buffer zones, and between £6.40 and £21.10 per ha when spread over the catchment area served by each zone (Leeds-Harrison et al. 1996). Such studies may provide a basis to implement incentives for conservation and restoration. However, there is a basic lack of long-term data sets on patterns, trends, and variability of riparian response to disturbance (Simenstad and Thom 1996). Further work is required to determine the cost of riparian conservation or restoration according to sites, land use, and soil type—particularly to integrate ecological considerations into a legal regime of property ownership and to develop public–private partnerships.

In the United States, the Conservation Reserve Enhancement Program (CREP) offers financial incentives to landowners who remove agricultural and pasture land from production in riparian areas (USDA 2003). This joint program of the U.S. Department of Agriculture (USDA) and state governments intends to establish more than 1 million ha of riparian buffer by engaging 10- to 15-year contracts with landowners. To qualify for assistance, the landowner must have cropped the land at least two of the past five years, and the land must be capable of normal cropping. The amount paid to landowners varies greatly as a function of local land rental rates, but the average cost of the program nationwide is approximately $170/ha/yr. In determining the amount to be paid to each landowner, the following factors are considered: (1) base rental rate, (2) cost of installation of conservation practices, (3) annual maintenance costs, and (4) any special incentives. Base rental rate is based on the location of the land and the three predominant soil types. The program will pay up to 50 percent of the cost of installing conservation practices and approximately $20/ha in annual maintenance costs. Special incentives are sometimes paid to landowners willing to install high-priority practices like riparian buffers.

The first and most successful implementation of the CREP was in the catchment of Chesapeake Bay, in the northeastern United States. As part of a larger effort to reduce nutrient loading to the bay, ~3,240 km of riparian forest buffers will be set aside in the catchment by the year 2010 through voluntary incentive-based programs. In this effort, the USDA has partnered with the Chesapeake Bay Foundation, Ducks Unlimited, the U.S. Fish and Wildlife Service, and the Maryland Department of Agriculture to implement the program. In addition to the basic CREP, a number of other cost-sharing programs are available to landowners (Lynch 2002). These include the Maryland State Buffer Incentive Program (providing a one-time $75/ha grant for up to 200 ha for tree planting and maintenance along streams and shorelines), the Maryland Agricultural Cost Share Program (providing cost share up to 87.5 percent of the cost of a conservation practice with a maximum of $20,000), the Maryland State Woodland Incentive Program (providing up to 50 percent cost share with a maximum of $5,000 for one-year or $15,000 for three-year projects involving tree planting, stand improvement, and management to private nonindustrial forest landowners), and a number of other USDA cost-sharing programs (the Stewardship Incentive Program, Wetland Reserve Program, Environmental Quality Incentives Program, and Wildlife Habitat Incentives Program).

Setting Priorities

Setting priorities for restoration of riparia includes two stages (Harris and Olson 1997). The first stage defines and classifies stream reaches as suitable for protection, for conservation, or for restoration. It involves collecting and analyzing data from topographic maps and aerial photographs and establishing reach-level reference conditions on the basis of existing regional studies (see Figure 10.5). The second stage is field sampling of reaches identified for restoration to determine the riparian plant communities present, associations between communities and floodplain landforms, and reference community conditions (see Figure 10.5). It requires data collection on geomorphology and vegetation, as well as types of plant community assemblages, while comparisons to reference conditions help define restoration needs. Reference conditions are defined specifically for each riparian segment through indicators such as (1) high percent cover of native tree and shrub communities, (2) low percent cover of urban or other irreversible land uses, (3) high percentage of floodplain–upland boundaries in native vegetation (an indicator of good connectivity between the floodplain and uplands), and (4) patchiness (number and size of patches), which can only be interpreted based on a knowledge of the causes of fragmentation. Fragmentation reflects the pressure of irreversible land uses or natural hydrologic and geomorphologic processes.

The combined results of the two stages provide a strategy for protecting and restoring riparian resources. Further site-specific information then allows implementation. The approach was applied to the 80-km San Luis Rey River below the Lake Henshaw

First stage: reach classification:

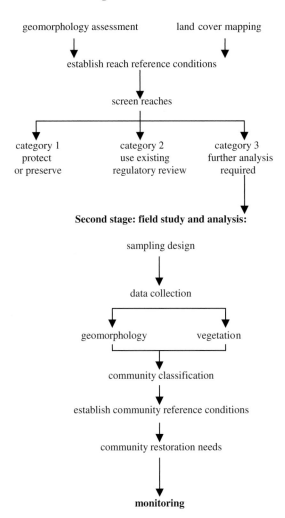

Figure 10.5 The two-stage system for prioritizing riparian restoration at the stream reach and community scales. After Harris and Olson 1997.

dam in semiarid southern California (Olson and Harris 1997). The two-stage process suggested that 32 km should be protected from further impacts as they met criteria for reference conditions at the stream reach scale (first category); 20 km were not suitable for restoration to reach-scale reference conditions, but there was a possibility for individual sites to be restored under existing regulatory review (second category); and 28 km were suitable for second stage evaluation of restoration needs (third category). Within the third category, two major departures from reference conditions were noted. First, tree-dominated communities were less extensive than they were historically. Second, numerous exotic plants had invaded some landforms, displacing natural communities. As a consequence, reestablishment of tree communities and removal of exotics were identified as goals for restoration of this category. A point to emphasize is that the project assessed "present" conditions and the restoration needs relative to reference conditions found at reach and community scales. It did not determine ecological functions to be restored or the desired landscape and community patterns, which remained the decision of people living in the catchment.

Assessing the Ecological Integrity of Riparia

The assessment of ecological integrity is a key component of any riparian restoration effort. It requires (1) classifying riparian assemblages, (2) identifying reference sites, and (3) assessing information on the history of resource development, hydrologic regime, soil and landforms, and vegetation (NRC 2002). Classification provides the framework that links a specific riparian area to an appropriate structural or functional grouping. The classifications described in Chapter 3 may be used, knowing that riparian conditions depend mainly on hydrologic and geomorphic forces (Brinson 1993) but may be significantly modified by biotic factors. Identifying reference sites is critical because they are (1) often desired goals for restoration, (2) common demonstration areas for scientists, managers, and the public, and (3) important for tracking natural variation relative to information derived from monitoring of the restoration site. The reference sites should encompass a range of conditions, from relatively unaltered sites subject to natural disturbances to sites significantly altered by human activities. Networks of reference sites capture natural variation among similar climates, stream orders, species assemblages, and disturbance regimes. Finally, assessing the history of resource development and the causes of degradation is required to understand present status and trends in riparia and to develop restoration strategies. For example, a condition for recovery of surface and subsurface water interactions may be to reestablish the natural stream-flow regimes (Rood and Mahoney 2000, Woessner 2000). Similarly, it is important to characterize floodplain soils and landforms to understand how they have changed under human activities and whether they can be conserved or restored.

Assessing the ecological integrity of riparia is, in essence, a formulation of an integrated statement regarding the current state and the factors that contribute to that state (Innis et al. 2000). It requires considering physical, chemical, and biological factors and understanding how these factors exist and interact in space and time, and why they exist as they do in their present state (see Sidebar 10.1). Many kinds of

---

Sidebar 10.1 Kissimmee River Restoration: Initial Response and Recovery

Restoration of the channelized Kissimmee River in central Florida is intended to reestablish ecological integrity of over 100 km² of a river/floodplain system. Evaluation of the success of this restoration project is being tracked by a suite of physical, chemical, biological, and functional indicators, with explicit expectations for both short- and long-term responses based on historical reference values for altered or degraded attributes (Toth et al. 2002).

The first phase of the restoration project (completed in 2001) reestablished flow through 12 km of river channel and re-inundated 4,400 ha of floodplain. Subsequent evaluation data indicate restoration of the Kissimmee River system is occurring through the predicted temporal sequence of response and recovery. Initial responses to re-established flow and floodplain inundation involved enhancement of key habitat characteristics such as dissolved oxygen regimes, river channel substrate and morphology, and littoral and floodplain vegetation communities. Reestablishment of flow through the remnant river channel (1) reduced cover of choking vegetation mats, including exotic species such as water hyacinth (*Eichhornia crassipes*) and water lettuce (*Pistia stratiotes*), by 50

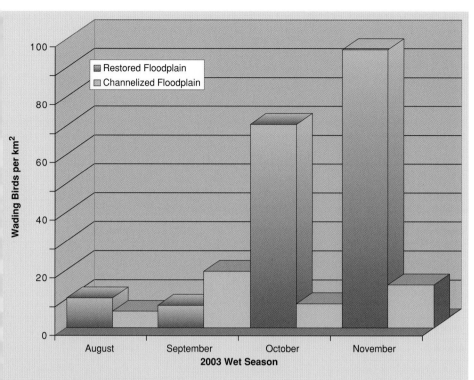

Figure SB10.1 Wading bird densities in channelized and restored sections of the Kissimmee River floodplain during the 2003 wet season. Data (G. Williams, unpublished data) show increased use of the floodplain three years after completion of the first phase of the restoration project.

percent; (2) flushed thick (up to 1 m) accumulations of organic matter and thereby reexposed the former sand substratum; (3) led to reformation of sand bars on channel meanders; and (4) increased chronically low dissolved oxygen levels (a major limiting factor for fish and their invertebrate food base) from <2 mg/L to >3 mg/L during wet season months (June to November). In the first year after the floodplain was re-inundated, cover of wetland plant species increased by 20 percent, while cover of wax myrtle *Myrica cerifera*, a native shrub species that had invaded the drained floodplain, declined from 35 to 8 percent.

Restoration of the river and floodplain is expected to lead to reestablishment of historical populations of over 300 fish and wildlife species. Initial signs of response and recovery of the food base have included re-colonization by bivalves (clams and mussels) and reestablishment of a filtering-collector guild of macroinvertebrates. During the first year of post-restoration avian monitoring, water-bird abundance more than doubled and included nine species of shorebirds that were not observed in baseline surveys of the channelized river (unpublished data). Even more dramatic increases were documented during the 2003 wet season, when wading bird densities reached 70 to 90 birds per km[2] on the restored floodplain (see Figure SB10.1). Details on the Kissimmee River restoration project can be found at www.sfwmd.gov/org/erd/krr/index.html.

Louis A Toth, Vegetation Management Division, South Florida Water Management District, West Palm Beach, Florida

wetland and riparian assessments have been proposed (see Table 10.5), but the usefulness of many of these techniques to resource managers is—sadly—questionable. Kusler and Niering (1998) found that managers are often dissatisfied with existing assessment approaches and seldom use formal rapid assessment techniques, that managers prefer qualitative assessments based on a "little common sense" over detailed models, and that they repeatedly qualify assessment methods as "unrealistic" and "out

Table 10.5 A List of Wetland and Riparian Assessment Methods

WETLANDS

Wetland Evaluation Technique (WET) (Adamus et al. 1991)

Parameters	Approach	Strengths	Limitations
Indicators: climate, landscape descriptors, hydrology, soils, vegetation, geomorphology, chemical environment, invertebrates.	Functional assessment. Uses Yes/No responses regarding ~80 qualitative indicators to categorize a group of 11 integrative functions and values as High, Medium, or Low.	• Simple approach. Integrates landscape conditions into evaluation. • Criteria are sensitive to different wetland classes.	• Allows focus to be limited to only those functions of social value. • Not aimed at assessing ecological integrity.

Hydrogeomorphic (HGM) Assessment (Brinson 1996, Hauer and Smith 1998, Whigham et al. 1999)

Parameters	Approach	Strengths	Limitations
Hydrologic and structural variables used in model. Specific choices dependent on function of interest.	Functional assessment. Models functions specific to an HGM subclass of wetland to determine whether functioning differently from reference wetland.	• Builds on HGM classification, which has been applied to many wetlands.	• Does not address ecological integrity. • Lack of quantification.

Wetlands Index of Biological Integrity (WIBI) (USEPA 1998)

Parameters	Approach	Strengths	Limitations
None indicated	IBI uses metrics that demonstrate an empirical, predictable response to human impacts to generate an overall score.	• Utilizes quantifiable metrics for statistical robustness. • Compares conditions to regional reference wetlands. • Focus is biological integrity.	• Not yet fully developed for wetland application. • Requires initial investment to develop and test metrics for an area.

RIPARIA

Integrated Riparian Evaluation Guide: Intermountain Region (USDA 1992)

Parameters	Approach	Strengths	Limitations
Aquatic habitat, soils, hydrology, geomorphology, vegetation, terrestrial habitat, riparian resource rating.	Semiquantitative, interdisciplinary approach to stratify, classify, and evaluate riparian areas. Uses three levels of evaluation: office procedures (level I), extensive fieldwork (II), and site-specific intensive fieldwork (III).	• Recognizes importance of watershed and landscape type characteristics. • Advises integration and synthesis of data into a written interpretation.	• Presentation is complicated. • Approach allows for inconsistency.

Riparian Evaluation and Site Assessment (RESA) (Fry et al. 1994)

Parameters	Approach	Strengths	Limitations
RE-flow regime, SA-vegetation, channel morphology, erosion, wildlife, local land use, groundwater recharge, surface water quality, recreation, upland conditions.	Semiquantitative. Based on land evaluation and site assessment. Adapted to rank riparian areas based on functions, values, and benefits they provide and to prescribe buffer widths for individual sites.	• Weighted scoring stresses most important features. • Approach is rapid and index generates an overall score.	• Does not prescribe use of reference conditions. • Focus is on maintaining valuable functions rather than ecological integrity. • Subjective scoring may lead to inconsistency.

Table 10.5 *Continued*

Assessment of Habitat of Wildlife Communities (Schroeder and Allen 1992)

Parameters	Approach	Strengths	Limitations
Percent vegetative cover type compared to 1956, channel and floodplain complexity, percent reach subject to human disturbance.	Generates a quantitative model whose output is a Habitat Suitability Index (HSI) for riparian vertebrates. Output predicts proportion 1956 wildlife species richness. Vegetation weighted over other attributes in the model.	• Provides temporal and spatial flexibility. • Concisely addresses three components of habitat integrity: hydrology, morphology, and vegetation.	• Criteria site-specific, so wide applicability uncertain. • Assumes 1956 vegetative communities undisturbed. • Does not look at historical state of other attributes (channel and floodplain complexity and human disturbance).

An Assessment of Riparian Environmental Quality by Using Butterflies and Disturbance Susceptibility Scores (Nelson and Andersen 1994)

Parameters	Approach	Strengths	Limitations
Disturbance Susceptibility Scores (DSS) for individual taxa ranked using species mobility, larval host-plant form and specificity, and riparian dependency.	Developed an index of butterfly riparian quality for vegetative habitat conditions. Index based on species richness and disturbance susceptibility scores for individual taxa. DSS based on semiquantitative scoring of species attributes related to riparian quality.	• Field components quick and inexpensive. • Relatively few, easily identifiable species.	• Not as effective in regions with less distinction between riparian and surrounding landscape. • Only one pair of sites was compared (disturbed and undisturbed).

Avian and Mammalian Guilds as Indicators of Cumulative Impacts in Riparian-Wetland Areas (Croonquist and Brooks 1991)

Parameters	Approach	Strengths	Limitations
Guilds based and scored for wetland dependency, trophic level, species status, habitat specificity, and seasonality (birds only).	Develops avian and mammalian response guilds based on life history and status for all Pennsylvania mammals and birds. Correlates responses to land use patterns at landscape level and investigates hierarchical patterns at the catchment level.	• Addresses landscape changes. Wide spatial coverage (catchment as well as riparian segments).	• Does not address alteration leading to the whole impact response. • Assessment endpoint is restoration for specific wildlife needs rather than ecological integrity.

Dragonflies (Insecta, Odonata) and the Ecological Status of Newly Created Wetlands (Chovanec and Raab 1997)

Parameters	Approach	Strengths	Limitations
Dragonflies. Also discussed, much more generally, other common indicator species: frogs, fish, and birds.	Assessed created wetlands using single indicator group. Sampled two created wetlands on the river Danube: one along man-made island and another bordering a new channel.	• Developed using frequent sampling over several years (could yield a fine degree of temporal resolution).	• Instead of collecting data at a reference site, used published data. • Clearly artificial structure could appear to have high integrity if they provide habitat.

A Qualitative Procedure for the Assessment of the Habitat Integrity (Kleynhans 1996)

Parameters	Approach	Strengths	Limitations
Water abstraction, flow modification, water quality, inundation, indigenous vegetation removal, exotic vegetation, encroachment, bank erosion.	Assesses relative significance of factors degrading river habitat quality. Scored in-stream and riparian separately, then combined for overall evaluation. Used aerial survey and ground-level observations.	• Provides a wide general survey of modifications that affect in-stream habitats. • Allows separate or combined evaluation of in-stream and riparian information.	• Use of largely remotely sensed data and descriptive classes limits sensitivity. • Many riparian attributes scored were based on function with respect to in-stream environment rather than stand-alone riparian integrity.

Table 10.5 *Continued*

The RCE: A Riparian, Channel, and Environmental Inventory for Small Streams in the Agricultural Landscape (Petersen 1992)

Parameters	Approach	Strengths	Limitations
Upland use, riparian zone width and completeness, vegetation, retention devices, channel structure and sediments, bank undercut.	Describes habitat quality of stream sites for comparison with other sites in same region. Rapid, semiquantitative scoring of 16 weighted metrics covering spatial scales from landscape to macrobenthos. Examines both in-stream and adjacent riparian zone, with emphasis on riparian metrics.	• Output easily understandable to nonscientists (color-coded maps reflect condition and management recommendations). • Quite rapid.	• Somewhat subjective. • High degree of correlation between many metrics. • Does not recommend the use of reference conditions.

SERCON: System for Evaluating Rivers for Conservation (Boon et al.1997)

Parameters	Approach	Strengths	Limitations
Scored: physical diversity, naturalness, representativeness, rarity, species richness, special features, and human impacts.	Hierarchical approach to assess conservation status of U.K. rivers. Uses site, evaluated catchment sections (ECS), or sub-catchment levels of spatial scale. Integrates riparian characteristics into scoring of ecological attributes of river as a whole.	• Results presented in a form easily understood by non-scientists. • Easy interpretation at different spatial scales. • Standardizes evaluations across U.K.	• Does not recommend the use of reference condition. • Criteria chosen for comprehensive rather than careful integration. • Riparian-related attributes would poorly represent the riparian integrity if stood alone.

From Innis et al. 2000.

of touch." Consequently, there remains a need to articulate the fundamental components of successful ecological assessments. Some attributes of riparian assessments that are absent or insufficiently developed in the methods reviewed by Innis et al. (2000) are ecosystem performance, reference condition, scale, and integration.

Ecosystem performance is appropriately measured by ecological integrity—that is, the capacity to support and maintain a balanced, integrated, and adaptive biological system having the full range of elements and processes expected in a region's natural habitat (Karr and Dudley 1981). Ecological integrity refers to reference conditions and naturalness, as well as to the dynamic nature of rivers and the presence of several riparian successional stages (i.e., a chronosequence). In addition, ecological integrity is clearly maintained by an appropriate disturbance regime. Of the assessment methods listed in Table 10.5, one addresses ecological integrity for wetlands and five for riparia.

Reference conditions describe a system's natural potential in its regional setting. They are based on "natural" areas when available or, most often, on historical information as a "best guess" about the ecological status of a "natural" site within the appropriate climate, geological, and landscape contexts. A range of reference conditions may be useful for comparing and testing indicators. Of the assessment methods listed in Table 10.5, two prescribe reference conditions for wetlands and two for riparian assessments.

Appropriate spatial and temporal *scales* should be addressed in connection with the question or issue being addressed. At the same time, it is important to consider natural spatio-temporal variation and information over multiple scales—for example, through a hierarchical approach (e.g., Frissell et al. 1986). Of the assessment methods listed in Table 10.5, only the Wetland Evaluation Technique considers seasonal or long-term dynamics. However, most methods can be adapted and applied over the long term for monitoring. Recall that temporal scales vary according to the approaches and that there is a range of time scales for restoration. For example, riparian restoration from pasture to native vegetation takes about 30 years, whereas modifying the path of groundwater flow may decrease inputs of nitrate to surface waters within some months (Downes et al. 1997).

Integration of information aids in determining underlying causes for any departure from desired conditions and for testing appropriate management actions. It is essential for articulating the state of an ecological system and implies an understanding of how chosen parameters and scales are related to each other as well as to the question or issue being addressed (Conquest and Ralph 1998). Indicators are useful integrators and, among them, living organisms are the optimal indicators of ecosystem integrity (Karr 1999). The issue of integration is generally addressed by the assessment methods in Table 10.5 but still requires development.

The dynamic nature of riparia, with continuous change and disruptions, complicates assessments related to the reestablishment of ecological integrity. Parameters are needed that indicate integrated biological responses to the physical environment. Five such indicators may be used to assess riparian integrity (Innis et al. 2000):

1. *Microclimate* and microclimatic gradients, indicators of conditions in adjacent terrestrial and aquatic systems and strongly interacting with biological communities.
2. *Patch heterogeneity*, an indicator of the integrity of the discharge regime, resulting in a mosaic of patches favoring the success of species with contrasting life history strategies or at different successional stages.
3. *Biodiversity*, of which one metric—the plant species richness—is positively correlated with flood frequency, productivity, and spatial heterogeneity (Pollock et al. 1998).
4. *Terrestrialization*, or the increasing relative abundance of terrestrial species in the riparian corridor, that may signal the degree to which a riparian forest has been impacted by water regulation.
5. *Seston*, or suspended particulate matter in streams, whose dynamics are significantly influenced by woody debris, canopy cover, and discharge regime and may integrate conditions in the surrounding riparian habitat (Elliott et al. 2004).

There are several key issues limiting the utility of riparian assessments, but there are also actions that can be taken to make assessment methods better available to practicing professionals (see Table 10.6). Among these, two are especially important and are emphasized many times in this book. First is a landscape perspective or "principle of place" (Dale et al. 2000), meaning that because every habitat has relationships with the surrounding landscape, specific data need to be interpreted within the larger spatial and temporal perspective. Second is an adequate consensus between scientists, managers, decision makers, and the public through open discussions about the inherent implications of restoration projects.

Table 10.6 Key Issues Limiting the Utility of Wetland and Riparian Assessments. Observed limitations lead to recommendations that present challenges to scientists and managers. These recommendations are of particular importance for a better availability of methods to practicing professionals, to begin with the retirement of out-of-date methods while strengthening and cross-calibrating useful ones

Limitation	Recommendation
• Scientific information may be improved	• Better attention to fundamental components
	• Recognize the "principle of place"
• Recognize gaps and limits in scientific knowledge, including uncertainty	• Formally field test techniques and indicators
• Deliver too little information for time and cost involved	• Develop rapid techniques
	• Test techniques for cost and practicality
• Not accurate enough for broad application	• Develop hierarchical techniques applicable at multiple levels of detail and expertise
• Inadequate consensus building for multi-agency and multicultural settings	• Openly discuss policy aspects of specific assessment methods
• Too many methods	• Retire out-of-date methods
	• Strengthen and cross-calibrate useful ones
• Inadequate training and support for users	• Consider a wide variety of users and situations
	• Provide regular training opportunities by developers

Innis et al. 2000, adapted from Karr 1999 and Kusler and Neiring 1998.

Specific Enhancements

Restoration of riparia needs to be strategic and systematic. As a consequence, specific goals should be included within a larger framework, preferably at the catchment level. However, resources are often too limited or uncertain, or it is not acceptable to invest in highly engineered restoration that requires long-term maintenance and that would be ultimately unsustainable. Thus, limited restoration measures or specific enhancements may be justified provided they are compatible with other measures. Limited measures might include implementing riparian silvicultural practices, reintroducing large woody debris, designing riparian buffer zones, excluding livestock, and optimizing riparia for biodiversity.

Implementing Riparian Silvicultural Practices

Specific ecological functions of riparia can be incorporated, to some degree, into silvicultural practices. Examples include bank stability, shading, and inputs of litter and woody debris. Effective silviculture in forested riparia prevents problems rather than only responding to problems (Palik et al. 2000). Consider the effects of riparian forests on stream bank integrity and channel stability (see Figure 10.6). It is most effective at the channel edge while, additionally, the provision of large woody debris and shade occurs within approximately half a tree height of the channel. Silvicultural harvest and planting options can thereby vary with distance from a stream and be responsive to ecological and landowner goals (see Figure 10.7).

Reintroducing Large Woody Debris

The reintroduction of large woody debris to rivers facilitates aquatic habitat restoration. This is particularly so where riparia have been exploited to the degree that they

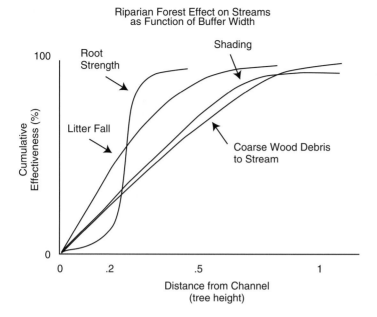

Figure 10.6 Generalized relationship indicating percentage of riparian ecological processes and functions occurring within indicated distances from the channel. After FEMAT 1993.

no longer deliver large wood, or the proper species, to neighboring streams. Considerable large woody debris has been introduced directly to many streams in order to restore habitat heterogeneity, biodiversity, and fish productivity, particularly in northwestern America, but with uncertain effects when focused on small spatial scales (Frissell and Ralph 1998). Reintroducing large wood to streams requires attention to processes rather than just structure, and a determination of the appropriate location, size, species, and amount of wood. It is also necessary to tailor the prescription toward the stream type and from one section of a stream to another depending on the physical environment.

The many functions of large woody debris are known to vary with the width of the stream (Beechie and Sibley 1997). For example, stream shade and effective large woody debris size are functions of the size of the trees relative to the size of the stream. Accordingly, restoration efforts are likely to be more beneficial in some places than in others. This is important to know when assessing whether trees are large enough to contribute logs that would form pools in the adjacent channels. The pool-forming size of wood is also a function of channel width, and riparian stands have recently been classified according to whether they had the capacity of forming salmon-holding pools in northwestern Washington State (Hyatt et al. 2004). Combining the classified stands with property ownership and threatened fish distribution led to a prioritization of riparian restoration location and strategies.

Clearly, it is important to assure a continued supply of large woody debris of appropriate size, volume, and species composition for the long-term integrity of streams (Naiman et al. 2002a). However it is also important to address the social and environmental trade-offs and constraints on introducing large woody debris to streams (see Sidebar 8.2 in Chapter 8). Wood accumulated and transported by floods can

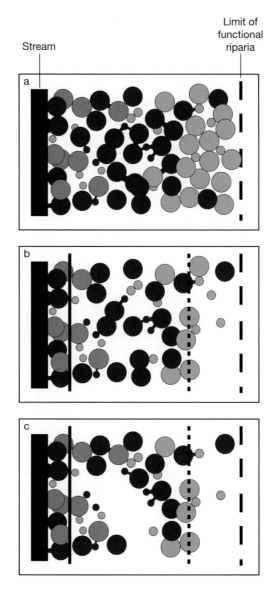

Figure 10.7 Harvest patterns in a riparian forest. (a) Uncut forest with riparian management area within the functional boundary. (b) Most of the trees are removed on the right and an uncut forest remains nearest the stream on the left; in the middle, few trees are removed such that the residual basal area is dispersed by cutting many small gaps. (c) As in (b) but the residual basal area in the middle part of the riparian area is clumped to open a single large gap. In (b) and (c), mature stand structure and riparian functions are less impacted by harvest nearest the stream. After Palik et al. 2000.

damage bridges and streets and is therefore frequently removed from rivers rather than being widely introduced. Millions of drifting trees and other driftwood have been removed from streams and rivers throughout North America and Europe to facilitate navigation and to reduce flooding over the last century (Maser and Sedell 1994). Compromises are necessary, based on a good knowledge of the pros and cons of wood introduction, as well as on public education (AFFA 2001).

Designing Riparian Buffer Zones

Various types of buffer zones have been proposed to protect the quality of running waters. The simplest ones are strips of herbaceous vegetation composed of one or several species along streams. Grasses, legumes, and/or forbs are used according to site conditions at widths varying from 15 to 50 m. Such buffers reduce sediment, nutrients, and contaminants from hillslope runoff; they also offer herbaceous habitat for wildlife. More complex, multiple species riparian buffer systems may be advisable in agricultural lands (Schultz et al. 1995, 2000; Iowa State University 1997). A widely used design consists of three distinct zones from upland to the stream (see Figure 10.8). The first zone is comprised of native forbs and grasses with a minimum width of 6 to 7 m and is used to slow overland flow of water, thus enhancing water infiltration and sediment deposition. The second zone is one or two rows of native shrubs having a minimum width of 3 to 4 m; it adds diversity and wildlife habitat and slows floodwaters when the banks are inundated. The third zone, next to the stream, includes four or five rows of fast-growing trees with some slower-growing ones interspersed between them. Its recommended minimum width is 10 m and it may provide bank stabilization, wildlife habitat, stream shading, nutrient removal, and selective timber harvest. Two zones may be constructed instead of three along the upper reaches of low-order streams (Shultz et al. 2000). In this case, shrubs replace the first two or three rows of trees next to the channel, or shrubs replace all the trees in the third zone.

Three-zone forest buffers are used in many agricultural regions of the humid eastern United States (Lowrance et al. 1985, 1995, 2000; Inamdar et al. 1999; Welsh 1991).

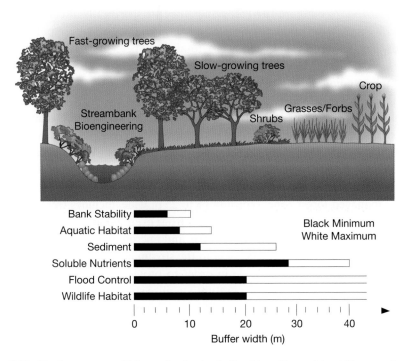

Figure 10.8 The three-zone, multiple species riparian buffer strip, with possible widths according to the function expected in management. After Schulz et al. 2000.

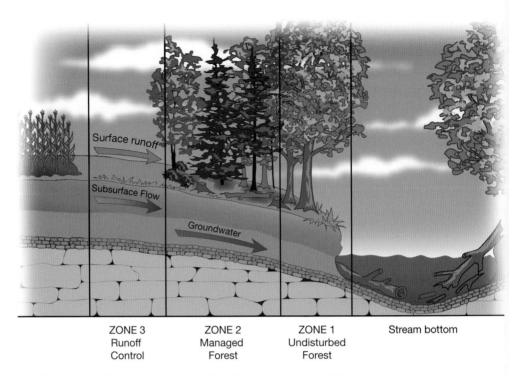

ZONE 3   ZONE 2   ZONE 1   Stream bottom
Runoff   Managed   Undisturbed
Control   Forest   Forest

Figure 10.9 The three-zone forest buffer. After Lowrance et al. 1995.

From upland to the stream, the zones are (1) a runoff control zone of grass, (2) a managed forest zone periodically harvested for timber, and (3) a permanent and undisturbed forest zone (see Figure 10.9). Together, they are specifically intended to provide water quality protection in the presence of adjacent sources of diffuse pollution from croplands, pastures, and urban settings. A total of 20 percent of the area contributing diffuse pollution is recommended for the three-zone forest buffer, and a width of 30 to 45 m depending on soil type and land use. The runoff control zone, at the upper edge of the riparia, is typically a minimum 6 m wide. It is usually covered with perennial cool season grasses, which are more effective when they form a dense cover of upright stems at ground level and have deep roots. This zone controls runoff in that it resists shallow overland flow, transforms concentrated flows into sheet flow, and reduces runoff velocity and sediment transport capacity. It promotes infiltration, thus improving the effectiveness of the adjacent managed forest zone in trapping pollutants. The managed forest zone, typically 15 to 25 m wide, includes trees and shrubs. It mainly removes and/or sequesters nutrients from overland and shallow subsurface flow, acting through infiltration, plant uptake, and microbial denitrification. The permanent and undisturbed forest zone next to the stream also includes native trees and shrubs. It is typically 5 to 10 m wide in order to provide habitat for terrestrial and aquatic wildlife, regulate stream temperature, stabilize stream banks, furnish leaf litter and large wood, and remove remaining diffuse pollutants.

Studies addressing the long-term performance of buffers with respect to water quality report that, in riparian zones established for decades, 84 to 90 percent of the sediment from uplands is trapped in the riparian buffer (Cooper et al. 1987). Further fluxes of nitrogen and phosphorus are reduced by 67 and 25 percent, respectively, in

Table 10.7 Some Reported Effectiveness of Buffer Zones for Sediment Reduction in the United States

GRASS

Reduction	Width	Reported By
More than 80 percent	4 to 9 m	Dillaha et al. (1989)
	9 m	Ghaffarzadeh et al. (1992)
	24 m	Chaubey et al. (1994)
	25 m	Young et al. (1980)
	61 m	Horner and Mar (1982)
Between 77 and 66 percent	3 to 6 m	Lee et al. (1999)
Between 60 and 30 percent	6 to 18 m	Daniels and Gilliam (1996)
Between 100 and 40 percent	20 m	Arora et al. (1996)
Below 50 percent	26 m	Schwer and Clausen (1989)

MIXED GRASS AND TREES

Reduction	Width	Reported By
Between 70 and 90 percent	7 to 16 m	Lee et al. (2000)

TREES

Reduction	Width	Reported By
Between 75 and 80 percent	30 m	Lynch et al. (1985)

After NRC 2002.

subsurface flow from croplands in Georgia (United States) when riparian buffers are in place (Lowrance et al. 1983). Nevertheless, the effectiveness of buffer zones for pollutant removal appears highly variable (see Table 10.7), suggesting a need to improve our knowledge on the utilization of riparian buffers as a landscape feature for water-quality protection, as well as for habitat enhancement. The National Conservation Buffer Workshop (Soil and Water Conservation Society 2001) identified several basic research needs related to riparian buffers. Among these are critical questions about the width required to meet site-specific pollutant reduction goals, the width required for each of the riparian subzones, the type of vegetation to be used and its impact on nutrient and pollutant sequestration as well as on habitat values, and the optimization of long-term nutrient and pollutant removal. Process-based riparian zone models such as the Riparian Ecosystem Management Model (REMM) may be useful here (Inamdar et al. 1999, Lowrance et al. 2000). This model simulates long-term sediment transport, plant uptake, nutrient transport, and denitrification in riparian systems. It helps in selecting plant species and widths required to meet site-specific goals for water quality (see Sidebar 10.2).

Prescribed Grazing

One goal of restoration may be to exclude livestock from riparian zones until biophysical functions are self-maintaining over the long term (see Table 10.8). Riparian vegetation, as well as bank and soil stability, differs between ungrazed and grazed areas—ungrazed areas generally exhibit uniformly better integrity than grazed areas (Belsky et al. 1999, McInnis and McIver 2001). Livestock grazing affects riparian vegetation by reducing reproductive output of mature plants and by increasing

Sidebar 10.2 Riparian Zones as Buffers

When terrestrial systems are disturbed, typically by human activities, riparian zones become important buffers, limiting changes in aquatic systems. Logically, the recognition of riparian zones as buffers goes back thousands of years, as people observed the movement of water in ancient landscapes. Scientifically, the recognition of the buffer role played by riparian ecosystems is quite recent, dating to a paper in *Science* by James Karr and Isaac Schlosser in 1978. Since then, numerous investigations in agricultural landscapes, managed forests, and urban environments have helped explain processes that provide buffer functions and under what conditions they exist.

By understanding processes active in riparian ecosystems, a few general principles have been identified that describe how riparian zones act as buffers. The general principles primarily describe the effects of water-transport mechanisms and biotic productivity on buffer function. In addition, understanding processes allows some generalizations about the roles of biotic and physical conditions in riparian buffer functions.

The first principle is that as the residence time of water in the riparian zone increases, the amount of pollutant removed increases. If water moves through the riparian zone in either channelized surface flow or enhanced groundwater flow, the less it acts as a buffer. Many of the processes that allow riparian zones to serve as buffers for water moving down slope depend on water residence time. In unmanaged landscapes, the amount of water bypassing riparian zones is probably negligible and only changes over very long time periods. As channels evolve over geologic time, the riparian zones evolve with them. In contrast, human-induced changes in the amount of bypass flow are rapid and allow little or no natural adaptation by the riparian zone. Runoff from impermeable surfaces, deliberate channeling of flow to remove water from upslope areas, and subterranean drainage are three examples of water movement that will largely negate the riparian buffer function. Even in areas where groundwater flow is naturally rapid, the nitrate removal function of riparian buffers is less effective than in areas of slower groundwater movement.

The second principle is that higher levels of biotic productivity will allow more filtering of pollutants than lower levels of productivity. Many buffering functions of riparian systems depend on nutrient uptake and the accumulation of soil organic matter. Nutrient uptake by vegetation leads to sequestering of nutrients both in plant biomass and in soil carbon stored on the surface or at depth due to the root biomass. Higher soil organic matter can lead to enhanced water retention. If water is retained effectively in the riparian buffer, it is more available for plant uptake and can be a source of nutrients for plants. Under a given climatic regime, higher levels of productivity will lead to higher levels of soil organic carbon. Many buffering processes known for riparian systems depend on soil organic matter. Enhanced levels of organic carbon can lead to enhanced denitrification and enhanced degradation of pesticides.

Until now, at least 90 percent of the research on riparian zones as buffers has been in "naturally occurring" riparian buffers in managed landscapes. These are distinguished from natural riparian systems because they are remnants of the natural systems that would have existed in less managed landscapes. For a variety of reasons—especially erosion and sedimentation from upslope areas, selective harvest and planting of vegetation, introduction of invasive exotic vegetation, and fire management—these "naturally occurring" remnants are often much different than the original natural riparian system. Because research suggested that the "naturally occurring" riparian buffers were effective in controlling nonpoint source pollution, there has been a strong push among conservationists and governmental natural resource agencies to reestablish riparian buffers, especially in agricultural and managed forest landscapes. In managed forest landscapes, riparian buffers are generally referred to as *streamside management zones* and often involve limited or no harvest of trees within a certain distance of the stream, depending on the slope of the surrounding catchment. In agricultural landscapes, riparian buffers generally replace either cultivated lands or permanent grazing lands adjacent to streams or other water bodies. In these cases, there are often drainage changes—ditches, subterranean drains, or both—that provide better drainage for agricultural lands. These areas present particular challenges for establishing riparian buffer functions. In many of the world's most productive agricultural areas, natural streams and wetland drainage-ways have been removed and replaced by ditches.

Unfortunately, there are not catchment-scale data to show that reestablishing riparian buffers can improve water quality. As riparian buffers are reestablished over large portions of a catchment, it is anticipated that there will be similar benefits to those measured at the smaller-scale studies of "naturally occurring" riparian buffers.

Richard Lowrance, USDA-Agricultural Research Service, Tifton, Georgia

Table 10.8 Livestock Impacts on Riparia and Streams

Decreased plant vigor
Decreased biomass
Alteration of species composition and diversity
Reduction or elimination of woody species
Elevated surface runoff
Erosion and sediment delivery to streams
Stream bank erosion and failure
Channel instability
Increased width to depth ratios
Degradation of aquatic species
Water quality degradation

After the Federal Interagency Working Group 1998.

mortality of seedlings and saplings. Livestock removal may also augment the water table, expand the hyporheic zone laterally, and allow the recovery of riparian soils, vegetation, and avifauna. These attributes have been observed in the northwestern Great Basin of the United States where key species of wet-meadow birds appeared in the third and fourth years following livestock removal (Elmore 1992, Dobkin et al. 1998). Tree canopies are the microhabitat stratum most widely used by riparian birds, suggesting that fenced plantings of endemic tree species, supplemented by native understory species, could be linked with existing vegetation to enhance landscape connectivity (Fisher and Goldney 1997).

Nevertheless, it is difficult to balance potential benefits of exclusion against the costs of construction and maintenance, lost forage, and possible impacts on other wildlife and recreation. Other site-specific management strategies are often necessary because livestock exclusion may be too costly. Such compromise strategies may include changing the season of use, reducing the stocking rate or grazing period, and resting the area from livestock use for several seasons (NRC 2002). In addition, as demonstrated by Elmore (1992), with proper levels of grazing, livestock can be present while stream systems are improving. Such an approach prescribes rather than eliminates grazing. In this perspective, riparian systems are evaluated for particular biophysical conditions as the basis for prescribing a suitable grazing strategy. For example, under the three-pasture rest rotation system, a given pasture is grazed early during the growing season the first year, grazed after upland grasses are seed-ripe the second year, and rested from livestock grazing the third year. This system provides total annual rest for each pasture on a regular basis and promotes plant vigor, seed production, seedling establishment, root production, and litter accumulation. However, because it is designed to meet the physiological needs of herbaceous plants, it is inappropriate for shrub-dominated riparia. Other grazing systems, such as winter grazing, early growing season grazing, deferred grazing, and rotational grazing, emphasize that adequate consideration of local site conditions is a prerequisite for an appropriate choice between grazing systems.

The influence of native ungulate herbivory on the recovery of riparian vegetation also may be important. Consider the results of Opperman and Merenlender (2000). They compared six deer exclosures with six upstream reference plots, along three streams with degraded riparian corridors in California. Livestock were excluded from both exclosures and reference plots. The exclosures had been in place for 15 years (three exclosures), for 6 years (one exclosure), and for 4 years (two exclosures). Their results suggest that herbivory by black-tailed deer (*Odocoileus hemionus columbianus*) substantially reduces the rate of recovery of species such as *Salix exigua, S. laevigata, S. lasiolepis, Alnus rhombifolia,* and *Fraxinus latifolia*. They discovered that the mean density of saplings was nearly 10 times greater (0.49 ± 0.15/m²) in deer exclosures as opposed to reference plots (0.05 ± 0.02/m²). Similarly, 65 percent of saplings were greater than 1 m high within exclosures, as opposed to only 3 percent within reference plots. Clearly, exclusionary fencing may be of importance for successful riparian restoration in regions with intense ungulate herbivory.

Optimizing Riparia for Biodiversity

Optimizing riparia for biodiversity is a difficult task for at least two reasons. First, most silvicultural and agricultural management strategies have focused mainly on water quality protection, not on riparian habitat values. As a result, restored buffers are often less structurally diverse than required for wildlife habitat. Second, riparia often are managed with single species as a target instead of a variety of native species. Nowadays, one needs to consider the interactions of organisms as well as spatio-temporal variability in plant and animal populations. Such an integrated management requires working both at site-specific and landscape scales.

The question most often asked by managers relates to the minimum width required to maintain the biodiversity of riparia. In reality, there will be a great variability in the width according to the taxa and sites under consideration (see Figure 10.10). This variability does not allow for the identification of one width that would maintain general biodiversity in riparia. Another difficulty in providing general rules for buffer widths is that the utilization of riparia often simplifies the habitat, reducing the chances for plant and animal diversity to recover. Emerging silvicultural techniques can provide the spatio-temporal heterogeneity necessary for maintaining diversified riparian plant and animal communities. Finally, the emphasis on width may be misleading, as that is only one aspect to be restored. Microtopography and physical heterogeneity may prove to be more important (Naiman et al. 2005) and, as discussed in Chapters 4 and 8, riparian biodiversity may result, in large part, from interactions between water-level fluctuations, microtopography, and microclimate (Brosofske et al. 1997, Pollock et al. 1998).

Conclusions

Restoration of riparian areas requires a holistic understanding of the causes of degradation, many of which reside outside riparia. One reason for adopting this perspective is that riparia are perfectly integrated into catchments where they connect uplands with aquatic systems and form corridors between low- and high-order streams. In addition, human activities throughout a drainage basin, and especially in the uplands, may affect riparia along all streams. For example, stopping the decline of riparian vegetation in the San Pedro Riparian National Conservation Area in Arizona would

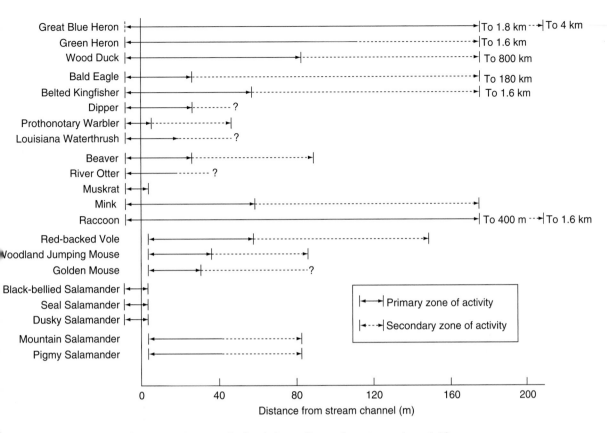

Figure 10.10 Distribution of some vertebrate species in relation to distance from stream channel. After Brinson et al. 1981.

have been impossible without limiting groundwater depletions in the uplands and recharging the regional aquifer. Thus, actions involving partnerships of various affected interests outside the riparia were necessary for riparian conservation and restoration (NRC 2002). As often emphasized, drainage basins should be the basic unit for the management of water resources, both above and below the ground, and riparian areas are integral parts of this basic unit (Wissmar and Beschta 1998). Therefore, any effort to restore riparia should be incorporated into a planning framework of hydrologically defined geographic areas.

Organizing catchment-scale information is a fundamental condition of success. An approach based on the concept of hydrogeologic equivalence has been proposed in that perspective (Bedford 1996). This concept assumes that water sources and flow paths determine the geographic distribution of riparian areas and wetlands, which is usually the case. Accordingly, different frequency distributions of wetlands and riparia that result from particular hydrogeologic settings can be obtained from maps and aerial photographs. These documents guide decisions about how to restore riparia and where to do so in the catchment. A similar approach has been proposed as a protocol to restore biophysical connectivity of entire river ecosystems (Stanford et al. 1996). Such a protocol can be adapted to the specific biophysical needs of selected

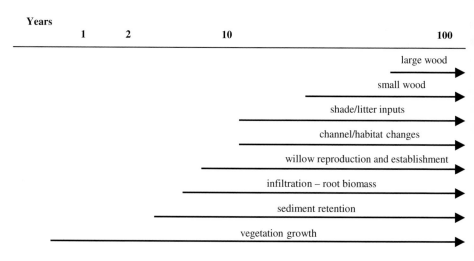

Figure 10.11 Projected recovery times of riparian processes and components in the interior Columbia River Basin following passive restoration (i.e., cessation of activities causing degradation or preventing recovery). After Beschta and Kauffman 2000.

riparia, thereby formalizing issues at the catchment scale. This allows the restoration of habitat heterogeneity, the maximization of migratory efficiency to allow recovery of meta-populations, the minimization of planting of nonnative vegetation, and the use of an adaptive management strategy.

Due to the position of riparia at the interface between terrestrial and aquatic environments, restoration efforts need to consider several specificities. First, changing the way land and water are used in a catchment may be a basic condition of successful restoration, although it is not easy to achieve. Second, the recovery of degraded riparia needs time just as environmentally unsustainable practices dominated for extended periods. This can be seen in the components of riparian areas associated with salmon habitats, which will take decades to recover in the Columbia River Basin (see Figure 10.11; Beschta and Kauffman 2000). Third, opposing opinions or conflicts often slow restoration efforts. Examples include balancing short-term versus long-term restoration goals, adopting small-scale versus large-scale perspectives, considering private lands versus public lands, and satisfying the water needs of riparian vegetation versus those of humans (NRC 2002). Fourth, it may be preferred to allocate efforts to low-risk, highly sustainable restoration, especially when public resources are limited. In some circumstances, it may not be feasible or efficient to restore a riparian area, and large investments in long-term maintenance would be required to achieve only compromised conditions. This is the case for many rivers in Europe due to the complexity of land ownership and tenure and to the high population density.

Globally, most options for improving riparian systems encompass a wide range of values, both individual and societal. It is thus essential that stakeholders—that is, scientists, land managers, regulators, and the public—confront their choices of where and how to initiate restoration efforts. It is also essential to ground these efforts on credible scientific information as well as on broadly inclusive participation of all stakeholders (Naiman et al. 1998a, Ward et al. 2001, NRC 2002). The full spectrum of perceptions must be discussed *openly* and *respectfully*, allowing for divergent views and positions. It must be remembered that such a process is time consuming and that full agreement or consensus is rare. Nevertheless, it is the only way to achieve a policy based on a substantive public support at the scale of the entire catchment (Committee of Scientists 1999). Such support is necessary for implementing a common vision that is based on science and on values, and one that addresses physical, biological, aesthetic, and spiritual benefits that future generations will value in riparia.

Synthesis

11

Overview

- This final chapter blends current knowledge of individual riparian characteristics with uses of riparian resources into a broad, holistic synthesis. We examine riparia as keystone components of catchments, reexamine their role as nodes of ecological organization, provide a unified perspective of riparian ecology, explore approaches for developing a future vision of riparia, and articulate principles to guide ecological management.

- It is nearly impossible to place monetary values on goods and services provided by riparia, whether for nature or for human societies. Nevertheless, enough is known about riparia to appreciate their value in providing clean water, flood control, habitat for plants and animals, and countless other benefits that help sustain human well-being.

- We review the 10 commonly assumed functions and uses of riparian systems (discussed in Chapter 1) related to biodiversity, productivity, nutrient retention, and other system-scale attributes. We conclude they are close to the originally stated tenets with their relative veracity depending on local environmental settings and on regional cultures.

- The *prevailing* perspective of riparia as systems shaped only by biophysical drivers is too limited in scope. A thorough system-scale understanding of riparia requires a broad integration of human and biophysical sciences. A more ample perspective must consider human societies as direct and indirect drivers of riparian form, where human influences vary depending on resources provided by riparia and on human perceptions of value.

- Sustainable riparian management relies on decisions based on knowledge of the system and on realistic predictions about the future. Consequently, understanding change and the associated uncertainties—whether natural or human-mediated—are vital aspects of riparian management decisions.

- A broad array of technical tools and approaches is available to understand, quantify, and model riparian systems. Emerging tools and approaches include advances in biotechnology, remote sensing, chemical and computer sciences, information management, and the social sciences—especially as related to resolving conflicts and developing shared visions of desired conditions.

- Ecological principles, when identified and properly applied, underpin the long-term maintenance of riparian goods and services. Knowledge, synthesized into basic principles, can guide ecologically sound riparian resource development.

Purpose

As a treatise on riparian systems, including those created or shaped by humans, our approach has been to use a landscape perspective to convey the broad spatial and temporal scales of processes shaping and supporting riparia. This requires integrating human perceptions and activities with biophysical patterns and processes. Our objective in this final chapter is to combine current knowledge about riparian characteristics into a broad, holistic synthesis.

Earlier chapters illustrated how plants and animals, and attendant ecological processes, are distributed in time and space along steep biophysical gradients in riparia. We also discussed how human cultures interact with nature to set management strategies, thereby affecting ecological processes and, finally, creating landscapes. The scope has been understandably broad as theory, physical drivers, typology, structure, ecological processes, environmental change, management, conservation, and restoration are melded into a coherent riparian story. The riparian story, of course, continues to evolve as new knowledge and perspectives are added at ever-increasing rates.

We have seen the importance of recognizing the multidimensionality of riparian systems. For example, the high degree of multidimensionality is manifest in a cascade of causes and effects: first, water saturation and disturbance gradients are driven by geomorphology and hydrodynamics; second, biophysical processes are shaped by water saturation gradients and disturbance regimes; and third, the components of biotic communities are adapted to specific patterns of saturation and disturbance arrayed in space and time along the gradients and regimes. According to our premise, the expression, persistence, and productivity of extant riparian communities are products of interactions between environmental history, groundwater flow paths, surface water dynamics, biotic interactions, and physical process domains. This is the basis for understanding natural and culturally modified riparian systems.

We have learned, among other things, that biodiversity and biotic production are often strong in riparia, especially when spatial patterns are examined in a landscape context, and that riparia are essential for effective catchment management (Décamps 1996). Nevertheless, riparia have only recently become the focus of concerted management efforts, and in much of the world they remain neither well managed nor well respected. Riparian systems, especially floodplains, are among the most altered landscape components on earth (Tockner et al. 2005). Stream regulation, urbanization, overgrazing, exotic species, and many other types of alterations fundamentally transformed riparian systems into vestiges of their former states. Yet, herein lies the conundrum since many human perceptions of the quality of life have a riparian basis, such as readily abundant resources and visual beauty (Carbiener 1983, Haslam 1991, Green and Tunstall 1992). It is our philosophy that in order to apply a riparian perspective to effective catchment management and human well-being, we must first elucidate the strong interactions between biophysical drivers, riparian characteristics, and healthy and appealing environments. Remaining uncertainties, especially those introduced by land use and climate change, hopefully will be resolved as cultural perceptions become more favorable toward the implementation of new knowledge and as science progresses to a point of making better predictions in ecology.

In the following sections, three related topics collectively provide a foundation for synthesis. We address riparia as keystone units of catchments, present a unified perspective of riparian ecology, and develop a future vision through technological innovation, restoration, and conservation.

Riparia as Keystone Units of Catchment Ecosystems

Nearly two decades ago, faced with limited support for research and catchment management, a small group of researchers and managers asked a seemingly easy question: Given limited resources, where within the catchment should they focus their activities so as to have the best knowledge-based return for their research and management investments? They reasoned that, even though riparian areas occupy a small percentage of the land area, they are disproportionately important because of their position at the intersection of land and water. They further reasoned the key position of riparia in landscapes and the unique processes operating within them led to disproportionately large values in terms of ecological goods and services. Over the course of the next decade this simple reasoning led to many discoveries and surprises that then stimulated a number of comprehensive, and complex, research programs (Naiman and Décamps 1990, Holland et al. 1991, Risser 1993, Pringle and Barber 2000).

What was discovered about riparia? They are often nodes of ecological diversity at the catchment scale; they can act as control points for the storage and transformation of essential nutrients; they are biophysically dynamic; they exhibit extraordinary resilience in response to disturbance; and they can be exceedingly useful in catchment management. Moreover, the basic biophysical properties of riparian systems appear to be, at one scale, similar throughout the world, while, at the site scale, they are remarkably adaptive to local conditions. While much has been learned, much also remains to be discovered about riparian soils, the roles of trees and herbivory in maintaining system integrity, emergent system-scale properties such as resilience, and other key issues (see Table 1.1 in Chapter 1). These are challenges for the future. But already it is clear that riparian systems are key to land and water restoration and management efforts in those areas where humans have intensively exploited riparian and other resources for centuries. In the developing and still wild parts of the world, riparian systems are key to sustainable development and conservation.

Applied ecological discoveries led to a riparian strategy for the management of catchments in many nations. As a result, regulations governing riparian conditions were regionally implemented as researchers informed the public and their representatives about the key ecological roles of riparia in maintaining the integrity of

Table 11.1 Functions of Riparia and Some of Their Key Relationships to Environmental Goods and Services

Functions	Indicators that Functions Exist	Effects of Functions	Goods and Services Provided
Hydrology and Sediment Dynamics			
Stores surface water (short term)	Floodplain connected to stream channel	Attenuates downstream flood peaks	Reduces damage from floodwaters
Maintains a high water table	Presence of flood-tolerant and drought-intolerant species	Maintains vegetation structure in arid climates	Contributes to regional biodiversity by providing habitat
Accumulates and transports sediments	Riffle-pool sequences, point bars, terraces	Contributes to fluvial geomorphology	Creates predictable yet dynamic channel and floodplain dynamics
Biogeochemistry and Nutrient Cycling			
Produces organic carbon	A balanced biotic community	Provides energy to maintain aquatic and terrestrial food webs	Supports populations of organisms
Contributes to overall biodiversity	High species richness of plants and animals	Reservoirs for genetic diversity	Contributes to biocomplexity
Cycles and accumulates chemical constituents	Healthy chemical and biological indicators	Intercepts nutrients and toxicants from runoff	Removes pollutants from runoff
Sequesters carbon in soil	Organic-rich soils	Contributes to nutrient and carbon retention	Potentially ameliorates global warming
Habitat and Food Web Maintenance			
Maintains streamside vegetation	Presence of shade-producing forest canopy	Shades streams during warm seasons	Maintains conditions for cool-water fish
Supports characteristic terrestrial vertebrate populations	Appropriate species having access to riparian areas	Allows daily and seasonal movements as well as annual migrations	Wildlife viewing and game hunting
Supports characteristic aquatic vertebrate populations	Fish migrations and population maintenance	Allows migratory fish to complete life cycles	Provides fish for food and recreation

Partially modified from NRC 2002.

catchment-scale systems. Riparian restoration became a research theme, a public activity, and even an industry in some areas. Riparian conservation efforts were bolstered by the inherent biodiversity found therein and were perhaps accentuated by the general decline of species and the simplification of ecological processes in other parts of the landscape.

Societies and cultures do not maintain and nurture what they do not value. This bit of wisdom has been integral to human behavior for countless generations. People tend to place little emphasis on protecting natural systems when they neither understand nor appreciate their value in providing services and goods (Daily 1997, Nassauer 1997). Ecosystem services encompass the conditions and processes through which natural systems maintain biodiversity and supply humans with goods, such as foods, fibers, fuels, and many chemicals and industrial products. In addition to the production of goods, ecosystem services provide actual life-support functions such as cleansing water and air, recycling and renewal, and they confer many intangible and cultural benefits as well.

Even though riparia provide many goods and services, both for nature as well as for human societies, placing monetary values on these goods and services is difficult. But even without specific dollar values, enough is known about riparia to appreciate their value (see Table 11.1). Engineering projects to replace services provided by riparia—if not completely beyond current technology—are prohibitively expensive if implemented at anything but a trivial scale. Unfortunately, the scientific and economic understanding of the dimensions and details of riparian goods and services remains shallow, although in certain geographic regions it has improved greatly in the last decade. Well-known examples are found along the Pacific Coast of North America, in South Africa, and in England and Australia.

On a global scale, riparian areas are valued by society because of their proximity to water, generally fertile soils, convenience as waste disposal sites, and opportunities for recreation (NRC 2002). Overall, whether in lowlands or uplands, riparia help shape hydrology and sediment dynamics, biogeochemistry and nutrient cycling, and habitat formation and maintenance—all of which are important ecosystem services with inherent "goods" and monetary values. Even though there do not appear to be any economic estimates for riparian goods and services, estimates for the goods and services provided by freshwater ecosystems certainly measure in the trillions of U.S. dollars (Postel and Carpenter 1997). Considering the importance of riparia in shaping the integrity of freshwater systems and in providing quality water, one might imagine that the global value of riparia would be of comparable magnitude.

Riparia as Nodes of Ecological Organization

At catchment and landscape scales, riparia act as strong organizers of ecological systems, especially in terms of spatial patterns. The distribution of plants and animals is shaped by the availability of water, food, and habitat—and these are features that riparia provide in abundance, even for many mobile species spending most of their time outside riparian zones. In wet regions this role may not seem obvious because of the dense network of riparia across the landscape. However, in arid and semiarid regions, the role of riparia in shaping the broader ecological organization is more obvious. For example, in the semiarid savannas of South Africa, the American South-west, and Australia, riparian corridors are perhaps the most obvious expression of widespread organization since so many organisms focus on them for habitat and nutri-tion. This applies as well to the regions of Europe, the Middle East, and North Africa that border the Mediterranean Sea. They are viewed as "hot spots" of activity because they sustain much of the mobile fauna in dry periods, maintain biodiversity, integrate terrestrial and aquatic systems, and act to shape the patterning of the landscape over broad spatial scales (Naiman and Décamps 1997). Riparia—and their connection to in-stream processes and to interchanges with the upland components of semiarid savanna landscapes—provide a powerful testament to the functional significance of riparian corridors in maintaining an ecological abundance that would not be expected given the scarcity of resources in the predominant components of the landscape.

A Unified Perspective of Riparian Ecology

Chapter 1 discussed 10 commonly assumed functions of riparian systems and offered some possible alternative explanations (see Table 1.1 in Chapter 1). Do these tenets hold in all riparian situations, or are the alternative explanations closer to reality? It appears that the *realities*, in a general sense, are close to the originally stated tenets but the *relative veracity of each tenet* depends on the local environmental setting and on the regional culture. If one examines the first tenet, one sees that, generally, riparian communities are areas of higher biodiversity and productivity, but the actual values vary according to local conditions. For example, riparian floodplains are generally more biologically diverse and productive than the riparian zones of headwater streams in the same region, and riparia in arid regions are relatively more biologically diverse and productive (with respect to adjacent uplands) than riparia in mesic regions—and there are differences between temperate and tropical regions as well (Décamps et al. 2004, Sabo et al. 2005). The same may be said for the tenet identifying riparia as nutrient filters. Owing to the dynamics of individual nutrients and local environmental conditions, riparia are not effective filters for all nutrients under all conditions. They are certainly important, but variable flow paths, flood regimes, ionic affinities of individual elements for particles, and ion-specific solution chemistries act to modify this general tenet. Similar arguments may be made for the remaining eight tenets, regardless of whether they address environmental, cultural, or system-level attributes.

The reality is that riparia are highly complex systems requiring a good level of interdisciplinary knowledge, including basic knowledge about social systems, to understand their intricate internal workings as well as their external interactions with adjacent systems. The general form of riparia is conceptually simple and readily appreciated, but the inherent dynamics are much more complex—much like modern engines that are becoming more varied in design, are increasingly sophisticated as technology advances, and are being used for a variety of purposes as human values and perceptions change. Recall that in the first chapter we introduced the concept of riparian biocomplexity as being shaped by biophysical drivers (i.e., Figure 1.8). This is only part of the system. In both the developing and developed regions of the world, now and into the future, human-mediated changes and human values and perceptions are significantly shaping riparian characteristics. This is why it is so important that those with a professional interest in riparia be well grounded in the human sciences as well as in the biophysical sciences.

Nevertheless, articulating a general functional and unified perspective of riparia remains perceptibly simple. Landscape characteristics and climate patterns determine the persistence of water saturation, as well as the chemical and material gradients, that collectively help shape the distinct aboveground and belowground attributes of riparia (see Figures 1.8 and 11.1). In essence, the spatial and temporal characteristics of disturbance regimes and saturation gradients structure the distribution of riparian biota. Biodiversity and production are maximized in riparia as a consequence of boundary dynamics—the juxtaposition of saturated and unsaturated zones, aerobic and anaerobic zones, terrestrial and wetland species, and strong bidirectional fluxes of materials. Riparia fundamentally are located at interaction zones that occur

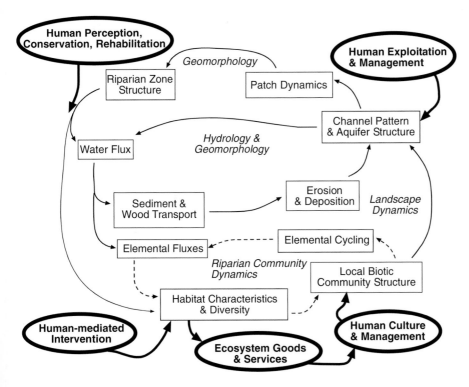

Figure 11.1 A conceptualization of the biophysical complexity—including dominant human dimensions—associated with riparia. Solid lines represent linkages driven predominantly by physical processes and dashed lines represent linkages driven predominantly by biogeochemical processes.

predictably in space and time within the physical setting of catchments. However, this *prevailing* perspective of riparia is too limited in scope for today's world. Our perspective is that, depending on the resources provided by riparia and perceptions of their value, individual riparia are culturally modified to serve the needs of human societies throughout the world. Riparia are shaped by political, institutional, and economic decisions, modified by management strategies and landowner values, further shaped by conservation initiatives, and reshaped by restoration efforts (see Figure 11.1). This adds a significant dimension to the prevailing perspective, but one that is essential for a fuller understanding of contemporary riparian systems.

Developing a Future Vision

What will the future bring? It is natural for scientists and managers to imagine what the future will be like, and use that vision to design the best courses for action. Riparian management relies on a process of decision making based on an understanding of the *whole* system to make predictions about the future. There are two basic approaches to considering the future in a system-level perspective: scenario development and forecasting, including prediction and projection. Selecting one or the other approach depends somewhat on the scale of the issue, and each has its own mechanism for addressing biophysical and socioeconomic change and uncertainty.

Scenario Development

Scenarios based on a range of probable conditions, and uncertainties, are especially useful for envisioning future environmental and human patterns. As a technique, scenario development has enjoyed numerous successes in helping build consensus about future directions for communities and regions because of its visual format built on existing data and trends, and the hope of actually achieving desired conditions once a community-scale consensus is reached (e.g., White et al. 1997, Hulse et al. 2000). Since the future is basically unpredictable, scenarios become powerful planning tools because they articulate different strategies for a range of possible outcomes. A further consideration is that since human well-being depends on consistent provision of ecosystem services, scenarios enhance the understanding of ecosystem services and change, and thereby improve management (Bennett et al. 2003).

There are three major benefits of scenario planning (Bennett et al. 2003). First, scenarios develop specific strategies for particular possible futures—for example, to identify policies that are adaptable across a range of futures. Second, they develop capacities for coping with surprises, such as investing in professional education. By describing the full range of possibilities, scenarios help planners develop the capacity to handle many different futures. Finally, scenarios scan the future for potential surprises and explore and map uncertainties, thereby creating an awareness that the future may differ from the present in important ways.

One of the more successful and comprehensive scenario planning efforts related to riparia occurred in the Willamette River catchment in Oregon (Hulse et al. 2002, Baker et al. 2004). The Willamette catchment, situated between the Coast Range and the Cascade Mountains, covers 29,728 km^2 of highly diverse geologic terrain and climatic conditions. Sustaining air quality, water supplies, anadromous fish (especially salmon, *Oncorhynchus*), and the livability of human communities are widespread concerns as the population is expected to nearly double (increasing from ~2 to ~4 million people) by the year 2050. This is the equivalent of adding three more cities the size of Portland, currently the state's largest city. The key challenge is to accommodate the expected human population growth with its attendant demands on natural resources while sustaining the catchment's natural resources. The initial step in meeting the challenge was the formation of a consortium to develop and promote a shared vision for enhancing the livability of the Willamette catchment (see Table 11.2). The basic approach was scenario development using the most advanced Geographic Information Systems (GIS) and graphics techniques, and much of that shared vision was focused on the river and its riparian zone.

Table 11.2 In Order to Develop Trajectories of Landscape Change, the Activities of the Willamette River Catchment Consortium Evolved Around Four Basic Questions

1. How have people altered the land, water, and biotic resources of the Willamette River catchment over the last 150 years since Euro-American settlement?
2. How might human activities alter the Willamette catchment over the next 50 years, considering a range of plausible policy options and land use changes?
3. What are the expected environmental consequences of these long-term landscape changes?
4. What types of management actions, in what geographic areas or types of ecosystems, are likely to have the greatest effect?

Modified slightly from Hulse et al. 2002.

The first task was to compile, produce, and analyze basic information on the phys-ical, biological, and human features of the catchment. This provided the basic infor-mation for projecting future change. Three alternative visions were then developed for the future based on implementation of current policies, on loosening of current poli-cies to allow a freer rein to market forces, and on conservation, which emphasized ecosystem protection and restoration as well as a balance between ecological, social, and economic considerations. Finally, the likely effects of these alternative futures and the long-term landscape changes on four areas of concern were evaluated, one of which was projected change in river channel structure and riparian vegetation.

The potential for riparian restoration was determined by a set of guiding criteria (e.g., amount of forested land in public or private ownership) and, as one might expect, it varied along the river. Analyses indicated that the middle and upper por-tions of the river had the best potential for future recovery. The middle reach of the Willamette River offered unique opportunities because it historically exhibited some of the most extensive floodplain forest, and more of the floodplain was vegetated by forests than any other reach of the river (see Figure 11.2). Fortunately, some of this land is owned by the state, and restoration on public lands represents the best oppor-tunity for a longer-term, ecologically sound solution.

The outcomes of the broader scenario planning effort are now informing commu-nity decisions regarding land and water use as results are discussed by stakeholder groups charged with developing a vision for the basin's future and a basin-wide restoration strategy (Baker et al. 2004). An interesting outcome of the three major scenarios examined was that, under an ecosystem protection and restoration scenario, most ecological indicators recovered 20 to 70 percent of the losses sustained since Euro-American settlement, even though water consumption increased by 43 percent over levels in 1990 AD. Water conservation measures were not sufficient to reverse recent trends of increasing water withdrawals for human use.

Forecasting, Prediction, and Projection

While ecology has many methods for envisioning the future—forecasting, prediction, and projection—these may be less efficient for decision making because of the complexity of riparian systems and an imperfect knowledge about how they work. Predictive modeling or forecasting is often used in ecological decision making and is appropriate for simulating well-understood systems over short time frames. However, most riparian management is underpinned by decisions affecting multi-decadal time periods for a system that is incompletely understood and capable of great change.

Ecological forecasting has been defined as the process of predicting the state of ecosystems, ecosystem services, and natural capital, with fully specified uncertainties, and is contingent on explicit scenarios for climate, land use, human population, tech-nologies, and economic activity (Clark et al. 2001). The time horizon for forecasting may extend up to 50 years, with the information content of a forecast being inversely proportional to forecast uncertainty. The most daunting source of uncertainty results from strong nonlinearity and stochasticity (randomness). Recently, however, there has been good progress in the capacity to accommodate stochasticity, particularly with the development of Markov chain Monte Carlo (MCMC) methods (Clark 2003).

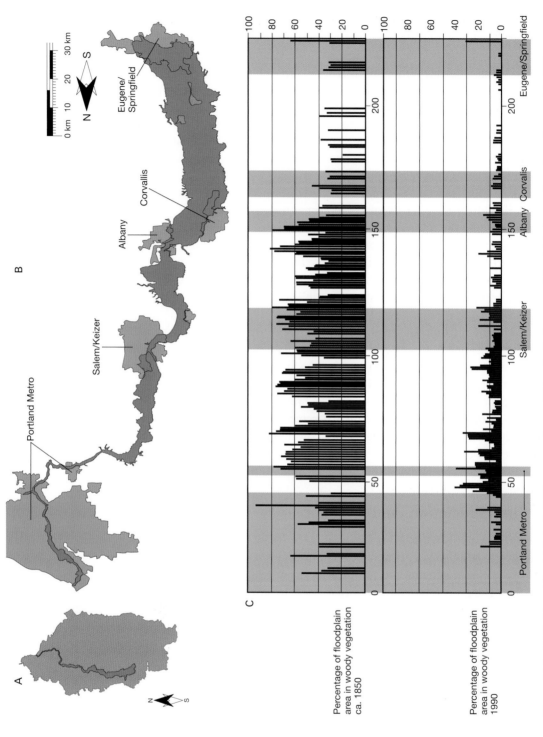

Figure 11.2 (A) Position of floodplain in the Williamatte catchment. (B) Longitudinal patterns of floodplain forests along the mainstem of the Willamette River, Oregon. (C) The percentage of floodplain area in woody vegetation (*ca.* 1850) is compared to the percentage for 1990 (adapted from Hulse et al. 2002).

These and other new approaches provide flexible methods for accommodating the variability that can stem from many sources, for incorporating uncertainties, and for data assimilation.

There are several guidelines to follow when forecasting future states of riparian systems. First, consider the "slow" variables that forewarn of ecological consequences years in advance. These may be climate trends that suggest regions are becoming wetter or dryer over decadal time scales, with attendant effects on water and sediment flux regimes (e.g., GWSP 2004). There are also many slow variables that constrain site-specific ecological processes. For example, successional change is constrained by climate and soils. If these change slowly relative to tree life spans, succession may be predictable using knowledge of physiology and ecological competition. Second, experimental and observational data that extend to landscapes or regions are a foundation for forecasting. Large experiments are critical because landscape processes are often unpredictable from site-specific studies. The feedbacks from riparian vegetation to future riparian states—for example, the role of large woody debris in forest initiation—can best be seen when the spatial extent of study exceeds an appropriate spatio-temporal threshold. Further, data networks and long-term monitoring provide key bases for forecasting. Invasions of riparia by exotic plants have benefited from historic records of species introductions and their vectors (e.g., cars, nursery shipments). Third, emerging technologies do not compensate for sparse data, but they promise to facilitate forecasting. Hydrologic forecasting and remote sensing, together with geophysical tomography, can provide high-resolution coverage of precipitation and the effects of dams and irrigation—which can build databases and assist in forecasting. Biogeochemical cycles, hydrology, and biodiversity forecasts require land inventory and census data. In combination with satellite-based information, habitat loss can be monitored, a predictor of extinction risk.

In general, most ecological models tend to systematically under-examine surprises and suppress uncertainties. Sudden changes in current trends, which are relatively common in riparian systems, are the hardest to predict, yet they carry the most risk and offer the greatest opportunities for management. Scenarios, on the other hand, are created specifically with the idea of paying attention to the unusual, the uncertain, and the surprising when making decisions (Bennett et al. 2003, Peterson et al. 2003). They focus attention on key shifts in current trends and other important opportunities and, together with forecasting, may provide the essential tools for developing a realistic future vision.

Emerging Perspectives, Tools, and Approaches

The last two decades have witnessed the rapid growth of riparian ecology from site-specific studies, mostly instigated by curiosity-driven research, to broad investigations emphasizing the roles of riparia in organizing landscapes, in catchment management, in restoration strategies, and in links to human cultural and economic systems (Gregory 1997). Along with this fundamental change in perspective has been the adaptation of a broad array of technical tools and approaches to understand, quantify, and model riparian systems. Together, these perspectives and tools constitute the intellectual frontier of riparian research and management.

Emerging Perspectives

Several emerging perspectives are at the core of this book: riparia in the landscape, riparia for catchment management, riparian restoration, and the links between riparian integrity and vibrant human cultural and economic systems. Although much remains to be learned about riparia in terms of structural features, ecological relationships, and processes shaping communities, riparia are today an important priority in catchment management. This stems mainly from discoveries about the structure and dynamics of riparian zones and their impacts on stream and catchment characteristics in the western United States, Canada, Australia, and Europe (e.g., Haycock et al. 1993, Naiman et al. 2000, Tockner et al. 2002). Restoration programs have begun to incorporate many of the dynamic riparian processes discussed in this book after realizing that riverine habitat conditions are closely linked with the integrity of riparian zones. One of the most difficult realizations for resource managers to accept was the need for periodic natural disturbances in the establishment of desired future conditions within catchments. This is illustrated in the following example.

In the northwestern United States, judicial injunctions in 1991 and 1992 against harvesting trees on federal government lands within the range of the northern spotted owl (*Strix occidentalis*) prompted then-President Clinton to chair the Forest Conference in 1993. A major outcome was the formation of an interdisciplinary scientific group, the Forest Ecosystem Management Assessment Team (FEMAT 1993), whose charge was to develop plans for both the long-term health of northwestern United States ecosystems and the associated human socioeconomic systems. The plans were to be scientifically sound, ecologically credible, and legally responsible, and to yield sustainable timber harvests while incorporating habitat protection for salmon, spotted owl, and other important species. A major outcome of this effort was an aquatic conservation strategy based on riparian management (Sedell et al. 1994).

This strategy established interim buffer requirements to ensure the long-term viability of aquatic and riparian species on federal lands within the range of the spotted owl. The interim buffer width was set at the height of a dominant riparian tree. However, implementation of such a system of streamside buffers in landscapes with high drainage densities would, in practice, create a dense network of riparian reserves that would fragment the exploited forest landscapes. As a consequence, activities like timber harvesting would become operationally difficult. The understanding developed over the last decade about natural catchment processes creating a mosaic of riparian conditions (Benda et al. 1998, 2004) provided the basis for applying a novel management paradigm that embraced disturbance history and natural disturbance. The FEMAT scientists therefore recommended catchment assessments be carried out as part of the management planning process and that the results of these assessments be used to shape management activities based on the unique characteristics of individual catchments (Naiman et al. 2000).

This approach illustrates how recent advances in understanding riparian processes and landscape dynamics can be applied to difficult management issues. Tailoring riparian buffers and timber management activities to catchment disturbance attributes results in a landscape-based plan that is designed to promote development of natural forests and healthier riparian zones. Maintaining the natural disturbance regime remains central to restoring diverse and productive aquatic and riparian systems in the northwestern United States. The use of disturbance history informa-

Table 11.3 A Summary of Five Major Attempts to Articulate Fundamental Principles for Maintaining the Biotic Vitality of Land and Freshwater Ecosystems. Principles are grouped and reformulated or modified from the original statements to make them more directly applicable to riparian systems. See Stanford et al. (1996), Naiman et al. (1998, 2002), Dale et al. (2000), and Jackson et al. (2001) for the original principles

Ecological Principles and Guidelines for Managing Riparia

1. Natural flow regimes shape the evolution of riparian biota and ecological processes.
   - Disturbances are essential, ubiquitous events whose effects strongly influence the ecological dynamics and spatial patterns of riparia.
   - Interactions between surface water and groundwater flows are essential to aquatic-riparian ecosystem integrity.
2. Individual and interacting networks of species have strong effects on ecosystem processes.
3. Riparian management demands unparalleled cooperation because the complexity of information and the scope of change exceed the capacity of any single group.
4. Appropriate human activities affecting riparia are fundamental ecological elements of the catchment.
5. Conserving and restoring biophysical properties of riparian zones improves all natural resource values and improves human well-being.

tion during catchment assessment therefore facilitated a landscape plan that is compatible with both conservation objectives and human uses.

The improved understanding of riparian ecology also played a role in the development of the concept of the "normative" river, which emphasizes the need to conserve, stabilize, enhance, and restore aquatic and riparian systems to a "normative" condition (see Table 11.3; Stanford et al. 1996). The term *normative* does not imply pristine, because it is often impossible to restore catchments to a state where anthropogenic disturbances are absent. Rather, it implies the restoration of ecosystem connections that permit as many natural processes to exist as possible, given other social and economic objectives. The normative river approach recently has been suggested as an alternative to current measures to restore fish and riparian wildlife in the Columbia River basin (Williams et al. 1999)—where salmon and other native plants and animals have steadily declined despite very costly restoration efforts.

Riparia are now managed for a wider variety of ecological functions than ever before, and their role in maintaining catchment "goods and services" has been given greater importance in regulatory frameworks. Today, the four emerging themes in riparian management—an emphasis on ecological function and natural riparian forest pattern, adoption of a landscape perspective of river networks, development of ecologically sound systems of restoring riparian system properties, and attention to social needs for riparian resources—are increasingly accepted and are becoming focal points around which conflicts between competing uses of riparian zones are resolved (Gregory 1997, Naiman et al. 2000).

Globally, many challenges remain. Even though each theme has been integrated into planning and implementing riparian management in many regions, there is a large discrepancy in the application of riparian management strategies across different land uses. At present, portions of the landscape devoted to forest management receive the most riparian protection, whereas riparian conservation in urban and industrial areas is usually limited to unnaturally narrow borders along stream banks. Additionally, restoration of highly altered riparian communities in heavily urbanized environments is often constrained by pavement and other structures preventing reestablishment of natural functions. Riparian protection on agricultural lands is usually intermediate between levels of protection given forested headwaters and urbanized floodplains. On a broad geographic scale, this has resulted in highly fragmented conditions in many catchments where islands of "healthy" riparian ecosystems are separated by long reaches of altered, dysfunctional river corridors. Building a network of functional, intact riparian zones—where appropriate—to reconnect these remaining areas remains one of the most significant challenges for natural resource management.

Many environmental and social approaches to catchment management have not proven to be effective beyond a decade or so. How might the success of management efforts be made more long lasting? One approach has been to develop a shared social-environmental vision of future riparian conditions (see Chapter 8). In an ideal sense, this may prove to be a nearly impossible task, although the process of identifying social-environmental endpoints aids communication and acts as an effective form of education about the diversity of cultural beliefs and values embedded in a catchment. For example, environmental endpoints may be related to the extent and condition of riparian forests, to acceptable levels of water quality and aquatic habitat, or to the persistence of viable populations of ecologically or culturally valuable plants and animals. Social endpoints may relate to the level of literacy about the structure and functioning of the catchment; the development of flexible (or adaptive) institutions that are able to respond to new and as yet unforeseen issues; the formulation of unique partnerships between industry, citizens, academia, and government; and the realization of levels of personal stewardship and responsibility needed for the long-term maintenance of a balanced riparian system. However, inherent in the implementation of a successful vision, is understanding the links between riparia and the human cultural and economic systems.

Integrated social-environmental models help explore alternative management scenarios and assist in identifying the links and in assessing desirable future conditions (Flamm and Turner 1994a, 1994b). An example of such a model is the Land-Use Change and Analysis System (LUCAS; Berry et al. 1996, Wear et al. 1996). LUCAS is a spatial simulation model operating at the catchment scale in which the probability of land cover being converted from one type to another depends on a variety of social, economic, and ecological factors (see Figure 11.3). Conditional transition probabilities are estimated empirically by comparing land cover, including riparian condition, at different times (e.g., from decade to decade), and these transition probabilities are then used to simulate potential future conditions (Turner et al. 1996).

Simulations begin with an initial map of land cover, and equations are used to generate a transition probability for each grid cell in the catchment map based on ownership type, elevation, slope, aspect, distance to road, distance to market, and population density. An integrated modeling approach permits the effects of a wide range of alternatives to be evaluated. For example, the effects of residential development in different locations within the basin or the effects of moving a large parcel of land into or out of intensive timber production can be examined. Linking projected land cover maps with effects on ecological indicators (such as riparian condition, species persistence, or water quality) allows the potential long-term implications of alternative human decisions to be compared.

LUCAS INTEGRATION MODULES

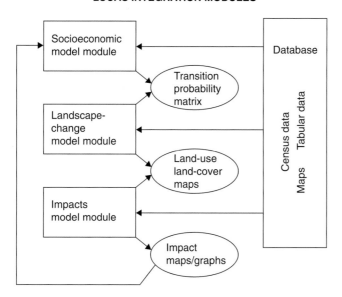

Figure 11.3 Integration of social, economic, and environmental aspects of catchment management can be accomplished with the Land Use Change Analysis System (LUCAS), a modeling environment (Berry et al. 1996, © 1996 IEEE).

Emerging Tools and Approaches

Techniques for examining riparia are expanding and improving as technical instruments, from microprobes to satellites, are improved. Living in an era of rapid technological advances is providing an abundance of electronic tools and chemical tracers for investigating riparian processes. Although one may be surprised to find a section on emerging tools and approaches in a synthesis chapter, we feel these provide key resources for establishing new riparian research avenues and knowledge generation for the future. Much of the future understanding of riparia will come inevitably from emerging tools and approaches. Although we cannot hope to examine them all here, there are several that promise to significantly enlighten us about riparia. We have grouped them into categories, representing those that are most useful for aboveground and belowground investigations, or for both.

Aboveground

Perhaps the most important new tools for assessing riparia are those coming from remote sensing (Hewitt 1990, Lefsky et al. 2002, Mertes 2002). Many instruments already are available for image acquisition from both commercial and government organizations (see Table 11.4). Further, digital data can be collected at spatial resolutions ranging from meters to kilometers and at spectral resolutions ranging from a few broad bands to hundreds of narrow bands on hyperspectral detectors such as the Airborne Visible/Infrared Spectrometer (AVIRIS; see Figure 11.4). As the number of instruments actively recording data has increased, the temporal resolution of acquisition for short-lived events such as floods has become finer. In addition, microwave

Table 11.4 Noncomprehensive List of Representative Remote-Sensing Instruments in Alphabetical Order

Acronym	Name	ER Region
AIRSAR	Airborne Synthetic Aperture Radar	Active microwave
ASTER	Advanced Spaceborne Thermal Emission and Reflection Radiometer	Optical, IR, thermal
AVHRR	Advanced Very High Resolution Radiometer	Optical, IR, thermal
AVIRIS	Airborne Visible/Infrared Imaging Spectrometer	Optical, IR, thermal
ERS	European Remote Sensing Satellite Synthetic Aperture Radar	Active microwave
FLIR	Airborne Forward-Looking Infrared	Thermal
Ikonos	Ikonos—A high spatial resolution satellite	Multispectral and panchromatic
IRS LISS	Indian Remote Sensing Linear Imaging Self Scanning Camera	Optical, IR
JERS	Japanese Earth Remote Sensing Synthetic Aperture Radar	Active microwave
LANDSAT	Landsat Multi-Spectral Scanner (MSS), Thematic Mapper (TM), Enhanced Thematic Mapper (ETM)	Optical, IR, thermal
LIDAR	Laser Radar or Light Detection and Ranging	Active microwave
MODIS	Moderate Resolution Imaging Spectrometer	Optical, IR, thermal
RADARSAT	Canadian Synthetic Aperture Radar	Active microwave
SEA WIFS	Sea-Viewing Wide-Field-of-View Sensor	Optical, IR
SMMR	Scanning Multichannel Microwave Radiometer	Passive microwave
SPOT HRV	Système Probatoire d'Observation de Terre, High Resolution Visible Sensor System	Optical, IR
SRTM	Shuttle Radar Topography Mission	Active microwave
TOPEX	Topex/Poseidon	Active microwave

The region of the electromagnetic spectrum (ER region) indicates the range of wavelengths covered by the instrument. IR denotes infrared waves.
From Mertes 2002.

and radar instruments can penetrate cloud cover, allowing for acquisition of information under all types of weather conditions. Finally, the radiometric properties of instruments are increasingly stable, and onboard calibration capabilities have allowed more accurate expression of the raw digital counts as meaningful physical values. Concurrent with the increased sophistication of the instrument array has been a rapidly expanding repertoire of image processing techniques (e.g., Mather 1999; see Figure 11.5).

The properties of riverine landscapes that have been most successfully measured with remote sensing include community- and habitat-level classification and connectivity of water bodies with optical and radar data. Radar and laser altimetric (LIDAR) data measured from *aircraft* provide water elevations at resolutions as fine as decimeters, as well as numerous vegetative characteristics (Lefsky et al. 2002). Several water properties are now routinely measured as absolute values (water surface elevation, temperature, surface sediment concentration, and algal concentration) with these methods. Additionally, new analyses of both passive and active radar data have led to measurements of inundation and wetness that are providing valuable insights into flooding dynamics and their effects on riverine landscapes. LIDAR is now an

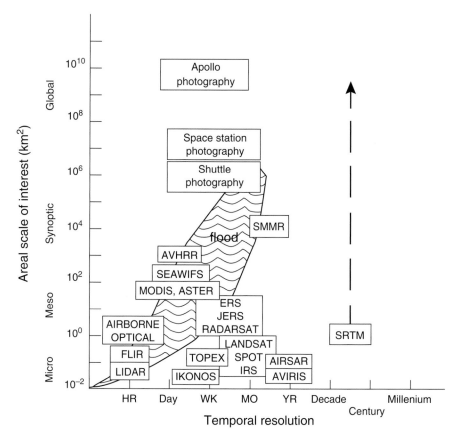

Figure 11.4 Spatial and temporal resolution of remote-sensing instruments. The abscissa shows the temporal resolution (i.e., the frequency of acquisition) of image data at the finest resolution for the listed instruments (see Table 11.4 for instrument names). The abscissa also represents the flood duration for catchments at the scales shown on the ordinate. The ordinate scale also represents the finest spatial resolution of the picture element (pixel) for each of the detectors. The arrow associated with SRTM suggests that although data were collected at finer than a km^2 of resolution, the product is globally comprehensive (from Mertes 2002).

extremely accurate tool for measuring topography, vegetation height, and cover, as well as more complex attributes of canopy structure and function, including leaf area index and aboveground biomass in three dimensions. Additionally, other types of satellite-derived data have proven useful in tracking changes in riparian land use and land cover (Apan et al. 2002).

An exciting recent development is the availability of Ikonos imagery, which is the first widely available digital *satellite* imagery with global coverage to have a high enough resolution to resolve riparian features (1 m panchromatic data and 4 m multispectral data). Ikonos data are enabling researchers to remotely map detailed vegetation communities in wetlands (Dechka et al. 2002) and to closely monitor changes in forest structure—detecting individual trees (Read et al. 2003). In remote areas where high-resolution aerial photography is unavailable, Ikonos data may also prove useful in delineating patterns of soil drainage (Peng et al. 2003).

From a ground-based perspective, Global Positioning Systems (GPS) increasingly are being used to understand the effects of flooding on the geomorphic composition

Wetness index at Lagarto river and lagoons

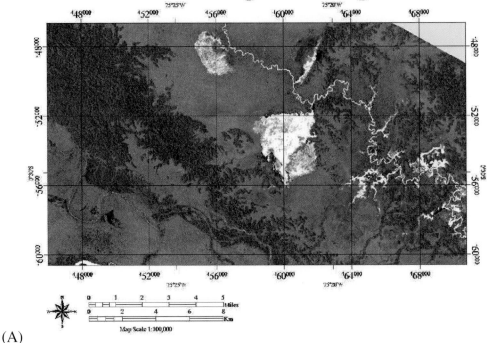

(A)

Greeness at Lagarto river and lagoons

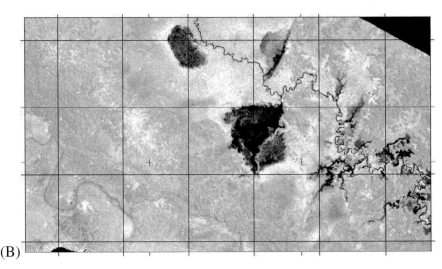

(B)

Figure 11.5 Satellite imagery is proving to be a valuable tool in the detection and mapping of riparia. It is especially useful in large-scale analyses and investigations of remote systems. These figures illustrate various techniques for visualizing and classifying aquatic and terrestrial systems using Landsat 7 ETM+ data. The study area is the Cuyabeno Wildlife Reserve in the Ecuadorian Amazon, A remote system of rivers and wetlands. Results of image analyses were checked by field visits to accessible portions of the reserve. Images (A) and (B) show wetness and greenness indices derived from Kauth-Thomas, or Tasseled Cap, transformation, a linear transformation of Landsat spectral bands into representative components. (C) Shows the soil fraction image derived for the same area using Spectral Mixture Analysis, in which boundaries between flooded and dry areas are better delineated (note enhanced dendritic pattern). (D) Shows the vegetation fraction image derived using Spectral Mixture Analysis. (E) Shows the classification obtained from combining three Kauth-Thomas indexes (brightness, wetness, and greeness). (F) Shows habitat classification obtained from the Spectral Mixture Analysis fraction images. This technique models individual pixel spectra as mixtures of spectrally pure endmembers. Different mixtures reflect different habitats or ecosystem types Images courtesy of A. Rosselli (2002).

Soil fraction at Lagarto river and lagoons

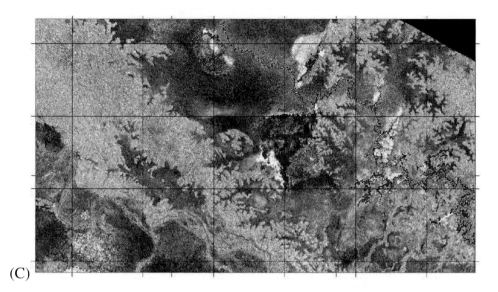

(C)

Vegetation fraction at Lagarto river and lagoons

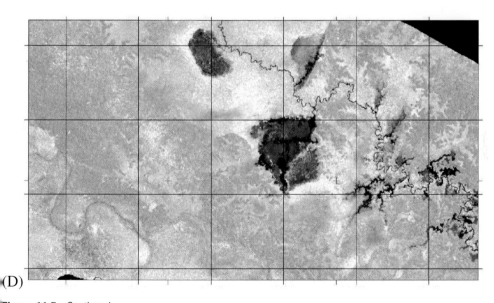

(D)

Figure 11.5 *Continued*

of rivers, the configuration of riparian patches, and their rate of turnover (Arscott et al. 2002). Using a GPS, major riparian features (water, gravel, vegetated patches and islands, and banks of the active floodplain) can be accurately mapped on the ground at different levels of discharge. Spatial data from the resulting maps can be analyzed using either MapInfo Professional (MapInfo Corporation, Troy, New York) or ArcView GIS (Environmental Systems Research Institute, Inc., Redlands, California).

Supervised classification of the brightness, greenness, and wetness derived from the tasseled cap transformation

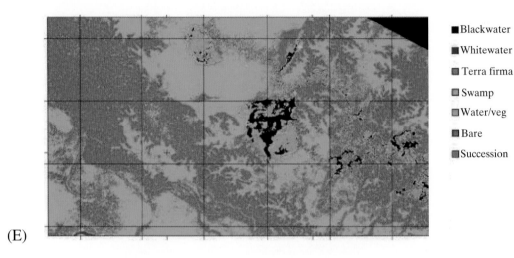

(E)

Classification of soil, vegetation, and water fraction images from feature space

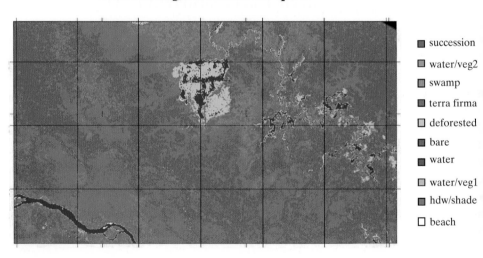

(F)

Figure 11.5 *Continued*

This approach can be effectively employed to better quantify the shifting mosaic steady-state model so often applied to floodplains.

Belowground

This is the most difficult environment to investigate without destroying or modify-ing the natural integrity of the sediment, soil, roots, and associated organisms and processes. Rapidly emerging geophysical methods, such as Ground Penetrating RADAR (GPR), have promise for profiling the textures and layering of riparian sediments to a depth of 10 to 20 m over extended areas. The principle of the GPR technique is similar to seismic and sonar reflection. A transmitting antenna emits

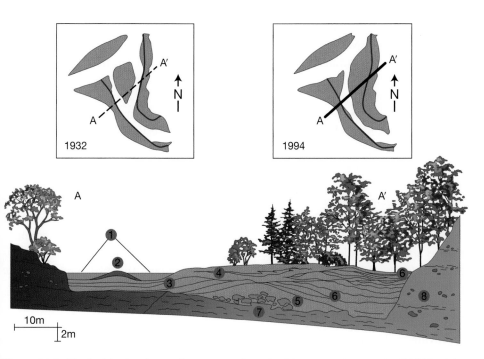

Figure 11.6 The fluvial subsurface environment, as determined from ground-penetrating RADAR, of the alluvial corridor of the lower Necker River, Switzerland. (1) Water surface, (2) a channel point bar, (3) point bar lateral accretion, (4) sand and intercalation of organic debris, (5) boulders forming the core of an ancient bar, (6) fills of former channel, (7) bedrock, (8) hillslope. From Naegeli et al. 1996.

short pulses of high-frequency electromagnetic energy into the ground. According to theory, reflections occur when the emitted energy waves meet boundaries between lithological units of contrasting dielectric properties. The total dielectric properties of the sediments are the sum of the volume fractions of gravel, water, and air times the respective dielectric properties. When the dielectric properties of gravel and pore water are constant in the saturated zone, differences in total dielectric properties are due to changes in the relative amount of pore water. In other words, the change in water saturation is strongly correlated to changes in porosity due to changes in sedimentary textures.

Although not widely employed because of equipment costs and the technical expertise needed, GPR has been applied effectively to mapping floodplain sediment textures (e.g., Naegeli et al. 1996, Poole et al. 1997). GPR is most suitable for deposits of low electrical conductivity such as gravel, which are the dominant sediments in many mountain rivers. In such environments, high-resolution continuous GPR profiling, used in combination with advanced data processing, can yield detailed information on the structure of fluvial gravel deposits (see Figure 11.6). This is a strong advancement over trench plots (e.g., Nanson et al. 1995), although both approaches have advantages.

Perhaps the most intellectually exciting and least known frontier in belowground riparian ecology is the investigation of root dynamics. As discussed in Chapter 5, root systems are important for anchorage and resource acquisition and may sequester more than half the carbon annually fixed by plants (Vogt et al. 1986). However, little is known about the mechanisms and timing of how nutrients reach riparian trees, how they move great distances laterally from the stream, and how quickly they contribute to growth. Mycorrhizae, a symbiotic association between fungi and plant roots (Allen 1991, Staddon et al. 2003), may serve an important intermediary in the transfer of nutrients from stream and groundwater to riparian vegetation. Transfer rates by mycorrhizal–root associations are unknown—and the mycorrhizal–root flux is a new research avenue for riparian ecologists. However, due to the inaccessibility of root systems, special techniques are required to investigate the standing stock, distribution, and turnover of roots, as well as to construct belowground elemental budgets. Traditionally, destructive techniques such as soil coring, in-growth cores, whole root system excavation, and trenching have been used to investigate root processes, while nondestructive techniques including rhizotrons and minirhizotrons have also been used (e.g., Riedacker 1974, Mahoney and Rood 1991, Pregitzer and Friend 1996, Johnson et al. 2001). Understanding root processes in riparian communities is a special challenge because of the dynamic and variable nature of the sediments in terms of turnover, wetness, range in texture, and the relatively young age of soils.

General Techniques

Improving and emerging technologies with wide applications to both aboveground and belowground riparian systems relate to genetic differentiation of populations, measurement of biogeochemical cycles, the tracing of key nutrients through system components, and the use of statistical innovations and Geographic Information Systems (GIS) for organizing, displaying, and analyzing spatial data.

The emergence and availability of molecular genetics techniques are greatly expanding our understanding of reproductive interactions between riparian populations, such as for cottonwood. Microsatellites, a class of highly polymorphic DNA markers, are well suited for studies of intraspecific population structure because they exhibit extensive variability and codominant Mendelian inheritance (Wright and Bentzen 1994). Thus, polymorphic microsatellite loci are routinely used to estimate variation in allelic frequencies within and among populations (Bradshaw 1996).

Along with the laboratory techniques are a variety of new statistical and mathematical tools for data analysis. The Markov chain method can be used to test for departures from Hardy-Weinberg equilibrium (HWE) and to determine whether genotypes at one locus are independent from genotypes at another locus (i.e., genotypic linkage disequilibrium or GLD; where loci are not in linkage equilibrium) for each pair of loci within each population (Raymond and Rousset 1995). Log-likelihood (G)-based tests can be used to determine whether genotypic distributions differ among populations at each locus, over all loci among all populations, and between all population pairs (Goudet et al. 1996). Population F-statistics (F, θ, and f), measures of the inbreeding of individuals relative to the total population and subpopulation, respectively, can be calculated. F also indicates the reduction in heterozygosity of individuals attributable to genetic drift. Graphical representation of pair-wise genetic distance (i.e., θ) between each sampled population is easily generated using multidimensional scaling (Kruskal and Wish 1978). Finally, genetic properties such as allelic richness (R_S), genetic diversity (H_S), heterozygosity (H_O), and other parameters can be estimated for each locus and population. Collectively, the molecular genetic techniques promise to unveil the reproductive mysteries between natural populations separated by distance and topography, as well as where propagule movement increasingly is compromised by dams or land use.

Advances in the measurement of biogeochemical cycles are coming, not so much from instrumentation as from the evolution and rediscovery of ideas. Specifically, the two most important are the chemical stoichiometry underlying ecosystem processes and the interactions between major biogeochemical cycles. Stoichiometry addresses the laws of definite proportions and the conservation of mass and energy; in other words, the relative proportion of key elements. It has had a long history in ecology because organisms are made up multiple elements that co-occur in fixed proportions depending on the biomolecule in which they occur (Sterner and Elser 2002, Hessen 2004). Stoichiometry may also be applied to the proportions of elements in different, ecologically relevant elemental pools, such as labile and refractory soil organic matter, dissolved and particulate aquatic pools, and so forth. By knowing the elemental stoichiometries of organisms mediating biogeochemical cycles and monitoring changes in elemental pools, one can infer specific processes in riparian systems. Appreciating elemental stoichiometry, and applying it to all aspects riparian systems, promises to greatly influence how scientists and managers address—and visualize—patterns and processes in riparia. The second advancement—interactions of the major biogeochemical cycles—is closely allied to stoichiometry (Melillo et al. 2003). The spatial distribution, cycles, and interactions of C, H, O, N, P, S, and approximately 25 other elements sustain life on earth by underpinning fundamental ecosystem processes. Understanding the stoichiometry of the cycles, and how community assemblages, changing disturbance regimes, and management activities affect it, is one of the great challenges for science.

The use of stable isotopes to investigate biogeochemical transformations in ecosystems is increasing rapidly because of their utility as tracers of ecological processes (Peterson and Fry 1987, Lajtha and Michener 1994). The elements C, N, S, H, and O all have more than one stable isotope, the isotopic compositions of natural materials can be measured with great precision with a mass spectrometer, and the analyses are becoming cheaper as more and better instruments are available. Isotopic compositions change in predictable ways as elements cycle through the biosphere, with perhaps C, N, and S being the most useful for ecological research because of inherent differences in isotopic ratios between major biosphere components (e.g., oceans and land).

Stable isotopes record two kinds of information. Where physical and chemical reactions fractionate stable isotopes, the resulting isotopic distributions reflect reaction conditions (process information). Stable isotope distributions also record information about the origins of samples (source information). The source sets an isotopic baseline that can be subsequently shifted by isotopic fractionation as the element cycles through the system. Stable isotopes have been especially useful in riparian investigations in providing information on the sources and fates of nitrogen in shaping riparian communities (e.g., Bilby et al. 2003), understanding nitrogen fixation (Martinelli et al. 1992), identifying plant energy sources to floodplain organisms (Forsberg et al. 1993), and elucidating the control of nitrogen export from catchments via headwater streams (Peterson et al. 2001).

Geographic Information Systems (GISs) are now in wide use for storing and analyzing spatial data, and abundant information can be found in the general literature.

However, the power of GISs for riparian applications may lie in combination with other technical instruments. Readily available geospatial data sets, which can be easily interrogated with a GIS, have considerable potential for predicting stream water chemistry on the basis of riparian characteristics using simple empirical relationships and multivariate statistics. In relatively unmodified catchments, the parent material and geochemistry of the riparian zone, when combined with a simple hydrologic flow path model, can accurately predict stream water chemistry at a range of flows (Q_{95} to > Q_5; 5 to 95 percent flow probabilities) and spatial scales (1 to 1,000 km^2; Smart et al. 2001). Such relatively simple tools have many potential benefits, particularly for assessing and managing catchments where acidification, erosion, and loss of diversity are potential issues.

The combination of remote sensing (e.g., aerial images), GIS techniques, and GPR data permits a three-dimensional analysis of riparia. The patchy mosaic of physical surfaces and vegetation associated with riparian corridors and floodplains are easily delineated from aerial photos and ground surveys of relative elevations and the spatial data organized into a GIS. Combined with GPR surveys of belowground structure, the results can suggest spatial positions of preferential groundwater flow paths (buried abandoned streambeds) and, collectively, be used to develop quantitative, three-dimensional characterizations of the surface and subsurface geomorphology for the entire riparian zone or floodplain. G. Poole and his colleagues (2002) developed this unique approach, using it to formulate heuristic models of the influence of floodplain geomorphology on the spatio-temporal patterns of surface and groundwater flow and exchange under dynamic hydrologic regimes.

Principles for the Ecological Management of Riparia

A continuing management challenge is to conduct activities within a framework of ecological principles. Properly applied, this approach helps maintain ecological integrity (i.e., goods and services). In the last decade, several groups have synthesized available knowledge into principles that can be used to guide ecologically sound land and water resource development (see Table 11.3). These deliberations are an important starting point for shaping ecological understanding into effective, long-term management strategies. Even though many human interventions alter flow regimes, the modifications need not ignore or completely violate fundamental ecological principles. Basic guidelines for effective catchment management and land use have been outlined (Naiman et al. 1998a, Dale et al. 2000) and innovative approaches for managing freshwater systems have been articulated (Jackson et al. 2001, Naiman et al. 2002c). The means exist for managing rivers and their riparian systems where water resource structures—such as dams and water diversions—are already in place (Stanford et al. 1996). In combination, these and other guidelines offer powerful approaches for maintaining the vitality of riparian systems at the landscape and catchment scales, providing that the recommendations are implemented and that the results are monitored. It is no longer sufficient to do *good* and *useful* science; the science must be *used*. Therein lies the fundamental challenge—and a principle.

More specifically, there are ecological principles addressing riparian integrity as well as inherent biogeochemical cycles and diversity. Some basic principles related to the integrity of riparian communities are:

1. The flow regime determines the successional evolution of riparian communities and associated ecological processes.
2. The riparian corridor serves as a pathway for the redistribution of organic and inorganic material influencing the riparian assemblages.
3. The riparian system is a transition zone between land and water and is disproportionately species-rich when compared to surrounding systems (Nilsson and Svedmark 2002).

Translating these principles into management directives requires information on how much water a river needs, and when and how it is delivered (i.e., flow variables described by magnitude, frequency, timing, duration, and rate of change). It also requires information on how various organisms are affected by habitat fragmentation, especially in terms of dispersal. Finally, it requires information about how effects of hydrologic alterations vary between different types of riparian systems and within specific locations of the catchment.

Basic principles addressing the biogeochemical cycle of nitrogen in riparia are complementary to those formulated for riparian organisms. These principles are:

1. The mode of nitrogen delivery affects riparian functioning.
2. Increasing contact between water and organic-rich soils or sediments increases nitrogen retention and processing.
3. Floods and droughts are important natural events that strongly influence pathways of nitrogen cycling (Pinay et al. 2002, McClain et al. 2003).

The challenge is to understand how site-specific alterations to riparian water regimes affect the cycling of nitrogen and other important elements. This is a nontrivial task as there are significant issues related to cumulative impacts, to the appropriate scale of investigation, and to separating natural variation from human changes.

Riparian biodiversity also is constrained by a small set of biophysical principles closely related to the flow regime (see Figure 11.7). These are summarized as follows:

1. Flow is a major determinant of physical habitat, which in turn is a major determinant of biotic composition.
2. Riparian species have evolved life history strategies primarily in direct response to natural flow regimes.
3. Maintenance of natural patterns of longitudinal and lateral connectivity is essential to the viability of many riparian species.
4. Altering flow and sediment regimes facilitates the invasion and success of nonnative species (Bunn and Arthington 2002).

Despite growing recognition of the collective principles and their importance, decision makers have difficulties in implementation—often because of economic considerations—and ecologists still struggle to predict specific biotic responses to altered

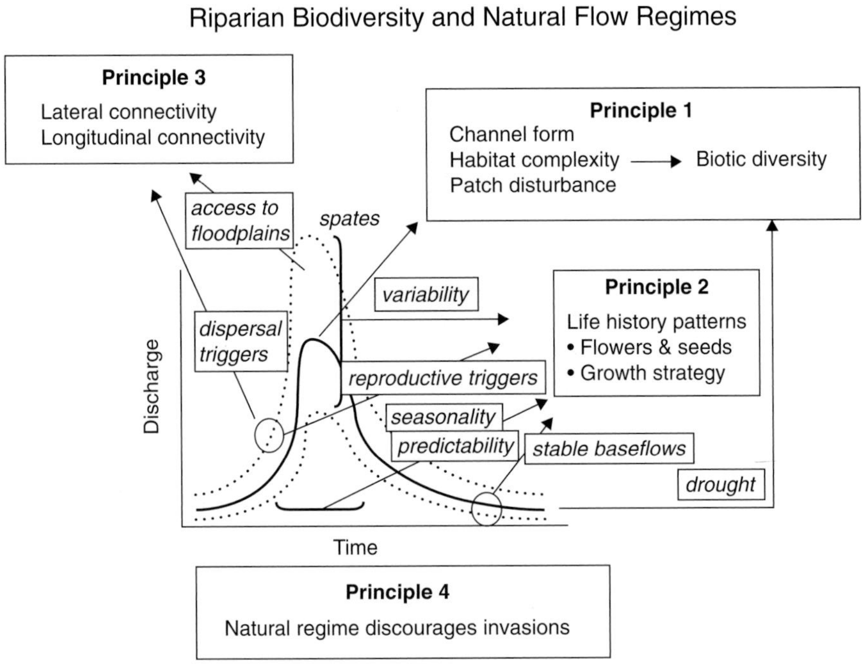

Riparian Biodiversity and Natural Flow Regimes

Figure 11.7 The natural flow regime influences riparian biodiversity via several interrelated mechanisms operating over different spatial and temporal scales. The relationship between biodiversity and the physical nature of the riparian habitat is driven primarily by large floods that shape channel form (Principle 1). However, droughts and base flows also play roles by limiting overall habitat availability. Many features of the flow regime influence life history patterns, especially the seasonality and predictability of the pattern (Principle 2). Some flows trigger longitudinal dispersal of riparian organisms and other large events inundate otherwise disconnected floodplain habitats (Principle 3). The native biota has evolved in response to the overall flow regime. Invasion by nonnative species are more likely to succeed at the expense of native biota if the former are better adapted to modified flow regimes (Principle 4). Modified from Bunn and Arthington 2002.

flow regimes. One obvious difficulty is the ability to distinguish the direct effects of modified flow from impacts associated with land use change that accompany water resource development, particularly in regions such as the Mediterranean where human cultures have followed one another for millennia (Décamps and Décamps 2002). All in all, it is painfully clear that riparian science needs to move quickly to a manipulative or experimental phase with the aim of restoration while measuring riparian system responses.

Global Environmental Change

The Earth's hydrologic cycle continues to undergo drastic alterations due to the combination of global climate change, growing population, and regional land use change which together have substantial cumulative effects on the global water system (see Chapter 7). Not surprisingly, the physical aspects of the global water cycle are best known while much less is known about water-mediated global biogeochemistry and still less about ecosystem states, biodiversity, and the social-economic impacts (GWSP 2004).

There is a high probability that global warming will modify the hydrologic cycle in the next few decades, with resulting changes in precipitation and evapotranspiration. The detection of such changes is complicated by additional effects from existing climate variability, such as fluctuations in oceanic conditions (e.g., Pacific Decadal Oscillation). In any case, there are recognizable regional trends in precipitation that, should they continue, will affect riparia over broad regions (Houghton et al. 2001). Additionally, extreme weather events such as typhoons and tropical depressions appear to have become more frequent and more intense (GWSP 2004), and their effects will certainly affect riparia.

Even though changes in surface runoff have been assessed only for individual regions, no uniform trends have been related to climate changes. However, runoff from many rivers has been heavily altered by dams, abstractions, transfers, and other water management activities, with deforestation possibly being the single most important pressure globally. Collectively, these alterations have already severely affected the global hydrologic cycle, and new large-scale water engineering schemes are planned or being implemented on all continents (Pearce 2003).

Anthropogenic loadings of biologically active elements, metals, and pollutants from point and nonpoint sources (e.g., agriculture, industry, cities, and long-range atmospheric deposition) have increased several-fold since the beginning of the industrial era, and will continue to increase. Globally, human-generated nitrogen currently totals about 210 Mt/yr of which agriculture accounts for some 86 percent, and this is expected to increase further in coming years (Seitzinger and Kroeze 1998, Galloway and Cowling 2002). Fortunately, only about a quarter of the additional nitrogen discharges into the coastal zone because of biotic retention, denitrification, and other riparian-related processes (Alexander et al. 2000). A significant challenge will be to model these transformations along the full continuum of landscape and fluvial systems (Billen and Garnier 2000). Additional water contaminants such as arsenic and fluoride also are emerging threats to water quality (WEHAB Working Group 2002), while the production of an estimated 1,500 km^3 of wastewater per year (WWDR 2003)—approximately a third of the volume of Lake Michigan—illustrates the increasing pressure on water quality. The vitality of aquatic and riparian ecosystems will continue to suffer from these pollutants, being the ultimate receptacles of much of the loads.

Trends in water temperature represent an additional issue because of temperature's effect on life history phenology, species distribution, and physiology (Gitay et al. 2001). There is good evidence for increasing water temperatures for European rivers, attributable to a mix of factors, in particular, increasing effluents, impoundments, and removal of riparian vegetation (Webb 1997). This is an issue for effective riparian management—both the proper species mix of riparian vegetation and the use of riparia for mitigating in-stream temperature regimes.

At the global scale, climate variability and land use change have brought about changes in water availability, which is demonstrated by water scarcity. As of 2002, 4 of every 10 people live in river basins experiencing water scarcity, and by 2025 at least 3.5 billion people will face water scarcity. The International Water Management Institute (IWMI; http://www.iwmi.org) has analyzed physical and economic water scarcity in various regions considering available water supplies and water demands by sector. These scarcity levels will be further exacerbated with anticipated changes in human population, changing food production systems, and intervention strategies for water resource utilization—with attendant and widespread effects on riparia.

The mechanisms of change for riparia will be as varied as the changes themselves. River fragmentation and land use changes continue to severely alter natural water regimes in quantity and timing through water abstraction and storage, channelization, river and habitat fragmentation, pollution, wetland drainage, and other drivers. Land use changes continue to manifest themselves through withdrawals of surface or subsurface water and consumption, which act in combination with socioeconomic and cultural practices at regional levels. In agriculture, humans are mobilizing large amounts of groundwater for irrigation, affecting baseflow conditions in many rivers. Globally, the impacts on freshwater and terrestrial ecosystems are being increasingly expressed in alterations to riparian habitat, diversity, structure, and functioning (Revenga et al. 2000, Poff et al. 2002).

Population growth and socioeconomic development are driving the rapid increase in water demand, especially from the industrial and household sectors. Humans use half of the runoff accessible globally (Postel 2000). Despite an average of over 6,000 m^3 of water available annually per capita, there are regions where water demand substantially exceeds renewable resources—and this leads to groundwater overuse (Postel 1999). Further, water pollution and a general lack of access to clean water have severe health consequences. As we saw in Chapter 7, 2.4 billion people are without adequate sanitation, while 1.2 billion people lack access to clean drinking water. It will be pragmatically difficult to maintain healthy riparian conditions under these human-related factors, even though good riparian management could significantly alleviate some conditions.

Collectively, global environmental changes will have extensive and long-lasting—if not permanent—effects on riparian systems. Fortunately, there are a number of international organizations, research groups, and private foundations attempting to address the more pressing issues through partnerships (see Table 11.5). Notable partnerships include the Global Water Systems Project (http://www.gwsp.org), the Millennium Ecosystem Assessment (http://www.millenniumassessment.org), and UNESCO's International Hydrologic Programme (http://www.unesco.org/water/ihp). Science examining the Earth's air, water, and land systems is generating abundant information from high-technology observing systems and modeling, either remotely sensed data or value-added products from climate, numerical weather prediction, or ecosystem model simulations. It has been estimated that by 2010, a constellation of 45 major planned satellites will provide 52 billion remotely sensed observations per day of continental land mass, oceans, and atmosphere; an increase of four orders of magnitude from the year 2000 (Lord 2001). Work is currently underway to assimilate data into numerical weather predictions, such as the LDAS (Land Data Assimilation System) administrated by the NOAA National Weather Service and the NASA in the United States, with many of the value-added data products being global in extent. Several real-time data sets are readily available, generated by the fusion of ground-based observations and model outputs including precipitation and other atmospheric variables necessary to estimate net water supply. A major theme of the global part-

Table 11.5 Key International Groups with Interests in Freshwater Systems and Their Internet Addresses

Acronym	Name	Internet Address
BAHC	Biospheric Aspects of the Hydrological Cycle	http://www.pik-postsdam.de/~bahc
ESSP	Earth System Science Partnership	http://www.ess-p.org
GWSP	Global Water System Project	http://www.gwsp.org
HELP	Hydrology for Environment, Life and Policy	www.nwl.ac.uk/ih/help/k/kl_recommendations.html
IGBP	International Geosphere Biosphere Programme	http://www.igbp.kva.se
IHDP	International Human Dimensions Programme on Global Environmental Change	http://www.ihdp.uni-bonn.de
IHP	International Hydrological Programme	http://www.unesco.org/water/ihp
IWMI	International Water Management Institute	http://www.iwmi.cgiar.org
SCOPE	Scientific Committee on Problems of the Environment	http://www.icsu-scope.org
UNESCO	United Nations Educational Scientific and Cultural Organization	http://www.unesco.org.water
WCD	World Commission on Dams	http://www.dams.org
WCRP	World Climate Research Programme	www.wmo.ch/web/wcrp
WEHAB	Water, Energy, Health, Agriculture and Biodiversity	http://www.johannesburgsummit.org/html/documents/wehab_papers.html

nerships is the use of this information in ecological and socioeconomic investigations, which fundamentally remain place-based as a series of regional case studies.

The emerging riparian-related challenges of the next decade are daunting in the face of changes in the global hydrologic system. Even though a water-related, global observation capability that can integrate environmental and human dimensions is largely lacking today, it may be available soon. The riparian-related challenges, however, can be met with the proper tools and the proper understanding of local riparia in the context of global changes. Developing an integrated observing capacity, consolidating data sets, working across disciplines, and having a set of indicators of water-related stress for riparia are key for resolving many of the issues.

Conclusions

This book has explored two major—and tightly linked—perspectives. First, the biophysical perspective examined how the physical environment is formed, how it sets up templates for the formation of the biotic riparian community, and the responses of the components and processes. Second, the cultural interventions examined the social context in which riparia exist, and how they might be conserved and restored or rehabilitated. Within a given climatic and geomorphic setting, it is fluvial dynamics, groundwater–surface water interactions, biogeochemical cycles, food webs, and cultural interventions that determine the characteristics of riparian systems.

In modern landscapes, however, the conservation and vitality of riparia depend on ecological and cultural sustainability as well (Décamps 2001). Ecological sustainability requires people to have knowledge about the environmental characteristics and functioning of riparian systems. Cultural sustainability demands people's attention and care toward these attributes. It is the interaction between ecological and cultural sustainabilities—indeed they are mutually supportive—that governs riparian management. The basic biophysical principles underpinning the origin and maintenance of healthy riparia will not change. In this context, human societies have the responsibility and power to make informed (or uninformed) choices about the future of riparia, all of which will have consequences for riparia and many of which will have important consequences for the human societies themselves.

Societies interpret their environment according to the way they manage it, and they manage their environment according to the way they interpret it.

—Berque (1994)

Bibliography

Abbe, T. B. and D. R. Montgomery. 1996. Large woody debris jams, channel hydraulics, and habitat formation in large rivers. *Regulated Rivers: Research & Management* 12:201–222.

Abbe, T. B. and D. R. Montgomery. 2003. Patterns and processes of wood debris accumulation in the Queets river basin, Washington. *Geomorphology* 51:81–107.

Abboud, M. 2002. Ibrahim River: A case study for investigating vegetation patterns and assessing riparian habitats. Masters Thesis, Faculty of Agricultural and Food Sciences, American University of Beirut, Lebanon.

Abboud, M. et al. 2004. Investigating vegetation patterns and riparian habitats: The Ibrahim River in Lebanon. University of Beirut, Lebanon.

Aber, J. D. and J. M. Melillo. 1981. Nitrogen immobilization in decaying hardwood leaf litter as a function of initial nitrogen and lignin content. *Canadian Journal of Botany* 60:2263–2269.

Aber, J. D., K. J. Nadelhoffer, P. Steudler, and J. M. Melillo. 1989. Nitrogen saturation in northern forest ecosystems. *BioScience* 39:378–386.

Abernethy, B. and I. D. Rutherfurd. 2000. Does the weight of riparian trees destabilize riverbanks? *Regulated Rivers* 16:565–576.

Acharya, C. N. 1935. Studies on the anaerobic decomposition of plant materials. III. Comparison of the course of decomposition of rice straw under anaerobic, aerobic and partially aerobic conditions. *Biochemical Journal* 29:1116–1120.

Acharya, G. 2000. Approaches to valuing the hidden hydrological services of wetland ecosystems. *Ecological Economics* 35:63–74.

Acreman, M. 1997. Principles of water management for people and the environment. IUCN (The World Conservation Union) Newsletter. *Wetlands Programme* 15:25–29.

Adamus, P. R. et al. 1991. Wetland Evaluation Technique (WET) Vol. I: Literature Review and Evaluation Rationale. WRP-DE-2. U.S. Army Corps of Engineers Waterways Experiment Station Vicksburg, Mississippi.

AFFA (Agriculture Forestry Fisheries Australia). 2001. Riparian management issues, sheets 1–7. www.rivers.go.au.

Alexander, M. 1977. Soil Microbiology. John Wiley & Sons: New York.

Alexander, R. B., R. A. Smith, and G. E. Schwarz. 2000. Effect of stream channel size on the delivery of nitrogen to the Gulf of Mexico. *Nature* 403:758–761.

Allanson, B. R., R. C. Hart, J. H. O'Keefe, and R. D. Roberts. 1990. Inland Waters of Southern Africa: An Ecological Perspective. Kluwer Academic Publishers: Dordrecht.

Allen, M. F. 1991. The Ecology of Mycorrhizae. Cambridge University Press: New York.

Ambroise, R., F. Bonneau, and V. Brunet-Vinck. 2000. Agriculteurs et Paysages, Dix Exemples de Projets de Paysage en Agriculture. Educagri: Dijon, France.

Amoros, C. and G. Bornette. 2002. Connectivity and biocomplexity in waterbodies of riverine floodplains. *Freshwater Biology* 47:761–776.

Amoros, C., J. C. Rostan, G. Pautou, and J. P. Bravard. 1987. The reversible process concept applied to the environmental management of large river systems. *Environmental Management* 11:607–617.

Anderson, D. C. and S. M. Nelson. 1999. Rodent use of anthropogenic and "natural" desert riparian habitat, lower Colorado River, Arizona. *Regulated Rivers: Research & Management* 15:377–393.

Angermeier, P. L. and J. R. Karr. 1983. Fish communities along environmental gradients in a system of tropical streams. *Environmental Biology of Fishes* 9:117–135.

Apan, A. A., S. R. Raine, and M. S. Paterson. 2002. Mapping and analysis of changes in the riparian landscape structure of the Lockyer Valley catchment, Queensland, Australia. *Landscape and Urban Planning* 59:43–57.

Aronson, J., C. Floret, E. Le Floc'h, C. Ovalle, and R. Pontanier. 1993. Restoration and rehabilitation of degraded ecosystems. I. A view from the south. *Restoration Ecology* 1:8–17.

Aronson, J. and J. van Andel. 2005. Challenges for ecological theory. *In* J. van Andel and J. Aronson, Eds. Restoration Ecology: A European Perspective. Blackwell Science: Oxford (in press).

Arora, K., S. K. Mickelson, J. L. Baker, D. P. Tierney, and C. J. Peter. 1996. Herbicide retention by vegetative buffer strips from runoff under natural rainfall. *Transactions of the American Society of Agricultural Engineers* 39:2155–2162.

Arscott, D. B., K. Tockner, D. van der Nat, and J. V. Ward. 2002. Aquatic habitat dynamics along a braided alpine river system (Taglimento River, Northeast, Italy). *Ecosystems* 5:802–814.

Arthington, A. H. 1998. Comparative evaluation of environmental flow assessment techniques: Review of holistic methodologies. LWRRDC (Land and Water Resources Research and Development Corporation) Occasional Papers 26/28, Canberra, Australia.

Arthington, A. H. et al. 2000. Environmental flow requirements of the Brisbane River downstream from Wivenhoe Dam. South East Queensland Water Corporation Ltd. and Centre for Catchment and Instream Research, Griffith University, Brisbane, Australia.

Arthington, A. H., S. O. Brizga, and M. J. Kennard. 1998. Comparative evaluation of environmental flow assessment techniques: Best practice framework. LWRRDC (Land and Water Resources Research and Development Corporation) Occasional Paper 25/98, Canberra, Australia.

Arthington, A. H. et al. 1992. Development of an holistic approach for assessing environmental flow requirements of riverine ecosystems. Pages 69–76 *in* J. J. Pigram and B. P. Hooper, Eds. Water Allocation for the Environment. Proceedings of an International Seminar and Workshop at the Centre for Water Policy Research, University of New England, Armidale, Australia.

Asner, G. P., T. R. Seastedt, and A. R. Townsend. 1997. The decoupling of terrestrial carbon and nitrogen cycles. *BioScience* 47:226–234.

Auble, G. T., J. M. Friedman, and M. L. Scott. 1994. Relating riparian vegetation to present and future streamflows. *Ecological Applications* 4:544–554.

Baker, J. P. et al. 2004. Alternative futures for the Willamette River basin, Oregon. *Ecological Applications* 14:313–324.

Baker, W. L. 1989. Classification of the riparian vegetation of the montane and sub-alpine zones in western Colorado. *Great Basin Naturalist* 9:214–228.

Baker, W. L. 1990. Species richness of Colorado riparian vegetation. *Journal of Vegetation Science* 1:119–124.

Baker, W. L. and G. M. Walford. 1995. Multiple stable states and models of riparian vegetation succession on the Animas River, Colorado. *Annuals of the Association of American Geographers* 85:320–338.

Bal, L. 1982. Zoological Ripening of Soils. Pudoc: Wageningen, The Netherlands.

Balian, E. 2001. Stem production dynamics of dominant riparian trees in the Queets River valley, Washington. Masters Thesis, University of Washington, Seattle.

Balian, E. and R. J. Naiman. 2005. Abundance and production of dominant riparian trees in the lowland floodplain of the Queets River, Washington. *Ecosystems* (in press).

Barbour, M. G., J. H. Burk, and W. A. Pitts. 1987. Terrestrial Plant Ecology. Benjamin/Cummings Publishing: Menlo Park, California.

Barnes, W. J. and E. Dibble. 1988. The effects of beaver in riverbank forest succession. *Canadian Journal of Botany* 66:40–44.

Baron, J. S. et al. 2002. Meeting ecological and societal needs for freshwater. *Ecological Applications* 12:1247–1260.

Barrington, M., D. Wolf, and K. Diebel. 2001. Analyzing riparian site capability and management options. *Journal of the American Water Resources Association* 37:1665–1679.

Barsoum, N. and F. M. R. Hughes. 1998. Regeneration response of black poplar to changing river levels. Pages 397–412 *in* Hydrology in a Changing Environment, Volume 1. British Hydrological Society and Wiley: Chichester, UK.

Barton, D. R., W. D. Taylor, and R. M. Biette. 1985. Dimensions of riparian buffer strips required to maintain trout habitat in southern Ontario streams. *North American Journal of Fisheries Management* 5:364–378.

Bartz, K. K. and R. J. Naiman. 2005. Effects of salmon-borne nutrients on riparian soils and vegetation in southwest Alaska. *Ecosystems* (in press).

Bayley, P. B. 1995. Understanding large river-floodplain ecosystems. *BioScience* 45:153–158.

Bazzaz, F. A. 1996. Plants in Changing Environments: Linking Physiological, Population and Community Ecology. Cambridge University Press: Cambridge.

Bechtold, J. S., R. T. Edwards, and R. J. Naiman. 2003. Biotic versus hydrologic control over seasonal nitrate leaching in a floodplain forest. *Biogeochemistry* 63:53–72.

Bechtold, J. S. and R. J. Naiman. 2005. Soil nitrogen mineralization potential across a riparian toposequence in a semi-arid savanna. *Soil Biology and Biogeochemistry* (in press).

Bedford, B. L. 1996. The need to define hydrologic equivalence at the landscape scale for freshwater wetland mitigation. *Ecological Applications* 6:57–68.

Beechie, T. J. and T. H. Sibley. 1997. Relationships between channel characteristics, woody debris, and fish habitat in northwestern Washington streams. *Transactions of the American Fisheries Society* 126:217–229.

Bégin, Y. and S. Payette. 1991. Population structure of lakeshore willows and ice-push events in subarctic Québec, Canada. *Holarctic Ecology* 14:9–17.

Bell, D. T., F. L. Johnson, and A. R. Gilmore. 1978. Dynamics of litter fall, decomposition, and incorporation in the streamside forest ecosystem. *Oikos* 30:76–82.

Bell, D., G. E. Petts, and J. P. Sadler. 1999. The distribution of spiders in the wooded riparian zone of three rivers in Western Europe. *Regulated Rivers: Research & Management* 15:141–158.

Belsky, A. J., A. Matzke, and S. Uselman. 1999. Survey of livestock influences on stream and riparian ecosystems in the western United States. *Journal of Soil and Water Conservation* 54:419–431.

Bencala, K. E. 1993. A perspective on stream-catchment connections. *Journal of North American Benthological Society* 12:44–47.

Benda, L. E., D. J. Miller, T. Dunne, G. H. Reeves, and J. K. Agee. 1998. Dynamic landscape systems. Pages 261–288 *in* R. J. Naiman and R. E. Bilby, Eds. River Ecology and Management: Lessons from the Pacific Coastal Ecoregion. Springer-Verlag: New York.

Benda, L. E. et al. 2004. The network dynamics hypothesis: How channel networks structure riverine habitats. *BioScience* 54:413–427.

Ben-David, M., T. A. Hanley, and S. M. Schell. 1998. Fertilization of terrestrial vegetation by spawning Pacific salmon: The role of flooding and predator activity. *Oikos* 83:47–55.

Bendix, J. and C. R. Hupp. 2000. Hydrological and geomorphological impacts on riparian plant communities. *Hydrological Processes* 14:2977–2990.

Benfield, E. F. 1997. Comparison of litterfall input to streams. *Journal of the North American Benthological Society* 16:104–108.

Benke, A. C. 1990. A perspective on America's vanishing streams. *Journal of the North American Benthological Society* 9:77–88.

Bennett, E. M. et al. 2003. Why global scenarios need ecology. *Frontiers in Ecology and the Environment* 1:322–329.

Berg, D. R. 1995. Riparian silvicultural design and assessment in the Pacific Northwest Cascade Mountains, USA. *Ecological Applications* 5:87–96.

Berkes, F. 1999. Sacred Ecology. Traditional Ecological Knowledge and Resource Management. Taylor & Francis: Philadelphia, Pennsylvania.

Berkes, F. and C. Folke, Eds. 1998. Linking Social and Ecological Systems: Management Practices and Social Mechanisms for Building Social Resilience. Cambridge University Press: Cambridge.

Bernhardt, E. S., G. E. Likens, D. C. Buso, and C. T. Driscoll. 2003. In stream uptake dampeas effects of major forest disturbance on watershed nitrogen export. *Proceedings of National Academy of Sciences* 100:10304–10308.

Berque, A., M. Conan, P. Donadieu, B. Lassus, and A. Roger. 1994. Cinq Propositions pour une Théorie du Paysage. Champ Vallon: Seyssel, France.

Berry, M., R. Flamm, B. Hazen, and R. MacIntyre. 1996. LUCAS: A system for modeling land-use change. *IEEE Computational Science and Engineering* 3:24–35.

Beschta, R. L. 1978. Long-term patterns of sediment production following road construction and logging in the Oregon coast range. *Water Resources Research* 14:1011–1016.

Beschta, R. L. 2003. Cottonwoods, elk, and wolves in the Lamar Valley of Yellowstone National Park. *Ecological Applications* 13:1295–1309.

Beschta R. L. and J. B. Kauffman. 2000. Restoration of riparian systems: Taking a broader view. Pages 323–328 *in* J. P. J. Wigington, Jr., and R. L. Beschta, Eds. Riparian Ecology and Management in Multi-Land Use Watersheds. American Water Resources Association: Middleburg, Virginia.

Biggs, B. J. F. et al. 1990. Ecological characterization, classification, and modeling of New Zealand rivers: An introduction and synthesis. *New Zealand Journal of Marine and Freshwater Research* 24:277–304.

Bilby, R. E., E. W. Beach, B. R. Fransen, J. K. Walter, and P. A. Bisson. 2003. Transfer of nutrients from spawning salmon to riparian vegetation in western Washington. *Transactions of the American Fisheries Society* 132:733–745.

Bilby, R. E. and P. A. Bisson. 1998. Function and distribution of large woody debris. Pages 324–346 *in* R. J. Naiman and R. E. Bilby, Eds. River Ecology and Management: Lessons from the Pacific Coastal Ecoregion. Springer-Verlag: New York.

Bilby, R. E., B. R. Franson, and P. A. Bisson. 1996. Incorporation of nitrogen and carbon from spawning coho into the trophic system of small streams: Evidence from stable isotopes. *Canadian Journal of Fisheries and Aquatic Sciences* 53:164–173.

Billen, G. and J. Garnier. 2000. Nitrogen transfers through the Seine drainage network: A budget based on the application of the "Riverstrahler" model. *Hydrobiologia* 410:139–150.

Bisson, P. A., J. L. Nielson, R. A. Palmason, and L. E. Gore. 1982. A system of naming habitat types in small streams, with examples of habitat utilization by salmonids during low stream flow. Pages 62–73 *in* N. B. Armantrout, Ed., Acquisition and Utilization of Aquatic Habitat Information. American Fisheries Society: Portland, Oregon.

Black, P. E. 1996. Watershed Hydrology. Ann Arbor Press: Chelsea, Michigan.

Blom, C. W. P. M. et al. 1990. Adaptations to flooding in plants from river areas. *Aquatic Botany* 38:29–47.

Blom, C. W. P. M. and L. A. C. J. Voesenek. 1996. Flooding: The survival strategies of plants. *Trends in Ecology and Evolution* 11:290–295.

Blum, M. D., M. J. Guccione, D. Wysocki, and P. C. Robnett. 2000. Late Pleistocene evolution of the Mississippi Valley, Southern Missouri to Arkansas. *Geological Society of America Bulletin* 112:221–235.

Boggs, K. and T. Weaver. 1994. Changes in vegetation and nutrient pools during riparian succession. *Wetlands* 14:98–109.

Bokhari, U. G. and J. S. Singh. 1974. Effects of temperature and clipping on growth, carbohydrate reserves and root exudation of western wheatgrass in hydroponic culture. *Crop Science* 14:790–794.

Boon, P. J., N. T. H. Holmes, P. S. Maitland, T. A. Rowell, and J. Davies. 1997. A system for evaluating rivers for conservation (SERCON): Development, structure and function. Pages 299–326 *in* P. J. Boon, and D. L. Howell, Eds. Freshwater Quality: Defining the Indefinable. Her Majesty's Stationary Office: Edinburg, United Kingdom.

Boon, P. J., J. Wilkenson, and J. Martin. 1998. The application of SERCON (System for Evaluating Rivers for Conservation) to a selection of rivers in Britain. *Aquatic Conservation* 8:597–616.

Bormann, F. H. and G. E. Likens. 1967. Nutrient cycling. *Science* 155:424–429.

Bormann, F. H., G. E. Likens, D. W. Fisher, and R. S. Pierce. 1968. Nutrient loss accelerated by clear-cutting of a forest ecosystem. *Science* 159:882–884.

Boulton, A. J. 1999. An overview of river health assessment: Philosophies, practice, problems and prognosis. *Freshwater Biology* 41:469–479.

Boyer, E. W., C. L. Goodale, N. A. Jaworski, and R. W. Howarth. 2002. Anthropogenic nitrogen sources and relationships to riverine nitrogen export in the northeastern USA. *Biogeochemistry* 57/58:137–169.

Braatne, J. H., S. B. Rood, and P. E. Heilman. 1996. Life history, ecology, and conservation of riparian cottonwoods in North America. Pages 57–85 *in* R. F. Stettler, H. D. Bradshaw, Jr., P. E. Heilman, and T. M. Hinckley, Eds. Biology of *Populus* and Its Implications for Management and Conservation. NRC Research Press, National Research Council of Canada: Ottawa, Ontario.

Bradshaw, H. D., Jr. 1996. Molecular genetics of *Populus*. Pages 183–199 *in* R. F. Stettler, H. D. Bradshaw, Jr., P. E. Heilman, and T. M. Hinckley, Eds. Biology of *Populus* and Its Implications for Management and Conservation. NRC Research Press, National Research Council of Canada: Ottawa, Ontario, Canada.

Bravard, G., C. Amoros, and G. Pautou. 1986. Impact of civil engineering works on the successions of communities in a fluvial system. A methodological and predictive approach applied to a section of the Upper Rhône River, France. *Oikos* 47:92–111.

Bray, J. R. and E. Gorham. 1964. Litter production in forests of the world. *Advances in Ecological Research* 2:101–157.

Briggs, M. K., M. K. Schmid, and W. L. Halvorson. 1997. Monitoring riparian ecosystems: An inventory of habitat along Rincon Creek near Tucson, AZ. Technical Report No. 58. US Geological Survey: Tucson, Arizona.

Brinson, M. M. 1990. Riverine forests. Pages 87–141 *in* A. E. Lugo, M. M. Brinson, and S. Brown, Eds. Forested Wetlands. Elsevier Science: Amsterdam, The Netherlands.

Brinson, M. M. 1993. A hydrogeomorphic classification for wetlands. Wetlands Research Program Technical Report WRP-DE-4. Final Report, U.S. Army Corps of Engineers Waterways Experiment Station, Vicksburg, Mississippi.

Brinson, M. M. 1996. Assessing wetland functions using HGM. *National Wetlands Newsletter* 18:10–16.

Brinson, M. M., B. L. Swift, R. C. Plantico, and J. S. Barclay. 1981. Riparian Ecosystems: Their Ecology and Status. FWS/OBS-81/17. Washington, DC: US Department of Interior, Fish and Wildlife Service, Biological Sciences Program.

Brizga, S. O. 1998. Methods addressing flow requirements for geomorphological purposes. Pages 8–46 *in* A. H. Arthington, and J. M. Zalucki, Eds. Comparative

Evaluation of Environmental Flow Assessment Techniques: Review of Methods. LWRRDC (Land and Water Resources Research and Development Corporation) Occasional Paper 27/98, Canberra, Australia.

Bromley, J., J. Brouwer, A. P. Barker, S. R. Gaze, and C. Valentin. 1997. The role of surface water redistribution in an area of patterned vegetation in a semi-arid environment, south-west Niger. *Journal of Hydrology* 198:1–29.

Brooks, K., P. Ffolliott, H. Gregersen, and L. DeBano. 1997. Hydrology and the Management of Watersheds. Iowa State University Press: Ames, Iowa.

Brosofske, K. D., J. Chen, R. J. Naiman, and J. F. Franklin. 1997. Effects of harvesting on microclimate from small streams to uplands in western Washington. *Ecological Applications* 7:1188–1200.

Brown, G. W. and J. T. Krygier. 1971. Clear-cut logging and sediment production in the Oregon coast range. *Water Resources Research* 7:1189–1198.

Brown, K. 2003. Integrating conservation and development: A case of institutional misfit. *Frontiers in Ecology and Environment* 1:479–487.

Brown, T. C. and T. C. Daniel. 1991. Landscape aesthetics of riparian environments: Relationship of flow quantity to scenic quality along a wild and scenic river. *Water Resources Research* 27:1787–1795.

Brussaard, L. et al. 1997. Biodiversity and ecosystem functioning in soil. *Ambio* 26:563–570.

Buijse, A. D. et al. 2002. Restoration strategies for river floodplains along large lowland rivers in Europe. *Freshwater Biology* 47:889–907.

Bunn, S. E. and A. H. Arthington. 2002. Basic principles and ecological consequences of altered flow regimes for aquatic biodiversity. *Environmental Management* 30:492–507.

Burgess, J., J. Clark, and C. M. Harrison. 2000. Knowledge in action: An actor network analysis of a wetland agri-environment scheme. *Ecological Economics* 35:35–45.

Burgess, S. S. O., M. A. Adams, N. C. Turner, and C. K. Ong. 1998. The redistribution of soil water by tree root systems. *Oecologia* 115:306–311.

Bush, D. E., N. L. Ingraham, and S. D. Smith. 1992. Water uptake in woody riparian phreatophytes of the southwestern United States: A stable isotope study. *Ecological Applications* 2:450–459.

Cadenasso, M. L. et al. 2003. An interdisciplinary and synthetic approach to ecological boundaries. *BioScience* 53:717–722.

Callicot, J. B. 1996. Do deconstructive ecology and sociobiology undermine Leopold's land ethic? *Environmental Ethics* 18:353–372.

Cals, M. J. R., R. Postma, A. D. Buijse, and E. C. L. Marteijn. 1998. Habitat restoration along the River Rhine in the Netherlands: Putting ideas into practice. *Marine and Freshwater Ecosystems* 8:61–70.

Carbiener, R. 1970. Un exemple de type forestier exceptionnel pour l'Europe occidentale: la forêt du lit majeur du Rhin au niveau du fossé rhénan. Intérêt écologique et biogéographique. Comparaison à d'autres forêts thermohygrophyles. *Vegetatio* 20:97–148.

Carbiener, R. 1983. Le Grand Ried d'Alsace. Ecologie et évolution d'une zone humide d'origine fluviale rhénane. *Bulletin d'Ecologie* 14:249–277.

Carbiener, R. and A. Schnitzler. 1990. Evolution of major pattern models and processes of alluvial forests of the Rhine in the rift valley (France/Germany). *Vegetatio* 88:115–129.

Carlyle, G. C. and A. R. Hill. 2001. Groundwater phosphate dynamics in a river riparian zone: Effects of hydrologic flowpaths, lithology, and redox chemistry. *Journal of Hydrology* 247:151–168.

Carpenter, S. R., S. G. Fisher, N. B. Grimm, and J. F. Kitchell. 1992. Global change and freshwater ecosystems. *Annual Review of Ecology and Systematics* 23: 119–139.

Carpenter, S. R. and M. G. Turner. 1998. At last: A journal devoted to ecosystem science. *Ecosystems* 1:1–5.

Case, T. J. 1990. Invasion resistance arises in strongly interacting species-rich model competition communities. *Proceedings of the National Academy of Sciences USA* 87:9610–9614.

Cattaneo, A., G. Salmoiraghi, and S. Gazzera. 1995. The rivers of Italy. Pages 479–505 *in* C. E. Cushing, K. W. Cummins, and G. W. Minshall, Eds. River and Stream Ecosystems. Elsevier Science: Amsterdam, The Netherlands.

Catterall, C. P., S. D. Piper, S. E. Bunn, and J. M. Arthur. 2001. Flora and fauna assemblages vary with local topography in a subtropical eucalypt forest. *Austral Ecology* 26:56–69.

Cech, T. V. 2003. Principles of Water Resources. History, Development, Management, and Policy. John Wiley & Sons: New York.

Cederholm, C. J., D. B. Houston, D. I. Cole, and W. J. Scarlett. 1989. Fate of coho salmon (*Oncorhynchus kisutch*) carcasses in spawning streams. *Canadian Journal of Fisheries and Aquatic Sciences* 46:1347–1355.

Chaubey, I., D. R. Edwards, T. C. Daniel, P. A. Moore, Jr., and D. J. Nichols. 1994. Effectiveness of vegetative filter strips in retaining surface-applied swine manure constituents. *Transactions of the American Society of Agricultural Engineers* 37:845–850.

Chen, J. et al. 1999. Microclimate in forest ecosystem and landscape ecology. *BioScience* 49:288–297.

Chestnut, T. J. and W. H. McDowell. 2000. C and N dynamics in the riparian and hyporheic zones of a tropical stream, Luquillo Mountains, Puerto Rico. *Journal of the North America Benthological Society* 19:199–214.

Chovanec, A. and R. Raab. 1997. Dragonflies (Insecta, Odonata) and the ecological status of newly created wetlands: Examples for long term bioindication programmes. *Limnologica* 27:381–392.

Christensen, B. 1992. Physical fractionation of soil and organic matter in primary particle size and density separates. *Advances in Soil Science* 20:1–90.

Christensen, N. L. et al. 1996. The report of the Ecological Society of America committee on the scientific basis for ecosystem management. *Ecological Applications* 6:665–691.

Church, M. 2002. Geomorphic thresholds in riverine landscapes. *Freshwater Biology* 47:541–557.

Cirmo, C. P. and J. J. McDonnell. 1997. Linking the hydrologic and biogeochemical controls of nitrogen transport in near-stream zones of temperate-forested catchments: A review. *Journal of Hydrology* 199:88–120.

Clark, D. A. et al. 2001a. Measuring net primary production in forests: Concepts and field methods. *Ecological Applications* 11:356–370.

Clark, D. A. et al. 2001b. Net primary production in tropical forests: An evaluation and synthesis of existing field data. *Ecological Applications* 11:371–384.

Clark, J. S. 2003. Uncertainty in ecological inference and forecasting. *Ecological Applications* 13:1349–1350.

Clark, J. S. et al. 2001. Ecological forecasts: An emerging imperative. *Science* 293:657–660.

Cleverly, J. R., S. D. Smith, A. Sala, and D. A. Devitt. 1997. Invasive capacity of *Tamarix ramosissima* in the Mojave Desert floodplain: The role of drought. *Oecologia* 111:12–18.

Clinton, S. M., R. T. Edwards, and R. J. Naiman. 2002. Subsurface metabolism and dissolved organic carbon dynamics in a floodplain terrace. *Journal of the American Water Resources Association* 38:619–631.

Cloe, W. W. and G. C. Garman. 1996. The energetic importance of terrestrial arthropod inputs to three warm-water streams. *Freshwater Biology* 36:105–114.

Clonts, H. A. and J. W. Malone. 1990. Preservation attitudes and consumer surplus in free-flowing rivers. Pages 301–315 *in* J. Vining, Ed. Social Science and Natural Resource Recreation Management. Westview Press: Boulder, Colorado.

Cohen, J. E. 1995. How Many People Can the Earth Support? W.W. Norton: New York.

Collier, K. J., S. Bury, and M. Gibbs. 2002. A stable isotope study of linkages between stream and terrestrial food webs through spider predation. *Freshwater Biology* 47:1651–1659.

Committee of Scientists. 1999. Sustaining the People's Lands: Recommendations for Stewardship of the National Forests and Grasslands into the Next Century. Report SD 426.U52, U.S. Department of Agriculture, Washington, DC.

Connell, J. H. 1978. Diversity in tropical rain forests and coral reefs. *Science* 199:1302–1310.

Connell, J. H. and W. P. Sousa. 1983. On the evidence needed to judge ecological stability or persistence. *American Naturalist* 121:798–824.

Conner, W. H., J. G. Gosselink, and R. T. Parrondo. 1981. Comparison of the vegetation of three Louisiana swamp sites with different flooding regimes. *American Journal of Botany* 68:320–331.

Conners, M. E. and R. J. Naiman. 1984. Particulate allochthonous inputs: Relationships with stream size in an undisturbed watershed. *Canadian Journal of Fisheries and Aquatic Sciences* 41:1473–1484.

Conquest, L. L. and S. C. Ralph. 1998. Statistical design and analysis considerations for monitoring and assessment. Pages 455–475 *in* R. J. Naiman, and R. E. Bilby, Eds. River Ecology and Management: Lessons from the Pacific Coastal Ecoregion. Springer-Verlag: New York.

Cooper, J. R., J. W. Gilliam, R. B. Daniels, and W. P. Robarge. 1987. Riparian areas as filters for agricultural sediment. *Journal of Soil Science Society America* 51:416–420.

Cordes, L. D., F. M. R. Hughes, and M. Getty. 1997. Factors affecting the regeneration and distribution of riparian woodlands along a northern prairie river: The Red Deer River, Alberta, Canada. *Journal of Biogeography* 24:675–695.

Correll, D. L. 1997. Buffer zones and water quality protection: General principles. Pages 7–20 *in* N. E. Haycock, T. P. Burt, K. W. T. Goulding, and G. Pinay, Eds. Buffer Zones: Their Processes and Potential in Water Protection. Quest Environmental: Harpenden, United Kingdom.

Costa, J. E. 1987. A comparison of the largest rainfall-runoff floods in the United States with those of the People's Republic of China and the world. *Journal of Hydrology* 96:101–115.

Costanza, R. and H. E. Daly. 1992. Natural capital and sustainable development. *Conservation Biology* 6:37–46.

Cowardin, L. M., V. Carter, F. C. Golet, and E. T. LaRoe. 1979. Classification of wetlands and deepwater habitats of the United States. FWS/OBS-79/31. U.S. Fish and Wildlife Service: Washington, DC.

Crockford, R. H. and D. P. Richardson. 2000. Partitioning of rainfall into throughfall, stemflow and interception: Effect of forest type, ground cover and climate. *Hydrological Processes* 14:2903–2920.

Croonquist, M. J. and R. P. Brooks. 1991. Use of avian and mammalian guilds as indicators of cumulative impacts in riparian-wetland areas. *Environmental Management* 15:701–714.

Cuffney, T. F. 1988. Input, movement and exchange of organic matter within a subtropical coastal blackwater river-floodplain ecosystem. *Freshwater Biology* 19:305–320.

Dahm, C. N. et al. 2002. Evapotranspiration at the land/water interface in a semi-arid drainage basin. *Freshwater Biology* 47:831–843.

Dahm, C. N., N. B. Grimm, P. Marmoier, H. M. Valett, and P. Vervier. 1998. Nutrient dynamics at the interface between surface waters and groundwaters. *Freshwater Biology* 40:427–451.

Dail, D. B., E. A. Davidson, and J. Chorover. 2001. Rapid abiotic transformation of nitrate in an acid forest soil. *Biogeochemistry* 54:131–146.

Daily, G. C., Ed. 1997. Nature's Services. Island Press: Washington, DC.

Dale, V. H. 1997. The relationship between land-use change and climate change. *Ecological Processes* 7:753–769.

Dale, V. H. et al. 2000. Ecological principles and guidelines for managing the use of land. *Ecological Applications* 10:639–670.

Dana, J. D. 1850. On denudation in the Pacific. *American Journal of Science* Series 2, 9:48–62.

Daniels, R. B. and J. W. Gilliam. 1996. Sediment and chemical load reduction by grass and riparian filters. *Soil Science Society of America Journal* 60:256–251.

Daniels, R. B. and J. W. Gilliam. 1997. Sediment and chemical load reduction by grass and riparian filters. *Soil Science Society of America Journal* 60:246–251.

Darby, S. 1999. Effect of riparian vegetation on flow resistance and flood potential. *Journal of Hydrological Engineering* 125:443–453.

Darveau, M., P. Beauchesne, L. Belanger, J. Huot, and P. Larue. 1995. Riparian forest strips as habitat for breeding birds in boreal forest. *Journal of Wildlife Management* 59:67–78.

Davidson, E. A. 1995. Spatial covariation of soil organic carbon, clay content, and drainage class at a regional scale. *Landscape Ecology* 10:349–362.

Davis, W. M. 1890. The rivers of northern New Jersey, with notes on the classification of rivers in general. *National Geographic Magazine* 2:82–110.

Dawson, T. E. and J. R. Ehleringer. 1991. Streamside trees that do not use stream water. *Nature* 350:335–337.

Dawson, T. E. and J. R. Ehleringer. 1993. Gender-specific physiology, carbon isotope discrimination, and habitat distribution in boxelder, *Acer negundo. Ecology* 74:798–815.

Dawson, T. E. and J. S. Pate. 1996. Seasonal water uptake and movement in root systems of phraeatophytic plants of dimorphic root morphology: A stable isotope investigation. *Oecologia* 107:13–20.

Décamps, H. 1984a. Biology of regulated rivers in France. Pages 495–514 *in* A. Lillehammer and S. J. Saltveit, Eds. Regulated Rivers. University of Oslo Press: Oslo, Norway.

Décamps, H. 1984b. Towards a landscape ecology of river valleys. Pages 163–178 *in* J. H. Cooley and F. G. Golley, Eds. Trends in Ecological Research for the 1980s. Plenum: New York.

Décamps, H. 1993. River margins and environmental change. *Ecological Applications* 3:441–445.

Décamps H. 1996. The renewal of floodplain forests along rivers: A landscape perspective. *Verhandlungen der Internationalen Vereinigung für Theoretische und Angewandte Limnologie* 26:35–59.

Décamps, H. 2001. How a riparian landscape finds form and comes alive. *Landscape and Urban Planning* 57:169–175.

Décamps, H. and O. Décamps. 2001. Mediterranean Riparian Woodlands. Medwet: La Tour du Valat, Le Sambuc, France.

Décamps, H., M. Fortuné, F. Gazelle, and G. Pautou. 1988. Historical influence of man on the riparian dynamics of a fluvial landscape. *Landscape Ecology* 1:163–173.

Décamps, H., J. Joachim, and J. Lauga. 1987. The importance for birds of the riparian woodlands within the alluvial corridor of the River Garonne, S.W. France. *Regulated Rivers: Research & Management* 1:301–316.

Décamps, H. and E. Tabacchi. 1994. Species richness in riparian vegetation along river margins. Pages 1–20 *in* P. S. Giller, A. G. Hildrew, and D. G. Rafaelli, Eds. Aquatic Ecology: Scale, Pattern and Process. Blackwell Scientific: London.

Décamps, H. et al. 2004. Riparian zones: Where biogeochemistry meets biodiversity in management practice. *Polish Journal of Ecology* 52:3–18.

Dechka, J., S. Franklin, M. Watmough, R. Bennett, and D. Ingstrup. 2002. Classification of wetland habitat and vegetation communities using multi-temporal Ikonos imagery in southern Saskatchewan. *Canadian Journal of Remote Sensing* 28:679–685.

DeFerrari, C. and R. J. Naiman. 1994. A multi-scale assessment of exotic plants on the Olympic Peninsula, Washington. *Journal of Vegetation Science* 5:247–258.

Dent, C. L. and N. B. Grimm. 1999. Spatial heterogeneity of stream water nutrient concentrations over successional time. *Ecology* 80:2283–2298.

Dent, C. L., N. B. Grimm, and S. G. Fisher. 2001. Multi-scale effects of surface–subsurface exchange on stream water nutrient concentrations. *Journal of the North American Benthological Society* 20:162–181.

Devol, A. H. and J. I. Hedges. 2001. Organic matter and nutrients in the mainstem Amazon River. Pages 275–306 *in* M. E. McClain, R. L. Victoria, and J. E. Richey, Eds. The Biogeochemistry of the Amazon Basin. Oxford University Press: Oxford.

Diamond, J. 1999. Guns, Germs, and Steel. W.W. Norton: New York.

Dickson, J. G., J. H.Williamson, R. N. Conner, and B. Ortego. 1995. Streamside zones and breeding birds in eastern Texas. *Wildlife Society Bulletin* 23:750–755.

Dietrich, W. E., C. J. Wilson, D. R. Montgomery, J. McKean, and R. Bauer. 1992. Channelization thresholds and land surface morphology. *Geology* 20:675–679.

Dillaha, T. A., R. B. Reneau, S. Mostaghimi, and D. Lee. 1989. Vegetative filter strips for agricultural nonpoint source pollution control. *Transactions of the American Society of Agricultural Engineers* 32:513–519.

Dixon, M. D. and W. C. Johnson. 1999. Riparian vegetation along the middle Snake River, Idaho: Zonation, geographical trends, and historical changes. *Great Basin Naturalist* 59:18–34.

Dixon, M. D., M. G. Turner, and C. Jin. 2002. Distribution of riparian tree seedlings on Wisconsin River sandbars: Controls at different spatial scales. *Ecological Monographs* 72:465–485.

Dobkin, D. S., A. C. Rich, and W. H. Pyle. 1998. Habitat and avifaunal recovery from livestock grazing in a riparian meadow system of the northwestern Great Basin. *Conservation Biology* 12:209–221.

Donahue, D. L. 1999. The Western Range Revisited: Removing Livestock from Public Lands to Conserve Native Biodiversity. University of Oklahoma Press: Norman, Oklahoma.

Downes, M. T., C. Howard-Williams, and L. A. Schipper. 1997. Long and short roads to riparian restoration: Nitrate removal efficiency. Pages 244–254 *in* N. E. Haycock, T. P. Burt, K. W. T. Goulding, and G. Pinay, Eds. Buffer Zones: Their Processes and Potential in Water Protection. Quest Environmental: Harpenden, United Kingdom.

du Toit, J. T., K. H. Rogers, and H. C. Biggs, Eds. 2003. The Kruger Experience. Island Press: Washington, DC.

Dudgeon, D. 2000. The ecology of tropical Asian streams in relation to biodiversity conservation. *Annual Review of Ecology and Systematics* 31:239–263.

Dumont, J. F., S. Lamotte, and F. Kahn. 1990. Wetland and upland forest ecosystems in the Peruvian Amazon: Plant species diversity in light of some geological and botanical evidence. *Forest Ecology and Management* 33/34:125–139.

Dunne, T. and L. Leopold. 1978. Water in Environmental Planning. W. H. Freeman and Company: New York.

Dunne, T., L. A. Mertes, R. H. Meade, J. E. Richey, and B. R. Forsberg. 1998. Exchanges of sediment between the flood plain and channel of Amazon River in Brazil. *Geological Society of America Bulletin* 110:450–467.

Dwire, K. A. and J. B. Kaufman. 2003. Fire and riparian ecosystems in landscapes of the western USA. *Forest Ecology and Management* 178:61–74.

Dynesius, M. and C. Nilsson. 1994. Fragmentation and flow regulation of river systems in the northern third of the world. *Science* 266:752–762.

Eaton, M. M. 1997. The beauty that requires health. Pages 85–106 *in* J. I. Nassauer, Ed. Placing Nature: Culture and Landscape Ecology. Island Press: Washington, DC.

Edwards, P. J. et al. 1999. A conceptual model of vegetation dynamics on gravel bars of a large Alpine river. *Wetlands Ecology and Management* 7:141–153.

Edwards, R. T. 1998. The hyporheic zone. Pages 399–429 *in* R. J. Naiman and R. E. Bilby, Eds. River Ecology and Management. Springer-Verlag: New York.

Ehleringer, J. R. and T. E. Dawson. 1992. Water uptake by plants: Perspectives from stable isotope composition. *Plant Cell Environment* 15:1073–1082.

Elliott, S. R., R. J. Naiman, and P. A. Bisson. 2004. The effect of riparian disturbance on physicochemical parameters of seston at summer baseflow. *Northwest Science* 78:150–157.

Ellis, L. M., M. C. Molles, Jr., and C. S. Crawford. 1999. Influence of experimental flooding on litter dynamics in a Rio Grande riparian forest, New Mexico. *Restoration Ecology* 7:193–204.

Elmore, W. 1992. Riparian responses to grazing practices. Pages 442–457 *in* R. J. Naiman, Ed. Watershed Management: Balancing Sustainability and Environmental Change. Springer-Verlag: New York.

Elwood, J. W., J. D. Newbold, A. F. Trimble, and R. W. Stark. 1981. The limiting role of phosphorus in a woodland stream ecosystem: Effects of P enrichment on leaf decomposition and primary producers. *Ecology* 62:146–158.

EU (European Union). 1992. Directive 92/43/EEC of the Council of May 12, 1992 on the conservation of natural habitats and of wild fauna and flora. Brussels, Belgium.

EU (European Union). 1999. Interpretation manual of European Union habitats. European Commission EUR 15/2. Brussels, Belgium.

EU (European Union). 2000. Directive 2000/60/EC of the European Parliament and of the Council of October 23, 2000 establishing a framework for community action in the field of water policy. Brussels, Belgium.

Everson, D. A. and D. H. Boucher. 1998. Tree species-richness and topographic complexity along the riparian edge of the Potomac River. *Forest Ecology and Management* 109:305–314.

Fabre, E. 1996. Relationships between aquatic hyphomycetes communities and riparian vegetation in three Pyrenean streams. *Comptes Rendu de l'Academie des Sciences Serie III Sciences de la Vie* 319:107–111.

Fagan, W. F., R. S. Cantrell, and C. Cosner. 1999. How habitat edges change species interactions. *The American Naturalist* 153:165–182.

Farina, A. and A. Belgrano. 2004. The eco-field: A new paradigm for landscape ecology. *Ecological Research* 19:107–110.

Farrier, D. 1995. Conserving biodiversity on private lands: Incentives for management or compensation for lost expectations? *Harvard Environmental Law Review* 19:303–408.

Federal Interagency Working Group. 1998. Stream Corridor Restoration. Principles, Processes, and Practices. National Technical Information Service, U.S. Department of Commerce: Washington, DC.

FEMAT (Forest Ecosystem Management Assessment Team). 1993. Forest ecosystem management: an ecological, economic, and social assessment. USDA Forest Service, U.S. Department of Commerce, National Oceanic and Atmospheric Administration, National Marine Fisheries Service, USDI Bureau of Land Management, Fish and Wildlife Service, National Park Service, Environmental Protection Agency: Portland, Oregon.

Fenchel, T. and T. H. Blackburn. 1979. Bacteria and Mineral Cycling. Academic Press: New York.

Fernald, A. G., P. J. Wigington, and D. H. Landers. 2001. Transient storage and hyporheic flow along the Willamette River, Oregon: Field measurements and model estimates. *Water Resources Research* 37:1681–1694.

Ferreira, L. V. and T. J. Stohlgren. 1999. Effects of river level fluctuation on plant species richness, diversity, and distribution in a floodplain forest in Central Amazonia. *Oecologia* 120:582–587.

Fetherston, K. L., R. J. Naiman, and R. E. Bilby. 1995. Large woody debris, physical process, and riparian forest development in montane river networks of the Pacific Northwest. *Journal of Geomorphology* 13:133–144.

Fielding, C. R. and J. Alexander. 2001. Fossil trees in ancient fluvial channel deposits: Evidence of seasonal and longer-term variability. *Palaeogeography, Palaeoclimatology, Palaeoecology* 170:59–80.

Findlay, S. 1995. Importance of surface-subsurface exchange in stream ecosystems. *Limnology and Oceanography* 40:159–164.

Firth, P. and S. G. Fisher, Eds. 1992. Global Climate Change and Freshwater Ecosystems. Springer-Verlag: New York.

Fisher, A. M. and D. C. Goldney. 1997. Use by birds of riparian vegetation in an extensively fragmented landscape. *Pacific Conservation Biology* 3:275–288.

Fitzpatrick, T. J. and M. F. L. Fitzpatrick. 1902. A study of the island flora of the Mississippi River near Sabula, Iowa. *Plant World* 5:198–201.

Flamm, R. O. and M. G. Turner. 1994a. Multidisciplinary modeling and GIS for landscape management. Pages 201–212 *in* V. A. Sample, Ed. Forest Ecosystem Management at the Landscape Level: The Role of Remote Sensing and Integrated GIS in Resource Management Planning, Analysis and Decision Making. Island Press: Washington, DC.

Flamm, R. O. and M. G. Turner. 1994b. Alternative model formulations of a stochastic model of landscape change. *Landscape Ecology* 9:37–46.

Fogel, R. 1985. Roots as primary producers in below-ground ecosystems. Pages 23–36 *in* A. H. Fitter, D. Atlinson, D. J. Read, and M. Usher, Eds. Ecological Interactions in Soil. British Ecological Society Special Publication 4. Blackwell: Oxford.

Folke, C. 1996. Conservation, driving forces, and institutions. *Ecological Applications* 6:370–372.

Fölster, J. 2000. The near-stream zone is a source of nitrogen in a Swedish forested catchment. *Journal of Environmental Quality* 29: 883–893.

Fonda, R. W. 1974. Forest succession in relation to river terrace development in Olympia National Park. *Ecology* 55:927–942.

Forman, R. T. T. 1995. Land Mosaics: The Ecology of Landscapes and Regions. Cambridge University Press: Cambridge.

Forman, R. T. T. and M. Gordon. 1981. Patches and structural components for a landscape ecology. *BioScience* 31:733–740.

Forman, R. T. T. and M. Godron. 1986. Landscape Ecology. John Wiley & Sons: New York.

Forsberg, B. R., C. A. R. M. Araujo-Lima, L. A. Martinelli, R. L. Victoria, and J. A. Bonassi. 1993. Autotrophic carbon sources for fish of the Central Amazon. *Ecology* 74:643–652.

Framenau, V. W., R. Manderbach, and M. Baehr. 2002. Riparian gravel banks of upland and lowland rivers in Victoria (south-east Australia): Arthropod community structure and life-history patterns along a longitudinal gradient. *Australian Journal of Zoology* 50:103–123.

Frederick, K. D. and P. H. Gleick. 1999. Water and global climate change: Potential impacts on U.S. water resources. Pew Center on Global Climate Change: Arlington, Virginia.

Friedman, J. M. and G. T. Auble. 2000. Floods, flood control, and bottomland vegetation. Pages 219–237 in E. E. Wohl, Ed. Inland Flood Hazards: Riparian and Aquatic Communities. Cambridge University Press: Cambridge.

Frissell, C. A., W. J. Liss, C. E. Warren, and M. D. Hurley. 1986. A hierarchical framework for stream habitat classification: Viewing streams in a watershed context. *Environmental Management* 10:199–214.

Frissell, C. A. and S. C. Ralph. 1998. Stream and watershed restoration. Pages 599–624 *in* R. J. Naiman, and R. E. Bilby, Eds. River Ecology and Management: Lessons from the Pacific Coastal Ecoregion. Springer: New York.

Fry, J., F. R. Steiner, and D. M. Green. 1994. Riparian evaluation and site assessment in Arizona. *Landscape and Urban Planning* 28:179–199.

Fryxell, J. M. 2001. Habitat suitability and source-sink dynamics of beavers. *Journal of Animal Ecology* 70:310–316.

Galloway, J. N. and E. B. Cowling. 2002. Reactive nitrogen and the world: 200 years of change. *Ambio* 31:64–71.

Gambrell, R. P., J. W. Gilliam, and S. B. Weed. 1975. Nitrogen losses from soils of the North Carolina coastal plain. *Journal of Environmental Quality* 4:317–323.

García-Gil, S. 1993. The fluvial architecture of the upper Buntsandstein in the Iberian Basin, central Spain. *Sedimentology* 40:125–143.

Gatto, M. and G. A. de Leo. 2000. Pricing biodiversity and ecosystem services: The never ending story. *BioScience* 50:347–355.

Gergel, S. E. and M. G. Turner. Eds. 2002 Learning Landscape Ecology. Springer-Verlag: New York.

Gerken, B. et al. 1991. Composition and distribution of Carabid communities along rivers and ponds in the region of Upper Weser (NW/NDS/FRG) with respect to protection and management of a floodplain ecosystem. *Regulated Rivers: Research & Management* 6:313–320.

Gessner, M. O. and E. Chauvet. 1994. Importance of stream microfungi in controlling breakdown rates of leaf litter. *Ecology* 75:1807–1817.

Ghaffarzadeh, M., C. A. Robinson, and R. M. Cruse. 1992. Vegetative filter strip effects on sediment deposition from overland flow. Agronomy abstracts, American Society of Agronomy: Madison, Wisconsin.

Gianessi, L. P. and C. M. Puffer. 1991. Herbicide use in the United States (revised April 1991). Resources for the Future: Washington, DC.

Gilbert, G. K. 1877. Report on the geology of the Henry Mountains. United States Geological and Geographic Survey, Rocky Mountain Region. U.S. Government Printing Office: Washington, DC.

Gilbert, G. K. 1914. The transportation of debris by running water. United States Geological Survey Professional Paper 86. U.S. Government Printing Office: Washington, DC.

Giller, P. S. and B. Malmqvist. 1998. The Biology of Stream and Rivers. Oxford University Press: New York.

Gilliam, J. W., R. B. Daniels, and J. F. Lutz. 1974. Nitrogen content of shallow ground water in the North Carolina coastal plain. *Journal of Environmental Quality* 3:147–151.

Gitay, H. et al. 2001. Ecosystems and their goods and services. Pages 235–342 *in* J. J. McCarthy, O. F. Canziani, N. A. Leary, D. J. Dokken, and K. S. White, Eds. Climate Change 2001: Impacts, Adaptations, and Vulnerability. Cambridge University Press: Cambridge.

Gleick, P. H., Ed. 1993. Water in Crisis. Oxford University Press: New York.

Gomez, D. M. and R. G. Anthony. 1998. Small mammal abundance in riparian and upland areas of five seral stages in Western Oregon. *Northwest Science* 72:293–302.

Gómez Peralta, D. 2000. Composición florística en el bosque ribereño de la Cuenca Alta San Alberto, Oxapampa—Perú. Ingeniero thesis. Facultad de Ciencias Forestales, Universidad Nacional Agraria La Molina: Lima, Peru.

Goodwin, C. N. 1999. Fluvial classification: Neanderthal necessity or needless normalcy? Pages 229–236 *in* D.S. Olson and J.P. Potyondy, Eds. Wildland Hydrology. American Water Resources Association: Middleburg, Virginia.

Goolsby, D. A., W. A. Battaglin, and E. M. Thurman. 1993. Occurrence and transport of agricultural chemicals in the Mississippi River basin, July through August 1993. United States Geological Survey Circular 1120-C, Denver, Colorado.

Gore, J. A. and F. D. Shields, Jr. 1995. Can large rivers be restored? *BioScience* 45:142–152.

Goudet, J., M. Raymond, T. De Meeus, and F. Rousset. 1996. Testing differentiation in diploid populations. *Genetics* 144:1933–1940.

Grace, J. B. 1999. The factors controlling species density in herbaceous plant communities: An assessment. *Perspectives in Plant Ecology, Evolution and Systematics* 2:1–28.

Graf, W. L. 1978. Fluvial adjustments to the spread of tamarisk in the Colorado plateau region. *Geological Society of America Bulletin* 98:1491–1501.

Graf, W. L. 1993. Landscapes, commodities, and ecosystems: The relationship between policy and science for American rivers. Pages 11–42. *in* Sustaining Our Water Resources. National Academy Press: Washington, DC.

Green, C. H. and S. M. Tunstall. 1992. The amenity and environmental value of river corridors in Britain. Pages 423–441 *in* P. J. Boon, P. Calow, and G. E. Petts, Eds. River Conservation and Management. John Wiley & Sons: Chichester, U.K.

Gregory, K. J. and R. J. Davis. 1993. The perception of riverscape aesthetics: An example from two Hampshire rivers. *Journal of Environmental Management* 39:171–185.

Gregory, S. V. 1997. Riparian management in the 21st century. Pages 69–85 *in* K. A. Kohm and J. F. Franklin, Eds. Creating a Forestry for the 21st Century: The Science of Ecosystem Management. Island Press: Washington, DC.

Gregory, S. V., K. Boyer, and A. Gurnell, Eds. 2003. The Ecology and Management of Wood in World Rivers. American Fisheries Society: Bethesda, Maryland.

Gregory, S. V., F. V. Swanson, W. A. McKee, and K. W. Cummins. 1991. An ecosystem perspective of riparian zones. *BioScience* 41:540–551.

Gren, I. M., K. H. Groth, and M. Sylvén. 1995. Economic values of Danube floodplains. *Journal of Environmental Management* 45:333–345.

Gresh, T. U., J. Lichatowich, and P. Schoonmaker. 2000. An estimation of historic and current levels of salmon production in the Northwest Pacific ecosystem: Evidence of a nutrient deficit in the freshwater systems of the Pacific Northwest. *Fisheries* 25:15–21.

Grette, G. B. 1985. The abundance and role of large organic debris in juvenile salmonid habitat in streams in second growth and unlogged forests. Masters Thesis, School of Fisheries, University of Washington, Seattle.

Gridley, N. and T. Prowse, Eds. 1993. Environmental Aspects of River Ice. National Hydrology Research Institute: Saskatoon, Canada.

Grier, C. C. and R. S. Logan. 1977. Old-growth *Pseudotsuga menziesii* communities of western Oregon watershed: Biomass distribution and production budgets. *Ecological Monographs* 47:373–400.

Grime, J. P. 1973. Competitive exclusion in herbaceous vegetation. *Nature* 242:344–347.

Grime, J. P. 1979. Plant Strategies and Vegetation Processes. J. Wiley & Sons: New York.

Groffman, P. M. et al. 2003. Down by the riverside: Urban riparian ecology. *Frontiers in Ecology and the Environment* 1:315–321.

Groffman, P. M. et al. 2002. Soil nitrogen cycle processes in urban riparian zones. *Environmental Science and Technology* 36:4547–4552.

Groisman, P. Y., R. W. Knight, and T. R. Karl. 2001. Heavy precipitation and high streamflow in the United States: Trends in the twentieth century. *Bulletin of the American Meteorological Society* 82:219–246.

Grove, A. T. and O. Rackham. 2001. The Nature of Mediterranean Europe: An Ecological History. Yale University Press: New Haven, Connecticut.

Gunderson, L. H. and C. S. Holling, Eds. 2002. Panarchy: Understanding Transformations in Human and Natural Systems. Island Press: Washington, DC.

Gurnell, A. M., P. Angold, and K. J. Gregory. 1994. Classification of river corridors: Issues to be addressed in developing an operational methodology. *Aquatic Conservation: Marine and Freshwater Ecosystems* 4:219–231.

Gurnell, A. M. and G. E. Petts. 2002. Island-dominated landscapes of large floodplain rivers: A European perspective. *Freshwater Biology* 47:581–600.

Gurnell, A. M. et al. 2001. Riparian vegetation and island formation along the gravel-bed Fiume Tagliamento, Italy. *Earth Surface Processes and Landforms* 26:31–62.

Gurnell, A. M. et al. 2000. Large wood retention in river channels: The case of the Fiume Tagliamento, Italy. *Earth Surface Processes and Landforms* 25:255–275.

GWSP (Framing Committee of the GWSP). 2004. The Global Water System Project: Science Framework and Implementation Activities. Earth System Science Partnership: Stockholm, Sweden (http://www.gwsp.org).

Hagan, P. T. 1996. Evaluating determinants of participation in voluntary riparian programs: A case study of Maryland's Buffer Incentive Program. Masters Thesis, University of Maryland, College Park, Maryland.

Hager, H. and H. Schume. 2001. The floodplain forests along the Austrian Danube. Pages 83–100 *in* E. Klimo, and H. Hager, Eds. The Floodplain Forests in Europe: Current Situation and Perspectives. Koninklijke Brill NV, Leiden, and European Forest Institute Research Report 10.

Hamilton, E. W. and D. A. Frank. 2001. Can plants stimulate soil microbes and their own nutrient supply? Evidence from a grazing tolerant grass. *Ecology* 82:2397–2402.

Hancock, C. N., P. G. Ladd, and R. H. Froend. 1996. Biodiversity and management of riparian vegetation in Western Australia. *Forest Ecology and Management* 85:239–250.

Hanley, T. A. and W. W. Brady. 1997. Understory species composition and production in old-growth western hemlock—Sitka spruce forests of southeastern Alaska. *Canadian Journal of Botany* 75:574–580.

Hanley, T. A. and T. Hoel. 1996. Species composition of old-growth and riparian Sitka spruce—Western hemlock forests in southeastern Alaska. *Canadian Journal of Forest Research* 26:1703–1708.

Hanson, G. C., P. M. Groffman, and A. J. Gold. 1994. Symptoms of nitrogen saturation in a riparian wetland. *Ecological Applications* 4:750–756.

Harding, J. S., E. F. Benfield, P. V. Bolstad, G. S. Helfman, and E. B. D. Jones III. 1998. Stream biodiversity: The ghost of land use past. *Proceedings of the National Academy of Sciences USA* 95:14843–14847.

Harmon, M. E. and J. F. Franklin. 1989. Tree seedlings on logs in *Picea-Tsuga* forests of Oregon and Washington. *Ecology* 70:48–59.

Harmon, M. E. et al. 1986. Ecology of coarse woody debris in temperate ecosystems. *Advances in Ecological Research* 15:133–302.

Harner, M. J. and J. A. Stanford. 2003. Differences in cottonwood growth between a losing and a gaining reach of an alluvial floodplain. *Ecology* 84:1453–1458.

Harper J. L. 1977. Population Biology of Plants. Academic Press: New York.

Harrington, W. A., A. J. Krupnick, and H. M. Peskin. 1985. Policies for nonpoint source water pollution control. *Journal of Soil and Water Conservation* 40:27–32.

Harris, R. and C. Olson. 1997. Two-stage system for prioritizing riparian restoration at the stream reach and community scales. *Restoration Ecology* 5:34–42.

Harris, R. R. 1988. Associations between stream valley geomorphology and riparian vegetation as a basis for landscape analysis in eastern Sierra Nevada, California, USA. *Environmental Management* 12:219–228.

Hartman, G. F. and J. C. Scrivener. 1990. Impacts of forestry practices on a coastal stream ecosystem, Carnation Creek, British Columbia. *Canadian Bulletin of Fisheries and Aquatic Sciences* 223:136–148.

Harvey, J. W. and K. E. Bencala. 1993. The effect of streambed topography on surface-subsurface water exchange in mountain catchments. *Water Resources Research* 29:89–98.

Harvey, J. W. and B. J. Wagner. 2000. Quantifying hydrologic interactions between streams and their subsurface hyporheic zones. Pages 3–44 *in* J. A. Jones and P. J. Mulholland, Eds. Streams and Ground Waters. Academic Press: San Diego.

Haslam, S. M. 1991. The Historic River. Cobden of Cambridge Press: Cambridge, United Kingdom.

Hauer, F. R., B. J. Cook, M. C. Gilbert, E. J. Clairain, Jr., and R. D. Smith. 2002. A regional guidebook for applying the hydrogeomorphic approach to assessing wetland functions of Riverine floodplains in the northern Rocky Mountains. Technical Report ERDC/EL TR-02-21. U.S. Army Corps of Engineers, Waterways Experiment Station: Vicksburg, Mississippi.

Hauer, F. R. and R. D. Smith. 1998. The hydrogeomorphic approach to functional assessment of riparian wetlands: Evaluating impacts and mitigation on river floodplains in the U.S.A. *Freshwater Biology* 40:517–530.

Haupt, H. F. and W. J. J. Kidd. 1965. Good logging practices reduce sedimentation in central Idaho. *Journal of Forestry* 63:664–670.

Hawkins, C. P. and J. R. Sedell. 1981. Longitudinal and seasonal changes in functional organization of macroinvertebrate communities in four Oregon streams. *Ecology* 62:387–397.

Hawkins, C. P. et al. 1993. A hierarchical approach to classifying stream habitat features. *Fisheries* 18:3–12.

Haycock, N. E., T. P. Burt, K. W. T. Goulding, and G. Pinay, Eds. 1997. Buffer Zones: Their Processes and Potential in Water Protection. Quest Environmental: Harpenden, United Kingdom.

Haycock, N. E. and G. Pinay. 1993. Groundwater nitrate dynamics in grass and popular vegetated riparian buffers during winter. *Journal of Environmental Quality* 22:273–278.

Haycock, N. E., G. Pinay, and C. Walker. 1993. Nitrogen retention in river corridors: European perspectives. *Ambio* 22:340–346.

Heal, O. W., P. W. Flanagan, D. D. French, and S. F. MacLean, Jr. 1981. Decomposition and accumulation of organic matter. Pages 587–633 *in* L. C. Bliss, O. W. Heal, and J. J. Moore, Eds. Tundra Ecosystems: A Comparative Analysis, Cambridge University Press: Cambridge.

Hedin, L. O., J. J. Armesto, and A. H. Johnson. 1995. Patterns of nutrient loss from unpolluted, old-growth temperate forests: Evaluation of biogeochemical theory. *Ecology* 76:493–509.

Hedin, L. O. et al. 1998. Thermodynamic constraints on nitrogen transformations and other biogeochemical processes at soil-stream interfaces. *Ecology* 79:684–703.

Heeg, J. and C. M. Breen. 1982. Man and the Pongolo Floodplain. Report 56, Council for Scientific and Industrial Research: Pretoria, South Africa.

Heilman, P. E., T. M. Hinckley, D. A. Roberts, and R. Ceulemans. 1996. Production physiology. Pages 450–489 *in* R. F. Stettler, H. D. Bradshaw, Jr., P. E. Heilman, and T. M. Hinckley, Eds. Biology of *Populus* and Its Implications for Management and Conservation. NRC Research Press: National Research Council of Canada, Ottawa, Ontario.

Helfield, J. M. and R. J. Naiman. 2001. Effects of salmon derived nitrogen on riparian forest growth and implications for stream productivity. *Ecology* 82:2403–2409.

Helfield, J. M. and R. J. Naiman. 2002. Salmon and alder as nitrogen sources to riparian forests in a boreal Alaskan watershed. *Oecologia* 133:573–582.

Helfield, J. M. and R. J. Naiman. 2005. Keystone interactions: Salmon and bear in riparian forests of Alaska. *Ecosystems* (in press).

Hershey, A. E. and G. A. Lamberti. 1998. Stream macroinvertebrate communities. Pages 169–199 *in* R. J. Naiman and R. E. Bilby, Eds. River Ecology and Management. Springer-Verlag: New York.

Hessen, D. O. 2004. Too much energy? *Ecology* 85:1177–1178.

Hewitt, M. J. 1990. Synoptic inventory of riparian ecosystems: The utility of Landsat Thematic Mapper data. *Forest Ecology and Management* 33–34:605–620.

Higler, L. W. G. 1993. The riparian community of north-west European streams. *Freshwater Biology* 29:229–241.

Hildebrand, G. V., T. A. Hanley, C. T. Robbins, and C. C. Schwartz. 1999. Role of brown bears (*Ursus arctos*) in the flow of marine nitrogen into a terrestrial ecosystem. *Oecologia* 121:546–550.

Hill, A. R. 1996. Nitrate removal in stream riparian zones. *Journal of Environmental Quality* 25:743–755.

Hill, A. R. 2000. Stream chemistry and riparian zones. Pages 83–110 *in* J. B. Jones and P. J. Mulholland, Eds. Streams and Ground Waters. Academic Press: San Diego, California.

Hill, A. R., K. J. Devito, S. Campagnolo, and K. Sanmugadas. 2000. Subsurface denitrification in a forest riparian zone: Interactions between hydrology and supplies of nitrate and organic carbon. *Biogeochemistry* 51:193–223.

Hill, A. R., C. F. Labadia, and K. Sanmugadas. 1998. Hyporheic zone hydrology and nitrogen dynamics in relation to the streambed topography of a N-rich stream. *Biogeochemistry* 42:285–310.

Hill, M., W. S. Platts, and R. L. Beschta. 1991. Ecological and geomorphological concepts for instream and out-of-channel flow requirements. *Rivers* 2:198–210.

Hillbricht-Ilkowska, A., L. Ryszkowski, and A. N. Sharpley. 1995. Phosphorus transfers and landscape structure: Riparian sites and diversified land use patterns. Pages 201–228 *in* H. Tiessen, Ed. Phosphorus in the Global Environment. John Wiley & Sons: New York.

Holland, M. M., P. G. Risser, and R. J. Naiman, Eds. 1991. The Role of Landscape Boundaries in the Management and Restoration of Changing Environments. Chapman and Hall: New York.

Holling, C. S. 1978. Adaptive Environmental Assessment and Management. University of British Columbia Press: Vancouver, Canada.

Holling, C. S. 1992. Cross-scale morphology, geometry and dynamics of ecosystems. *Ecological Monographs* 62:447–502.

Holling, C. S. 2001. Understanding the complexity of economic, ecological, and social systems. *Ecosystems* 4:390–405.

Holmes, T. P. 1988. The offsite impact of soil erosion on the water treatment industry. *Lands Economics* 64:356–367.

Hood, G. W. and R. J. Naiman. 2000. Vulnerability of riparian zones to invasion by exotic vascular plants. *Plant Ecology* 148:105–114.

Hornberger, G. M., J. P. Raffensperger, P. L. Wiberg, and K. N. Eshleman. 1998. Elements of Physical Hydrology. The Johns Hopkins University Press: Baltimore, Maryland.

Horner, R. R. and B. W. Mar. 1982. Guide for Water Quality Impact Assessment of Highway Operations and Maintenance. Report WA-RD-39.14, Washington Department of Transportation: Olympia, Washington.

Horton, R. E. 1945. Erosional development of streams and their drainage basins: Hydrophysical approach to quantitative morphology. *Bulletin of the Geological Society of America* 56:275–370.

Hosner, J. F., A. L. Leaf, R. Dickson and J. B. Hart, Jr. 1965. Effects of varying soil moisture on the nutrient uptake of four bottomland tree species. *Soil Science Society America Proceedings* 29:313–316.

Houghton, J. T. et al. 2001. Climate Change 2001: The Scientific Basis. Intergovernmental Panel on Climate Change: Working Group I. Cambridge University Press: Cambridge.

Hudson-Edwards, K. A., C. Schell, and M. G. Macklin. 1999. Mineralogy and geochemistry of alluvium contaminated by metal mining in the Rio Tinto area, southwest Spain. *Applied Geochemistry* 14:1015–1030.

Huet, M. 1949. Aperçu des relations entre la pente et les populations piscicoles des eaux courantes. *Schweizerische Zeitschrift für Hydrologie* 11:322–351.

Huet, M. 1954. Biologie, profils en long et en travers des eaux courants. *Bulletin Français de Pisciculture* 175:41–53.

Hughes, F. M. R. 1997. Floodplain biogeomorphology. *Progress in Physical Geography* 21:501–529.

Hughes, F. M. R. 2002. Can we manage rivers to benefit alluvial woodlands? Pages 187–196 *in* B. C. van Dam and S. Bordacs, Eds. Genetic Diversity in River Populations of European Black Poplar: Implications for Riparian Ecosystem Management. Proceedings of an International Symposium held in Szekszard, Hungary, May 16–20, 2001. Szekszard, Hungary.

Hughes, F. M. R., Ed. 2003. The Flooded Forest: Guidance for Policy Makers and River Managers in Europe on the Restoration of Floodplain Forests. FLOBAR2 (Floodplain Biodiversity and Restoration), Department of Geography, University of Cambridge, Cambridge.

Hughes, F. M. R. and B. Rood. 2003. Allocation of river flows for restoration floodplain forest ecosystems: A review of approaches and their applicability in Europe. *Environmental Management* 32:12–33.

Hughes, F. M. R. et al. 1997. Woody riparian species response to different soil moisture conditions: laboratory experiments on *Alnus incana* (L.) Moench. *Global Ecology and Biogeography Letters* 6:247–256.

Hulse, D., J. Eilers, K. Freemark, C. Hummon, and D. White. 2000. Planning alternative future landscapes in Oregon: Evaluating effects on water quality and biodiversity. *Landscape Journal* 19:1–19.

Hulse, D., S. V. Gregory, and J. Baker, Eds. 2002. Willamette River Basin Planning Atlas: Trajectories of Environmental and Ecological Change. Oregon University Press: Corvallis, Oregon.

Huntly, N. and R. Inouye. 1988. Pocket gophers in ecosystems: Patterns and mechanisms. *BioScience* 38:786–793.

Hupp, C. R. 1982. Stream-grade variation and riparian-forest ecology along Passage Creek, Virginia. *Bulletin of the Torrey Botanical Club* 109:488–499.

Hupp, C. R. and W. R. Osterkamp. 1985. Bottomland vegetation distribution along Passage Creek, Virginia, in relation to fluvial landforms. *Ecology* 66:670–681.

Hupp, C. R. and W. R. Osterkamp. 1996. Riparian vegetation and fluvial geomorphic processes. *Geomorphology* 14:277–295.

Huppert, D. and S. Kantor. 1998. Economic perspectives. Pages 572–595 *in* R. J. Naiman and R. E. Bilby, Eds. River Ecology and Management: Lessons from the Pacific Coastal Ecoregion. Springer-Verlag: New York.

Huston, M. A. 1979. A general hypothesis of species diversity. *The American Naturalist* 113:81–101.

Huston, M. A. 1994. Biological Diversity: The Coexistence of Species on Changing Landscapes. Cambridge University Press: Cambridge.

Hutchinson, H. B. and E. H. Richards. 1921. Artificial farmyard manure. *Journal of the Ministry of Agriculture (London)* 28:398–411.

Hyatt, T. L., T. Z. Waldo, and T. J. Beechie. 2004. A watershed scale assessment of riparian forests, with implications for restoration. *Restoration Ecology* 12: 175–183.

Hynes, H. B. N. 1970. The Ecology of Running Waters. University of Toronto Press: Toronto, Ontario, Canada.

Hynes, H. B. N. 1975. The stream and its valley. *Verhandlungen der Internationalen Vereinigung für Theoretische und Angewandte Limnologie* 19:1–15.

Hynes, H. B. N. 1983. Groundwater and stream ecology. *Hydrobiologia* 100:93–99.

Illies, J. and L. Botosaneanu. 1963. Problèmes et méthodes de la classification et de la zonation écologique des eaux courantes, considérées surtout du point de vue faunistique. *Mitteilungen der Internationalen Vereinigung für theoretische und angewandte Limnologie* 12:1–57.

Inamdar, S. P. et al. 1999. Riparian ecosystem management model (REMM): I. Testing of the hydrologic component for a coastal plain riparian system. *Transactions of the American Society of Agricultural Engineers* 42:1679–1689.

Innis, S. A. 2003. Water use in trees across a riparian-upland gradient in a semi-arid savanna landscape. Masters Thesis, University of Washington, Seattle.

Innis, S. A., R. J. Naiman, and S. R. Elliott. 2000. Indicators and assessment methods for measuring the ecological integrity of semi-aquatic terrestrial environments. *Hydrobiologia* 422/423:111–131.

Iowa State University. 1997. Stewards of Our Streams: Buffer Strip Design, Establishment, and Maintenance. Publication 1626b, Iowa University State Extension, Ames, Iowa.

Irons, J. G., M. W. Oswood, and J. P. Bryant. 1988. Consumption of leaf detritus by a stream shredder influence of tree species and nutrient status. *Hydrobiologia* 160:53–62.

Iversen, T. M. 1973. Decomposition of autumn-shed beech leaves in a springbrook and its significance for the fauna. *Archiv für Hydrobiologie* 72:305–312.

Jackson, J. B. 1994. A Sense of Place, A Sense of Time. Yale University Press: New Haven, Connecticut.

Jackson, J. K. and S. G. Fisher. 1986. Secondary production, emergence, and export of aquatic insects of a Sonoran desert stream. *Ecology* 67:629–638.

Jackson, R. B. et al. 2001. Water in a changing world. *Ecological Applications* 11:1027–1045.

Jacobs, T. C. and J. W. Gilliam. 1983. Nitrate loss from agricultural drainage waters: Implications for nonpoint source control. Report No. 209. Water Resources Research Institute: University of North Carolina, Raleigh.

James, L. A. 1991. Incision and morphologic evolution of an alluvial channel recovering from hydraulic mining sediment. *Geological Society of America Bulletin* 103:723–736.

Jansson, R., C. Nilsson, M. Dynesius, and E. Andersson. 2000a. Effects of river regulation on river-margin vegetation: a comparison of eight boreal forest rivers. *Ecological Applications* 10:203–224.

Jansson, R., C. Nilsson, and B. Renöfält. 2000b. Fragmentation of riparian floras in rivers with multiple dams. *Ecology* 81:899–903.

Jasanoff, S. et al. 1997. Conversations with the community: AAAS (American Association for the Advancement of Science) at the millennium. *Science* 278:2066–2067.

Johansson, M., C. Nilsson, and E. Nilsson. 1996. Do rivers function as corridors for plant dispersal? *Journal of Vegetation Science* 7:593–598.

Johnson, B. R. and K. Hill. 2002. Introduction: Toward landscape realism. Pages 1–26 *in* B. R. Johnson and K. Hill, Eds. Ecology and Design: Frameworks for Learning. Island Press: Washington, DC.

Johnson, M. G., D. T. Tingey, D. L. Phillips, and M. J. Storm. 2001. Advancing fine root research with minirhizotrons. *Environmental and Experimental Botany* 45:263–289.

Johnson, N., C. Revenga, and J. Echeverria. 2001. Managing water for people and nature. *Science* 292:1071–1072.

Johnson, R. L., R. J. Alig, E. Moore, and R. J. Moulton. 1997. NIPF: Landowner's view of regulation. *Journal of Forestry* 95:23–28.

Johnson, S. L. and J. A. Jones. 2000. Stream temperature responses to forest harvest and debris flows in western Cascades, Oregon. *Canadian Journal of Fisheries and Aquatic Sciences* 57 (Supplement 2):30–39.

Johnson, W. C. 1994. Woodland expansion in the Platte River, Nebraska: Patterns and causes. *Ecological Monographs* 64:45–84.

Johnson, W. C. 2002. Riparian vegetation diversity along regulated rivers: Contribution of novel and relict habitats. *Freshwater Biology* 47:749–759.

Johnston, C. A. and R. J. Naiman. 1990a. Aquatic patch creation in relation to beaver population trends. *Ecology* 71:1617–1621.

Johnston, C. A. and R. J. Naiman. 1990b. Browse selection by beaver: Effects on riparian forest composition. *Canadian Journal of Forest Research* 20:1036–1043.

Johnston, C. A., J. Pastor, and R. J. Naiman. 1993. Effects of beaver and moose on boreal forest landscapes. Pages 237–254 *in* R. Haines-Young, D. R. Green, and S. H. Cousin, Eds. Landscape Ecology and Geographic Information Systems. Taylor & Francis: London.

Johnston, N. T., J. S. MacDonald, K. J. Hall, and P. J. Tschaplinski. 1997. A preliminary study of the role of sockeye salmon (*Oncorhynchus nerka*) carcasses as carbon and nitrogen sources for benthic insects and fishes in the "Early Stuart" stock spawning streams, 1050 km from the Ocean. British Columbia Ministry of Environment, Lands and Parks, Victoria, British Columbia.

Jones, J. B. and R. M. Holmes. 1996. Surface-subsurface interactions in stream ecosystems. *Trends in Ecology and Evolution* 11:239–242.

Jonsson, B. G. 1997. Riparian bryophyte vegetation in the Cascade mountain range, Northwest USA: Patterns at different spatial scales. *Canadian Journal of Botany* 75:744–761.

Junk, W. J. 1989. Flood tolerance and tree distribution in the central Amazon floodplains. Pages 47–64 *in* L. B. Holme-Nielson, I. C. Nielson, and H. Basley, Eds. Tropical Forests: Botanical Dynamics, Speciation, and Diversity. Academic Press: Orlando, Florida.

Junk, W. J., Ed. 1997. The Central Amazon Floodplain: Ecology of a Pulsing System. Springer Verlag: Berlin.

Junk, W. J., P. B. Bayley, and R. E. Sparks. 1989. The flood pulse concept in river-floodplain systems. Pages 110–127 *in* D. P. Dodge, Ed. Proceedings of the International Large River Symposium. Canadian Special Publication of Fisheries and Aquatic Sciences 106.

Junk, W. J. and M. T. F. Piedade. 1997. Plant life in the floodplain with special reference to herbaceous plants. Pages 147–185 *in* W. J. Junk, Ed. The Central Amazon Floodplain. Ecological Studies. Springer-Verlag: Berlin.

Juracek, K. E. and F. A. Fitzpatrick. 2003. Limitations and implications of stream classification. *Journal of the American Water Resources Association* 39:659–670.

Kabat, P. et al., Eds. 2004. Vegetation, Water, Humans and the Climate: A New Perspective on an Interactive System. Springer-Verlag: Berlin and Heidelberg.

Kalff, J. 2002. Limnology. Prentice-Hall: Upper Saddle River, New Jersey.

Kalliola, R. and M. Puhakka. 1988. River dynamics and vegetation mosaicism: A case study of the River Kamajohka, northernmost Finland. *Journal of Biogeography* 15:703–719.

Karl, T. R. and R. W. Knight. 1998. Secular trends of precipitation amount, frequency, and intensity in the USA. *Bulletin of the American Meteorological Society* 79:231–241.

Karl, T. R., R. W. Knight, D. R. Easterling, and R. Q. Quayle. 1996. Indices of climate change for the United States. *Bulletin of the American Meteorological Society* 77:279–292.

Karr, J. R. 1996. Ecological integrity and ecological health are not the same. Pages 100–113 *in* P. C. Schulze, Ed. Engineering within Ecological Constraints. National Academy Press: Washington, DC.

Karr, J. R. 1999. Seeking Suitable Endpoints: Biological Monitoring for Wetland Assessment. Report to U.S. Environmental Protection Agency, Region 10, Seattle, Washington.

Karr, J. R. 2002. What from ecology is relevant to design and planning? Pages 133–164 *in* B. R. Johnson, and K. Hill, Eds. Ecology and Design: Frameworks for Learning. Island Press: Washington, DC.

Karr, J. R., and D. R. Dudley. 1981. Ecological perspectives on water quality goals. *Environmental Management* 5:55–68.

Karr, J. R. and I. J. Schlosser. 1978. Water resources and the land-water interface. *Science* 201:229–134.

Kattenberg, A. et al. 1996. Climate models: Projections of future climate. Pages 285–357 *in* J. T. Houghton, L. G. Meira Filho, B. A. Callander, N. Harris, A. Kattenberg, and K. Maskell, Eds. Climate Change 1995: The Science of Climate Change. Contribution of Working Group I to the Second Assessment Report of the Intergovernmental Panel on Climate Change. Cambridge University Press: Cambridge.

Kaushik, N. K. and H. B. N. Hynes. 1971. The fate of dead leaves that fall into streams. *Archiv für Hydrobiologie* 68:465–515.

Keddy, P. A. 1989. Competition, Population and Community Biology. Chapman and Hall: New York.

Keiter, R. B. 1998. Ecosystems and the law: Toward an integrated approach. *Ecological Applications* 8:332–341.

Keller, C. M. E., C. S. Robbins, and J. S. Hatfield. 1993. Avian communities in riparian forests of different widths in Maryland and Delaware. *Wetlands* 13:137–144.

Keller, E. A., and F. J. Swanson. 1979. Effects of large organic material on channel form and fluvial processes. *Earth Surface Processes* 4: 361–380.

Kellman, M. and R. Tackaberry. 1993. Disturbance and tree species coexistence in tropical riparian forest fragments. *Global Ecological and Biogeography Letters* 3:1–9.

Kelsey, K. A. and S. D. West. 1998. Riparian wildlife. Pages 235–258 *in* R. J. Naiman and R. E. Bilby, Eds. River Ecology and Management: Lessons from the Pacific Coastal Ecoregion. Springer-Verlag: New York.

Kielland, K., J. P. Bryant, and R. W. Ruess. 1997. Moose herbivory and carbon turnover of early successional stands in interior Alaska. *Oikos* 80:25–30.

Kilen, M. 2000. Mississippi garbage-man makes a splash. Des Moines Register: Des Moines, Iowa, August 6, 2000, 1E–2E.

King, D. M., P. T. Hagan, C. C. Bohlen. 1997. Setting priorities for riparian buffers. University of Maryland Center for Environmental and Estuarine Studies. Technical Contribution 96–160, College Park, Maryland.

King, J. M., C. Brown, and H. Sabat. 2003. A scenario-based approach to environmental flow assessments for rivers. *River Research and Application* 19:619–639.

King, J. M. and D. Louw. 1998. Instream flow assessment for regulated rivers in South Africa using the building block methodology. *Aquatic Ecosystem Health and Management* 1:109–124.

King, S. L. and B. D. Keeland. 1999. Evaluation of reforestation in the lower Mississippi River alluvial valley. *Restoration Ecology* 7:348–359.

Klapproth, J. C. and J. E. Johnson. 2000. Understanding the Science Behind Riparian Forest Buffers: Effects on Plant and Animal Communities. University of Virginia Cooperative Extension. Publication 420–152. Blacksburg, Virginia.

Klapproth, J. C. and J. E. Johnson. 2001a. Understanding the Science Behind Riparian Forest Buffers: Benefits to Communities and Landowners. Virginia Cooperative Extension. Publication 420–153. Blacksburg, Virginia.

Klapproth, J. C. and J. E. Johnson. 2001b. Understanding the Science Behind Riparian Forest Buffers: Factors Influencing Adoption. Virginia Cooperative Extension. Publication 420–154. Blacksburg, Virginia.

Klapproth, J. C. and J. E. Johnson. 2001c. Understanding the Science Behind Riparian Forest Buffers: Planning, Establishment, and Maintenance. Virginia Cooperative Extension. Publication 420–155.

Klapproth, J. C. and J. E. Johnson. 2001d. Understanding the Science Behind Riparian Forest Buffers: Resources for Virginia Landowners. Virginia Cooperative Extension. Publication 420–156. Blacksburg, Virginia.

Kleynhans, C. J. 1996. A qualitative procedure for the assessment of the habitat integrity of the Luvuvuhu River (Limpopo System, South Africa). *Journal of Aquatic Ecosystem Health* 5:41–54.

Klimo, E. and H. Hager, Eds. 2001. The Floodplain Forests in Europe: Current Situation and Perspectives. European Forest Institute Research Report 10, Koninklijke Brill NV, Leiden.

Klopatek, J. M. and R. H. Gardner. Eds. 1999. Landscape Ecological Analysis: Issues and Applications Springer-Verlag: New York.

Knopf, F. L. 1986. Changing landscapes and the cosmopolitism of the eastern Colorado avifauna. *Wildlife Society Bulletin* 14:132–142.

Knopf, F. L. and F. B. Samson. 1994. Scale perspectives on avian diversity in western riparian ecosystems. *Conservation Biology* 8:669–676.

Knox, J. C. 1993. Large increases in flood magnitude in response to modest changes in climate. *Nature* 361:430–432.

Kolb, T. E., S. C. Hart, and R. Amundson. 1997. Boxelder water source and physiology at perennial and ephemeral stream sites in Arizona. *Tree Physiology* 17:151–160.

Kondolf, G. M. 1995. Geomorphological stream channel classification in aquatic habitat restoration: Uses and limitations. *Aquatic Conservation: Marine and Freshwater Ecosystems* 5:127–141.

Kondolf, G. M. 1998. Development of flushing flows for channel restoration on Rush Creek, California. *Rivers* 6:183–193.

Kovalchik, B. L. 2001. Classification and management of aquatic, riparian, and wetland sites on the national forests of eastern Washington (Part 1: The Series Descriptions). Technical Report. U.S. Department of Agriculture, Forest Service: Colville, Washington.

Kozlowski, T. T., P. J. Kramer, and S. G. Pallardy. 1991. The Physiological Ecology of Woody Plants. Academic Press: New York.

Kruskal, J. B. and M. Wish. 1978. Multidimensional Scaling. Sage Publications: Beverly Hills, California.

Kukalak, J. 2003. Impact of mediaeval agriculture on the alluvium in the San River headwaters (Polish Eastern Carpathians). *Catena* 51:255–266.

Kunkel, K. E., S. A. Changnon, and J. R. Angel. 1994. Climatic aspects of the 1993 Upper Mississippi River Basin flood. *Bulletin of the American Meteorological Society* 75:811–822.

Kusler, J. and W. Niering. 1998. Wetland assessment: Have we lost our way? *National Wetlands Newsletter* 20:9–20.

Lachat, B. 1994. Guide de Protection des Berges de Cours d'Eau en Techniques Végétales. Ministère de l'Environnement: Paris, France.

Lajtha K. and R. H. Michener, Eds. 1994. Stable Isotopes in Ecology and Environmental Science. Blackwell: London.

Lake, P. S. 2000. Disturbance, patchiness, and diversity in streams. *Journal of the North American Benthological Society* 19:573–592.

Langlade, L. R. and O. Décamps. 1995. Accumulation de limon et colonisation végétale d'un banc de galets. Comptes Rendus de l'Académie des Sciences. 318:1073–1082.

Langston, N. E. 1998. People and nature: Understanding the changing interactions between people and ecological systems. Pages 25–76 *in* S. I. Dodson, T. F. H. Allen, S. R. Carpenter, A. R. Ives, R. L. Jeanne, J. F. Kitchell, N. E. Langston, and M. G. Turner. Ecology. Oxford University Press: New York.

Langston, N. E. 2004. Where Land and Water Meet: A Western Landscape Transformed. Universtiy of Washington Press: Seattle, Washington.

Lassettre N. 2003. Process Based Management of Large Woody Debris at the Basin Scale. Ph.D. Dissertation, Department of Environmental Planning, University of California, Berkeley, California.

Lassus, B. 1998. The Landscape Approach. University of Pennsylvania Press: Philadelphia, Pennsylvania.

Lavelle, P. 1997. Faunal activities and soil processes: adaptive strategies that determine ecosystem function. *Advances in Ecological Research* 27:93–132.

Le Floch, S. 2002. Les «ramiers»: Un espace riverain inaccessible de la Garonne? *Ethnologie Française* 32:719–726.

Le Floch, S. and D. Terrasson. 1995. Enjeux écologiques et sociaux autour d'un paysage rural: Le développement de la populiculture dans les Basses Vallées Angevines. *Natures Sciences Sociétés* 3:129–143.

Léger, L. 1909. Principes de la méthode rationnelle du peuplement des cours d'eau à salmonidés. *Travaux de la Laboratoire de Pisciculture à Grenoble* 1:533–568.

Lee, D., T. A. Dillaha, and J. H. Sherrard. 1999. Modeling phosphorus transport in grass buffer strips. *Journal of Environmental Engineering* 115:409–427.

Lee, K. H., T. M. Isenhart, R. C. Schultz, and S. K. Mickelson. 2000. Multi-species riparian buffer system in Central Iowa for controlling sediment and nutrient loss during simulated rain. *Journal of Environmental Quality* 29:1200–1205.

Lee, K. N. 1993. Compass and Gyroscope: Integrating Science and Politics for the Environment. Island Press: Washington, DC.

Lee, R. G. et al. 1992. Integrating sustainable development and environmental vitality: A landscape ecology approach. Pages 499–521 *in* R. J. Naiman, Ed. Watershed Management: Balancing Sustainability and Environmental Change. Springer-Verlag: New York.

Leeds-Harrison, P. B. et al. 1996. Buffer Zones in Headwater Catchments. Report on MAFF (Ministry of Agriculture, Fisheries and Food)/English Nature Buffer Zones Project CSA 2285. Cranfield University, Silsoe, United Kingdom.

Lefsky, M. A., W. B. Cohen, G. G. Parker, and D. J. Harding. 2002. LIDAR remote sensing for ecosystem studies. *BioScience* 52:19–30.

Leonard, S. et al. 1992. Riparian area management: Procedures for ecological site inventory—With special reference to riparian-wetland sites. Technical Report 1737-7. U.S. Department of the Interior, Bureau of Land Management: Denver, Colorado.

Leopold, A. 1949. A Sand County Almanac, and Sketches Here and There. Oxford University Press: New York.

Leopold, L. B. and M. G. Wolman. 1957. River channel patterns: Braided, meandering and straight. United States Geological Survey Professional Paper 282-B. U.S. Government Printing Office: Washington, DC.

Leopold, L. B., M. G. Wolman, and J. P. Miller. 1964. Fluvial Processes in Geomorphology, W. H. Freeman: San Francisco, California.

Lespez, L. 2003. Geomorphic responses to long-term land use changes in Eastern Macedonia (Greece). *Catena* 51:181–208.

Lettenmaier D. P., A. W. Wood, R. N. Palmer, E. F. Wood, and E. Z. Stakhiv. 1999. Water resources implications of global warming: A U.S. regional perspective. *Climatic Change* 43:537–579.

Lettenmaier, D. P., E. F. Wood, and J. R. Wallis. 1994. Hydro-climatological trends in the continental U.S., 1948–88. *Journal of Climate* 7:586–607.

Levin, S. A. 1992. The problem of pattern and scale in ecology. *Ecology* 73:1943–1983.

Levine, C. M. and Stromberg J. C. 2001. Effects of flooding on native and exotic plant seedlings: Implications for restoring southwestern riparian forests by manipulating water and sediment flows. *Journal of Arid Environments* 49:111–131.

Levine, J. M. 2000. Complex interactions in a streamside plant community. *Ecology* 81:3431–3444.

Lewin, J. 2001. Alluvial systematics. Pages 19–41 *in* D. Maddy, M. G. Macklin, and J. C. Woodward, Eds. River Basin Sediment Systems: Archives of Environmental Change. A. A. Balkema: Lisse, The Netherlands.

Lewis, B. J. and T. Slider. 1999. The social dimension of ecosystem management. Pages 265–295 *in* H. K. Cordell and J. C. Bergstrom, Eds. Integrating Social Science and Ecosystem Management. Sagamore Press: Champaign, Illinois.

Lichtenberg, E. and B. V. Lessley. 1992. Water quality, cost-sharing, and technical assistance: Perceptions of Maryland farmers. *Journal of Soil and Water Conservation* 47:260–264.

Likens, G. E. 1984. Beyond the shoreline: A watershed-ecosystem approach. *Verhandlungen Internationale Vereinigung für Theoretische und Angewandte Limnologie* 22:1–22.

Likens, G. E. 1991. Human-accelerated environmental change. *BioScience* 41:130. The Heinz Center Report. 2002. The State of the Nation's Ecosystems. Cambridge University Press: Cambridge.

Likens, G. E. and F. H. Bormann. 1974. Linkages between terrestrial and aquatic ecosystems. *BioScience* 24:447–456.

Likens, G. E., F. H. Bormann, and N. M. Johnson. 1969. Nitrification: Importance to nutrient losses from a cutover, forested ecosystem. *Science* 163:1205–1206.

Likens, G. E., F. H. Bormann, N. M. Johnson, D. W. Fisher, and R. S. Pierce. 1970. Effects of forest cutting and herbicide treatment on nutrient budgets in the Hubbard Brook watershed-ecosystem. *Ecological Monographs* 40:23–47.

Linkins, A. E. and J. Neal. 1982. Soil cellulase, chitinase, and protease activity in *Eriphorum vaginatum* tussock tundra at Eagle Summit, Alaska. *Holarctic Ecology* 5:135–139.

Lins, H. F. and J. R. Slack. 1999. Streamflow trends in the United States. *Geophysical Research Letters* 26:227–230.

Litton, R. B. 1977. River landscape quality and its assessment. Pages 46–54 *in* River Recreation Management and Research: Proceedings of a Symposium. U.S. Department of Agriculture, Forest Service Publication GTR-NC628. St Paul, Minnesota.

Lock, P. A. and R. J. Naiman. 1998. Effects of stream size on bird community structure in coastal temperate forests of the Pacific Northwest, USA. *Journal of Biogeography* 25:773–782.

Lord, S. 2001. NOAA data assimilation activities. Presented at: Coordinated Enhanced Observation Program International Workshop. February 2001. NASA/Goddard Space Flight Center: Greenbelt, Maryland.

Lowe-McConnell, R. H. 1987. Ecological Studies in Tropical Fish Communities. Tropical Biology Series. Cambridge University Press: Cambridge.

Lowrance, R., J. K. Sharp, and J. M. Sheridan. 1986. Long-term sediment deposition in the riparian zone of a coastal plain watershed. *Journal of Soil and Water Conservation* 41:266–271.

Lowrance, R. et al. 1984. Riparian forests as nutrient filters in agricultural watersheds. *BioScience* 34:374–377.

Lowrance, R. R. et al. 1995. Water Quality Functions of Riparian Forest Buffer Systems in the Chesapeake Bay Watershed. U.S. Chesapeake Bay Program: Annapolis, Maryland.

Lowrance, R. R. et al. 2000. REMM: The riparian ecosystem management model. *Journal of Soil and Water Conservation* 55:27–36.

Lowrance, R. R., R. Leonard, and J. M. Sheridan. 1985. Managing riparian ecosystems to control nonpoint pollution. *Journal of Soil and Water Conservation* 40:87–97.

Lowrance, R. R., R. L. Todd, and L. E. Asmussen. 1983. Waterborne nutrient budgets for the riparian zone on an agricultural watershed. *Agriculture, Ecosystems and Environment* 10:371–384.

Ludwig, D., M. Mangel, and B. Haddad. 2001. Ecology, conservation, and public policy. *Annual Review of Ecology and Systematics* 32:481–517.

Ludwig, J., F. Meixner, B. Vogel, and J. Förstner. 2001. Soil-air exchange of nitric oxide: An overview of processes, environmental factors, and modeling studies. *Biogeochemistry* 52:225–257.

Luken, J. O. and R. W. Fonda. 1983. Nitrogen accumulation in a chronosequence of red alder communities along the Hoh River, Olympic National Park, Washington. *Canadian Journal of Forest Resources* 13:1228–1237.

Lynch, J. A., E. S. Corbett, and K. Mussallem. 1985. Best management practices for controlling nonpoint source pollution on forested watersheds. *Journal of Soil and Water Conservation* 40:164–167.

Lynch, L. 1997. Closed geese season brings economic chill to eastern shore's winter. *Economics Viewpoints* 2:9–11.

Lynch, L. 2002. Riparian Buffer Financial Assistance Opportunities. Fact Sheet 769 of the Maryland Cooperative Extension. Available at http://www.riparianbuffers.umd.edu.

Lynne, G. D., J. S. Shonkwiler, and L. R. Rola. 1988. Attitudes and farmer conservation behavior. *American Journal of Agriculture and Economics* 70:12–19.

MacCall, A. D. 1996. Too little, too late: Treating the problem of inaction. *Ecological Applications* 6:368–369.

Machtans, C. S., M. A. Villard, and S. J. Hannon. 1996. Use of riparian buffer strips as movement corridors by forest birds. *Conservation Biology* 10:1366–1377.

Maddy, D., M. G. Macklin, and J. C. Woodward. 2001. River Basin Sediment Systems: Archives of Environmental Change. A. A. Balkema: Lisse, The Netherlands.

Magette, W. L., R. B. Brinsfield, R. E. Palmer, and J. D. Wood. 1989. Nutrient and sediment removal by vegetated filter strips. *Transactions of the American Society of Agricultural Engineering* 32:6663–6667.

Magnuson, J. J. 1990. Long-term ecological research and the invisible present. *BioScience* 40:495–501.

Mahoney J. M. and S. B. Rood. 1991. A device for studying the influence of declining water table on poplar growth and survival. *Tree Physiology* 8:305–314.

Mahoney, J. M. and S. B. Rood. 1992. Response of a hybrid poplar to water table decline in different substrates. *Forest Ecology and Management* 54:141–156.

Mahoney, J. M. and S. B. Rood. 1998. Streamflow requirements for cottonwood seedling recruitment: An integrative model. *Wetlands* 18:634–645.

Malanson, G. P. 1993. Riparian Landscapes. Cambridge University Press: Cambridge.

Malard, F., K. Tockner, M. J. Dole-Olivier, and J. V. Ward. 2002. A landscape perspective of surface–subsurface hydrological exchanges in river corridors. *Freshwater Biology* 47:621–640.

Malik, A. S., B. A. Larson, and M. Ribaudo. 1994. Economic incentives for agricultural nonpoint source pollution control. *Water Resources Bulletin* 30:471–479.

Mangel, M. et al. 1996. Principles for the conservation of wild living resources. *Ecological Applications* 6:338–362.

Maridet, L., J. G. Wasson, M. Philippe, C. Andros, and R. J. Naiman. 1998. Riparian and morphological controls in structuring the macroinvertebrate stream community. *Archiv für Hydrobiologie* 144:61–85.

Marinucci, A. C., J. E. Hobbie, and J. V. K. Helfrich. 1983. Effects of litter nitrogen on decomposition and microbial biomass in *Spartina alterniflora*. *Microbial Ecology* 9:27–40.

Marsh, G. P. 1965. Man and Nature. Charles Scribner: New York.

Martin, T. L., J. T. Trevors, and N. K. Kaushik. 1999. Soil microbial diversity, community structure and denitrification in a temperate riparian zone. *Biodiversity and Conservation* 8:1057–1078.

Martinelli, L. A., R. L. Victoria, P. C. O. Trivelin, A. H. Devol, and J. E. Richey. 1992. ^{15}N natural abundance in plants of the Amazon River floodplain and potential atmospheric N-fixation. *Oecologica* 90:591–596.

Maser, C. and J. R. Sedell. 1994. From the Forest to the Sea: The Ecology of Wood in Streams, Rivers, Estuaries, and Oceans. St. Lucie Press: Delray Beach, Florida.

Mason, C. F. and S. M. MacDonald. 1982. The input of terrestrial invertebrates from tree canopies to a stream. *Freshwater Biology* 12:305–311.

Masonis, R. J. and F. L. Bodi. 1998. River law. Pages 553–571 *in* R. J. Naiman and R. E. Bilby, Eds. River Ecology and Management. Springer-Verlag: New York.

Mather, P. M. 1999. Computer Processing of Remotely-Sensed Images: An Introduction. John Wiley & Sons: New York.

Mathooko, J. M. and S. T. Kariudi. 2000. Disturbances and species distribution of the riparian vegetation of a rift valley stream. *African Journal of Ecology* 38:123–129.

Matson, P., K. A. Lohse, and S. J. Hall. 2002. The globalization of nitrogen deposition: consequences for terrestrial ecosystems. *Ambio* 31:113–119.

McCarthy, J. J., O. F. Canziani, N. A. Leary, D. J. Dokken, and K. S. White, Eds. 2001. Climate Change 2001: Impacts, Adaptation, and Vulnerability. Cambridge University Press: Cambridge.

McClain, M. E., J. E. Richey, and T. P. Pimentel. 1994. Groundwater nitrogen dynamics at the terrestrial-lotic interface of a small catchment in the Central Amazon Basin. *Biogeochemistry* 27:113–127.

McClain, M. E. and J. E. Richey. 1996. Regional-scale linkages of terrestrial and lotic ecosystems in the Amazon Basin: A conceptual model for organic matter. *Archiv für Hydrobiologie* 113 (Supplement):111–125.

McClain, M. E., R. E. Bilby, and F. J. Triska. 1998. Nutrient cycles and responses to disturbance. Pages 347–372 *in* R. J. Naiman and R. E. Bilby, Eds. River Ecology and Management. Springer-Verlag: New York.

McClain M. E., R. L. Victoria, and J. E. Richey, Eds. 2001. The Biogeochemistry of the Amazon Basin. Oxford University Press: New York.

McClain M. E. et al. 2003. Biogeochemical hot spots and hot moments at the interface of terrestrial and aquatic ecosystems. *Ecosystems* 6:301–312.

McClain, M. E. and R. E. Cossio. 2003. The use of riparian environments in the rural Peruvian Amazon. *Environmental Conservation* 30:242–248.

McCully, P. 1996. Silenced Rivers: The Ecology and Politics of Large Dams. Zed Books: London and New Jersey.

McCully, P. 2001. Silenced Rivers: The Ecology and Politics of Large Dams. Enlarged and updated edition. Zed Books: London.

McDowell, W. H., W. B. Bowden, and C. E. Asbury. 1992. Riparian nitrogen dynamics in two geomorphologically distinct tropical rain forest watersheds: Subsurface solute patterns. *Biogeochemistry* 18:53–75.

McHale, M. R., M. J. Mitchell, J. J. McDonnell, and C. P. Cirmo. 2000. Nitrogen solutes in an Adirondack forested watershed: Importance of dissolved organic nitrogen. *Biogeochemistry* 48:165–184.

McInnes, P. F., R. J. Naiman, J. Pastor, and Y. Cohen. 1992. Effects of moose browsing on vegetation and litter of the boreal forest, Isle Royale, Michigan, USA. *Ecology* 73:2059–2075.

McInnis, M. L. and J. McIver. 2001. Influence of off-stream supplements on streambanks of riparian pastures. *Journal of Range Management* 54:648–652.

McIntyre, S., S. Lavorel, J. Landsberg, and T. D. A. Forbes. 1999. Disturbance response in vegetation: Towards a global perspective on functional traits. *Journal of Vegetation Science* 10:621–630.

Meave, J. and M. Kellman. 1994. Maintenance of rain forest diversity in riparian forests and tropical savannas: Implications for species conservation during Pleistocene drought. *Journal of Biogeography* 21:121–135.

Meehan, W. R. 1991. Influences of Forest and Rangeland Management on Salmonid Fishes and Their Habitats. American Fisheries Society Special Publication 19. Bethesda, Maryland.

Melillo, J. M., C. B. Field, and B. Moldan, Eds. 2003. Interactions of the Major Biogeochemical Cycles. SCOPE 61. Island Press: Washington, DC.

Melillo, J. M., R. J. Naiman, J. D. Aber, and K. N. Eshleman. 1983. The influence of substrate quality and stream size on wood decomposition dynamics. *Oecologia* 58:281–285.

Melillo, J. M., R. J. Naiman, J. D. Aber, and A. E. Linkins. 1984. Factors controlling mass loss and nitrogen dynamics of plant litter decaying in northern streams. *Bulletin of Marine Science* 35:341–356.

Mensforth, L. J., P. J. Thorburn, S. D. Tyerman, and G. R. Walker. 1994. Sources of water used by riparian *Eucalyptus camaldulensis* overlying highly saline groundwater. *Oecologia* 100:21–28.

Mertes, L. A. K. 1997. Documentation and significance of the perirheic zone on inundated floodplains. *Water Resources Research* 33:1749–1762.

Mertes, L. A. K. 2002. Remote sensing of riverine landscapes. *Freshwater Biology* 47:799–816.

Metz, B., O. Davidson, R. Swart, and J. Pan. 2001. Climate Change 2001: Mitigation. Contribution of Working Group III to the Third Assessment Report of the Intergovernmental Panel on Climate Change. Cambridge University Press: Cambridge.

Metzger J. P. and H. Décamps. 1997. The structural connectivity threshold: A hypothesis in conservation biology at the landscape scale. *Acta Oecologica* 18:1–12.

Meybeck, M. 2001. Global alteration of riverine geochemistry under human pressure. Pages 97–113 *in* E. Ehlers and X. Kraft, Eds. Understanding the Earth System: Compartments, Processes and Interactions. Springer: Heidelberg.

Meybeck, M. 2003. Global analysis of river systems: From earth system controls to anthropocene syndromes. *Philosophical Transactions of the Royal Society of London. Series B* 358:1935–1955.

Meybeck, M. and C. J. Vörösmarty. 2004. The integrity of river and drainage basins systems: Challenges from environmental change. Pages 297–480 *in* Kabat, P., M. Claussen, P. A. Dirmeyer, J. H. C. Gfash, L. Bravo de Guenni, M. Meybeck, R. A. Pielke, Sr., C. J. Vörösmarty, R. W. A. Hutjes and S. Lütkenmeier, Eds. Vegetation, Water, Humans and the Climate: A New Perspective on an Interactive System. Springer-Verlag: Berlin and Heidelberg.

Meyer, J. L., M. J. Sale, P. J. Mulholland, and N. L. Poff. 1999. Impacts of climate change on aquatic ecosystem functioning and health. *Journal of the American Water Resources Association* 35:1373–1386.

Meyer, W. B. 1995. Past and present land use and land cover in the USA. *Consequences*, Spring 1995:25–33.

Meyers, L. H. 1989. Riparian area management: Inventory and monitoring riparian areas. Technical Report 1737-3. US Department of the Interior, Bureau of Land Management: Denver, Colorado.

Michael, D. N. 1995. Barriers and bridges to learning in a turbulent human ecology. Pages 461–485 *in* L. H. Gunderson, C. S. Holling, and S. S. Light, Eds. Barriers and Bridges to the Renewal of Ecosystems and Institutions. Columbia University Press: New York.

Mighall, T. M., S. Timberlake, S. H. E. Clark, and A. E. Caseldine. 2002. A palaeoenvironmental investigation of sediments from the prehistoric mine of Copa Hill, Cwmystwyth, mid-Wales. *Journal of Archaeological Science* 29:1161–1188.

Miller, J. R. and J. B. Ritter. 1996. An examination of the Rosgen classification of natural rivers. *Catena* 27:295–299.

Miller, J. R., J. A. Wiens, N. T. Hobbs, and D. M. Theobald. 2003. Effects of human settlement on bird communities in lowland riparian areas of Colorado (USA). *Ecological Applications* 13:1041–1059.

Milly, P. C. D., R. T. Wetherald, K. A. Dunne, and T. L. Delworth. 2002. Increasing risk of great floods in a changing climate. *Nature* 415:514–517.

Minshall, G. W. 1988. Stream ecosystem theory: A global perspective. *Journal of the North American Benthological Society* 7:263–288.

Minshall, G. W. et al. 1983. Interbiome comparison of stream ecosystem dynamics. *Ecological Monographs* 53:1–25.

Mitsch, W. J. and J. G. Gosselink. 1993. Wetlands. Van Nostrand Reinhold: New York.

Montgomery, D. R. 1999. Process domains and the river continuum. *Journal of the American Water Resources Association* 35:397–410.

Montgomery, D. R. et al. 1996. Distribution of bedrock and alluvial channels in forested mountain drainage basins. *Nature* 381:578–589.

Montgomery, D. R. and J. M. Buffington. 1997. Channel reach morphology in mountain drainage basins. *Geological Society of America Bulletin* 109:596–611.

Montgomery, D. R. and J. M. Buffington. 1998. Channel processes, classification, and response. Pages 13–42 *in* R. J. Naiman and R. E. Bilby, Eds. River Ecology and Management. Springer-Verlag: New York.

Montgomery, D. R. and W. Dietrich. 1992. Channel initiation and the problem of landscape scale. *Science* 255:826–830.

Montgomery, D. R., G. E. Grant, and K. Sullivan. 1995. Watershed analysis as a framework for implementing ecosystem management. *Water Resources Bulletin* 31:369–386.

Montgomery, D. R. and L. H. MacDonald. 2002. Diagnostic approach to stream channel assessment and monitoring. *Journal of the American Water Resources Association* 38:1–16.

Montgomery, D. R. and H. Piégay. 2003. Wood in rivers: Interactions with channel morphology and processes. *Geomorphology* 51:1–5.

Moring, J. B. and K. W. Stewart. 1994. Habitat partitioning by the wolf spider (Araneae, Lycosidae) guild in streamside and riparian vegetation zones of the Conejos River, Colorado. *Journal of Arachnology* 22:205–217.

Morisawa, M. 1968. Streams: Their Dynamics and Morphology. McGraw-Hill: New York.

Mouchot, M-C., T. Alföldi, D. de Lisle, and G. McCullough. 1991. Monitoring water bodies of the Mackenzie Delta by remote sensing methods. *Arctic* 44:21–28.

Moulin, B. and H. Piégay. 2004. Characteristics and temporal variability of large woody debris stored in the reservoir of Genissiat, Rhône: Elements for river basin management. *River Research and Applications* 20:79–97.

Movle, P. B. and H. W. Li. 1979. Community ecology and predator-prey relationships in warmwater streams. Pages 171–181 *in* R. H. Stroud and H. E. Clepper, Eds. Predator-Prey Systems in Fisheries Management. Sport Fishing Institute, Washington, DC.

Mulholland, P. J. 1992. Regulation of nutrient concentrations in a temperate forest stream: Roles of upland, riparian and instream processes. *Limnology and Oceanography* 37:1512–1526.

Munn, T., A. Whyte, and P. Timmerman. 1999. Emerging environmental issues: A global perspective of SCOPE (Scientific Committee on Problems of the Environment). *Ambio* 28:464–471.

Murphy, M. L. and K. V. Koski. 1989. Input and depletion of woody debris in Alaska streams and implications for streamside management. *North American Journal of Fisheries Management* 9:427–436.

Murray, N. L. and D. F. Stauffer. 1995. Nongame bird use of habitat in central Appalachian riparian forests. *Journal of Wildlife Management* 59:78–88.

Nadler, C. T. and S. A. Schumm. 1981. Metamorphosis of South Platte and Arkansas rivers, eastern Colorado. *Physical Geography* 2:95–115.

Naegeli, M. W., P. Huggenberger, and U. Uehlinger. 1996. Ground penetrating radar for assessing sediment structures in the hyporheic zone of a prealpine river. *Journal of the North American Benthological Society* 15:353–366.

Naiman, R. J., Ed. 1992. Watershed Management: Balancing Sustainability and Environmental Change. Springer-Verlag: New York.

Naiman, R. J. 1998. Biotic stream classification. Pages 97–119 *in* R. J. Naiman and R. E. Bilby, Editors. River Ecology and Management: Lessons from the Pacific Coastal Ecoregion. Springer-Verlag: New York.

Naiman, R. J., E. V. Balian, K. K. Bartz, R. E. Bilby, and J. J. Latterell. 2002. Dead wood dynamics in stream ecosystems. Pages 23–48 *in* W. F. Laudenslayer, Jr., P. J. Shea, B. Valentine, C. P. Weatherspoon, and T. E. Lisle, Eds. Proceedings of the Symposium on the Ecology and Management of Dead Wood in Western Forests. General Technical Report GTR-PSW-181. U.S. Department of Agriculture, Forest Service, Pacific Southwest Research Station, Albany, California.

Naiman, R. J. et al. 1992a. Fundamental elements of ecologically healthy watershade in the Pacific Northwest coastal ecoregion. Pages 127–188 *in* R. J. Naiman, Ed. Watershed Management: Balancing Sustainability and Environmental Change. Springer-Verlag: New York.

Naiman, R. J. and R. E. Bilby. 1998. River ecology and management in the Pacific Coastal Ecoregion. Pages 1–10 *in* R. J. Naiman and R. E. Bilby, Eds. River Ecology and Management: Lessons from the Pacific Coastal Ecoregion. Springer-Verlag: New York.

Naiman, R. J. et al. 2005. Origins, patterns, and importance of heterogeneity in riparian systems. *In* G. Lovett, C. G. Jones, M. G. Turner, and K. C. Weathers, Eds. Ecosystem Function in Heterogeneous Landscapes. Springer-Verlag: New York. (In Press)

Naiman, R. J., R. E. Bilby, and P. A. Bisson. 2000. Riparian ecology and management in the Pacific coastal rain forest. *BioScience* 50:996–1011.

Naiman, R. J., R. E. Bilby, D. E. Schindler, and J. M. Helfield. 2002b. Pacific Salmon, nutrients, and the dynamics of freshwater and riparian ecosystems. *Ecosystems* 5:399–417.

Naiman, R. J., P. A. Bisson, R. G. Lee, and M. G. Turner. 1998a. Watershed management. Pages 642–661 *in* R. J. Naiman and R. E. Bilby, Eds. River Ecology and Management. Springer-Verlag: New York.

Naiman, R. J. et al. 2003. Interactions between species and ecosystem characteristics. Pages 221–241, *in* J. T. du Toit, K. H. Rogers, and H. Biggs, Eds. The Kruger Experience: Ecology and Management of Savanna Heterogeneity. Island Press: Washington, DC.

Naiman, R. J. et al. 2002c. Legitimizing fluvial systems as users of water: An overview. *Environmental Management* 30:455–467.

Naiman, R. J. and H. Décamps, Eds. 1990. The Ecology and Management of Aquatic-Terrestrial Ecotones. Paris: UNESCO, Park Ridge: Parthenon.

Naiman, R. J. and H. Décamps. 1997. The ecology of interfaces: Riparian zones. *Annual Review of Ecology and Systematics* 28:621–658.

Naiman, R. J., H. Décamps, and M. Pollock. 1993. The role of riparian corridors in maintaining regional biodiversity. *Ecological Applications* 3:209–212.

Naiman, R. J., K. L. Fetherston, S. McKay, and J. Chen. 1998b. Riparian forests. Pages 289–323 *in* R. J. Naiman and R. E. Bilby, Eds. River Ecology and Management: Lessons from the Pacific Coastal Ecoregion. Springer-Verlag: New York.

Naiman, R. J., C. A. Johnston, and J. C. Kelley. 1988. Alteration of North American streams by beaver. *BioScience* 38:753–762.

Naiman, R. J., D. G. Lonzarich, T. J. Beechie, and S. C. Ralph. 1992. General principles of classification and the assessment of conservation potential in rivers. Pages 93-123 *in* P. Boon, P. Calow, and G. Petts, Eds. River Conservation and Management. John Wiley & Sons: Chichester, United Kingdom.

Naiman, R. J., J. J. Magnuson, and P. L. Firth. 1998c. Integrating cultural, economic and environmental requirements for fresh water. *Ecological Applications* 8:569–570.

Naiman, R. J., J. J. Magnuson, D. M. McKnight, and J. A. Stanford, Eds. 1995. The Freshwater Imperative: A Research Agenda. Island Press: Washington, DC.

Naiman, R. J., J. M. Melillo, and J. E. Hobbie. 1986. Ecosystem alteration of boreal forest streams by beaver (*Castor canadensis*). *Ecology* 67:1254–1269.

Naiman, R. J., J. M. Melillo, M. A. Lock, T. E. Ford, and S. R. Reice. 1987. Longitudinal patterns of ecosystem processes and community structure in a subarctic river continuum. *Ecology* 68:1139–1156.

Naiman, R. J., G. Pinay, C. A. Johnston, and J. Pastor. 1998. Beaver-induced influences on the long-term bioglochamical characteristics of boreal forest drainage metworks. *Ecology* 75:905–921.

Naiman, R. J. and K. H. Rogers. 1997. Large animals and the maintenance of system-level characteristics in river corridors. *BioScience* 47:521–529.

Naiman, R. J. and M. G. Turner. 2000. A future perspective on North America's freshwater ecosystems. *Ecological Applications* 10:958–970.

Nakano, S., H. Miyasaka, and N. Kuhara. 1999. Terrestrial-aquatic linkages: Riparian arthropod inputs alter trophic cascades in a stream food web. *Ecology* 80: 2435–2441.

Nakano, S. and M. Murakami. 2001. Reciprocal subsidies: Dynamic interdependence between terrestrial and aquatic food webs. *Proceedings of the National Academy of Sciences* 98:166–170.

Nanson, G. C., M. Barbetti, and G. Taylor. 1995. River stabilization due to changing climate and vegetation during the late Quaternary in western Tasmania, Australia. *Geomorphology* 13:145–158.

Nassauer, J. I. 1992. The appearance of ecological systems as a matter of policy. *Landscape Ecology* 6:239–250.

Nassauer, J. I. 1995. Culture and changing landscape structure. *Landscape Ecology* 10:229–237.

Nassauer, J. I. 1997. Cultural sustainability: Aligning esthetics and ecology. Pages 65–83 *in* J. I. Nassauer, Ed. Placing Nature: Culture and Landscape Ecology. Island Press: Washington, DC.

Nassauer, J. I., Ed. 1997. Placing Nature: Culture and Landscape Ecology. Island Press: Washington, DC.

Nelson, P. N., J. A. Baldock, and J. M. Oades. 1993. Concentration and composition of dissolved organic carbon in streams in relation to catchment soil properties. *Biogeochemistry* 19:27–50.

Nelson, S. M. and D. C. Andersen. 1994. An assessment of riparian environmental quality by using butterflies and disturbance susceptibility scores. *The Southwestern Naturalist* 39:137–142.

Nepstad, D. C. et al. 1994. The role of deep roots in the hydrological and carbon cycles of Amazonian forests and pastures. *Nature* 372:666–669.

Neumeister, H., A. Krüger, and B. Schneider. 1997. Problems associated with the artificial flooding of floodplain forests in an industrial region in Germany. *Global Ecology and Biogeography Letters* 6:197–209.

Nierenberg, T. R., and D. E. Hibbs. 2000. A characterization of unmanaged riparian areas in the central Coast Range of western Oregon. *Forest Ecology and Management* 129:195–206.

Nilsson, C. 1986. Changes in plant community composition along two rivers in northern Sweden. *Canadian Journal of Botany* 64:589–592.

Nilsson, C. 1992. Conservation management of riparian communities. Pages 352–372 *in* L. Hansson, Ed. Ecological Principles of Nature Conservation. Elsevier Applied Science: London.

Nilsson, C. and K. Berggren. 2000. Alteration of riparian ecosystems caused by river regulation. *BioScience* 50:783–792.

Nilsson, C. and M. Dynesius. 1994. Ecological effects of river regulation on mammals and birds: A review. *Regulated Rivers* 9:45–53.

Nilsson, C. et al. 1994. A comparison of species richness and traits of riparian plants between a main river channel and its tributaries. *Journal of Ecology* 82:281–295.

Nilsson, C., G. Grelsson, M. Dynesius, M. E. Johansson, and U. Sperens. 1991. Small rivers behave like large rivers: Effects of postglacial history and plant species richness along riverbanks. *Journal of Biogeography* 18:533–541.

Nilsson, C., G. Grelsson, M. Johansson, and U. Sperens. 1989. Patterns of plant species richness along riverbanks. *Ecology* 70:77–84.

Nilsson, C., R. Jansson, and U. Zinco. 1997. Long-term response of river-margin vegetation to water level regulation. *Science* 276:798–800.

Nilsson, C. et al. 1993. Processes structuring riparian vegetation. *Current Topics in Botanical Research* 1:419–431.

Nilsson, C. and M. Svedmark. 2002. Basic principles and ecological consequences of changing water regimes: riparian plant communities. *Environmental Management* 30:468–480.

Nowak, P. J. and P. F. Korsching. 1983. Social and institutional factors affecting the adoption and maintenance of agricultural BMPs. Pages 349–373 *in* F. W. Schaller and G. W. Bailey, Eds. Agricultural Management and Water Quality. Iowa State University Press: Ames, Iowa.

NRC (National Research Council). 1992. Restoration of Aquatic Systems: Science, Technology and Public Policy. National Academy Press: Washington, DC.

NRC (National Research Council). 1998. New Strategies for America's Watersheds. National Academy Press: Washington, DC.

NRC (National Research Council). 2000. Clean Coastal Waters: Understanding and Reducing the Effects of Nutrient Pollution. National Academy Press: Washington, DC.

NRC (National Research Council). 2002. Riparian Areas. National Academy Press: Washington, DC.

Oakley, A. L. et al. 1985. Riparian zones and freshwater wetlands. Pages 57–80 *in* E. R. Brown, Ed. Management of Wildlife and Fish Habitats in Forests of Western Oregon and Washington. Pacific Northwest Research Station of the United States Forest Service, Department of Agriculture, Technical Report PNW-R6-F&WL-192-1985.

Odum, W. E., P. W. Kirk, and J. C. Zieman. 1979. Non-protein nitrogen compounds associated with particles of vascular plant detritus. *Oikos* 32:363–367.

O'Keefe, T. C. and R. T. Edwards. 2002. Evidence for hyporheic transfer and storage of marine-derived nutrients in sockeye streams in southwest Alaska. *American Fisheries Society Symposium* 33:99–107.

Oleksyn, J. and P. B. Reich. 1994. Pollution, habitat destruction, and biodiversity in Poland. *Conservation Biology* 8:943–960.

Oliver, C. D. and B. C. Larson. 1996. Forest Stand Dynamics. John Wiley & Sons: New York.

Olson, C. and R. Harris. 1997. Applying a two-stage system to prioritize riparian restoration at the San Luis Rey River, San Diego County, California. *Restoration Ecology* 5:43–55.

Omernik, J. and R. Bailey. 1997. Distinguishing between watersheds and ecoregions. *Journal of the American Water Resources Association* 33:935–949.

Opperman, J. J. and A. M. Merenlender. 2000. Deer herbivory as an ecological constraint to restoration of degraded riparian corridors. *Restoration Ecology* 8:41–47.

Osterkamp, W. R. and C. R. Hupp. 1984. Geomorphic and vegetative characteristics along three northern Virginia streams. *Geological Society of American Bulletin* 95:1093–1101.

Ostrom, E., J. Burger, C. B. Field, R. B. Norgaard, and D. Policansky. 1999. Revisiting the commons: Local lessons, global challenges. *Science* 284:278–282.

Pabst, R. J. and T. A. Spies. 1999. Structure and composition of unmanaged riparian forests in the coastal mountains of Oregon, USA. *Canadian Journal of Forest Research* 29:1557–1573.

Paetzold, A., C. J. Schubert, and K. Tockner. 2005. Aquatic-terrestrial linkages along a braided river: Riparian arthropods feeding on aquatic insects. *Ecosystems* (in press).

Paine, R. T. 1980. Food webs: Linkage strength and community infrastructure. *Journal of Animal Ecology* 49:666–685.

Palik, B. J., J. C. Zasada, and C. W. Hedman. 2000. Ecological principles for riparian silviculture. Pages 233–254 *in* E. S. Verry, J. W. Hornbeck, and C. A. Dollof, Eds. Riparian Management in Forests of the Continental Eastern United States. Lewis Publishers: New York.

Parolin, P. 2001. Morphological and physiological adjustments to waterlogging and drought in seedlings of Amazonian floodplain trees. *Oecologia* 128:326–335.

Pastor, J., J. Aber, C. McClaugherty, and J. Melillo. 1984. Aboveground production and N and P cycling along a nitrogen mineralization gradient on Blackhawk Island, Wisconsin. *Ecology* 65:256–268.

Pastor, J., B. Dewey, R. J. Naiman, P. F. McInnes, and Y. Cohen. 1993. Moose browsing and soil fertility in the boreal forests of Isle Royale National Park. *Ecology* 74:467–480.

Pastor, J., R. J. Naiman, B. Dewey, and P. McInnes. 1988. Moose, microbes, and the boreal forest. *BioScience* 38:770–777.

Pautou, G. 1984. L'organisation des forêts alluviales dans l'axe rhodanien entre Genève et Lyon: Comparaison avec d'autres systèmes fluviaux. *Documents de Cartographie Ecologique* 27:43–64.

Pautou, G. and J. Girel. 1986. La végétation de la basse plaine de l'Ain: Organisation spatiale et évolution. *Documents de Cartographie Ecologique* 29:75–108.

Pautou, G., J. Girel, and J. L. Borel. 1992. Initial repercussions and hydroelectric developments in the French Upper Rhone valley: A lesson for predictive scenarios propositions. *Environmental Management* 16:231–242.

Pawelko, K. A., E. B. Drogin, A. R. Graefe, and D. P. Huden. 1996. Examining the nature of river recreation visitors and their recreational experiences in the Delaware River. Pages 43–49 *in* C. P. Dawson, compiler. Proceedings of the 1995 Northeastern Recreation Research Symposium. U.S. Department of Agriculture, Forest Service Publication GTR-NE-218. Saratoga Springs, New York.

Pearce, F. 2003. Replumbing the planet. *New Scientist* 178:30

Peckham, S. D. 1999. A reformulation of Horton's laws for large river networks in terms of statistical self-similarity. *Water Resources Research* 35:2763–2777.

Pedroli, B., G. de Blust, K. van Looy, and S. van Rooij. 2002. Setting targets in strategies for river restoration. *Landscape Ecology* 17 (suppl. 1):5–18.

Pellerin, B. A. et al. 2004. Role of wetlands and developed land use on dissolved organic nitrogen concentrations and DON/TDN in northeastern U.S. rivers and streams. *Limnology and Oceanography* 49:910–918.

Peng, W., D. Wheeler, J. Bell, and M. Krusemark. 2003. Delineating patterns of soil drainage class on bare soils using remote sensing analyses. *Geoderma* 115:261–279.

Penka, M., M. Vyskot, E. Klimo, and F. Vašíček. 1991. Floodplain Forest Ecosystem. II. After Water Management Measures. Elsevier: Amsterdam, The Netherlands.

Perrin, C. J., K. S. Shortreed, and J. G. Stockner. 1984. An integration of forest and lake fertilization: Transport and transformations of fertilizer elements. *Canadian Journal of Fisheries and Aquatic Sciences* 41: 253–262.

Peterjohn, W. T. and D. L. Correll. 1984. Nutrient dynamics in an agricultural watershed: Observations on the role of the riparian forest. *Ecology* 65:1466–1475.

Peters, R. L. 1989. Effects of global warming on biological diversity. Pages 82–95 *in* D. E. Abrahamson, Ed. The Challenge of Global Warming. Island Press: Washington, DC.

Petersen, R. C., Jr. 1992. The RCE: A riparian, channel, and environmental inventory for small streams in the agricultural landscape. *Freshwater Biology* 27:295–306.

Peterson, B. J. and B. Fry. 1987. Stable isotopes in ecosystem studies. *Annual Review of Ecology and Systematics* 18:293–320.

Peterson, B. J. et al. 2001. Control of nitrogen export from watersheds by headwater streams. *Science* 292:86–90.

Peterson, B. J. et al. 2001. Control of nitrogen export from watersheds by headwater streams. *Science* 292:86–90.

Peterson, D. L. and G. L. Rolfe. 1982. Nutrient dynamics and decomposition of litterfall in floodplain and upland forests of central Illinois. *Forest Science* 28:667–681.

Peterson, G. D., G. Cumming, and S. R. Carpenter. 2003. Scenario planning: A tool for conservation in an uncertain world. *Conservation Biology* 17:358–366.

Pettit, N. E. and R. H. Froend. 2001. Variability in flood disturbance and the impact on riparian tree recruitment in two contrasting river systems. *Wetlands Ecology and Management* 9:13–25.

Pettit, N. E., R. H. Froend, and R. H. Davies. 2001. Identifying the natural flow regime and the relationship with riparian vegetation for two contrasting Western Australian rivers. *Regulated Rivers: Research and Management* 17:201–215.

Pettit, N. E. and R. J. Naiman. 2005. Heterogeneity and the role of large woody debris piles in the post-flood recovery of riparian vegetation in a semi-arid river ecosystem. *Oecologia* (submitted).

Pettit, N. E., R. J. Naiman, K. H. Rogers, and J. E. Little. 2005. Post flooding distribution and characteristics of large woody debris piles along the semi-arid Sabie River, South Africa. *River Research and Applications* 21:27–38.

Petts, G. E., Ed. 1989. Historical Change of Large Alluvial Rivers: Western Europe. John Wiley & Sons: Chichester, United Kingdom.

Phillips, J. D. 1999. Edge effects in geomorphology. *Physical Geography* 20:53–66.

Pickett, S. T. A. and P. S. White. 1985. The Ecology of Natural Disturbance and Patch Dynamics. Academic Press: Orlando, Florida.

Piedade, M. T. F., M. Worbes, and W. J. Junk. 2001. Geoecological controls on elemental fluxes in communities of higher plants in Amazonian floodplains. Pages 209–234 *in* M. E. McClain, R. L. Victoria, and J. E. Richey, Eds. The Biogeochemistry of the Amazon Basin. Oxford University Press: Oxford.

Piégay, H. and J. P. Bravard. 1997. Response of a Mediterranean riparian forest to a 1 in 400 year flood, Ouveze River, Drome, Vaucluse, France. *Earth Surface Processes and Landforms* 22:31–43.

Pinay, G., V. J. Black, A. M. Planty-Tabacchi, B. Gumiero, and H. Décamps. 2000. Geomorphic control of denitrification in large river floodplain soils. *Biogeochemistry* 50:163–182.

Pinay, G., J. C. Clément, and R. J. Naiman. 2002. Basic principles and ecological consequences of changing water regimes on nitrogen cycling in fluvial ecosystems. *Environmental Management* 30:481–491.

Pinay, G., H. Décamps, E. Chauvet, and E. Fustec. 1990. Functions of ecotones in fluvial systems. Pages 141–169 *in* R. J. Naiman and H. Décamps, Eds. The Ecology and Management of Aquatic-Terrestrial Ecotones. Parthenon Publishing Group: Carnforth, United Kingdom.

Pinay, G. and R. J. Naiman. 1991. Short-term hydrologic variations and nitrogen dynamics in beaver-created meadows. *Archiv für Hydrobiologie* 123:187–205.

Pinay, G., C. Ruffinoni, S. Wondzell, and F. Gazelle. 1998. Change in groundwater nitrate concentration in a large river floodplain: Denitrification, uptake, or mixing? *Journal of the North American Benthological Society* 17:179–189.

Piorkowski, R. J. 1995. Ecological effects of spawning salmon on several central Alaskan streams. Thesis, University of Alaska, Fairbanks.

Planty-Tabacchi, A. M., E. Tabacchi, R. J. Naiman, C. Deferrari, and H. Décamps. 1996. Invasibility of species-rich communities in riparian zones. *Conservation Biology* 10:598–607.

Platteeuw, M., N. Geilen, J. de Jonge, and M. H. I. Schropp. 2001. Reconstruction measures in the floodplains of the Rhine and Meuse. Pages 53–70 *in* H. A. Wolters, M. Platteeuw, and M. M. Schoor, Eds. Guidelines for Rehabilitation and Management of Floodplains: Ecology and Safety Combined. NRC (The Netherlands Centre for River Studies) Publication 09–2001, Delft, The Netherlands.

Platts, W. S. 1987. Methods for evaluating riparian habitats with applications to management. U.S. Department of Agriculture, Forest Service. General Technical Report INT-221. Ogden, Utah.

Poesen, J. W. A., and J. M. Hooke. 1997. Erosion, flooding and channel management in Mediterranean environments of southern Europe. *Progress in Physical Geography* 21:159–199.

Poff, N. L. 1992a. Why disturbances can be predictable: A perspective on the definition of disturbance in streams. *Journal of the North American Benthological Society* 11:86–92.

Poff, N. L. 1992b. Regional hydrologic response to climate change: An ecological perspective. Pages 88–115 *in* P. Firth and S. G. Fisher, Eds. Global Climate Change and Freshwater Ecosystems. Springer-Verlag: New York.

Poff, N. L. et al. 1997. The natural flow regime: A paradigm for conservation and restoration of river ecosystems. *BioScience* 47:769–784.

Poff, N. L. et al. 2003. River flows and water wars: Emerging science for environmental decision making. *Frontiers in Ecology and the Environment* 1:298–306.

Poff, N. L., M. M. Brinson, and J. W. Day, Jr. 2002. Aquatic Ecosystems and Global Climate Change. Potential Impacts on Inland Freshwater and Coastal Wetland Ecosystems in the United States. Prepared for the Pew Center on Global Climate Change: Arlington, VA. http://www.pewclimate.org/projects/aquatic.cfm.

Poff, N. L. and J. V. Ward. 1989. Implications of stream flow variability and pre-dictability for lotic community structure: A regional analysis of streamflow patterns. *Canadian Journal of Fisheries and Aquatic Sciences* 48:1805–1818.

Polis, G. A., W. B. Anderson, and R. D. Holt. 1997. Toward an integration of landscape and food web ecology: The dynamics of spatially subsidized food webs. *Annual Review of Ecology and Systematics* 28:289–319.

Pollock, M. M. 1998. Biodiversity. Pages 430–452 *in* R. J. Naiman and R. E. Bilby, Eds. River Ecology and Management. Springer-Verlag: New York.

Pollock, M. M. et al. 1994. Beaver as engineers: Influences on biotic and abiotic characteristics of drainage basins. Pages 117–126 *in* C. G. Jones and J. H. Lawton, Eds. Linking Species and Ecosystems. Chapman and Hall: New York.

Pollock, M. M., R. J. Naiman, and T. A. Hanley. 1998. Plant species richness in riparian wetlands: A test of biodiversity theory. *Ecology* 79:94–105.

Poole, G.C. 2002. Fluvial landscape ecology: addressing uniqueness within the river discontinuum. *Freshwater Biology* 47:641–660.

Poole, G. C., R. J. Naiman, J. Pastor, and J. A. Stanford. 1997. Uses and limitations of ground penetrating RADAR in two riparian systems. Pages 140–148 *in* J. Gibert, J. Mathieu, and F. Fournier, Eds. Groundwater/Surface Water Ecotones: Biological and Hydrological Interactions and Management Options. Cambridge University Press: Cambridge.

Poole, G. C., J. A. Stanford, C. A. Frissell, and S. W. Running. 2002. Three-dimensional mapping of geomorphic controls on flood-plain hydrology and connectivity from aerial photos. *Geomorphology* 48:329–347.

Posey, D. A. 1985. Indigenous management of tropical forest ecosystems: The case of Kayapo Indians of the Brazilian Amazon. *Agroforestry Systems* 3:139–158.

Post, D. M. 2002. The long and short of food-chain length. *Trends in Ecology and Evolution* 17:269–277.

Postel, S. L. 1997. Last Oasis. Norton & Company: New York

Postel, S. L. 1998. Water for food production: Will there be enough in 2025? *BioScience* 48:629–637.

Postel, S. L. 1999. Pillars of Sand. W.W. Norton: New York.

Postel, S. L. 2000. Entering an era of water scarcity: The challenges ahead. *Ecological Applications* 10:941–948.

Postel, S. L. and S. Carpenter. 1997. Freshwater ecosystem services. Pages 195–214 *in* G. C. Daily, Ed. Nature's Services. Island Press: Washington, DC.

Postel, S. L., G. C. Daily, and P. R. Ehrlich. 1996. Human appropriation of renewable freshwater. *Science* 271:785–788.

Postel, S. L. and B. Richter. 2003. Rivers for Life: Managing Water for People and Nature. Island Press: Washington, DC.

Powell, J. W. 1875. Exploration of the Colorado River of the West and its Tributaries. U.S. Government Printing Office: Washington, DC.

Power, M. E. and W. E. Dietrich. 2002. Food webs in river networks. *Ecological Research* 17:451–471.

Pregitzer, K. S. and A. L. Friend. 1996. The structure and function of *Populus* root systems. Pages 331–354 *in* R. F. Stettler, H. D. Bradshaw, Jr., P. E. Heilman, and T. M. Hinckley, Eds. Biology of *Populus* and its Implications for Management and Conservation. NRC Research Press: National Research Council of Canada, Ottawa, Ontario, Canada.

Primack, R. B. 1993. Essentials of Conservation Biology. Sinauer: Sunderland, Massachusetts.

Pringle, C. M. 1997. Exploring how disturbance is transmitted upstream: Going against the flow. *Journal of the North American Benthological Society* 16:425–438

Pringle, C. M. 2000. Threats to U.S. public lands from cumulative hydrologic alterations outside of their boundaries. *Ecological Applications* 10:971–989.

Pringle, C. M. 2001. Hydrologic connectivity and the management of biological reserves: A global perspective. *Ecological Applications* 11:981–998.

Pringle, C. M. and M. Barber. 2000. The land-water interface: Science for a sustainable biosphere. *Ecological Applications* 10:939–940.

Pringle, C. M., M. C. Freeman, and B. J. Freeman. 2000. Regional effects of hydrologic alterations on riverine macrobiota in the New World: tropical-temperate comparisons. *BioScience* 50:807–823.

Pringle, C. M. et al. 1988. Patch dynamics in lotic ecosystems: The stream as a mosaic. *Journal of the North American Benthological Society* 7:503–524.

Puckridge, J. T., F. Sheldon, K. F. Walker, and A. J. Boulton. 1998. Flow variability and the ecology of large rivers. *Marine and Freshwater Research* 49:55–72.

Pyšek, P. and K. Prach. 1994. How important are rivers for supporting plant invasions? Pages 19–26 *in* L. C. de Waal, L. E. Child, P. M. Wade, and J. H. Brock, Eds. Ecology and Management of Invasive Riverside Plants. John Wiley & Sons: Chichester, United Kingdom.

Quinn, J. M. et al. 2001. Riparian zone classification for management of stream water quality and ecosystem health. *Journal of the American Water Resources Association* 37:1509–1515.

Raedeke, K. J., Ed. 1988. Streamside Management: Riparian Wildlife and Forestry Interactions. Contribution Number 59. Forest Resources Institute, College of Forest Resources, University of Washington, Seattle.

Rapport, D. J. and W. G. Whitford. 1999. How ecosystems respond to stress. *BioScience* 49:193–203.

Raymond, M. and F. Rousset. 1995. GENEPOP (Version 1.2): Population genetics software for exact tests and ecumenicism. *Journal of Heredity* 86:248–249.

Read, J., D. Clark, E. Venticinque, and M. Moreira. 2003. Application of merged 1-m and 4-m resolution satellite data to research and management in tropical forests. *Journal of Applied Ecology* 40:592–600.

Reeves, G., P. A. Bisson, and J. M. Dambacher. 1998. Fish communities. Pages 200–234 *in* R. J. Naiman and R. E. Bilby, Eds. River Ecology and Management, Springer-Verlag: New York.

Reich, P. B., D. F. Grigal, J. D. Aber, and S. T. Gower. 1997. Nitrogen mineralization and productivity in 50 hardwood and conifer stands on diverse soils. *Ecology* 78:335–347.

Reid, L. M. 1998. Cumulative watershed effects and watershed analysis. Pages 476–501 *in* R. J. Naiman and R. E. Bilby, Eds. River Ecology and Management. Springer-Verlag: New York.

Remsen, J. V. and T. A. Parker. 1983. Contribution of river-created habitats to bird species richness in Amazonia. *Biotropica* 15:223–231.

Resh, V. H. et al. 1988. The role of disturbance in stream ecology. *Journal of the North American Benthological Society* 7:433–455.

Revenga, C., J. Brunner, N. Henninger, K. Kassem, and R. Payne. 2000. Pilot Analysis of Global Ecosystems (PAGE): Freshwater Systems. World Resources Institute: Washington, DC.

Richards, E. H. and A. G. Norman. 1931. The biological decomposition of plant materials versus some factors determining the quantity of nitrogen immobilized during decomposition. *Biochemical Journal* 25:1769–1778.

Richards, J. F. 1990. Land transformation. Pages 163–178 *in* B. L. Turner III, W. C. Clark, R. W. Kates, J. F. Richards, J. T. Mathews, and W. B. Meyer, Eds. 1990. The Earth as Transformed by Human Action. Cambridge University Press: Cambridge.

Richards, J. H. and M. M. Cadwell. 1987. Hydraulic lift: Substantial nocturnal water transport between soil layers by *Artemisis tridentate* roots. *Oecologia* 73:486–489.

Richter, B. D. and H. E. Richter. 2000. Prescribing flood regimes to sustain riparian ecosystems along meandering rivers. *Conservation Biology* 14:1467–1478.

Rickard, J. 1993. Warthog (*Phacochoerus aethiopicus*, Pallas) foraging patterns in stands of wild rice (*Oryza longistaminata*, A. Chev and Roehr) in the Nyl River flood plain. Thesis, University of Witwatersrand, Johannesburg, South Africa.

Riedacker, A. 1974. Un nouvel outil pour l'étude des racines et de la rhizosphère: le minirhizotron. *Annales des Sciences Forestières* 31:129–134.

Ripple, W. J., and R. L. Beschta. 2003. Wolf reintroduction, predation risk, and cottonwood recovery in Yellowstone National Park. *Forest Ecology and Management* 184:299–313.

Risser, P. G., Ed. 1993. Ecotones. *Ecological Applications* 3:369–445.

Risser, P. G., J. R. Karr, and R. T. T. Forman. 1984. Landscape ecology: Directions and approaches. Special Publication 2. Illinois Natural History Survey: Champaign, Illinois.

Robertson, A., A. M. Jarvis, C. J. Brown, and R. E. Simmons. 1998. Avian diversity and endemism in Namibia: Patterns from the Southern Africa Bird Atlas Project. *Biodiversity and Conservation* 7:495–511.

Rodieck, J. E. 2002. Landscape and Urban Planning cover for 2003. *Landscape and Urban Planning* 62:1–2.

Rodriguez-Iturbe, I. and A. Rinaldo. 1997. Fractal River Basins. Cambridge University Press: New York.

Rogers, K. H. 1997. Inland wetlands. Pages 322–347 *in* R. Cowling and D. Richardson, Eds. Vegetation of Southern Africa. Cambridge University Press: Cambridge.

Rood, S. B., J. H. Braatne, and F. M. R. Hughes. 2003a. Ecophysiology of riparian cottonwoods: Stream flow dependency, water relations and restoration. *Tree Physiology* 23:1113–1124.

Rood, S. B., A. R. Kalischuk, M. L. Polzin, and J. H. Braatne. 2003b. Branch propagation, not cladoptosis, permits dispersive, clonal reproduction of riparian cottonwoods. *Forest Ecology and Management* 186:227–242.

Rood, S. B. and J. M Mahoney. 1990. Collapse of riparian poplar forests downstream from dams in western prairies: Probable causes and prospects for mitigation. *Environmental Management* 14:451–464.

Rood, S. B. and J. M. Mahoney. 2000. Revised instream flow regulation enables cottonwood recruitment along the St Mary River, Alberta, Canada. *Rivers* 7:109–127.

Rood, S. B. et al. 2003b. Flows for floodplain forests: Successful riparian restoration along the lower Truckee River, Nevada, USA. *BioScience* 7:647–656.

Rosales, J., G. Petts, and J. Salo. 1999. Riparian flooded forests of the Orinoco and Amazon basins: A comparative review. *Biodiversity and Conservation* 8:551–586.

Rosenberg, D. M., P. McCully, and C. M. Pringle. 2000. Global-scale environmental effects of hydrological alterations: Introduction. *BioScience* 50:746–751.

Rosgen, D. L. 1994. A classification of natural rivers. *Catena* 22:169–199.

Rosselli, A. 2002. Aquatic Habitat Identification and Mapping in the Western Amagon Using Landsat 7 Enhanced Thematic Mapper Plus (ETM+). Master Thesis, Florida International University, Miami.

Rot, B. W., R. J. Naiman, and R. E. Bilby. 2000. Stream channel configuration, landform, and riparian forest structure in the Cascade Mountains, Washington. *Canadian Journal of Fisheries and Aquatic Sciences* 57:699–707.

Roux, A. L., J. P. Bravard, C. Amoros, and G. Pautou. 1989. Ecological changes of the French upper Rhône River since 1750. Pages 323–350 *in* G. E. Petts, Ed. Historial Changes of Large Alluvial Rivers: Western Europe. John Wiley & Sons: Chichester, United Kingdom.

Rundel, P. W. and S. B. Sturmer. 1998. Native plant diversity in riparian communities of the Santa Monica Mountains, California. *Madroño* 45:93–100.

Sabater, S. et al. 2003. Nitrogen removal by riparian buffers along a European climatic gradient: Patterns and factors of variation. *Ecosystems* 6:20–30.

Sabo, J. L. and M. E. Power. 2002a. Numerical response of lizards to aquatic insects and short-term consequences for terrestrial prey. *Ecology* 83:3023–3036.

Sabo, J. L. and M. E. Power. 2002b. River-watershed exchange: Effects of riverine subsidies on riparian lizards and their terrestrial prey. *Ecology* 83:1860–1869.

Sabo, J. L. et al. 2005. Riparian zones increase regional species diversity by harboring different, not more species. *Ecology* (in press).

Sadler, J. P., D. Bell, and A. Fowles. 2004. The hydroecological controls and conservation value of beetles on exposed riverine sediments in England and Wales. *Biological Conservation* 118:41–56.

Salafsky, N. and E. Wollenberg. 2000. Linking livelihoods and conservation: A conceptual framework and scale for assessing the integration of human needs and biodiversity. *World Development* 28:1421–1438.

Salo, J. 1990. External processes influencing origin and maintenance of inland waterland ecotones. Pages 37–64 *in* R. J. Naiman and H. Décamps, Eds. Ecology and Management of Aquatic-Terrestrial Ecotones. UNESCO, Paris and Parthenon Publishing Group: Carnforth, United Kingdom.

Salo, J. et al. 1986. River dynamics and the diversity of Amazon lowland forest. *Nature* 322:254–258.

Sanchez, P. P., P. M. Rivas, and A. Cadena. 1996. Biological diversity of a community of Chiroptera and its relation with the structure of the gallery forest habitat, Macarena Mountains, Colombia. *Caldasia* 18:343–353 (in Spanish).

Sanzone, D. M. et al. 2003. Carbon and nitrogen transfer from a desert stream to riparian predators. *Oecologia* 134:238–250.

Schama, S. 1995. Landscape and Memory. Fontana Press: London.

Schmidt, J. C., R. H. Webb, R. A. Valdez, G. R. Marzolf, and L. E. Stevens. 1998. Science and values in river restoration in the Grand Canyon. *BioScience* 48:735–747.

Schmidt, W. 1989. Plant dispersal by motor cars. *Vegetatio* 80:147–152.

Schneider, R. L. and R. R. Sharitz. 1988. Hydrochory and regeneration in a bald cypress-water tupelo swamp forest. *Ecology* 69:1055–1063.

Schneider, S. H. and T. L. Root. 1998. Climate change. Pages 89–116 *in* M. J. Mac, P. A. Opler, P. Doran, and C. Haecker, Eds. Status and Trends of Our Nation's Biological Resources. Volume 1. National Biological Service: Washington, DC.

Schnitzler, A. 1994. Conservation of biodiversity in alluvial hardwood forests of the temperate zone: The example of the Rhine valley. *Forest Ecology and Management* 68:385–398.

Schnitzler, A., R. Carbiener, and M. Trémolières. 1992. Ecological segregation between closely related species in the flooded forests on the upper Rhine plain. *New Phytologist* 121:203–301.

Schreiner, E. G., K. A. Krueger, P. J. Happe, and D. B. Houston. 1996. Understory patch dynamics and ungulate herbivory in old-growth forests of Olympic National Park, Washington. *Canadian Journal of Forest Research* 26:255–265.

Schroeder, R. L. and A. W. Allen. 1992. Assessment of Habitat of Wildlife Communities on the Snake River, Jackson, Wyoming. Resource Publication 190, U.S. Department of the Interior, Fish and Wildlife Service: Washington, DC.

Schuldt, J. A. and A. E. Hershey. 1995. Effect of salmon carcass decomposition on Lake Superior tributary streams. *Journal of the North American Benthological Society* 14:259–268.

Schulz, R. C. et al. 2000. Riparian forest practices. Pages 189–281 *in* North American Agroforestry: An Integrated Science and Practice. American Society of Agronomy: Madison, Wisconsin.

Schulz, R. C. et al. 1995. Design and placement of a multi-species riparian buffer strip system. *Agroforestry Systems* 29:201–226.

Schumm, S. A. 1977. The Fluvial System. John Wiley & Sons: New York.

Schumm, S. A. and R. Hadley. 1961. Progress in the application of landform analysis in studies of semi-arid erosion. *U.S. Geological Survey Circular 437*, Washington, DC.

Schwer, C. B. and J. C. Clausen. 1989. Vegetative filter treatment of dairy milkhouse wastewater. *Journal of Environmental Quality* 18:446–451.

Scott, M. L., J. M. Friedman, and G. T. Auble. 1996. Fluvial processes and the establishment of bottomland trees. *Geomorphology* 14:327–339.

Scott, M. L., P. B. Shafroth, and G. T. Auble. 1999. Responses of riparian cottonwoods to alluvial water table declines. *Environmental Management* 23:347–358.

Scott, R. L., W. J. Shuttleworth, D. C. Goodrich, and T. Maddock III. 2000. The water use of two dominant vegetation communities in a semiarid riparian ecosystem. *Agricultural and Forest Meteorology* 105:241–256.

Seastedt, T. R., R. A. Ramundo, and D. C. Hayes. 1988. Maximization of densities of soil animals by foliage herbivory: Empirical evidence, graphical and conceptual models. *Oikos* 51:243–248.

Sedell, J. R. and J. L. Froggett. 1984. Importance of streamside forests to large rivers: The isolation of the Williamette River, Oregon, USA, from its floodplain by snagging and streamside forest removal. *Verhandlungen der Internationalen Vereinigung für Theoretische und Angewandte Limnologie* 22:1828–1834.

Sedell, J. R., G. H. Reeves, and K. M. Burnett. 1994. Development and evaluation of aquatic conservation strategies. *Journal of Forestry* 92:28–31.

Segelquist, C. A., M. L. Scott, and G. T. Auble. 1993. Establishment of *Populus deltoides* under simulated alluvial groundwater declines. *American Midland Naturalist* 130:274–285.

Seitzinger, S. and C. Kroeze, 1998. Global distribution of nitrous oxide production and N inputs in freshwater and coastal marine ecosystems. *Global Biogeochemical Cycles* 12:93–113.

SER (Society for Ecological Restoration) and Policy Working Group. 2002. The SER Primer on Ecological Restoration. www.ser.org [Web site accessed June 2004].

Shafroth, P. B., G. T. Auble, and M. L. Scott. 1995. Germination and establishment of the native plains cottonwood (*Populus deltoides* Marshall subsp. Monilifera) and the exotic Russian olive (*Eleagnus angustifolia* L.). *Conservation Biology* 9:1169–1175.

Shafroth, P. B., G. T. Auble, J. C. Stromberg, and D. T. Patten. 1998. Establishment of woody vegetation in relation to annual patterns of streamflow, Bill William River, Arizona. *Wetlands* 18:557–590.

Shannon, M. A. 1998. Social organizations and institutions. Pages 529–552 *in* R. J. Naiman and R. E. Bilby, Eds. River Ecology and Management. Springer-Verlag: New York.

Shear, T. H., T. J. Lent, and S. Fraver. 1996. Comparison of restored and mature bottomland hardwood forests of Southwestern Kentucky. *Restoration Ecology* 4:111–123.

Shelly, D. 2000. Comparative vertebrate fauna survey of the Paroo, Cobham, and Gumbalara landsystems in the western division of New South Wales. *Australian Zoologist* 31:470–481.

Sher, A. A., D. L. Marshall, and S. A. Gilbert. 2000. Competition between native *Populus deltoides* and invasive *Tamarix ramosissima* and the implications for reestablishing flooding disturbance. *Conservation Biology* 14:1744–1754.

Shirley, H. L. 1929. The influence of light intensity and light quality upon the growth and survival of plants. *American Journal of Botany* 16:354–390.

Shirley, H. L. 1945. Light as an ecological factor and its measurement. *Botanical Review* 1:497–532.

Shure, D. J., M. R. Gottschalk, and K. A. Parsons. 1986. Litter decomposition processes in a floodplain forest. *American Midland Naturalist* 115:314–327.

Sigafoos, R. S. 1964. Botanical evidence of floods and flood plain deposition. U.S. Geological Survey Professional Paper 485-A, Washington, DC.

Silver, W. L., D. J. Herman, and M. K. Firestone. 2001. Dissimilatory nitrate reduction to ammonium in upland tropical forest soils. *Ecology* 82:2410–2416.

Simenstad, C. A. and R. M. Thom. 1996. Functional equivalency trajectories of the restored Gog-Le-Hi-Te estuarine wetland. *Ecological Applications* 6:38–56.

Sinsabaugh, R. L., E. F. Benfield, and A. E. Linkins. 1981. Cellulase activity associated with the decomposition of leaf litter in a woodland stream. *Oikos* 36:184–190.

Sioli, H. E. 1984. The Amazon. Limnology and Landscape Ecology of Mighty Tropical River and its Basin. Dr. W. Junk Publishers. The Hague: Dordrecht.

Slater, F. M., P. Curry, and C. Chadwell. 1987. A practical approach to the examination of the conservation status of vegetation in river corridors in Wales. *Biological Conservation* 43:259–263.

Smart, R. P. et al. 2001. Riparian zone influence on stream water chemistry at different spatial scales: A GIS-based modeling approach, an example for the Dee, NE Scotland. *The Science of the Total Environment* 280:173–193.

Smith, S. D., A. B. Wellington, J. A. Nachlinger, and C. A. Fox. 1991. Functional responses of riparian vegetation to streamflow diversions in the eastern Sierra Nevada. *Ecological Applications* 1:89–97.

Smits, A. J. M., P. Laan, R. H. Their, and G. van der Velde. 1990. Root aerenchyma, oxygen leakage patterns and alcoholic fermentation ability of the roots of some nymphaeid and isoetid macrophytes in relation to the sediment type of their habitat. *Aquatic Botany* 38:3–17.

Snelder, T. H. and B. J. F. Biggs. 2002. Multiscale river environment classification for water resources management. *Journal of the American Water Resources Association* 38:1225–1239.

Snyder, E. B. and J. A. Stanford. 2001. Review and Synthesis of River Ecological Studies in the Yakima River, Washington, with Emphasis on Flow and Salmon Habitat Interactions. Final Report to U.S. Department of Interior, Yakima, Washington.

Snyder, K. A. and D. G. Williams. 2000. Water sources used by riparian trees varies among stream types on the San Pedro River, Arizona. *Agricultural and Forest Meteorology* 105:227–240.

SOAC (Systems Operations Advisory Committee). 1999. Report on Biologically-Based Flows for the Yakima River Basin. Final Report to U.S. Department of Interior, Yakima, Washington.

Söderqvist, T., W. J. Mitsch, and R. K. Turner. 2000. Valuation of wetlands in a landscape and institutional perspective. *Ecological Economics* 35:1–6.

Soil and Water Conservation Society. 2001. Realizing the Promise of Conservation Buffer Technology. Ankeny, Iowa.

Soulé, M. E., Ed. 1986. Conservation Biology: The Science of Scarcity and Diversity. Sinauer: Sunderland, Massachusetts.

Sparks, R. E., P. B. Bayley, S. L. Kohler, and L. L. Osborne. 1990. Disturbance and recovery of large floodplain rivers. *Environmental Management* 14:699–709.

Staddon, P. L., C. B. Ramsey, N. Ostle, P. Ineson, and A. H. Fitter. 2003. Rapid turnover of mycorrhizal fungi determined by AMS microanalysis of ^{14}C. *Science* 300:1138–1140.

Stanford, J. A. and G. C. Poole. 1996. A protocol for ecosystem management. *Ecological Application* 6:741–744.

Stanford, J. A. and J. V. Ward. 1988. The hyporheic habitat of river ecosystems. *Nature* 335:64–66.

Stanford, J. A. and J. V. Ward. 1992. Management of aquatic resources in large catchments: Recognizing interactions between ecosystem connectivity and environmental disturbance. Pages 91–124 *in* R. J. Naiman, Ed. Watershed Management. Springer-Verlag: New York.

Stanford, J. A. and J. V. Ward. 1993. An ecosystem perspective of alluvial rivers: Connectivity and the hyporheic corridor. *Journal of the North American Benthological Society* 12:48–60.

Stanford J. A. et al. 1996. A general protocol for restoration of regulated rivers. *Regulated Rivers: Research and Management* 12:391–413.

Steel, E. A., R. J. Naiman, and S. D. West. 1999. Use of woody debris piles by birds and small mammals in a riparian corridor. *Northwest Science* 73:19–26.

Steinberger, K. H. 1996. The spider fauna of riparian habitats of the River Lech (Northern Tyrol, Austria): (Arachnida: Araneae). *Berichte des Naturwissenschaftlich Medizinischen Vereins in Innsbruck* 83:187–210 (in German).

Steinblums, I. J., H. A. Froehlich, and J. K. Lyons. 1984. Designing stable buffer strips for stream protection. *Journal of Forestry* 82:49–52.

Sterner, R. W. and J. J. Elser. 2002. Ecological Stoichiometry. Princeton University Press: Princeton, New Jersey.

Stettler, R. F., H. D. Bradshaw, Jr., P. E. Heilman, and T. M. Hinckle, Eds. 1996. Biology of *Populus* and its implications for Management and Conservation. NRC Research Press: National Research Council of Canada, Ottawa.

Stevens, L. E. et al. 2001. Planned flooding and Colorado River riparian trade-offs downstream from Glen Canyon Dam, Arizona. *Ecological Applications* 11:701–710.

Stevenson, F. J. 1982. Humus Chemistry. John Wiley & Sons: New York.

Stohlgren, T. J., K. A. Bull, Y. Otsuki, C. A. Villa, and M. Lee. 1998. Riparian zones as havens for exotic plant species in the central grasslands. *Plant Ecology* 138:113–125.

Stohlgren, T. J. et al. 1999. Exotic plant species invade hot spots of native plant diversity. *Ecological Monographs* 69:25–46.

Strahler, A. N. 1957. Quantitative analysis of watershed geomorphology. *Transactions of the American Geophysical Union* 38:913–920.

Strahler, A. N. 1964. Quantitative geomorphology of drainage basins and channel networks. Section 4-2 *in* V.T. Chow, Ed. Handbook of Applied Hydrology. McGraw-Hill: New York.

Strayer, D. L., M. E. Power, W. F. Fagan, S. T. A. Pickett, and J. Belnap. 2003. A classification of ecological boundaries. *BioScience* 53:723–729.

Stromberg, J. C. 1998. Dynamics of Fremont cottonwood (*Populus fremontii*) and salt cedar (*Tamarix chinensis*) populations along the San Pedro River, Arizona. *Journal of Arid Environments* 40:133–155.

Stromberg, J. C. 2001. Influence of stream flow regime and temperature on growth rate of the riparian tree, *Plantanus wrightii*, in Arizona. *Freshwater Biology* 46:227–239.

Stromberg, J. C. 2001. Restoration of riparian vegetation in the south-western United States: Importance of flow regimes and fluvial dynamism. *Journal of Arid Environments* 49:17–34.

Stromberg, J. C. and M. K. Chew. 2002. Flood pulses and restoration of riparian vegetation in the American Southwest. Pages 11–49 *in* B. A. Middleton, Ed. Flood

Pulsing in Wetlands: Restoring the Hydrological Balance. John Wiley & Sons: New York.

Stromberg, J. C. and D. T. Patten. 1995. Instream flow and cottonwood growth in the eastern Sierra Nevada of California USA. *Regulated Rivers* 12:1–12.

Stromberg, J. C., R. Tiller, and B. D. Richter. 1996. Effects of groundwater decline on riparian vegetation of semiarid regions: The San Pedro, Arizona. *Ecological Applications* 6:113–134.

Stumm, W. and J. J. Morgan. 1996. Aquatic Chemistry, 3rd edition. John Wiley & Sons: New York.

Suberkropp, K. and E. Chauvet. 1995. Regulation of leaf breakdown by fungi in streams: Influences of water chemistry. *Ecology* 76:1433–1445.

Suberkropp, K. and M. J. Klug. 1976. Fungi and bacteria associated with leaves during processing in a woodland stream. *Ecology* 57:707–719.

Suberkropp, K. F. 1998. Microorganisms and organic matter decomposition. Pages 120–143 *in* R. J. Naiman and R. E. Bilby, Eds. 1998. River Ecology and Management. Springer-Verlag: New York.

Surell, A. 1841. Étude sur les Torrents des Hautes-Alpes. Paris, France.

Swales, S. and J. H. Harris. 1995. The expert panel assessment method (EPAM): A new tool for determining environmental flows in regulated rivers. Pages 125–134 *in* D. M. Harper and A. J. D. Ferguson, Eds. The Ecological Basis for River Management. John Wiley & Sons: Chichester, United Kingdom.

Swanson, F. J., S. V. Gregory, J. R. Sedell, and A. G. Campbell. 1982. Land-water interactions: The riparian zone. Pages 267–291 *in* R. L. Edmonds, Ed. Analysis of Coniferous Forest Ecosystems in the Western United States. Hutchinson Ross: Stroudsburg, Pennsylvania.

Swanson, S., R. Miles, S. Leonard, and K. Genz. 1988. Classifying rangeland riparian areas: The Nevada task force approach. *Journal of Soil and Water Conservation* 43:259–263.

Sweeney, B. W., J. K. Jackson, J. D. Newbold, and D. H. Funk. 1992. Climate change and the life histories and biogeography of aquatic insects in eastern North America. Pages 143–176 *in* P. Firth and S. G. Fisher, Eds. Global Climate Change and Freshwater Ecosystems. Springer-Verlag: New York.

Swift, B. L. 1984. Status of riparian systems in the United States. *Water Resources Bulletin* 20:223–228.

Tabacchi, E. et al. 2000. Impacts of riparian vegetation on hydrological processes. *Hydrological Processes* 14:2959–2976.

Tabacchi, E. and A.-M. Planty-Tabacchi. 2000 Riparian plant community composition and the surrounding landscape: Functional significance of incomers. Pages 11–16 *in* P. J. Wigington, Jr. and R. L. Beschta, Eds. Riparian Ecology and Management in Multi-Land Use Watersheds. Proceedings of the American Water Resources Association's 2000 Summer Conference, Corvallis, Oregon.

Teller, J. T. 1990. Volume and routing of late-glacial runoff from the southern Laurentide ice sheet. *Quaternary Research* 34:12–23.

Tenny, F. G. and S. A. Waksman. 1929. Composition of natural organic materials and their decomposition in the soil: IV The nature and rapidity of decomposition of the various organic complexes in different plant materials under aerobic conditions. *Soil Science* 28:55–84.

405

Terborgh, J. and K. Petren. 1991. Development of habitat structure through succession in an Amazonian floodplain forest. Pages 28–46 *in* S. S. Bell, E. D. McCoy, and H. R. Mushinsky, Eds. Habitat Structure: The Physical Arrangement of Objects in Space. Chapman and Hall: London.

Thébaud, C. and M. Debussche. 1991. Rapid invasion of *Fraxinus ornus* L. along the Hérault River system in southern France: The importance of seed dispersal by water. *Journal of Biogeography* 18:7–12.

Thieme, D. M. 2001. Historic and possible prehistoric human impacts on flood-plain sedimentation, North Brach of the Susquehanna River, Pennsylvania, United States. Pages 375–403 *in* D. Maddy, M. G. Macklin, and J. C. Woodward, Eds. River Basin Sediment Systems: Archives of Environmental Change. A. A. Balkema: Lisse, The Netherlands.

Thorburn, P. J. and G. R. Walker. 1994. Variations in stream water uptake by *Eucalyptus camaldulensis* with differing access to stream water. *Oecologia* 100:293–301.

Thorne, C. R. 1990. Effects of vegetation on river bank erosion and stability. Pages 123–144 *in* C. R. Thorne, Ed. Vegetation and Erosion. John Wiley & Sons: Chichester, United Kingdom.

Tilman, D. 1982. Resource competition and community structure. Princeton University Press: Princeton, New Jersey.

Tobias, C. R., S. A. Macko, I. C. Anderson, E. A. Canuel, and J. W. Harvey. 2001. Tracking the fate of a high concentration groundwater nitrate plume through a fringing marsh: A combined groundwater tracer and in situ isotope enrichment study. *Limnology and Oceanography* 46:1977–1989.

Tockner, K. et al. 2005. Flood plains: Critically threatened ecosystems. *In* N. V. C. Polunin, Ed. Aquatic Ecosystems: Trends and Global Prospects. Cambridge University Press, Cambridge, United Kingdom (in press).

Tockner K., F. Malard, U. Uehlinger, and J. V. Ward. 2002. Nutrient and organic matter in a glacial river-floodplain system (Val Roseg, Switzerland). *Limnology and Oceanography* 47:266–277.

Tockner, K., F. Malard, and J. V. Ward. 2000. An extension of the flood pulse concept. *Hydrological Processes* 14:2861–2883.

Tockner, K., F. Schiemer, and J. V. Ward. 1998. Conservation by restoration: The management concept for a river-floodplain system on the Danube River in Austria. *Aquatic Conservation: Marine and Freshwater Ecosystems* 8:71–86.

Tockner, K., and J. A. Stanford. 2002. Riverine floodplains: Present state and future trends. *Environmental Conservation* 29:308–330.

Tockner, K. and J. V. Ward. 1999. Biodiversity along riparian corridors. *Archiv für Hydrobiologie*. 115 (Supplement):293–310.

Tockner, K. et al. 2003. The Tagliamento River: A model ecosystem of European importance. *Aquatic Sciences* 65:239–253.

Tockner, K., J. V. Ward, J. Kollmann, and P. J. Edwards, Eds. 2002. Riverine landscapes. *Freshwater Biology* 47:497–907.

Toth, L. A., J. W. Koebel, Jr., A. G. Warne, and J. Chamberlain. 2002. Implications of reestablishing prolonged flood pulse characteristics of the Kissimmee River and floodplain ecosystem. Pages 191–221 *in* B. A. Middleton, Ed. Flood Pulsing in Wetlands: Restoring the Natural Hydrological Balance. John Wiley & Sons: New York.

Townsend, C. N. 1989. The patch dynamics concept of stream community ecology. *Journal of the North American Benthological Society* 8:36–50.

Townsend, C. R. 1996. Concepts in river ecology: Pattern and process in the catchment hierarchy. *Archiv für Hydrobiologie* 113 (Supplement):3–21.

Trémolières, M., J. M. Sánchez-Pérez, A. Schnitzler, and D. Schmitt. 1998. Impact of river management history on the community structure, species composition and nutrient status in the Rhine alluvial hardwood forest. *Plant Ecology* 135:59–78.

Trimble, S. W. 1997. Stream channel erosion and change resulting from riparian forests. *Geology* 25:467–469.

Triska, F. J., J. H. Duff, and R. J. Avanzino. 1990. Influence of exchange flow between the channel and hyporheic zone on nitrate production in a small mountain stream. *Canadian Journal of Fisheries and Aquatic Sciences* 47:2099–2111.

Triska, F. J., A. P. Jackman, J. H. Duff, and R. J. Avanzino. 1994. Ammonium sorption to channel and riparian sediments: a transient storage pool for dissolved inorganic nitrogen. *Biogeochemistry* 26:67–83.

Triska, F. J., V. C. Kennedy, R. J. Avanzino, G. W. Zellweger, and K. E. Bencala. 1989. Retention and transport of nutrients in a third-order stream in northwestern California: Hyporheic processes. *Ecology* 70:1893–1905.

Troll, C. 1939. Luftbildplan und okologische Bodenforsching. *Zeitschrift Gesellschaft für Erdkunde* 241–298.

Turner III, B. L. et al. 1990. The Earth as Transformed by Human Action. Cambridge University Press: Cambridge.

Turner M. G. 1998. Landscape ecology. Pages 77–122 *in* S. I. Dodson, T. E. H. Allen, S. R. Carpenter, A. R. Ives, R. J. Jeanne, J. F. Kitchell, N. E. Langston, and M. G. Turner, Eds. Ecology: Oxford University Press: New York.

Turner, M. G., S. R. Carpenter, E. J. Gustafson, R. J. Naiman, and S. M. Pearson. 1998. Land use. Pages 37–61 *in* M. J. Mac, P. A. Opler, P. Doran, and C. Haecker, Eds. Status and Trends of Our Nation's Biological Resources. Volume 1. National Biological Service: Washington, DC.

Turner, M. G., D. N. Wear, and R. O. Flamm. 1996. Land ownership and land-cover change in the Southern Appalachian Highlands and the Olympic Peninsula. *Ecological Applications* 6:1150–1172.

U.S. Department of Commerce. 1997. Demographic state of the nation: 1997. Special Studies Series P23–193. Bureau of the Census: Washington, DC.

U.S. Department of the Interior, Fish and Wildlife Service and U.S. Department of Commerce Bureau of the Census. 1996 National Survey of Fishing, Hunting, and Wildlife-Associated Recreation. U.S. Fish and Wildlife Service Report FHW/96 NAT, Arlington, Virginia.

U.S. Fish and Wildlife Service and Hoopa Valley Tribe. 1999. Trinity River Flow Evaluation. Report SH222.C3 U654, U.S. Department of the Interior, Arcata, California.

Uetz, G. W., K. L. van der Laan, G. F. Summers, P. A. K. Gibson, and L. L. Getz. 1979. The effects of flooding on floodplain arthropod distribution, abundance and community structure. *American Midland Naturalist* 101:286–299.

UNEP (United Nations Environment Programme) World Conservation Monitoring Centre. 2000. European Forest and Protected Areas: Gap Analysis. Cambridge, United Kingdom.

UNESCO (United Nations Educational, Scientific and Cultural Organization). 2003. Division of Ecological Sciences and Man and the Biosphere Web site http://www.unesco.org/mab/EE/home.htm [Web site accessed June 2004].

USDA (United States Department of Agriculture). 1992. Integrated riparian evaluation guide: Intermountain region. Technical Report Number PB-94-115953/XAB, U.S. Forest Service: Ogden, Utah.

USDA (U.S. Department of Agriculture). 2003. The Conservation Reserve Program 26th Signup. Available at http://www.fsa.usda.gov/conservation.

USDA Soil Survey Staff. 1998. Keys to Soil Taxonomy, 8th Edition. Pocahontas Press: Blacksburg, Virginia.

USEPA (U.S. Environmental Protection Agency). 1998. Wetland Bioassessment Fact Sheet. EPA 843-F-98-001. Office of Wetlands, Oceans, and Watersheds, Office of Water, USEPA, Washington, DC.

Vadas, R. L. and J. E. Sanger. 1997. Lateral zonation of trees along a small Ohio stream. *Ohio Journal of Science* 97:107–112.

Valett, H. M., J. A. Morrice, C. N. Dahm, and M. E. Campana. 1996. Parent lithology, surface-groundwater exchange, and nitrate retention in headwater streams. *Limnology and Oceanography* 41:333–345.

van Andel, J. and J. Aronson, Eds. 2005. Restoration Ecology: A European Perspective. Blackwell Science: Oxford.

van Breemen, N. et al. 2002. Where did all the nitrogen go? Fate of nitrogen inputs to large watersheds in the northeastern USA. *Biogeochemistry* 57/58:267–293.

van Cleve, K., L. A. Viereck, and R. L. Schlentner. 1971. Accumulation of nitrogen in alder (*Alnus*) ecosystems near Fairbanks, Alaska. *Arctic and Alpine Research* 3:101–114.

van Cleve, K., J. Yarie, R. Erickson, and C. T. Dyrness. 1993. Nitrogen mineralization and nitrification in successional ecosystems on the Tanana River floodplain, interior Alaska. *Canadian Journal of Forest Research* 23:970–978.

van Coller, A.L. 1993. Riparian vegetation of the Sabie River: Relating spatial distribution patterns to characteristics of the physical environment. M.Sc. Thesis, University of the Witwatersrand, Johannesburg, South Africa.

van Coller, A. L., K. H. Rogers, and G. L. Heritage. 1997. Linking riparian vegetation types and fluvial geomorphology along the Sabie River within Kruger National Park, South Africa. *African Journal of Ecology* 35:194–212.

van Coller, A. L., K. H. Rogers, and G. L. Heritage. 2000. Riparian vegetation-environment relationships: Complimentarity of gradients versus patch hierarchy approaches. *Journal of Vegetation Science* 11:337–350.

Van den Brink, F. W. B., A. van der Velde, A. D. Buijse, and A. G. Klink. 1996. Biodiversity in the Lower Rhine and Meuse river-floodplains: Its significance for ecological management. *Netherlands Journal of Aquatic Ecology* 30:129–149.

van der Berg, M. W. and T. van Hoof. 2001. The Maas terrace sequence at Maastricht, SE Netherlands: Evidence for 200 m of late Neogene and Quaternary surface uplift. Pages 45–86 *in* D. Maddy, M. G. Macklin, and J. C. Woodward, Eds. River Basin Sediment Systems: Archives of Environmental Change. A. A. Balkema: Lisse, The Netherlands.

van der Valk, A. G., S. D. Swanson, and R. F. Nuss. 1983. The response of plant species to burial in three types of Alaskan wetlands. *Canadian Journal of Botany* 61:1150–1164.

Van Dijk, G. M., E. C. L. Marteijn, and A. Schulte-Wülwer-Leidig. 1995. Ecological rehabilitation of the River Rhine: Plans, progress, and perspectives. *Regulated Rivers* 11:377–388.

Vandenberghe, J. 1995. Timescales, climate and river development. *Quaternary Science Reviews* 14:631–638.

Vannote, R. L., G. W. Minshall, K. W. Cummins, J. R. Sedell, and C. E. Cushing. 1980. The river continuum concept. *Canadian Journal of Fisheries and Aquatic Sciences* 37:130–137.

Veneklaas, E. J. 1991. Litterfall and nutrient fluxes in two montane tropical rain forests, Columbia. Journal of Tropical Ecology 7:319–336.

Verneaux, J. 1973. Cours d'eau de Franche-Comté (massif du Jura). Recherches écologiques sur le réseau hydrographique du Doubs—Essai de biotypologie. *Annales Scientifiques de l'Université de Besançon, Zoologie* 3:79–90.

Viljoen, J. A. and E. B. Fred. 1924. The effect of different kinds of wood and of wood pulp cellulose on plant growth. *Soil Science* 17:199–208.

Virgos, E. 2001. Relative value of riparian woodlands in landscapes with different forest cover for medium-sized Iberian carnivores. *Biodiversity and Conservation* 10:1039–1049.

Vita-Finzi, C. 1969. The Mediterranean Valleys: Geological Changes in Historical Times. Cambridge University Press: Cambridge.

Vitousek, P. M. 1984. Litterfall, nutrient cycling, and nutrient limitation in tropical forests. *Ecology* 65:285–298.

Vitousek, P. M., S. Hattenschwiler, L. Olander, and S. Allison. 2002. Nitrogen and Nature. *Ambio* 31:97–101.

Vitousek, P. M. and J. R. Sanford. 1986. Nutrient cycling in moist tropical forest. *Annual Review of Ecology and Systematics* 17:137–167.

Vogt, K. A., C. C. Grier, and D. J. Vogt. 1986. Production, turnover, and nutrient dynamics of above- and below-ground detritus of world forests. *Advances in Ecological Research* 15:303–377.

von Hagen, B., S. Beebe, P. Schoonmaker, and E. Kellogg. 1998. Nonprofit organizations and watershed management. Pages 625–641 *in* R. J. Naiman and R. E. Bilby, Eds. River Ecology and Management. Springer-Verlag: New York.

Vörösmarty, C. J. and D. Sahagian. 2000. Anthropogenic disturbance of the terrestrial water cycle. *BioScience* 50:753–765.

Vörösmarty, C. J. et al. 1997. The storage and aging of continental runoff in large reservoir systems of the world. *Ambio* 26:210–19.

Walker, B. et al. 2002. Resilience management in social-ecological systems: A working hypothesis for a participatory approach. *Conservation Ecology* 6:14 [online] URL: http://www.consecol.org/vol6/iss1/art14.

Walker, L. 1989. Soil nitrogen changes during primary succession on a floodplain in Alaska, USA. *Arctic and Alpine Research* 21:341–349.

Wallace, J. B., S. L. Eggert, J. L. Meyer, and J. R. Webster. 1997. Multiple trophic levels of a forest stream linked to terrestrial litter inputs. *Science* 277:102–104.

Wallace, J. B., S. L. Eggert, J. L. Meyer, and J. R. Webster. 1999. Effects of resource limitation on a detrital-based ecosystem. *Ecological Monographs* 69:409–442.

Ward, J. K., T. E. Dawson, and J. R. Ehleringer. 2002. Responses of *Acer negundo* genders to inter-annual differences in water availability determined from carbon isotope ratios of tree ring cellulose. *Tree Physiology* 22:339–346.

Ward, J. V. 1989. The four dimensional nature of lotic ecosystems. *Journal of the North American Benthological Society* 8:2–8.

Ward, J. V. 1998. Riverine landscapes: Biodiversity patterns, disturbance regimes, and aquatic conservation. *Biological Conservation* 83:269–278.

Ward, J. V. and J. A. Stanford. 1983. The serial discontinuity concept of lotic ecosystems. Pages 29–42 *in* T. D. Fontaine and S. M. Bartell, Eds. Dynamics of Lotic Ecosystems. Ann Arbor Science: Ann Arbor, Michigan.

Ward, J. V. and J. A. Stanford. 1995. The serial discontinuity concept: Extending the model to floodplain rivers. *Regulated Rivers: Research and Management* 10:159–168.

Ward, J. V. and K. Tockner. 2001. Biodiversity: Towards a unifying theme for river ecology. *Freshwater Biology* 46:807–819.

Ward, J. V., K. Tockner, D. B. Arscott, and C. Claret. 2002. Riverine landscape diversity. *Freshwater Biology* 47:517–539.

Ward, J. V., K. Tockner, U. Uehlinger, and F. Malard. 2001. Understanding natural patterns and processes in river corridors as the basis for effective river restoration. *Regulated Rivers*: *Research and Management* 17:311–323.

Ward, J. V. et al. 1999. A reference river system for the Alps: The "Fiume Tagliamento," *Regulated Rivers: Research and Management* 15:63–75.

Ward, J. V. et al. 2000. Potential role of island dynamics in river ecosystems. *Verhandlungen der Internationalen Vereinigung für Theoretische und Angewandte Limnologie* 27:2582–2585.

Wardle, D. A. and R. D. Bardgett. 2004. Human-induced changes in large herbivorous mammal density: The consequences for decomposers. *Frontiers in Ecology and the Environment* 2:145–153.

Waring, R. H. and W. H. Schlesinger. 1985. Forest Ecosystems: Concepts and Management. Academic Press: Orlando, Florida.

Warren, A. and J. French. 2001. Relations between nature conservation and the physical environment. Pages 1–5 *in* A. Warren and J. R. French, Eds. Habitat Conservation: Managing the Physical Environment. John Wiley & Sons: Chichester, United Kingdom.

Warren, C. E.1979. Toward classification and rationale for watershed management and stream protection. United States Environmental Protection Agency Report Number EPA—600/3-79-059. United States Environmental Protection Agency: Corvallis, Oregon.

Wasson, J. G. 1989. Eléments pour une typologie fonctionnelle des eaux courants: 1. Revue critique de quelques approches existantes. *Bulletin d'Ecologie* 20:109–127.

WCD (World Commission on Dams). 2000. Dams and Development. Earthscan Publications: London.

Wear, D. N., M. G. Turner, and R. O. Flamm. 1996. Ecosystem management with multiple owners: Landscape dynamics in Southern Appalachian watershed. *Ecological Applications* 6:1173–1188.

Webb, B. W. 1997. Trends in stream and river temperature. Pages 81–102 *in* N. E. Peters, O. P. Bricker, and M. M. Kennedy, Eds. Water Quality Trends and Geochemical Mass Balance. John Wiley & Sons: New York.

Webster, J. R. and J. L. Meyer. 1997. Stream organic matter budgets. *Journal of the North American Benthological Society* 16:3–161.

Webster, J. R., J. B. Wallace, and E. F. Benfield. 1995. Organic processes in streams of the eastern United States. Pages 117–187 *in* C. E. Cushing, K. W. Cummins, and G. W. Minshall, Eds. Ecosystems of the World 22. River and Stream Ecosystems. Elsevier: Amsterdam.

WEHAB Working Group. 2002. A Framework for Action on Water and Sanitation (August), United Nations, http://www.johannesburgsummit.org/html/documents/summit_docs/wehab_papers/wehab_water_sanitation.pdf [Site accessed April 3, 2003].

Welcomme, R. L. 1985. River Fisheries. FAO Fisheries Technical Paper 262. Food and Agriculture Organization (FAO) of the United Nations: Rome, Italy.

Welch, E. B., J. M. Jacoby, and C. W. May. 1998. Stream quality. Pages 69–94 *in* R. J. Naiman and R. E. Bilby, Eds. River Ecology and Management. Springer-Verlag: New York.

Welsh, D. 1991. Riparian Forest Buffers: Function and Design for Protection and Enhancement of Water Resources. USDA-FS Publication NA-PR-07-91, U.S. Department of Agriculture, Forest Service: Radnor, Pennsylvania.

Whigham, D. F. et al. 1999. Hydrogeomorphic (HGM) assessment: A test of user consistency. *Wetlands* 19:560–569.

Whitaker, D. M. and W. A. Montevecchi. 1997. Breeding bird assemblages associated with riparian, interior forest, and nonriparian edge habitats in a balsam fir ecosystem. *Canadian Journal of Forest Research* 27:1159–1167.

White, D. et al. 1997. Assessing risks to biodiversity from future landscape change. *Conservation Biology* 11:349–360.

White, P. S. and S. T. A. Pickett. 1995. Natural disturbance and patch dynamics: An introduction. Pages 3–13 *in* S. T. A. Pickett and P. S. White, Eds. The Ecology of Natural Disturbance and Patch Dynamics. Academic Press: San Diego, California.

Whiting, P. J. 1998. Floodplain maintenance flows. *Rivers* 6:160–170.

Wigley, T. B. and T. H. Roberts. 1997. Landscape level effects of forest management on faunal diversity in bottomland hardwoods. *Forest Ecology and Management* 90:141–154.

Wigley, T. M. L. 1999. The science of climate change: Global and U.S. perspectives. Pew Center on Global Climate Change: Arlington, Virginia.

Williams, D. D. 1984. The hyporheic zone as a habitat for aquatic insects and associated arthropods. Pages 430–455 *in* V. H. Resh and D. M. Rosenberg, Eds. Ecology of Aquatic Insects. Praeger: New York.

Williams, K. S. 1993. Use of terrestrial arthropods to evaluate restored riparian wood-lands. *Restoration Ecology* 1:107–116.

Williams, R. N. et al. 1999. Scientific issues in the restoration of salmonid fishes in the Columbia River. *Fisheries* 24:10–19.

Wilson, L. G. 1967. Sediment removal from flood water by grass filtration. *Transactions of the American Society of Agricultural Engineers* 10:35–37.

Winter, T. C., J. W. Harvey, O. L. Franke, and W. M. Alley. 1998. Ground water and surface water: A single resource. U.S. Geological Survey Circular 1139, Washington, DC.

Wissmar, R. C. and R. R. Beschta. 1998. Restoration and management of riparian ecosystems: A catchment perspective. *Freshwater Biology* 40:571–585.

Wissmar, R. C. and P. A. Bisson, Eds. 2003. Strategies for Restoring River Ecosystems: Sources of Variability and Uncertainty in Natural and Managed Systems. American Fisheries Society: Bethesda, Maryland.

Wissmar, R. C., J. H. Braatne, R. L. Beschta, and S. B. Rood. 2003. Variability of riparian ecosystems: Implications for restoration. Pages 107–127 *in* R. C. Wissmar, and P. A. Bisson, Eds. Strategies for Restoring River Ecosystems: Sources of Variability and Uncertainty in Natural and Managed Systems. American Fisheries Society: Bethesda, Maryland.

Woessner, W. W. 2000. Stream and fluvial plain ground water interactions: Rescaling hydrogeologic thought. *Ground Water* 38:423–429.

Wolman, M. G. 1954. A method of sampling coarse river-bed material. *Transactions of the American Geophysical Union* 35:951–956.

Wright, J. M. and P. Bentzen. 1994. Microsatellites: Genetic markers for the future. *Reviews in Fish Biology and Fisheries* 4:384–388.

Wright, J. P., C. G. Jones, and A. S. Flecker. 2002. An ecosystem engineer, the beaver, increases species richness at the landscape scale. *Oecologia* 132:96–101.

Wu, J. et al. 2004. The Three Gorges Dam: An ecological perspective. *Frontiers in Ecology and the Environment* 2:241–248.

WWDR (World Water Development Report of UNESCO). 2003. Available at http://www.unesco.org/water/wwap/wwdr/index.shtml.

WWF (World Wide Fund for Nature). 1999. Europe's Living Rivers: An agenda for Action. WWF European Freshwater Programme, Copenhagen, Denmark.

WWF (World Wide Fund for Nature). 2000. Wise Use of Floodplains: Policy and Economic Analysis of Floodplain Restoration in Europe. WWF European Freshwater Programme, Copenhagen, Denmark.

Xiong, S. and C. Nilsson. 1997. Dynamics of leaf litter accumulation and its effects on riparian vegetation: A review. *The Botanical Review* 63:240–264.

Xiong, S. and C. Nilsson. 1999. The effects of plant litter on vegetation: A meta-analysis. *Journal of Ecology* 87:984–994.

Xiong, S., C. Nilsson, M. E. Johansson, and R. Jansson. 2001. Responses of riparian plants to accumulations of silt and plant litter: The importance of plant traits. *Journal of Vegetation Science* 12:481–490.

Yodzis, P. 1986. Competition, mortality and community structure. Pages 480–491 *in* J. M. Diamond and T. J. Case, Eds. Community Ecology. Harper & Row: New York.

Young, R. A., T. Huntrods, and W. Anderson. 1980. Effectiveness of vegetated buffer strips in controlling pollution from feedlot runoff. *Journal of Environmental Quality* 9:483–487.

Zavaleta, E. 2000. Valuing ecosystem services lost to *Tamarix* invasion in the United States. Pages 261–300 *in* H. A. Mooney and R. J. Hobbs, Eds. Invasive Species in a Changing World. Island Press: Washington DC.

Zavitkovski, J. and R. D. Stevens. 1972. Primary productivity of red alder ecosystems. *Ecology* 53:235–242.

Zedler, J. B. 2000. Progress in wetland restoration ecology. *Trends in Ecology and Evolution* 15:402–407.

Zhang, X., K. D. Harvey, W. D. Hogg, and R. R. Yuzyk. 2001. Trends in Canadian streamflow. *Water Resources Research* 37:987–998.

Ziemer, R. R. and T. E. Lisle. 1998. Hydrology. Pages 43–68 *in* R. J. Naiman and R. E. Bilby, Eds. River Ecology and Management. Springer-Verlag: New York.

Index

A
aboveground biomass, 112
aboveground communities, 97–104
aboveground fauna diversity, 122–23
aboveground tools, 341–46
Acer negundo, 129
adaptive environmental, 254–55
aerated tissue, 84
aerenchyma, 84–85
aerobic–anaerobic boundaries, 54
agencies, 241, 243; *see also names of specific
    agencies*
agents of change, *see* disturbance and agents
    of change
Airborne Visible/Infrared Spectrometer
    (AVIRIS), 341
air spaces, 84
Alces alces, 182
allochthonous organic matter, 175
alluvial complexes, 25–26, 30
alluvial valleys, 63
Alnus incana, 148
alpha diversity, 117
alternative forest harvest management
    strategies, 257
Amazon River, 4–5
    floodplain of, 44–45
    interaction with floodplain, 39
    riparian forests of, 116
    scale of basin of, 22
ammonia–nitrate reactions, 54
ammonium (NH_4^+), 132, 134
anadromous salmon, 304
anaerobic decay, 152
ancient water development events, 235
anemochory, 88
animals, *see also names of specific types of
    animals*
    components of food webs and
        biogeochemical cycles, 161

diversity among, 121–23
functional grouping of large animal
    interactions, 180–83
importance of, 273–74
in riparian soils, 96–97
soil, 96
anoxic environments, 84
anthropogenic disturbance, 189, 198
anthropogenic environmental change, 206
anthropogenic stress, 205–7
application, of ecological information, 55–57
aquatic environment, 274–75
aquatic insects, 177
aquatic processes, 50–51
aquatic systems, 174–79
aquent suborder, 74
arachnida, 122
Architectural Ensemble, 30
ArcView GIS, 345
areanormalized fluxes, 174
arid regions, 331
Arkansas River, 204
arthropods, 176–79
artificial structures, 27
asexual reproduction, 88, 89–90
Aspect II (morphologic description), 69–70
assessment methods, wetlands, 310
assessments, differentiating, 56
Australia, flow allocation methodology of, 300
avalanches, 146–47
AVIRIS (Airborne Visible/Infrared
    Spectrometer), 341
avoider, 82–83
avulsion, 26

B
BAI (basal area growth), 138, 139–40
banks
    erosion of, 147
    stability of, 294–95

basal area, 112–14
basal area growth (BAI), 138, 139–40
basin water balances, temperature changes effect on, 26
Bear River, mining effects on, 28
beavers, 8, 181–82
bed load transport, 36
bedrock geology, 23
bedrock valley, 62
behavioral and motivational sources, economic value of, 238
belowground biomass, 112
belowground communities, 97–104, 183
belowground tools, 346–48
benthic aquatic herbivorous arthropods, 177
benthic macroinvertebrates, 175
beta diversity, 117–18
between-microhabitat diversity, 117
Big Spring Creek catchment, 248
biocomplexity, 76
biodiversity, 250, 294–95, 313
    conserving riparia for, 271–74
    optimizing riparia for, 322
    theories of, 115, 120
biogeochemical cycles, 161, 349
    natural, 251
    of nitrogen, principles for, 351
biogeographic factors, 73
biological attributes, effect of spatial scale on, 70
biological denitrification, 134
biological diversity, 114–23
    diversity theory and measurement, 115–18
    faunal diversity, 121–23
    vegetative diversity, 118–21
biological integrity, 244
biologically active elements, 353
biologically based approach, 50
biological processes, 20, 157
biomass, 112–14
biomass loss, 155
biophysical characteristics of riparian soils
    fauna, 96–97
    moisture, 94–96
    organic matter, 93–94
biophysical connectivity and riparian functions, 159–88
    energy flows and food webs, 173–79
    large animal connections, 179–86
        functional grouping of large animal interactions, 180–83
        pacific salmon influences on riparian ecosystems, 183–86

nutrient flows, 163–73
    particle-nutrient considerations, 172–73
    riparian zones as buffers against high in-stream nutrient levels, 168–72
    riparian zones as buffers against nutrient pollution from upland runoff, 164–68
    overview, 159–61
    patch dynamics and landscape perspective of catchments, 161–63
biophysical heterogeneity, 91–92
biotic classification, 73–76
    plants, 74–75
    soils, 73–74
    wildlife, 75–76
biotic decomposition, 175
biotic functions of riparia, 125–58
    decomposition dynamics, 149–55
        decomposition of riparian litter, 154–55
        factors controlling immobilization of nitrogen, 151–52
        mechanisms of nitrogen immobilization, 152
        nutrient dynamics during decay, 151
        principles of, 149–51
    information fluxes, 155–56
    microclimate, 156–58
    nutrient fluxes, 131–35
    overview, 125–27
    production ecology, 135–48
        growth and metabolism of riparian trees, 135–37
        litterfall, 143–44
        mortality rates, 145–47
        root production, 147–48
        timing of growth and rates of net primary production, 137–43
    water use and flux, 127–31
biotic influences, 76
biotic patterns, 80
biotic processes, 132
biotic vitality of land, principles for maintaining, 339
biotic zonation patterns, 51
bird species, 8, 122–23
black-tailed deer, 322
BLM (Bureau of Land Management), 242
BLM (Interior's Bureau of Land Management), 56
bole wood production, 138, 140
bottomland forest, 304–5
bottom-up approach, 300
boundary conditions, 54
boundary traits, 53
boxelder, 129

Breonadia, 85
broad geomorphic characterization, 66–67
broad-level classification, 68
buffering diffuse pollution, 294–95
buffer strips, 257
buffer widths, 322
buffer zones, 278–79
    against high in-stream nutrient levels,
        168–72
    against nutrient pollution from, 164–68
    designing, 317–19
    effects on reduction of inputs from
        surface runoff, 280
    reported effectiveness for sediment
        reduction, 319
Bureau of Land Management (BLM), 242
Bureau of Reclamation (USBR), 241
buried paleochannels, 26

C
Ca^{2+} (calcium), 230
calcium (Ca^{2+}), 230
candidate mechanisms, 165
Canis lupus, 182
Castor canadensis, 8, 181–82
catchment ecosystems, 329–31
catchment-level classification, 61
catchment management
    environmental and social approaches to,
        340
    guidelines for, 350
    process linked to and river management,
        246–53
        catchments and rivers benefit from
            riparian management, 251–53
        riparian benefits from catchment
            management, 246–48
        riparian benefits from river
            management, 248–51
catchments, 3
    benefit from riparian management, 251–53
    form, 23–25
    history, 25–28
    patch dynamics and landscape perspective
        of, 161–63
    patterns, 118–19
    scale, 246
    size and shape of, 22, 23
catchments and physical template, 19–48
    geomorphic processes and process
        domains, 32–39
        channel processes, 34–37
        floodplain and channel interactions,
            38–39
        headwater erosion, 33–34

goal of, 20–22
and hierarchical patterns of geomorphic
        features, 22–32
    catchment form and channel networks,
        23–25
    catchment history, 25–28
    hydrologic connectivity and surface
        water–groundwater exchange, 39
    dynamics of linked surface–subsurface
        hydrologic system, 45–47
    surface connectivity and flooding, 40–45
    overview, 19–20
catchment-scale patterns, 119
categories of change, 192–97
    human demography, 192
    resource use, 192–96
    social organization, 197
    technology development, 197
cell collapse, 84
cellulase activity, 150
cellulose degradation, 154
changes, agents of, *see* disturbance and
        agents of change
channel classification, hierarchical levels of,
        62
channeled colluvial valleys, 62
channel fragmentation, data on, 196
channel heads, 22
channel interactions, 38–39
channel networks, 23–25
channel processes, 34–37
channel-reach classification, 37
chaotic environments, 81
chemical information, 155
chemoautotrophic, catabolic processes
        occurring within riparian zones, 135
Chesapeake Bay Preservation Act of 1988,
        285
classifications, 50, 56
classification systems, 76–77
Clean Water Act, 72, 243, 285
Clementsian model, 106
climate change, 218–24
clonal growth, 89–90
collapse, 84
collectors, 175
cological science, ecological subdisciplines to
        use, 262
Colorado River, 210
Columbia River, 4
combined approach, 300
communication needs, 266–67
community-level patterns, 73
community patterns, riparian, 76
community perspective, 259

complex interconnected corridors, 8
complex surface, 4
conditioned litter, 175
Congo river, 5
connectedness property, 263–64
connectivity, 156
conservation biology, 269–90
    conserving riparia for
        biodiversity, 271–74
        ecosystem services, 274–75
        hydrologic effects, 276–77
    emergence of new conservation legislation,
        284–86
    goal of, 270–71
    human benefits from riparian
        conservation, 282–84
    overview, 269–70
    riparian conservation
        for long term, 286–88
        in management context, 277–82
conservation legislation, 284–86
conservation management
    predicting how riparian zone will react to,
        287
    and restoration, 293
Conservation Reserve Enhancement
    Program (CREP), 305–6
contrasting taxa, lateral patterns among,
    100
Convention on Biological Diversity, 271–72
convex hillslopes, 70
Corderie Royale, restoration of, 252
Corps or USACE (U.S. Army Corps of
    Engineers), 241
cottonwood trees, 296
    role of sediment characteristics in, 108
    root production of seeds, 148
    seedling, 7
    sexual reproduction, establishment and
        growth of, 139
CREP (Conservation Reserve Enhancement
    Program), 305–6
crevasse splays, 26, 38
cross-sectional profile, 67
cultural perspective, 238–41, 259
cultural setting, 11–12
cut-and-fill alluviation, 6, 10
cut-and-fill process, 38
cycles, 132–35

D
dams, 208–10, 235, 250
Danube River, 4, 210
decay rate, cause-and-effect relations
    between litter quality, 149

decomposition dynamics, 149–55
    factors controlling immobilization of
        nitrogen, 151–52
    mechanisms of nitrogen immobilization,
        152
    nutrient dynamics during decay, 151
    principles of, 149–51
    riparian litter, 154–55
deer, black-tailed, 322
deforestation, 27
delineative criteria, broad-level classification,
    68
delta diversity, 117–18
dendritic networks, 9
denitrification, 165–66
density, 112–14
depositional zone, 33
derived arthropods, 176–77
derived wood, 11
design professions, 263
detrital nitrogen, 153
development pathways, 294
dielectric properties, 347
differentiation diversity, 118
dilute, in rivers, 7
discharge, 40–41, 224
dissemination and information collection,
    243–45
dissolved organic carbon (DOC), 171–72
distance attribute of riparian corridors, 156
distress syndrome, 189, 205
disturbance and agents of change, 189–232
    categories of change, 192–97
        human demography, 192
        resource use, 192–96
        social organization, 197
        technology development, 197
    consequences of global climate and land
        use changes, 218–32
        climate change, 218–24
        land use change, 225–32
        riparia adapting to, 224–25
    disturbance ecology responses to stress,
        205–7
    ecological consequences of flow
        regulation, 207–18
        alterations to energy and nutrient
            budgets, 215–17
        basic ecological principles, 217–18
        effects on native species and processes,
            212–15
        extent of flow regulation, 210–12
        theory addressing flow regulation by
            dams, 208–10
    goal of, 190–91

overview, 189–90
riparian disturbances, 197–205
  basic concepts and approaches to
    understanding history, 198–201
  defining anthropogenic disturbance,
    198
  historical examples of riparian
    alterations, 201–4
  pervasive human-mediated changes,
    204–5
disturbance ecology, responses to stress,
  205–7; *see also* riparian disturbances
diversity measurement, 118–21
diversity theory, 115–18
DOC (dissolved organic carbon), 171–72
Dolly Varden, 177
down-cutting, 34
downwelling surface water, 170
drainage, 22, 23, 93–94
droughts, 119–20
dry regions, plant production in, 141
dune-ripple reaches, 64
dynamic equilibrium model, 115, 116–17

E
Earth's hydrologic cycle, 352–53
ecological boundaries, 53
ecological engineering, 294
ecological forecasting, prediction, and
  projection, 335
ecological information, application of, 55–57
ecological integrity of riparia, 308–14
ecological knowledge, 288
ecological management, 350–52
ecological mechanisms, 55
ecological organization, 331
ecological principles, 217–18, 351
ecological problem solving, 287–88
ecological processes in rivers, 11
ecological restoration, 292, 294
ecological subdisciplines, 262
ecological theory, 123
ecology, production, *see* production ecology
Ecology of Running Waters, The, 51
economic valuation, riparia, 237–38
ecoregions, 59
Ecosystem Analysis, 201
ecosystems
  effect of spatial scale on, 70
  emerging, 294
  flow allocation approaches for
    rehabilitation of, 297
  pacific salmon influences on, 183–86
  performance, 312
  processes, 157

ecosystem services
  conserving riparia for, 274–75
  importance of, 330–31
El Niño Southern Oscillation (ENSO)
  phenomenon, 25
emotional landscapes, 11
Endangered Species Act, 243, 285
endogenous forces, 53
endurer category, 82
enduring classification system, attributes of,
  50, 77–78
energy budgets, 215–17
energy flows, 173–79
ENSO (El Niño Southern Oscillation)
  phenomenon, 25
Entisol orders, 73–74
environmentally arbitrary boundaries, 55
environmental protection and conservation,
  285
environmental regulations, 243
EPA (U.S. Environmental Protection
  Agency), 242
epigean channel networks, 4
epsilon diversity, 117–18
erosion
  on banks, 147
  power of, 26
  on rivers, 5
erosional zone, 32
ethical values, 239, 241
European countries
  alluvial forests in, 293
  development of river regulation on rivers,
    202
  primeval flooded forests, 237
  range of changes on rivers, 201
  regulations affecting river and floodplain
    management, 286
European grey alder, 148
European Union Habitat Directive, 284–85
European Water Framework Directive
  (WFD), 285, 286
evapotranspiration, 127–28, 277
existence value, 238
exogenous forces, 53
exogenous nutrient supply, 150, 151–52

F
factual landscapes, 11
farmers, 284
Federal Emergency Management Agency
  (FEMA), 242
Federal Energy Regulatory Commission
  (FERC), 242
Federal Interagency Working Group, 248, 274

Federal Land Planning and Management Act, 243
Federal Land Policy and Management Act, 285
federal-level regulations, 285
Federal Power Act, 243, 285
FEMA (Federal Emergency Management Agency), 242
FEMAT (Forest Ecosystem Management Assessment Team), 338
FERC (Federal Energy Regulatory Commission), 242
filtering capability, 164
filtering function, 274
filtration sites, 278–79
financial incentives, 305–6
fires, 145
fish production, 294–95
Fiume Tagliamento, 107–8
Flathead River, 4
flooded forest, Europe, 237
floodplains, 38–39
    ecology, 17
    ecosystem, 297
    maintenance flows, 298
    recruitment box model, 299
    terraces, 26
floods, 40–45, 145
    adaptations to in riparian plants, 83
    biomass loss during, 155
    magnitude, frequency, timing, and duration, 43–44
    management, 249
    stress on plants, 85
    waves, 41–42
flow networks, 8–9
flow regime, 36
flow regulation, 189, 196
    ecological consequences, 207–18
        alterations to energy and nutrient budgets, 215–17
        basic ecological principles, 217–18
        effects on native species and processes, 212–15
        extent of flow regulation, 210–12
        theory addressing flow regulation by dams, 208–10
    effects on native species and processes, 212–15
    extent of, 210–12
flushing flows, 298
fluvent soil, 74
fluvial actions, 76
fluvial geomorphologists, 59
flux, 127–31

food chains, 174
food webs, 161, 173–79
forecasting, 335–37
Forest Ecosystem Management Assessment Team (FEMAT), 338
forested buffers, 284
forest harvest, 227, 257
forests, 120
    of Amazon River and Orinoco River, 116
    flooded, 237
    practices regulation, 243
    upland, 141
Form units, 30–31
F, θ, and ƒ (population F-statistics), 348
freshwater ecosystems, principles for maintaining, 339
functional grouping, 180–83
fungal decomposition, 175
fungi and plant roots, 348

G
gamma diversity, 117–18
Garonne River, 4–5, 8, 225–26
gender-specific ecophysiology, 129
genetic information, 155
geographic boundaries, 302
Geographic Information Systems (GISs), 334, 349–50
geology, 20–21
geomorphic
    effect of spatial scale on geomorphic influences, 70
    provinces, 60–61
    templates, 77
    variables, 21
geomorphic characterization, 67–69
geomorphic classifications, 51, 58
    hierarchical classification, 59–65
    hydrogeomorphic (HGM) approach, 72
    process domain concept (PDC), 70–72
    Rosgen's classification, 65–72
geomorphic features, hierarchical patterns of, 22–32
    catchment form and channel networks, 23–25
    catchment history, 25–28
    in catchments, 28–32
geomorphic processes and process domains, 32–39
    channel processes, 34–37
    floodplain and channel interactions, 38–39
    headwater erosion, 33–34
geomorphic-vegetation units, 74–75
geomorphology, influence on lateral extent, 7
geophysical tomography, 337

GISs (Geographic Information Systems), 334, 349–50
glides, 64
global climate changes, 218–24, 224–25
global environmental change, 352–55
Global Positioning Systems (GPS), 343–45
global warming, 353
Global Water Systems Project, 354
G (Log-likelihood)-based tests, 348
goals, defining, 303–4
GPR (Ground Penetrating RADAR), 346–47, 350
GPS (Global Positioning Systems), 343–45
graded rivers, 34
grazing, prescribed, 319–22
greenhouse gases, 204–5, 218–19
grey alder, root production of, 148
Ground Penetrating RADAR (GPR), 346–47, 350
groundwater, 128
grouping, functional, 180–83
Gwynns Falls Trail, 252

H
habitat distribution, 129
habitat diversity, 294–95
Habitats Directive, 284–85
Habitat Spatial Boundaries, 61
Hadejia River, 237
Hardy-Weinberg equilibrium (HWE), 348
Harris' classification system, 75
headwalls, 62
headwaters, 5, 33–34
health, defined, 294
Hemlock roots, 148
heterogeneity, 76–77, 91, 114–15
heterogeneous environment, 117
HGM (hydrogeomorphic) approach, 58, 72
hierarchical classifications, 51, 58, 59–65, 77–78
hierarchical patch dynamics perspective, 10
highest litterfall rates, 143
high in-stream nutrient levels, 168–72
Hippopotamus amphibious, 180, 182
historical context, riparian typology, 51–52
Histosol orders, 73–74
hollows, 62, 70
human attitudes, toward riparia, 190
human benefits, from riparian conservation, 282–84
human dimension of riparian management, 257–67
    communication needs, 266–67
    shared socioenvironmental visions, 260–61
    social-ecological systems, 261–66

human-mediated changes, 204–5, 207
HWE (Hardy-Weinberg equilibrium), 348
hydraulic gradients, 45–46
hydrochory, 89
hydrogeomorphic (HGM) approach, 58, 72
hydrographs, 41
hydrological context, 4–7
hydrologic alterations, 189
hydrologic connectivity, and surface water–groundwater exchange, 39
    dynamics of linked surface–subsurface hydrologic system, 45–47
    surface connectivity and flooding, 40–45
hydrologic cycle, Earth's, 352–53
hydrologic effects, 276–77
hydrologic flows, 161
hydrologic forecasting, 337
hydrologic influences, 76
hydrologic regimes, 295–301
hydrologic restoration, 295–96
hydrologic variables, 21
hydropower regulation, 243
hyphomycete fungi, 175
hypogean, 4
hyporheic processes, 170
hyporheic zone, 45, 168

I
Ibrahim River, Lebanon, 283
ICSU (International Council for Science), 244
Ikonos imagery, 343
Immobilization I, 153–54
Immobilization II, 154
Inceptisol orders, 73–74
information collection, 243–45
information fluxes, 155–56
initial litter quality, 151
inorganic nitrogen, 150
insects, 112, 177
in-stream ecological processes, 52
in-stream nutrient levels, high, 168–72
integration, 313
integrity, 294
interactive subzone, 171–72
interconnected corridors, complex, 8
Interior's Bureau of Land Management (BLM), 56
intermediate disturbance hypothesis, 115
International Council for Science (ICSU), 244
international groups, for freshwater systems, 355
invader category, 82
invasive species, 230–32

inventory diversity, 118
invertebrates, 96–97, 112
islands, formation and disintegration of, 119
isotopes, 349
Italian Tagliamento River, 275

J
Justinian Code (533 AD), 235

K
K⁺ (potassium), 230
Kissimmee River, 308–9
knee roots, 84
knowledge, 244

L
laboratory techniques, 348–49
lag-times, 200
laissez faire, 293
land
    changes in alter fluxes of greenhouse
        gases, 204–5
    human effect on, 192–93
    land cover change, 225
    use changes, 224–32
Land Data Assimilation System (LDAS),
    354
Land Evaluation and Site Assessment
    (LESA), 75
Land Mosaics, 239
landowners, riparian, 284
landscapes, 255, 283
    context, 9–11
    mosaic, 161
    perspective, 11–12, 239
        of catchments, 161–63
        of riparian systems, 10
landslides, 146–47
Land-Use Change and Analysis System
    (LUCAS), 340–41
Langston, Nancy E., 237
lateral dimension, 7
lateral distribution, 98
lateral patterns, 119
lateral slopes, 248
lateral zonation, 97
latitude and litter production, 143
LDAS (Land Data Assimilation System),
    354
leaching phase, 151
leaf decomposition, 272–73
leaf litter, 109, 144, 273
legacies, 200
Léger's classification, 51
Lena River, 4–5

Leopold, Aldo, 241
LESA (Land Evaluation and Site
    Assessment), 75
LIDAR (radar and laser altimetric) data,
    342–43
life history strategies, 81–82
light, 156, 157
lignin degradation, 154
lignin-rich materials, 149
linkage number, 23
linked surface–subsurface hydrologic system,
    45–47
lithotopographic units, 30
litter, 144, 273; *see also* woody debris
        accumulations, 109–10
    conditioned litter, 175
    decomposition of plant litter with
        inorganic nitrogen, 150
    effect on plant reproduction, 111
    fluxes, 174–76
    leaf litter, 109
    litterfall, 143–44
    microbially colonized litter, 175
    plants, 109, 143
    production, 143
    quality, 149–50, 151
    riparian, 154–55
litterfall, 143–44
livestock grazing, 319–22
living resources, seven principles for
    conservation of, 271
living vegetation, 11
local agencies, 243
Log-likelihood (*G*)-based tests, 348
longitudinal dimension, 7, 118–19
longitudinal profile, 67
longitudinal slopes, 248
longitudinal zonation, 100–104
Lorzing, Han, 11
lower-basin rivers, 103
lowest litterfall rates, 143
LUCAS (Land-Use Change and Analysis
    System), 340–41
lysigeny, 84

M
macrofauna, 96
Madison River, 7
Magdalena River, 5
magnesium (Mg²⁺), 230
magnitude, flood, 43–44
Magnuson, J. J., 200
maintenance flows, 297
Major World Biomes, 143
mammal diversity, 123

management, riparian, *see* riparian
    management
man-made landscapes, 11
MapInfo Professional, 345
marine-derived nutrients, 184–86
Markov chain method, 348
Markov chain Monte Carlo (MCMC)
    methods, 335–37
mass wasting, 33–34
matrix, 161
MCMC (Markov chain Monte Carlo)
    methods, 335–37
meander development, 38
measurable biophysical features, 55
MedWet (Mediterranean wetlands), 244
Mekong River, 4–5
Melaleuca, 85
mesic conditions, 119–20
mesic regions, 141
mesofauna, 96
metals, 353
Mg^{2+} (magnesium), 230
microbial activity, 152, 153–54
microbial biomass, 153
microbial communities, 96, 183
microbial ecology, 151
microbially colonized litter, 175
microclimate, 156–58, 313
microclimatic environment, 157
microfauna, 96–97
microhabitat diversity, 117
microhabitat scales, 30, 59
microsatellite, 348
mid-order streams function, 103
migrating birds, 8
Milankovich cycles, 25
Millennium Ecosystem Assessment, 354
Mineralization I & II, 154
mining, effects of on rivers, 27
Mississippi River
    evolution of, 26–27
    factors influencing reforestation of, 296
    flood in 1993, 44
    range of changes on, 201–2
Missouri-Mississippi River, 4
mitigation, 294
Moisie River, 24–25
moisture
    availability, 154–55
    environmental variables, 156
    in riparian soils, 94–96
molecular genetics techniques, 348
molecular oxygen, 150–51
monitoring, restoration, 304–5
moose, 182

morphological adaptations, 82–87
morphologic description, channel, 66–67
morphologic description (Aspect II), 69–70
mortality rates, 145–47
mosaic, landscape, 161
multi-metric index, 75
multiple flow methodology, 300
multiple quasi-stable states, 107
multistate water management agencies, 243
Murray River, 210
mycorrhizae, 348

N
narrows, 156
narrow valley, 62
Nassauer, Joan, 11
National Conservation Buffer Workshop,
    319
National Forest Management Act, 243, 285
National Marine Fisheries Service (NMFS),
    242
National Park Organic Act of 1916, 270–71
National Park Service (NPS), 242
National Wild and Scenic Rivers Act, 243
Nationwide Rivers Inventory, 193–94
natural capital, 294
natural-cultural systems, 14
natural disturbances, 198–99
natural hydrogeomorphic processes, 198
Natural Resources Conservation Service
    (NRCS), 242
natural riparian systems, 51
net nitrogen, 153–54
net primary production, 137–43
net primary productivity (NPP), 135–36
NGOs (nongovernmental organizations), 255
NH_4^+ (ammonium), 132, 134
Nile River, 5, 210
nitrate inputs, 167
nitrate mass balance, 166
nitrate runoff, 164
nitrifying bacteria, 171
Nitrobacter, 171
nitrogen, 132, 164–65, 228–30
    accumulation of, 153–54
    compounds enriched with, 153
    exogenous nutrient supply, 151–52
    factors controlling immobilization of,
        151–52
    immobilization, 152
    inorganic, decomposition of plant litter
        with, 150
    mineralization, 93
    observed during immobilization, 151
    removal of, 167

Nitrosomonas, 171
NMFS (National Marine Fisheries Service), 242
NO$_3^-$, 169–70, 171
nongovernmental organizations (NGOs), 255
nonprofit organizations, 255
nonuse value, 238
normative river, 339
North American cottonwoods, *see* cottonwood trees
northern spotted owl, 338
novel habitats, 250–51
NPP (net primary productivity), 135–36
NPS (National Park Service), 242
NRCS (Natural Resources Conservation Service), 242
Nr (reactive nitrogen), 229
nutrients, 151–52
    budgets, 215–17
    dynamics during decay, 151
    enrichment, 228–30
    exogenous nutrient supply, 150
    filters, 332
    flows, 163–73
        particle-nutrient considerations, 172–73
        riparian zones as buffers against high in-stream nutrient levels, 168–72
        riparian zones as buffers against nutrient pollution from upland runoff, 164–68
    fluxes, 131–35
    levels, 217
    understanding removal and retention of, 164

O

Odocoileus hemionus columbianus, 322
OM (soil organic matter), 93–94
Oncorhynchus, 183–86, 304
operation (contemporary), 53
order-of-magnitude change, 46
organic matter, 93–94, 149
Orinoco River, 5, 116
orthophospate (PO$_4^{3-}$), 171
overbank flow, 42–43
overstory vegetation, 74
oxidation-reduction conditions, 134, 172–73
oxygen, molecular, 150–51
oxygen tension, 150–51

P

Pacific Decadal Oscillation, 353
pacific salmon, 183–86
paleochannels, buried, 26
paperbark tree, 85

Parana river, 5
particle-nutrient considerations, 172–73
passive restoration, 293
patches, 161
    causal, 53
    consequential, 53
    dynamics, 161–63
    heterogeneity, 313
PDC (process domain concept), 58, 70–72
pebble count method, 69
pedological processes, 97
perceived landscapes, 11
Phacochoerus aethiopicus, 180–81
phosphorus, 132, 151, 173, 229
phreatophytes, 127–28
physical environmental variables, 100
physical processes, 20, 80
physical template, *see* catchments and physical template
physiological adaptations, 82–87
physphysically based geomorphic approach, 50
phytosociological terminology, 97
Picea sitchensis, 82
placer mining, 27
plane-bed reaches, 64
plants, 74–75
    communities, 249–50
    demography, 145
    large seeds, 111
    litter, 109, 143
    reproduction, litter effect on, 111
    riparian, 82–87
    roots and fungi, 348
pleistocene droughts, 120
pneumatophores, 84
PO$_4^{3-}$ (orthophospate), 171
point diversity, 117
pollution, 172–73, 353; *see also* litter; woody debris
    buffering diffuse pollution, 294–95
    nutrient, 164–68
polymorphic microsatellite loci, 348
ponds, 181–82
pool-riffle reaches, 64
poplar trees, 266
population expansion, 192
population *F*-statistics (F, θ, and *f*), 348
Populus, 7, 88, 131, 296
Populus fremontii, 131
Populus x canadensis, 173
Pô River, 4
porosity, 156
potassium (K$^+$), 230
Potomac River, 41–42

precipitation
    changes in, 222–23
    increased variability, 223–24
    influence on lateral extent, 7
predatory spider communities, 112
predictable environments, 81
prediction, ecology methods for, 335–37
Pregracke, Chad, 237
press disturbances, 198
primeval flooded forest, Europe, 237
principles of decomposition dynamics,
        149–51
process-based riparian zone models, 319
process domain concept (PDC), 58, 70–72
process domains, 32–39
process domains, and geomorphic processes,
        32–39
    channel processes, 34–37
    floodplain and channel interactions, 38–39
    headwater erosion, 33–34
production ecology, 135–48
    growth and metabolism of riparian trees,
        135–37
    litterfall, 143–44
    mortality rates, 145–47
    root production, 147–48
    timing of growth and rates of net primary
        production, 137–43
projection, 335–37
propagules, dispersal of, 8
Prosopis velutina, 131
protection and restoration, 293
proximate controls, 57
psamment soil, 74
pulse disturbances, 198

Q
quasi-stable conditions, 106–7
Queets River, 140
Querus robur, 150

R
radar and laser altimetric (LIDAR) data,
        342–43
rain forest, 112–13
ramp disturbances, 198
rapids, 64
RCE (riparian, channel, and environmental
        inventory), 75
reach classification, 306, 307
reach-scale patterns, 38
reactive nitrogen (Nr), 229
reciprocal energy, 177–79
reciprocal information, 288
reclamation, 294

Red Deer River, 107
red oak leaves, 150
redox conditions, 170
refuges, 119–20
regeneration flows, 297
regional scales, 59, 243
rehabilitation, 294
relict habitats, 250–51
relict (site), 53
relief ratio, 23
REMM (Riparian Ecosystem Management
        Model), 319
remote-sensing instruments, 342
reproductive adaptations, 82, 85
reproductive strategies, 87–91
RESA (Riparian Evaluation and Site
        Assessment), 75
resilience property, 263
resistance, 156
resister category, 82–83
resource allocation, 81
resource competition, in heterogeneous
        environment, 115
resource use, 192–96
restoration, 291–326
    assessing ecological integrity of riparia,
        308–14
    developing restoration plan, 301–8
        defining goals and objectives, 303–4
        financial incentives, 305–6
        getting organized, 302
        identifying problems and opportunities,
            302–3
        implementing, monitoring, and
            evaluating, 304–5
        setting priorities, 306–8
    general principles and definitions, 294–95
    goal of, 292–93
    overview, 291–92
    programs, 338
    returning to more natural hydrologic
        regimes, 295–301
    specific enhancements, 314–22
        designing riparian buffer zones, 317–19
        implementing riparian silvicultural
            practices, 314
        optimizing riparia for biodiversity,
            322
        prescribed grazing, 319–22
        reintroducing large woody debris,
            314–16
revegetation of riverbanks, 257–59
Rhine River, 215
rhizosphere, 84, 85
Rhône River, 4, 201, 249–50

riffles, 64
riparia
    alterations, 201–4
    assessments, 314
    boundaries, 53
    corridors, 156
    defined, 1
    direct changes to, 193–94
    discoveries about, 329–30
    educational efforts, 244
    inventory approaches, 56
    placing monetary value on, 330
riparian, channel, and environmental
        inventory (RCE), 75
riparian communities, 91–114
    density, basal area, and biomass, 112–14
    general distributions of aboveground and
        belowground communities, 97–104
    identification of riparian zones based on
        soils and vegetation type, 91–92
    patterns, 76
riparian-derived arthropods, 176–77
riparian-derived wood, 11
riparian disturbances, 197–205
    basic concepts and approaches to
        understanding history, 198–201
    defining anthropogenic disturbance, 198
    historical examples of riparian alterations,
        201–4
    pervasive human-mediated changes, 204–5
Riparian Doctrine, 235
Riparian Ecosystem Management Model
        (REMM), 319
Riparian Evaluation and Site Assessment
        (RESA), 75
riparian functions, biophysical connectivity
        and, see biophysical connectivity and
        riparian functions
Riparian Landscapes, 239
riparian management, 233–68
    goals of, 234
    highly specific process of, 253–59
        adaptive environmental, 254–55
        appropriate management, 257–59
        revegetation of riverbanks, 257–59
        sustainable riparian management,
            255–56
        timber harvest practices, 257
    human dimension of, 257–59
        communication needs, 266–67
        shared socioenvironmental visions,
            260–61
        social-ecological systems, 261–66
    overview, 233–34
    process linked to catchment and, 246–53

catchments and rivers benefit from
        riparian management, 251–53
    riparian benefits from catchment
        management, 246–48
    riparian benefits from river
        management, 248–51
recent and evolving concern, 235–45
    economic valuation of riparia, 237–38
    information collection and
        dissemination, 243–45
    social and cultural perspective, 238–41
    suitable management institutions,
        241–43
riparian science, 16
riparian silvicultural practices, 314
riparian systems, 2–4, 51
riparian tree species, 140
riparian typology, 49–78
    attributes of enduring classification
        system, 77–78
    biotic classification, 73–76
        plants, 74–75
        soils, 73–74
        wildlife, 75–76
    emerging classification concepts, 57–58
    geomorphic classification, 58
        hierarchical classification, 59–65
        hydrogeomorphic (HGM) approach, 72
        process domain concept (PDC), 70–72
        Rosgen's classification, 65–72
    historical context, 51–52
    overview, 49
    purpose of, 49–51
    theoretical basis for classification, 53–57
    treating complexity and heterogeneity in
        classification systems, 76–77
riparian vegetation, 98, 128
riparian wildlife community, 102–3
riparian zones, 12–14, 156
    based on soils and vegetation type, 91–92
    as buffers, 320
        against high in-stream nutrient levels,
            168–72
        against nutrient pollution from upland
            runoff, 164–68
    effects of changing climate on, 220–21
    predicting how will react to conservation
        management, 287
    rivers by dams and ecological
        consequences for, 195
riverbank erosion, 279–80
river basin, 22
river energy, 34–37
River Habitat Survey, 75
riverine hierarchical scaling system, 29

riverine organisms, 8
River Lech, 122
rivers, *see also names of specific rivers*
    benefit from riparian management, 251–53
    classification of, 51
    corridor protection, 243
    direct changes to, 193–94
    European, regulations on, 202
    flow, 249–50
    habitat scale, 30
    inventories, 66
    management of, 248–51
    reach scale, 30
    regulations on European rivers, 202
    segment scale, 30
    small river restoration, 296
rooting zone, 84
root production, 147–48
root systems, 277, 348
Rosgen's classification, 58, 65–72
round-shaped catchments, 23
Rubus, 85
runoff control zone, 318
runoff regimes, changes in, 222

S
S^{-2} (sulfide), 167
safe sites, 82
Salix gooddingii, 129, 131
Salix spp., 82
salmon, 183–86, 304
saltcedar, 296
Salvelinus malma, 177
sanctuary, riparia, 271
San Pedro River, 128, 131
São Franciso River, 5
Saskatchewan River, 4
scales, 312–13
scenario development, 334–35
schizogeny, 84
Scientific Committee on Water Research
    (SCOWAR), 244
SCOWAR (Scientific Committee on Water
    Research), 244
SDC (serial discontinuity concept), 208
sediment-bound pollutants, 172–73
sediments, 35–36, 172–73
    moisture in, 94–95
    reported effectiveness of buffer zones for
        sediment reduction, 319
seeds
    dispersal, 88–89, 299
    survival, 109
semiarid regions, 331
semiterrestrial environment, 275

Senegal River, 210
SER and Policy Working Group 2002, 292,
    293, 294
SERCON (System for Evaluating Rivers for
    Conservation), 75
serial discontinuity concept (SDC), 208
seston, 313
sexual reproduction, 88, 139
shade-tolerant conifers, 145
shared socioenvironmental visions, 260–61
shredders, 175
Sierra Nevada rivers, 27
Sierra Vista sub-catchment, 128
silt accumulations, 111
silvicultural practices, 314
site patterns, 118–19
site (relict), 53
sitka spruce, 82
smaller-scale patterns, 74
small river restoration, 296
snapshots, biophysical, 56
snowmelt runoff, 223
SO_4^{-2} (sulfate), 167
social-ecological systems, 261–66, 287, 294
social-environmental models, 340
social organization, 197
social perspective, 238–41
social relations, 259
socioenvironmental conditions, 243–44
soil, 73–74, 94–95; *see also* biophysical
    characteristics of riparian soils
    biophysical characteristics of riparian soils
        fauna, 96–97
        moisture, 94–96
        organic matter, 93–94
    diversity of soil organisms, 121–22
    erosion of, 281
    soil microbe activity, 157
    texture of, 93–94, 166
    upland, 73
Soil and Water Conservation Society 2001,
    319
soil organic matter (OM), 93–94
solar radiation, 157
soluble-reactive-phosphorus (SRP), 173
southern forested wetlands, 296
South Platte River, 203–4
spatial dimensions, 7
spatial orientation, 156
spatial scales, 62, 312–13
Spatio-Temporal Scales, 60
species–habitat relationships, 73
species richness, factors controlling, 120–21
spider communities, 112, 122
SRP (soluble-reactive-phosphorus), 173

S (sulfur), 132
stable environments, 81
stable isotopes, 349
state agencies, 243
State of Washington Forest Practices
  Regulations, 78
stem exclusion stage, 104
stem production, 136–37, 139–40
step-pool reaches, 64
stoichiometry, 349
Strata sets, 31–32
streams
  condition, 66–67
  corridor, 303
  energy, 34–37
  events or processes controlling habitat,
    60
  livestock impacts on riparia and, 321
  riparian zones sources or sinks of nitrogen
    to, 169–70
  woodland, effects of depriving of leaf
    litter, 273
stress, disturbance ecology responses to,
  205–7
Strix occidentalis, 338
structural patterns, 79–124
  biological diversity, 114–23
    diversity theory and measurement,
      115–18
    faunal diversity, 121–23
    vegetative diversity, 118–21
  distribution, structure, and abundance of
    riparian communities, 91–114
    biophysical characteristics of riparian
      soils, 92–97
    density, basal area, and biomass, 112–14
    distributions of aboveground and
      belowground communities, 97–104
    identification of riparian zones, 91–92
    successional and seasonal community
      patterns, 104
  life history strategies, 81–82
  morphological and physiological
    adaptations of riparian plants,
    82–87
  overview, 79–81
  reproductive strategies, 87–91
structure of riparian communities, see
  riparian communities
subsurface channel, 4
sulfate (SO_4^{-2}), 167
sulfide (S^{-2}), 167
sulfur (S), 132
surface connectivity, 40–45
surface runoff, 280

surface–subsurface hydrologic system,
  linked, 39
surface subzone, 171
surface water–groundwater exchanges, and
  hydrologic connectivity, 39
  dynamics of linked surface–subsurface
    hydrologic system, 45–47
  surface connectivity and flooding, 40–45
sustainable riparian management, 255–56
swales, 62
synthesis, 327–56
  developing future vision, 333–50
    emerging perspectives, tools, and
      approaches, 337–41
    forecasting, prediction, and projection,
      335–37
    scenario development, 334–35
  global environmental change, 352–55
  goal of, 328–29
  overview, 327–28
  principles for ecological management of
    riparia, 350–52
  riparia as keystone units of catchment
    ecosystems, 329–31
  unified perspective of riparian ecology,
    332–33
System for Evaluating Rivers for
  Conservation (SERCON), 75
system-scale management, 234

T
Tamarix, 296
taxonomic class, 104
technology development, 197
tectonic activity, 20–21
temperature, 150, 152, 157, 353
  changes, affect on basin water balances,
    26
  environmental variables, 156
  regimes, 222, 225–28
templates, 77
temporally dimension, 7
temporal models, 142–43
temporal scales, 61, 312–13
tenets, 17, 332
terminal electron-accepting, 135
terraces, floodplain, 26
terrestrial components, 9
terrestrialization, 212, 214, 223, 313
terrestrial processes, 50–51
theoretical basis for classification,
  53–57
thermal regime, 220–21
Thiobacillus denitrificans, 171
three-pasture rotation system, 278, 321

three-zone forest buffers, 317–18
timber harvest practices, 257
timing, flood, 43–44
too little, too late syndrome, 290
tools and approaches
    developing future vision, 341–50
        aboveground tool, 341–46
        belowground tool, 346–48
        general techniques, 348–50
    emering, 337–41
top-down approach, 300
topographic divides, 3
topographic highs, 45
transfer zone, 32–33
transient storage, 168
trees, *see also names of specific trees*
    deep-rooted trees, 277
    riparian tree species, 140
Trinity River, California, 295
Triple Divide Peak, 4
TWINSPAN analysis, 75
typology, riparian, *see* riparian typology

U
ultimate controls, 57
unchanneled valleys, 62
understory initiation stage, 104
understory vegetation, 74
UNEP-World Conservation Monitoring
    Centre 2000, 293
UNESCO's International Hydrologic
    Programme, 354
unpredictable environments, life history
    strategies of organisms in, 81
upland forests, 141
upland runoff, 164–68
upland soil, 73
U.S. Army Corps of Engineers (Corps or
    USACE), 241
USBR (Bureau of Reclamation), 241
USDA Forest Service's Integrated Riparian
    Evaluation Guide, 75
USDA Soil Survey Staff 1998, 73
U.S. Department of Agriculture's Forest
    Service, 56
U.S. Department of the Interior and
    Department of Commerce 1996,
    282
U.S. Environmental Protection Agency
    (EPA), 242
U.S. Fish and Wildlife Service (USFWS),
    242, 295
U.S. Forest Service (USFS), 242
USFS (U.S. Forest Service), 242
U.S. Geological Survey (USGS), 241

V
valley segments, 62
várzea floodplains, 138
vegetation, 276–77
    communities, 108–9
    distribution, 7
    diversity of, 118–21
    influences, 276
    lateral distribution of, 98
    moderator of cut-and-fill alluviation,
        10
    overstory, 74
    patterns, 74
    reducing risk of erosion, 279–80
    succession, 104–11
    understory, 74
    water sequestered by, 128
vertical dimensions, 7, 118
vocal information, 155
Volga-Kama River, 210

W
warthogs, 180–81
water
    ancient water development events,
        235
    availability, 353–54
    consumption, 194–95
    water resources development, principles
        for, 245
water–groundwater systems, 45
water-mediated erosion, 6
water-related agencies, 243
Watershed Analysis, 201
Watershed Partnership, 248
watersheds, 3, 22
Waterton River, 4
wealth property, 263
WEHAB Working Group 2002, 353
Wetland Evaluation Technique, 313
wetlands
    assessment methods of, 310–12
    classification scheme, 72
    key issues limiting utility of, 314
    regulations, 243
wet regions, 331
WFD (European Water Framework
    Directive), 285, 286
wildlife, 75–76, 273
wildlife community, 102–3
Willamette River, 201, 334–35
willow trees, 82, 108, 258
wind, seed dispersal by, 88
wind speed, environmental variables,
    156

windstorms, 145
Wisconsin River, 109
wolves, 182
woodlands, 273, 281
woody debris, 107–8, 272
   large, 239, 314–16
   removal of, 202
   in riparian corridors, 11
   in rivers, issues and challenges, 240

Y
Yangtze River, 4–5
Yellow River, 4–5
Yellowstone National Park, 270

Z
Zambezi River, 5
zoochory, 89
zoological ripening of soils processes, 97